Emerging Water Pollutants: Concerns and Remediation Technologies

Edited by

Shaukat Ali Mazari
Department of Chemical Engineering
Dawood University of Engineering and Technology
Karachi 74800
Pakistan

Nabisab Mujawar Mubarak
Department of Chemical Engineering
Faculty of Engineering and Science
Curtin University
Malaysia

&

Nizamuddin Sabzoi
School of Engineering
RMIT University
Melbourne 3000
Australia

Emerging Water Pollutants: Concerns and Remediation Technologies

Editors: Shaukat Ali Mazari, Nabisab Mujawar Mubarak & Nizamuddin Sabzoi

ISBN (Online): 978-981-5040-73-9

ISBN (Print): 978-981-5040-74-6

ISBN (Paperback): 978-981-5040-75-3

First published in 2022.

need for a court order if at any point you breach any terms of this License Agreement. In no event will any delay or failure by Bentham Science Publishers in enforcing your compliance with this License Agreement constitute a waiver of any of its rights.

3. You acknowledge that you have read this License Agreement, and agree to be bound by its terms and conditions. To the extent that any other terms and conditions presented on any website of Bentham Science Publishers conflict with, or are inconsistent with, the terms and conditions set out in this License Agreement, you acknowledge that the terms and conditions set out in this License Agreement shall prevail.

Bentham Science Publishers Pte. Ltd.
80 Robinson Road #02-00
Singapore 068898
Singapore
Email: subscriptions@benthamscience.net

BENTHAM SCIENCE

CONTENTS

PREFACE

With the increasing population, industrial growth and consumer needs, water is getting adulterated with several new contaminants of concern. Nanomaterials, pharmaceuticals & personal care products, endocrine disrupting compounds, artificial sweeteners, surfactants, *etc.* are some of the emerging sources of water pollutants. Several of these emerging water pollutants introduce noxious effects on humans and the ecosystem. These emerging pollutants need to be removed from wastewater sources. Some of the most investigated technologies for the removal of emerging pollutants from wastewater include catalytic processes including oxidation, membrane process, adsorption, osmotic processes, integrated processes, *etc*. This book cincludes of thirteen chapters. The first five chapters discuss and analyze the sources of emerging water pollutants, their toxicities, and legislations available to monitor and regulate their emissions. The next three chapters (6-8) are on the risk assessment of emerging pollutants, their fate and life cycle assessment. The last five chapters are on the remediation technologies for wastewater treatment containing emerging pollutants. This book is equally good for academia, industry professionals and students for state-of-the-art learning on emerging water pollutants and their remediation methods.

Chapter 1 provides an extensive review on classification of various emerging pollutants reported worldwide along with their physio-chemical properties and potential environmental effects. Moreover, the legislative policy regulations formulated worldwide for the monitoring of various emerging pollutants are also discussed.

Chapter 2 discusses the state-of-the-art global applications of pharmaceuticals and personal care products, the mechanism of water pollution by pharmaceuticals and personal care products, possible biohazards, and negative impacts on the environment. Besides that, various types of pharmaceuticals and personal care products, the most applied chemical compounds in pharmaceuticals and personal care products have been discussed in this chapter. The chapter also presents a future perspective for reducing the pharmaceuticals and personal care products' contamination of surface water with cutting-edge technologies along with wastewater treatment.

Chapter 3 focuses on the release of emerging pollutants from the food and packaging industries. This chapter interconnects the use of chemicals and fertilizers for the production and prevention of food crops and food. Also, the use of drugs for livestock and poultry for the food raw materials are highlighted. Furthermore, this chapter summarizes all aspects of emerging water pollutants from the food and packaging industry and highlights the source-sink relation of emerging water pollutants as well.

Chapter 4 introduces nanomaterials and their release to the environment and the relevant concerns. In addition, the effect of various nanomaterials in the aquatic environment is discussed, including the behavior and toxicity of nanomaterials to the aquatic ecosystem. It is important to identify, assess and reduce the environmental impact of these nanomaterials. The chapter further highlights environment-friendly and advantageous use of engineered nanomaterials for a sustainable world.

Chapter 5 highlights the industrial sources of emerging water pollutants, their possible routes to waterbodies, and their respective impacts on the environment and ecosystem. Furthermore, the concerns of emerging contaminants from industrial operations and processes have also been discussed. Furthermore, an insight into the challenges in minimizing water pollution is also focused on the shared benefit.

Chapter 6 discusses the risk assessment of the emerging water pollutants. The environmental risk assessment of emerging pollutants can help to indicate potential risks associated with these substances, highlighting the importance of the identification of their hazard, dose response, exposure assessment and finally their risk characterization. The chapter provides useful insights into the recent findings related to the pollutants' effects and their assessment approaches.

The major focus of chapter 7 is the fate of emerging water pollutants. The chapter introduces emerging pollutants of concerns, and identifies their routes to sources where they end up. More importantly chapter highlights the end sink of pollutants with potential threats and risk evaluation for plants and human health through the end sink.

Chapter 8 briefs on the lifecycle assessment of emerging water pollutants. This includes the process of lifecycle assessment for various emerging pollutants, including the release and accumulation to examine the impacts and associated risks to water quality, the aquatic environment, and ultimately on the human beings. Furthermore, a deep insight into lifecycle has been provided that may help to understand the flow of pollutants in environment and fill the existing knowledge gaps.

Chapter 9 introduces various conventional wastewater technologies such as disinfection, desalination, coagulation, chemical precipitation, filtration, adsorption, *etc*. Some of these technologies have potential uses for the removal of emerging water pollutants. Furthermore, the chapter discusses the limitations of these processes and their advantages over the others.

Chapter 10 briefly describes advanced membrane processes for the removal of emerging water pollutants. Various membranes, such as nanofiltration, reverse osmosis, membrane bioreactors have been discussed and evaluated for the removal of emerging water pollutants in detail. Furthermore, challenges related to membrane technologies such as their efficiency, prevention from fouling, stability when using cleaning agents, permeability, cost, and energy reduction have been discussed as well.

Chapter 11 discusses osmotic and filtration processes for the removal of emerging water pollutants. A detailed account has been provided for the pressure driven membrane filtration processes including micro-filtration, nano-filtration, ultra-filtration, reverse osmosis, forward osmosis, osmotically driven membrane filtration processes such as pressure retarded osmosis, *etc*. Apart from their technicalities, the performance evaluation of these osmotic processes has also been given.

Chapter 12 brings a detailed account of various catalytic processes for the removal of emerging water pollutants. Some of these processes include ozonation, electrocatalysis process including electrocatalytic oxidation, electro-Fenton process, photoelectro-Fenton process, photocatalysis, reduction by hydrodehalogenation, *etc*. Moreover, the features, mechanisms, and potential applications of catalytic processes in the treatment of emerging water pollutants are discussed in detail.

Shaukat Ali Mazari
Department of Chemical Engineering
Dawood University of Engineering and Technology
Karachi 74800
Pakistan

Nabisab Mujawar Mubarak
Department of Chemical Engineering
Faculty of Engineering and Science
Curtin University
98009
Miri Sarawak
Malaysia

Nizamuddin Sabzoi
School of Engineering
RMIT University
Melbourne 3000
Australia

List of Contributors

Aansa Rukya Saleem	Department of Earth and Environmental Sciences, Bahria, University, Islamabad, Pakistan
A. Masudi	Clean Energy Research Centre, Korea Institute of Science and Technology, P.O. Box 131, Cheongryang, Seoul 136-791, Republic of Korea
Abdul Sattar Jatoi	Department of Chemical Engineering, Dawood University of Engineering and Technology, Karachi 74800, Pakistan
Abdul Qayoom Memon	Department of Chemical Engineering, Dawood University of Engineering and Technology, Karachi 74800, Pakistan
Abdul Karim Shah	Department of Chemical Engineering, Dawood University of Engineering and Technology, Karachi 74800, Pakistan
Akanksha Verma	Department of Physics and Materials Science and Engineering, Jaypee Institute of Information Technology, A-10, Sector 62, Noida 201309, India
Arbab Tufail	Strategic Water Infrastructure Laboratory, School of Civil, Mining and Environmental Engineering, University of Wollongong, Wollongong, NSW 2522, Australia
Asif Shah	Department of Metallurgy and Materials Engineering Dawood, University of Engineering and Technology, Karachi, Pakistan
Atta Muhammad	Department of Chemical Engineering, Dawood University of Engineering and Technology, Karachi 74800, Pakistan
Atta Muhammad	Department of Chemical Engineering, Dawood University of Engineering and Technology, Karachi 74800, Pakistan
Audil Rashid	Department of Botany, University of Gujrat, Pakistan
Brian Moon	Plamica Labs, Batten Hall, 125 Western Ave, Allston, MA 02163, USA
Behzad Ataie-Ashtiani	Department of Civil Engineering, Sharif University of Technology, P.O. Box 11155-9313, Tehran, Iran
Ezzat Chan Abdullah	Department of Chemical Process Engineering, Malaysia-Japan International, Institute of Technology (MJIIT) Universiti Teknologi Malaysia (UTM), Jalan Sultan Yahya Petra, 54100 Kuala Lumpur, Malaysia
Fahad Saleem Ahmed Khan	Department of Chemical Engineering, Faculty of Engineering and Science Curtin University, 98009 Miri Sarawak, Malaysia
Ghulam Mujtaba	Department of Energy & Environment Engineering, Dawood University of Engineering & Technology, Karachi, Pakistan
Hafiza Salma	Institute of Agro-Industry and Environment, The Islamia University of Bahawalpur, Punjab, Pakistan
Joydeb Babu Pranta	Department of Civil Engineering, Stamford University, Bangladesh

Kamran Manzoor	Institute of Environmental Sciences and Engineering (IESE), School of Civil and Environmental Engineering (SCEE), National University of Sciences and Technology (NUST), Sector H-12, Islamabad, Pakistan
Mohammad Khalid	Graphene & Advanced 2D Materials Research Group (GAMRG), School of Science and Technology, Sunway University No. 5 Jalan University Bandar Sunway 47500 Subang Jaya, Selangor, Malaysia
Muhammad Umair	Institute of Soil & Environmental Sciences, University of Agriculture, Faisalabad 38000, Pakistan
Muhammad Zia ur Rehman	Institute of Soil & Environmental Sciences, University of Agriculture, Faisalabad 38000, Pakistan
Mujahid Ali	Institute of Soil & Environmental Sciences, University of Agriculture, Faisalabad, 38000, Pakistan
Manoj Tripathi	Department of Physics and Materials Science and Engineering, Jaypee Institute of Information Technology, A-10, Sector 62, Noida 201309, India
Marziyeh Jannesari	Institute for Nanoscience and Nanotechnology, Sharif University of Technology, P.O. Box 14588 89694, Tehran, Iran
Mohammadhossein Taghipour	Department of Materials Engineering, University of Tabriz, P.O. Box 51666-16471, Tehran, Iran
Muhammad Saud Baig	Department of Energy & Environment Engineering, Dawood University of Engineering & Technology, Karachi, Pakistan
Muhammad Rizwan	U.S.-Pakistan Center for Advanced Studies in Water, Mehran University of Engineering & Technology, Jamshoro, Pakistan
Nazia Hossain	School of Engineering, RMIT University, Melbourne VIC 3001, Australia
Nimra Khalid	Institute of Agro-Industry and Environment, The Islamia University of Bahawalpur, Punjab, Pakistan
Nizamuddin Sabzoi	School of Engineering, RMIT University, Melbourne 3000, Australia
N.W.C. Jusoh	Department of Chemical and Environmental Engineering (ChEE), Malaysia-Japan International Institute of Technology (MJIIT), Universiti Teknologi Malaysia (UTM) Kuala Lumpur, Jalan Sultan Yahya Petra, 54100 Kuala Lumpur, Malaysia
N.F. Jaafar	School of Chemical Sciences, Universiti Sains Malaysia, 11800 USM Penang, Malaysia
Nisa Waqar-Un	Center for Interdisciplinary Research in Basic Sciences (SA-CIRBS), International Islamic University, Islamabad, Pakistan
Nazneen Bangash	Department of Biosciences, COMSATS University, Islamabad, Pakistan
Pranta Barua	Department of Electronic Materials Engineering, Kwangwoon University, Seoul 1897, South Korea
P.Y. Liew	Department of Chemical and Environmental Engineering (ChEE), Malaysia-Japan International Institute of Technology (MJIIT), Universiti Teknologi Malaysia (UTM) Kuala Lumpur, Jalan Sultan Yahya Petra, 54100 Kuala Lumpur, Malaysia

Rama Rao Karri   Petroleum and Chemical Engineering Faculty of Engineering, Universiti Teknologi Brunei, Bandar Seri Begawan BE1410, Brunei Darussalam

Rashmi Walvekar   School of Energy and Chemical Engineering, Department of Chemical Engineering, Xiamen University Malaysia, Jalan Sunsuria, Bandar Sunsuria, 43900 Sepang, Selangor, Malaysia

Rashid Abro   Department of Chemical Engineering, Dawood University of Engineering and Technology, Karachi 74800, Pakistan

Shoaib Ahmed   Department of Chemical Engineering, Faculty of Engineering and Science Curtin University, 98009 Miri Sarawak, Malaysia

Shaukat Ali Mazari   Department of Chemical Engineering, Dawood University of Engineering and Technology, Karachi 74800, Pakistan

Sabzoi Nizamuddin   School of Engineering, RMIT University, Melbourne 3000, Australia

Sana Rana   Institute of Agro-Industry and Environment, The Islamia University of Bahawalpur, Punjab, Pakistan

Shuakat Ali Mazari   Department of Chemical Engineering, Dawood University of Engineering and Technology, Karachi 74800, Pakistan

Samia Qadeer   Department of Environmental Science, Pir Mehr Ali Shah-Arid Agriculture University, Rawalpindi, Pakistan

Shoaib Ahmed   Department of Chemical Engineering, Dawood University of Engineering and Technology, Karachi 74800, Pakistan

Sher Jamal Khan   Institute of Environmental Sciences and Engineering (IESE), School of Civil and Environmental Engineering (SCEE), National University of Sciences and Technology (NUST), Sector H-12, Islamabad, Pakistan

Shabnam Taghipour   Department of Civil Engineering, Sharif University of Technology, P.O. Box 11155-9313, Tehran, Iran

Shabnam Taghipour   Department of Chemical and Biological Engineering, The Hong Kong University of Science and Technology, Clear Water Bay, Kowloon, Hong Kong

Seiyed Mossa Hosseini   Physical Geography Department, University of Tehran, P.O. Box 14155-6465, Tehran, Iran

Siraj Ahmed   Department of Energy & Environment Engineering, Dawood University of Engineering & Technology, Karachi, Pakistan

Sheeraz Ahmed   Institute of Environmental Engineering, Mehran University of Engineering & Technology, Jamshoro, Pakistan

Tahir Hayat Malik   Department of Environmental Sciences, COMSATS University Islamabad, Abbottabad Campus, Pakistan

Talat Ara   Department of Environmental Sciences, International Islamic University, Islamabad, Pakistan

Wajid Umar   Doctoral School of Environmental Science, Szent Istvan University, Gödöllő, Hungary

Yie Hua Tan   Department of Chemical Engineering, Faculty of Engineering and Science Curtin University, 98009 Miri Sarawak, Malaysia

Zahoor Ahmad	Department of Botany, University of Central Punjab Bahawalpur Campus, Punjab, Pakistan
Zubair Hashmi	Department of Chemical Engineering, Dawood University of Engineering and Technology, Karachi 74800, Pakistan
Zahra Zahra	Department of Civil & Environmental Engineering, University of California-Irvine, Irvine, CA 92697, USA
Zunaira Habib	Institute of Environmental Sciences and Engineering (IESE), School of Civil and Environmental Engineering (SCEE), National University of Sciences and Technology, Islamabad, Pakistan
Zahoor Ahmad	Department of Botany, University of Central Punjab, Bhawalpur, Pakistan
Naveed Ahmed	U.S.-Pakistan Center for Advanced Studies in Water, Mehran University of Engineering & Technology, Jamshoro, Pakistan

CHAPTER 1

Emerging Water Pollutants, their Toxicities, and Global Legislations

Shoaib Ahmed[1,6], Fahad Saleem Ahmed Khan[1], Nabisab Mujawar Mubarak[2,*], Yie Hua Tan[1], Rama Rao Karri[2], Mohammad Khalid[3], Rashmi Walvekar[4], Ezzat Chan Abdullah[5], Shaukat Ali Mazari[6] and Sabzoi Nizamuddin[7]

[1] *Department of Chemical Engineering, Faculty of Engineering and Science, Curtin University, 98009, Miri Sarawak, Malaysia*

[2] *Petroleum and Chemical Engineering, Faculty of Engineering, Universiti Teknologi Brunei, Bandar Seri Begawan BE1410, Brunei Darussalam*

[3] *Graphene & Advanced 2D Materials Research Group (GAMRG), School of Science and Technology, Sunway University, No. 5, Jalan University, Bandar Sunway, 47500 Subang Jaya, Selangor, Malaysia*

[4] *School of Energy and Chemical Engineering, Department of Chemical Engineering, Xiamen University Malaysia, Jalan Sunsuria, Bandar Sunsuria, 43900 Sepang, Selangor, Malaysia*

[5] *Department of Chemical Process Engineering, Malaysia-Japan International Institute of Technology (MJIIT) Universiti Teknologi Malaysia (UTM), Jalan Sultan Yahya Petra, 54100 Kuala Lumpur, Malaysia*

[6] *Department of Chemical Engineering, Dawood University of Engineering and Technology, Karachi 74800, Pakistan*

[7] *School of Engineering, RMIT University, Melbourne 3000, Australia*

Abstract: Emerging pollutants (EPs) in the environment have become a significant source of pollution and cause of serious concern for the ecosystem and human health. Although during the recent decades, extensive research has been performed worldwide for the detection and analysis of EPs, continuous refinement, and development of specific analytical techniques; a great number of undetected EPs still need to be investigated in different components of the ecosystem and biological tissues. Therefore, this chapter provides extensive reviews of several emerging pollutants reported around the globe along with their physiochemical properties and potential ecological impacts. Moreover, formulated legislations and policy regulations for the monitoring of EPs are also discussed in this chapter.

** **Corresponding author Nabisab Mujawar Mubarak:** Petroleum and Chemical Engineering, Faculty of Engineering, Universiti Teknologi Brunei, Bandar Seri Begawan BE1410, Brunei Darussalam; E-mail:mubarak.mujawar@utb.edu.bn*

Keywords: Antibiotics, Concerns of emerging pollutants, Emerging pollutants, Emerging water pollutants, Legislations, Personal-care-products, Pesticides, Pollutant toxicity.

INTRODUCTION

Over the recent years, owing to the uncontrolled progress in multiple human activities, such as transport, agriculture, industrialization, and urbanization, the world has experienced antagonistic consequences. The change in living standards and unsustainable consumer demand have enhanced the air pollution load, for example, particulate matter, SO_2, NOx, greenhouse gases, and ozone. On the other hand, the water sources are adulterated with numerous manmade chemicals, heavy metals, nondegradable materials, oil spills, nutrients, landfill leachates, *etc* [1, 2]. Emerging contaminants include a variety of anthropogenic chemicals, for example, active pharmaceutical compounds, personal care products, pesticides, and numerous industrial chemicals, which are widely used in the world [3, 4]. It has been estimated that global production of anthropogenic chemicals between 1930 and 2000 reached from 1 million to 400 million tons per year [5]. Hence, increasing chemical demands and their uses have become a reason for several ecological impacts worldwide.

According to the environmental protection agencies, these EPs are newly detected in the environment. Therefore, their impacts are not completely determined. Thus, they are often considered un-regulated trace contaminants [6]. Moreover, conventional water and wastewater treatment facilities cannot completely remove these pollutants. Therefore, numerous studies reported their occurrence in drinking as well as surface and groundwater [5, 7, 8]. Furthermore, various metabolites and transformation by-products of EPs have also been reported by researchers, which are quite harmful and biologically active and become a reason for various adverse effects [9].

The majority of studies related to EPs occurrence and toxicities have been conducted in developed regions of the world. The circumstances are worse in developing countries where the occurrence and concentration of EPs are high owing to less efficient wastewater treatment plants, unskilled personnel, large population size, and disposal of international expired chemicals near rivers, *etc* [10]. The release of these chemicals into the river also pollutes the surrounding environment, including animals and aquatic life [11, 12]. Based on recent research studies, the improved administration and methodology must be formulated and implemented [13]. However, for precise regulations and monitoring of EPs to determine the permissible limit of these pollutants in the environment, there is a dire need for a complete understanding of fate and toxicities [14]. Ecotoxicology study is a major concern of the EPs because most pollutants are persistent and possess bioaccumulation [15].

The ecotoxicity of different emerging toxicants has been evaluated previously on sentinel species recommended by the U.S.-Environmental Protection Agency (USEPA) and the European Union (EU) for the safety potential of xenobiotics. To determine exposure limits of specific pollutants in the environment and decide the safety, various models and experimental protocols have been developed previously by researchers. The two widely used are whole organism battery tests and cell-lines decipher. The whole-organism battery test determines ecological exposure obtained by performing significant co-relation with realistic scenarios. At the same time, laboratory-based information obtained by cell lines deciphers certain limitations in extrapolating the adverse effects information regarding organisms of higher levels, such as humans, and is considered the bottleneck in ecotoxicity assessment.

Based on numerous emerging pollutants in an ecosystem, this chapter provides the classification of potential emerging contaminants, such as active pharmaceutical chemicals, personal care products, pesticides, and emerging industrial chemicals found in the environment. The knowledge of physicochemical properties of pollutants helps determine the fate and transport of specific pollutants, which are also discussed. Moreover, the ecotoxicological impacts of several EPs and available legislations formulated for the monitoring and handling of several EPs in the world are also discussed in this chapter.

PATHWAYS OF EPS IN THE ECOSYSTEM

EPs enter the ecosystem through different paths, including industrial/municipal wastewater, hospital wastewater, treated/partially treated effluent from the treatment plant, farmyard and agricultural runoff released into surface water, and application of sludge and biomass [16, 17]. For example, some pharmaceuticals are not completely metabolites by animals and humans and are released into the environment by urine and feces [8]. The classes of different EPs and their metabolites recently detected in other components of the environment around the globe are described in Table **1**. Most of these pollutants are not completely removed by conventional water and wastewater treatment facilities. Therefore, when released into rivers, lakes, and coastal water, partially or untreated water pollute the water with trace concentration of EPs. Thus, when used for horticulture, irrigation, or other non-potable purposes, water from these natural sources intensifies the concentration of these pollutants from parts per trillion to parts per million, thus deteriorating the quality of surface water, groundwater, and soil [18]. Other possible sources of EPs pollution in the environment might include agricultural runoff, recreational activities, swimming, veterinary medicines, and the discharge of various disposable goods [19, 20]. Typical pathways of EPs pollution from different sources are described in Fig. (1).

Table 1. Description of various classes of emerging pollutants.

Pollutants	Major Sources	Description	Ref.
PhACs	• Human or animal excreta • Wastewater • Treated Sewage Sludge • Medical Waste • Landfill leachate	• Impact human health and water supply in ecosystems. Research on its adverse effects has been recently initiated. • Extensive usage in veterinary and human medicine, such as anti-diabetics, antibiotics, and analgesics, among others. • Incomplete removal with water treatment. Most pharmaceuticals are not metabolites completely.	[21, 22]
PCPs	• Sewage Treatment Plant • Cosmetics • Dental-care products • Household items	• Therefore, some PCPs are not metabolites and are released into freshwater and underground water resources. • The main source of these pollutants is domestic wastewater treatment plants because most pollutants are resistant to conventional wastewater treatment facilities.	[3]
Metabolite	• Agricultural wastewater • Human and Animal excreta • Domestic wastewater	• Few PhACs are partially or completely transformed in the liver, containing several enzymes. • These compounds are found in rivers or streams because most pollutants are left after residual treatment water. • Some of the persistent nature of metabolites from agricultural activities (use of pesticides) do not undergo natural degradation, ultimately being released into rivers and streams.	[5, 22]
Endocrine disruptors	• Wastewater treatment plants • Human and Animals excreta	• The resemblance with natural hormones causes negative impacts on the endocrine system of animals and humans. • Estrogens found in freshwater bodies are natural estrogens, estriol, 17α-ethinylestradiol, 17β-estradiol	[23]
Pesticides	• Agricultural activities • Pesticide manufacturing plants	• Chemicals used to control the disease and spread of pests in crops; in horticulture, forestry, agriculture, *etc.*, these compounds, at the time of their application, reach the soil through irrigation water or rainwater, ultimately reaching the ground as well as surface waters.	[8, 24]
Fire Retardants	• Construction, electronic, and furniture industries	• These chemicals are soluble in water and used in various consumer products to meet fire retardations. It was due to these chemicals' solubility properties detected in water bodies.	[2]

(Table 1) cont.....

Pollutants	Major Sources	Description	Ref.
Food Additives	• Food industry • Domestic wastewater	• Synthetic organic compounds, such as antioxidants, sweeteners, *etc.*, owing to the higher water solubility of sweeteners, are widely detected in wastewater. In contrast, antioxidants are due to insolubility found in underground water.	[25]
UV Filters	• Domestic wastewater • Cosmetics industry • Paints and Plastic industries	• Endocrine disruption impacts aquatic and terrestrial organisms by developing brain and reproductive organs that interfere with the thyroid axis.	[26]
Hydrocarbons	• Exhaust fumes of vehicle • Petrochemical industries • Coal combustions • Oil spillage	• Intense toxicity, carcinogenic and mutagenic • Fire release various types of hydrocarbons, such as polycyclic aromatic hydrocarbons. They enter the ecosystem *via* industrial effluent; fossil fuel combustion reactions pollute the surface and underground water.	[27, 28]

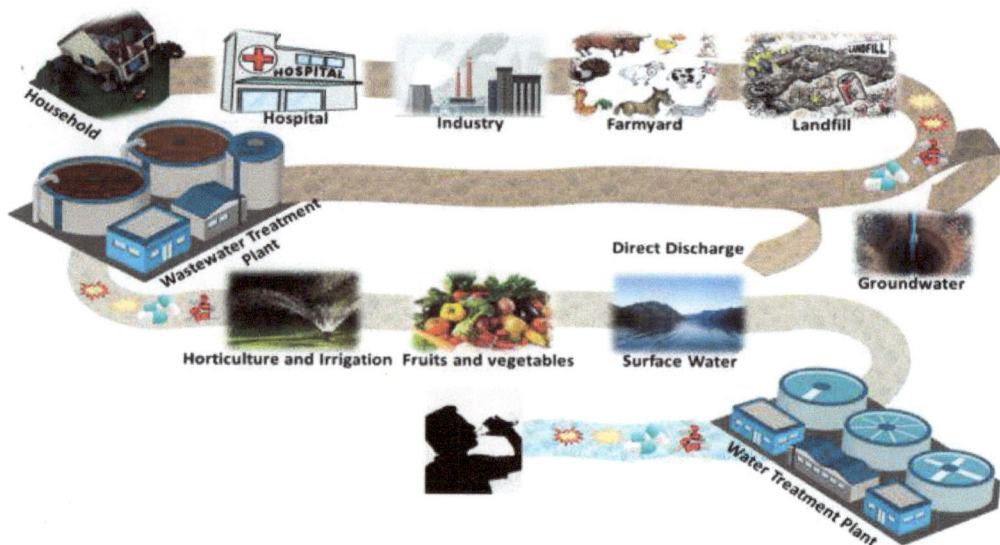

Fig. (1). Pathways and Potential Sources of Emerging Pollutants [29].

CLASSIFICATION OF EMERGING POLLUTANTS

Emerging pollutants are broadly classified into 4 major groups: pharmaceuticals, personal care products, pesticides, and numerous industrial chemicals based on their sources and usage. The fate and transport of EPs are strongly dependent upon the physicochemical properties of the contaminants, including the octanal-water partition coefficient (K_{ow}), distribution coefficient (D), water solubility, and dissociation constant (pKa), as well as the metabolic capability of microbes in the ecosystem. Various groups and subgroups of EPs and their physicochemical properties are described in Table **2**.

Table 2. Summary of physiochemical properties of representative EPs detected in water.

Group	Sub-groups	Pollutants	Molecular Formula	pKa	Log K_{ow}	Log D	Water Solubility (mg/l) at 25°C
PhACs	Antibiotics	Ethromycine	$C_{37}H_{67}NO_{13}$	8.9	3.06	1.55	4.2
		Amoxicillin	$C_{16}H_{19}N_3O_5S$	3.2	0.87	-	3430
		Tylosin	$C_{46}H_{77}NO_{17}$	7.73	1.633	-	211
		Penicillin	$C_{16}H_{18}N_2O_4S$	2.76	1.83	-	210
		Cephalexin	$C_{16}H_{17}N_3O_4S$	0.65	5.2	-	10,000
		Ofloxacin	$C_{18}H_{20}FN_3O_4$	5.8	-2.0	-0.25	28000
		Tetracycline	$C_{22}H_{24}N_2O_8$	3.30	−1.37	-	231
		Sulfamethoxazole	$C_{10}H_{11}N_3O_3S$	1.7	0.89	0.45	610
		Oxytetracycline	$C_{22}H_{24}N_2O_9$	3.37	-0.97	-	313
	Analgesic and anti-inflammatory drugs	Acetaminophen	$C_8H_9NO_2$	9.5	0.46	0.23	14,000
		Ibuprofen	$C_{13}H_{18}O_2$	4.47	3.97	1.44	21
		Naproxen	$C_{14}H_{14}O_3$	4.2	3.18	0.34	15.9
		Mefenamic Acid	$C_{15}H_{15}NO_2$	3.8	5.12	2.04	200
		Fenoprofen	$C_{15}H_{14}O_3$	4.21	3.9	0.38	0.128
		Ketoprofen	$C_{16}H_{14}O_3$	4.29	3.12	0.41	51
		Indometacin	$C_{19}H_{16}ClNO_4$	3.8	4.23	0.75	-
		Diclofenac	$C_{14}H_{11}C_{12}NO_2$	4.08	4.51	1.59	23.7

(Table 2) cont.....

Group	Sub-groups	Pollutants	Molecular Formula	pKa	Log K_{ow}	Log D	Water Solubility (mg/l) at 25°C
	Antiepileptic drugs	Primidone	$C_{12}H_{14}N_2O_2$	12.2	0.91	0.83	600 at 37°C
		Carbamazepine	$C_{15}H_{12}N_2O$	13	2.45	2.58	17.7
	lipid regulators	Clofibric acid	$C_{10}H_{11}ClO_3$	3.35	2.57	1.08	582.5
		Acebutolo	$C_{18}H_{28}N_2O_4$	-	1.71	-	259
		Gemifibrozil	$C_{15}H_{22}O_3$	4.45	4.77	2.22	27.8
		Bezafibrate	$C_{19}H_{20}ClNO_4$	3.44	4.25	0.69	2
		Fenofibric acid	$C_{17}H_{15}ClO_4$	-	4	-	9
		Pravastatin	$C_{23}H_{36}O_7$	4.2	3.1	-1.21	-
	β-blockers Contrast	Propranolol	$C_{16}H_{21}NO_2$	9.6	3.48	1.15	79.4
		Atenolol	$C_{14}H_{22}N_2O_3$	9.6	0.16	-	26.5 (37°C)
		Metoprolo	$C_{15}H_{25}NO_3$	9.49	1.88	0.61	16900
		Sotalol	$C_{12}H_{20}N_2O_3S$	-	0.24	-	5500
	Contrast media	Iopromide	$C_{18}H_{24}I_3N_3O_8$	13	2.10	-	366.6
		Iopamidol	$C_{17}H_{22}I_3N_3O_8$	10.7	-	-	1216 at 37°C
		Iohexol	$C_{19}H_{26}I_3N_3O_9$	11.7	3.05	-	-
	Hormones	Estriol	$C_{18}H_{24}O_3$	-	2.45	-	441
		Estrone	$C_{18}H_{22}O_2$	10.3	3.13	3.6 at pH9	147
		17-α ethinylestradiol	$C_{20}H_{24}O_2$	10.4	4.01	4.12at pH9	10
		17B-estradiol	$C_{18}H_{24}O_2$	10.3	3.9	2.53	1.51
	Cytostatic drugs	Cyclophosphamide	$C_7H_{15}Cl_2N_2O_2P$	0.5	0.97	-	15000
	Psychostimulants	Caffeine	$C_8H_{10}N_4O_2$	5.3	-0.04	-	20,000
		Paraxanthine	$C_7H_8N_4O_2$	-	-	-	1000

(Table 2) cont.....

Group	Sub-groups	Pollutants	Molecular Formula	pKa	Log K_{ow}	Log D	Water Solubility (mg/l) at 25°C
PCPs	Insect repellent	N, N-diethyl-m-toluamide	$C_{12}H_{17}NO$	< 2	2.8	-	11000
	Preservatives	Propyl-paraben	$C_{10}H_{12}O_3$	8.5	3.04	-	960
		Methyl-paraben	$C_8H_8O_3$	-	-	1.86 at pH6	3690
	Anti-microbial agents/ Disinfectants	Triclosan	$C_{12}H_7C_{l3}O_2$	7.8	5.34	5.28	10
		Triclocarban	$C_{13}H_9C_{l3}N_2O$	11.4	4.90	-	-
	Sunscreens	Oxybenzone	$C_{14}H_{12}O_3$	7.6	3.79	-	< 1
	Cosmetics	Salicylic acid	$C_7H_6O_3$	2.97	-	-	2480
		Tonalide	$C_{18}H_{26}O$	-	5.7	-	1.25
		Galaxolide	$C_{18}H_{26}O$	-	5.9	-	1.75
	UV filters	Homosalate	$C_{16}H_{22}O_3$	-	6.16	-	37
		Benzophenone-3	$C_{14}H_{12}O_3$	-	3.79	-	68.56
		4-methoxycinnamate	$C_{11}H_{12}O_3$	-	5.80	-	0.15
		Octyl methoxycinnamate	$C_{18}H_{26}O_3$	-	3.75	-	7.226
		4-methylbenzylidene camphor	$C_{18}H_{22}O$	-	5.28	-	0.20
Pesticides	Fungicides	Chlorpyrifos	$C_{22}H_{17}ClN_2$	-	-2.63 at 20°C	-	7850 at 20°C
		Dimethoate	$C_5H_{12}NO_3PS_2$	-	0.78	-	25000
		Tebuconazole	$C_{16}H_{22}ClN_3O$	-	-	-	36.0
		Pyraclostrobin	$C_{19}H_{18}ClN_3O_4$	-	3.69	-	1.9 at 20°C
		Propiconazole	$C_{15}H_{17}Cl_2N_3O$	1.09	3.72	-	150 at 20°C

(Table 2) cont.....

Group	Sub-groups	Pollutants	Molecular Formula	pKa	Log K_{ow}	Log D	Water Solubility (mg/l) at 25°C
	Herbicides	Atrazine	$C_8H_{14}ClN_5$	1.7	2.6	2.08	33
		Isoproturon	$C_{12}H_{18}N_2O$	-	2.87	-	70
		Diuron	$C_9H_{10}C_{12}N_2O$	2.68	-	-	37
		Bentazone	$C_{10}H_{12}N_2O_3S$	3.3	-0.46	-	500
		Oxyfluorfen	$C_{15}H_{11}ClF_3NO_4$	-	4.47	-	0.116
		Clomazone	$C_{12}H_{14}ClNO_2$	-	2.5	-	1100
		Glyphosate	$C_3H_8NO_5P$	5.44	-	-	12000
		Simazine	$C_7H_{12}ClN_5$	-	2.18	-	4006
	Insecticides	Diazinon	$C_{12}H_{21}N_2O_3PS$	2.6	3.8	-	40 at 20 °C
		Pirimicarb	$C_{11}H_{18}N_4O_2$	-	1.4	-	970
Industrial Chemicals	Plasticizers	Bisphenol A	$C_{15}H_{16}O_2$	-	9.6	3.32	381
		dibutyl phthalate	$C_{16}H_{22}O_4$	-	4.27	-	9.9
		di(2-ethylhexyl) phthalate	$C_{24}H_{38}O_4$	-	-	-	7.73
		dimethyl phthalate	$C_{10}H_{10}O_4$	-	1.53	-	24.70
		diisobutyl phthalate	$C_{16}H_{22}O_4$	-	9.9	-	4.27
	Flame Retardants	Tri(chloropropyl) Phosphate	$C_9H_{18}C_{13}O_4P$	-	1.44	-	7000
		Tri(2-chloroethyl) Phosphate	$C_9H_{15}O_6P$	-	2.59	-	6570
	Food additives	Butylhydroxytoluene	$C_{15}H_{24}O$	-	6.27	-	0.06
	Hydrocarbons	Benzopyrene	$C_{20}H_{12}$	-	6.04	-	0.0038
		Perfluorooctane sulfonate (PFOS)	$C_8HF_{17}O_3S$	-3.27	4.51	-	750
		Perfluorooctanoic acid (PFOA)	$C_8HF_{15}O_2$	0.50	6.44	-	13600
The data were compiled from [4, 5, 8, 22, 31, 43, 50, 55, 60 - 81].							

Pharmaceutical Active Compounds

Pharmaceutical active compounds are an imperative class of emerging pollutants because of their widespread occurrence in water bodies, drinking water contamination, and toxic effects on human health and ecosystems [30]. In recent years, numerous countries have reported the presence of different subgroups of PhACs in finished drinking water, including antibiotics, analgesic and anti-inflammatory drugs, antiepileptic drugs, β-blockers Contrast, Harmons and others [29, 31]. Different pollutants of subgroups of pharmaceutically active compounds and physicochemical properties are described in Table **2**. Most of these pollutants are reported at an elevated level in treated water based on the type of treatment facilitated [32 - 34]. A trace amount of a few pollutants (ng/l) such as carbamazepine, acetaminophen, ibuprofen, and clofibric acid has been found in treated drinking water in several countries. It is observed that the treatment system utilizing surface water shows a higher concentration of these contaminants than the system that uses groundwater [35]. Furthermore, it was also observed that a few of the drinking water treatment methods such as coagulation/flocculation, disinfection and filtration could efficiently remove the parent pollutants of some PhACs. Still, these pollutants sometimes react with disinfectants and form disinfection by-products (DBPs), which are difficult to remove [36, 37].

In addition to excessive use of pharmaceuticals by humans, they are also used in fish farming, livestock, and poultry. A wide range of drugs is frequently used for animals to reduce infection and diseases. Although around three thousand chemicals are used as pharmaceutical ingredients, only a few are investigated in the environment. Regulatory agencies and scientists are always questioned about the adverse effects of these contaminants on the ecosystem at minimum trace concentration (ng/l). Even though several pharmaceutical pollutants can potentially affect humans, animals and ecosystems adversely, most environmental concentrations are observed well below the lowest observed effect concentrations (LOECs). However, few PhACs such as diclofenac, ciprofloxacin, ethinylestradiol, carbamazepine, fluoxetine, and clofibric acid have been found in water at a level well above the LOECs [38].

Previously, the two major disasters occurred owing to the release of trace amounts of PhACs in the environment. First, about 40 million vultures died in Pakistan from consuming diclofenac, an anti-inflammatory drug from cattle carcasses [39]. This event was significantly surprising because about 90% of the vulture population was poisoned by diclofenac. Therefore, the event was named as "worst case of wildlife poisoning ever". The second happened in Ontario, Canada, where the collapse and feminization of a wild fish owing to the release of trace concentration of estrone (5 to 6 ng/l) into lake water were investigated and found responsible [40].

Furthermore, hormones are also one of the most important subgroups of PhACs, owing to their potential androgenic and estrogenic effects on animals [41]. A wide range of synthetic and natural hormones are often released into the environment by wastewater treatment and agriculture runoff [32]. The U.S. environmental protection agency published a list of 9 widely reported hormones in the background in new CCL-4 (Ethinylestradiol, norethindrone, aquiline, estrone, equilenin, mestranol, 17β estradiol, 17α-estradiol) (UEPA, 2020).

Personal Care Products

Personal care products (PCPs) include various types of goods used in our daily life, such as surfactants, perfumes, insect repellents, sunscreens, UV filters, disinfectants, *etc*. Unlike pharmaceuticals, PCPs do not undergo any metabolic changes because these products are applied externally, therefore, released into the environment in their original forms [25]. The most widely detected PCPs and subgroups and physicochemical properties are described in Table **2**. However, due to intense urbanization and excess use of these products, they are widely detected in surface and groundwater. Moreover, most of these pollutants are persistent and have bioaccumulation potential [42]. It is observed that when N, N-diethyle-m-toluamide, a widely used ingredient of insect repellent, is applied, only 20 percent of the total chemical is consumed by the screen, and the remaining 80 percent is discharged into water [43]. Polycyclic tonalide and galaxolide musk are the two most used ingredients in fragrance manufacturing. In Europe alone, around 5000 tons of these chemicals were produced in 2004. it has been reported that about 77 percent of used musk was released into water [25].

Furthermore, disinfectants such as triclosan are commonly used ingredients in toothpaste, shampoo, and soap and are often detected in the environment [44]. The use of this chemical was banned in Minnesota in 2017. Similarly, in Europe, the excess use of this chemical was recently limited (EU, 2014), and a ban on other use is being considered. In the United States, an extensive study was conducted in 2002 to investigate lakes and rivers affected by wastewater. About 50 percent of the samples were polluted with triclosan chemicals [45]. One of the major concerns of triclosan contamination is that it has characteristics of active contribution to antibiotic resistance. It is validated by one of the studies performed to identify triclosan resistance bacteria in the aquatic environment [46]. Triclosan contamination equally affects marine life, including crustaceans, algae, and fish. Moreover, this pollutant also has endocrine-disrupting, cytotoxic, and genotoxic properties. Additionally, it is also reported that triclosan may transform into hazardous by-products during water and wastewater treatment [47].

UV filters and sunscreens are another sub-group of personal care products widely used in manufacturing cosmetic products such as hairsprays, moisturizers, skincare, hair dyes, and lipsticks. These chemicals also manufacture non-cosmetic products such as furniture, washing powder, plastics, and carpet [48]. The most often used UV-filters are camphor, 4-methylbenzylidene, benzophenone-3, and homosalate *etc* [48]. These pollutants enter the aquatic environment through bathing, washing and swimming [20]. Human and animal exposure pattern to these contaminants overlaps through the food chain.

Pesticides

Due to rapid urbanization, the extensive use of various pesticides in agriculture, forestry, and amenities such as airports, sports grounds, public parks, and industrial sites increases [49]. The widely used sub-groups of pesticides include insecticides, herbicides, and fungicides to protect the crops from pests and disease [50]. When applied to the crops, these pesticides enter freshwater bodies by irrigation runoff water, spray drift, or wash water. Moreover, during the application of pesticides on crops, some amounts of chemicals reach the soil, eventually polluting the unground water after seepage [51].

Industrial Chemicals

The extensive use of different plasticizers, including bisphenol A, DBP, DEHP *etc.*, for improving the rheology and plasticity of gypsum, clays, concrete, and plastics has created a variety of harmful pollutants. Among these pollutants, most plasticizers are a potential endocrine disruptor. For example, Bisphenol A, an ingredient of polycarbonate plastics, has been used in wide applications, including flame retardants, containers of beverages and foods, electronic components, paper coatings, eyeglass lenses, building materials and dental sealants. They are the highest manufactured chemicals around the globe, with an annual production of around 11.5 billion pounds in 2012 and are on the increase with a 4.6% annual rate and are expected to reach 10.6 million metric tons by 2022 [52]. This chemical enters the water cycle by releasing wastewater from industries, landfill leachates and other domestic waste [35]. According to the United States environmental protection agencies, around 1.5 lac pounds of BPA was released into the environment in 2010 (UEPA 2010).

Flame retardants are another subgroup of emerging industrial chemicals widely produced globally. The most commonly manufactured flame retardants include tri(chloropropyl) Phosphate and tri(2-chloroethyl) Phosphate [53]. These chemicals are used in various applications such as thermoplastics, thermostats, electronics and furniture coatings [54]. Owing to the hydrophobic properties, a trace amount of these chemicals are often found in sediments than in the aquatic environment [55]. Moreover, their persistent nature might transport them away from the point source for a long distance and uptaken by vegetables and marine life when released [56]. Therefore, various studies have reported their occurrence in food chains. Similarly, numerous other chemicals are widely detected, including different food additives and hydrocarbons.

Electronic Waste

Since the abrupt digitalization, the production of electrical and electronic products has increased significantly. The waste from electronic devices contains metals and organic and inorganic materials, some of which are toxic and hazardous. The valuable portion of the electronic waste is recovered and recycled. However, some of its portions still go into wastewater [57]. The major portion of electronic equipment is the metals, several of these metals and metalloids are reported to develop various forms of cancer. For example, As, Be, Cd, and Cr(VI) is reported by the International Agency for Research on Cancer (IARC) as carcinogenic to humans. Similarly, Pb, antimony trioxides, Co, organic Hg compounds, and Ni are suspected to be carcinogenic [58, 59].

TOXICITY AND ECOLOGICAL RISK OF EPS

Various EPs and their metabolites in the environment can pose a serious threat to human health, animals, and aquatic species, including non-targeted organisms such as bacteria, algae, and plants [82]. The unsustainable production of numerous synthetic pharmaceuticals and chemicals creates various pollutants in an aquatic environment, which becomes a reason for several adverse effects [83]. Over usage and release of ample amounts of antibiotics by humans and animals create numerous types of antibiotic resistance in the aquatic environment, posing a serious threat to public health. According to the data released by the Center for Disease Control and Prevention (CDC), particularly in the EU and U.S., about 25 thousand and 23 thousand deaths occurred due to the presence of antibiotic resistance, respectively. In addition, around 2 million individuals developed drug resistance in the U.S [84]. Some researchers predict that by 2050, 10 million deaths will be caused by exposure to these antibiotic resistances, and cancers as a leading cause of mortality globally [85]. The principal reason for this issue includes poor hygiene and sanitation, antibiotics prescriptions, and insufficient facilities in a laboratory for infection detection accurately and quickly [86]. Various health effects in humans due to different emerging pollutants in water are shown in Fig. (2). At the same time, the summary of ecotoxicology effects of widely reported emerging pollutants is described in Table **3**.

Table 3. Summary of potential ecotoxicology effects of various EPs.

Categories	Representative Pollutants	Adverse Health Effects of Representative Pollutants				Reference
		Human	Animals	Aquatic Species	Other	
Antibiotic	Sulfamethoxazole	Growth inhibition of humans was observed when mixed with 12 other antibiotics	-	Chronic, acute toxicity at lower mg/l; Mutagenic; chronic impacts on the aquatic environment	Antimicrobial resistance	[89, 90]
	Ciprofloxacin	-	Chronic hepatotoxicity in animals	Abnormal development and histopathological changes observed in fish.	-	[91, 92]
	Penicillin	Cytotoxic effects	-	Demag antioxidant and immune response in fish	Cause resistance in bacterial pathogens, the primary reason for altered microbial community structure in the ecosystem, ultimately affecting the higher food chain.	[64]
	Roxithromycin, Tylosin	-	Liver or kidney disease in pets	-	Growth inhibition of algae (Pseudo kirchneriellasubcapitata)	[20]
Harmons	Estrone, 17-β estradiol, 17-α ethinylestradiol	A higher level can cause breast cancer in women	Abnormalities in domestic animals; change in teat length is observed	Abnormal growth of testicular in male; fathead minnows at lower concentration (ng/l); can cause feminism on chronic exposure.	Sunflower, potato plant, seedings; impacts flowering, root, shoot growth with estrone.	[3, 87, 88]
Analgesic and anti-inflammatory drugs	Acetaminophen	Enhance estrogenicity in adrenal cells modify steroidogenic	At high concentration, it causes chronic hepatoxicity in mice (cirrhosis and hepatocyte necrosis)	Embryonic development is affected in zebrafish; at ppm level, survival of fish and daphnia are affected	-	[44, 93]
	Diclofenac	A higher dosage for a long time may cause ulcers in the stomach	Couse kidney failure in birds	Gill alteration and renal lesions were observed in the seawater species (rainbow trout)	Depression and tiredness were observed in birds; higher doses may reach in mortality in birds (pigeons)	[3, 39, 94]
	Naproxen	-	Toxicity and genotoxicity	Renal and Gastrointestinal effects in zebrafish	Bacteria, rotifers, algae, and microcrustaceans	[95, 96]
β-blockers Contrast	Propranolol	-	Slow heart rate, low blood pressure, bronchospasm in pets	Reduction of viable eggs of Japanese medaka fish was observed (Oryiaslatipes)	Inhibit shivering in birds	[97, 98]

(Table 3) cont.....

Categories	Representative Pollutants	Adverse Health Effects of Representative Pollutants				Reference
		Human	**Animals**	**Aquatic Species**	**Other**	
Contrast media	Iopromide	Low possibility of human toxicity	In chloraminated and chlorinated drinking water, iodo-DBPs are formed, which are genotoxic and cytotoxic to mammalian	At about 100 ppm, no chronic toxicity is observed of the degradation product on zebrafish	Notmicrocidal to bacteria	[99]
Psychostimulants	Caffein	-	Pets may lose muscle control and have tremors or seizures	Endocrine disruption of goldfish (Carassiusauratus) Renal	Distress, hyperactivity, and cardiac arrest can cause in birds.	[100]
UV-filters	4-MBC	After applying dermal UV-filter mixtures, thyroid hormones and the reproductive system might be affected.	Alteration in steroid hormones and gonadal weight production in male rats	Estrogenic activity potency fish-MCF-7 cell lines; mixture with other UV-filters may risk aquatic environment; antiestrogenic activity using fish ERα;	mortality decreased; Reproduction increased; in worm Lumbriculus variegates	[101]
Antimicrobial Agents	Triclosan	Risk of asthma, allergies, food sensitization	Cancer development on mouse liver	A decrease in sperm count is observed in fish.	Growth inhibition of algae (Pseudokirchneriellasubcapitata) Proven	[44]
Herbicides	Atrazine	Kidney failure and congestive heart failure	Enhance incidence of estrogen-dependent mammary cancers in rodents	Osmolality, elevated plasma cortisol; thyroxine, major physiological stress in fish	Reduced growth and feed intake in birds	[49, 78, 102, 103]
	Simazine	Reproductive disorder and hear failure	Cachexia was observed in the dog at a higher dose	-	-	[49, 104]
Fungicides	Dimethoate	Diarrhea, blurred vision, respiratory disorder	Skin irritation and eye irritation in the rabbits	-	Cholinesterase inhibition observed in the hen	-
Plasticizer	Bisphenol A, DBP	Probability of carcinogenic impact, EDC	EDC, synaptic, miotic aneuploidy	Reproductive and Estrogenic impacts	Possible EDC in bird	[25]
Fire retardants	Tri(chloropropyl) Phosphate	thyroid cancer, heart disease high exposure might affect fertility.	Central nervous system excitation, including seizures in dogs (Pitbull, German Shepherd)	Spinal deformities in fish (Killifish); Central nervous system (CNS) development	Reduce heart rate of chicken embryo	[105 - 109]

(Table 3) cont.....

Categories	Representative Pollutants	Adverse Health Effects of Representative Pollutants				Reference
		Human	Animals	Aquatic Species	Other	
Per-fluorinated compounds	PFOS	Possibility of low sperm count; and thyroid disease.	Compact bodyweight and cholesterol, neonatal mortality, liver weight increase, carcinogenic to rodents	Mussel mortality observed	-	[110]

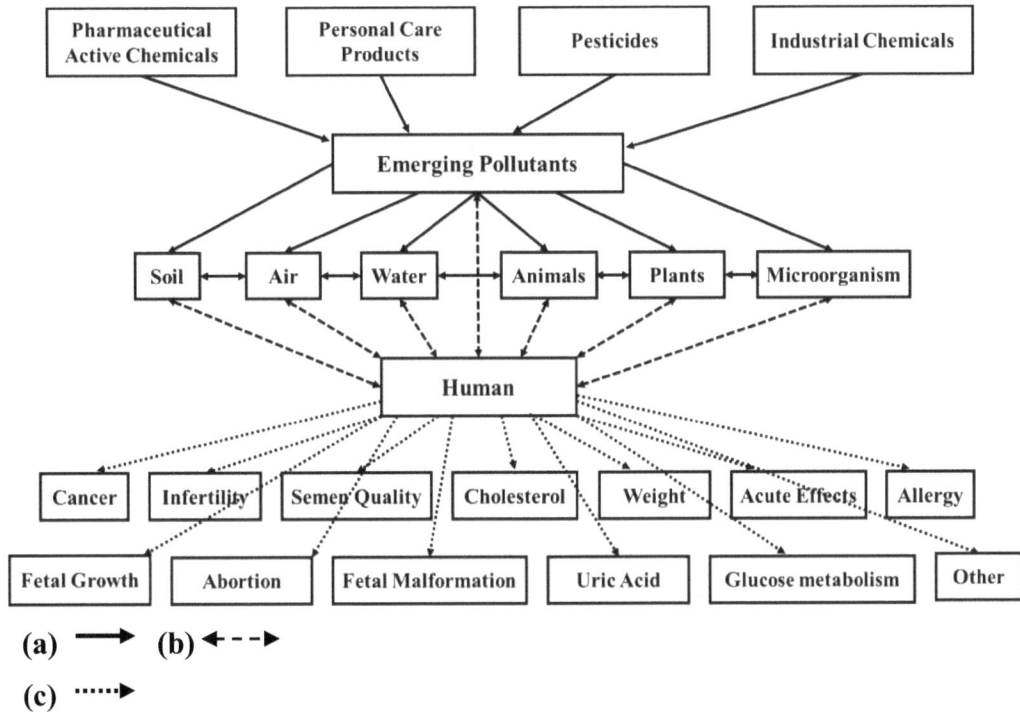

Fig. (2). Effects of different emerging pollutants. (a) Represents the other groups of emerging contaminants. (b) Represents the interaction between humans, plants, animals, microorganisms, air, soil, water and emerging pollutants. (c) Describes the potential adverse effects of emerging contaminants on humans [48].

Other problematic pollutants are the compounds of harmons, commonly prescribed these days, the leading cause of disease in animals and aquatic life. One of the studies performed to predict adverse health effects associated with widely reported harmons (Estrone, 17-β estradiol, 17-α ethinylestradiol) predicted the altered sexual development, presence of intersex species, changed mating behavior, *etc.* in marine life. Higher intersexuality was found in walleyes (Sander vitreus vitreus) and wild roach (Rutilus) in rivers [87]. Chronic exposure to 17-α ethinylestradiol results in the feminization of various fish species [40]. In addition, different health effects in other domestic animals are also reported in studies, such as shortness of teat length [88].

LEGISLATIONS

To develop a sustainable and protected environment for future generations, strict policy regulations are essential. It is evident from the previous practices that intense urbanization and industrialization created numerous harmful emerging pollutants in our environment, which are difficult to be removed. The studies on the occurrence of EPs in the ecosystem, their toxicities, and removal have been conducted worldwide. Yet, most studies include case studies in Europe, the United States, Canada, and some developed regions of Asia [111]. There is a lack of information about emerging pollutants in the aquatic environment [112]. Therefore, most policies and regulations are formed in developed regions [13]. Currently, available policy regulations for mitigating EPs are as follows:

- EU water policy (Directive, 2000) has been framed to describe and prioritize the high-risk pollutants [113]
- 33 priority compounds based on their ecological quality standards have been confirmed as per directive 2008/105/EPs [114]
- A watch list of chemicals of emerging concern for monitoring in Europe has been formulated in the field of water policy in a decision 2015/495/EU [115]
- Fate and occurrence of EPs in the watch list show fewer reports in the list, 27 examinations of transformation by-products, and the study of unspiked water matrices.

In emerging pollutants, such as PhACs, PCPs, pastiches, and various industrial contaminants, fate and transport mechanisms are not discovered at diverse levels of biological organization for determining the effective risk assessment policies [49]. In some developed countries like the U.S., legislation for plastic beads was recently formulated to ban microplastics because of confirmed adverse effects on human health and the aquatic environment. Similarly, other developed countries like New Zealand, Canada, and Kenya also implemented strict regulations on plastic beads [116]. The United States' drug and food administration center for drug evaluation and research regulates over-the-counter widely prescribed drugs, including genetic and therapeutics. In addition, this department also considers new drugs which are widely prescribed and used for humans and animals for policy formulation and safety efficacy.

Similarly, in Japan, to evaluate various pharmaceuticals used for veterinary and human health formulated, the Organization of Pharmaceuticals and Research worked under the umbrella of their ministry [117]. Furthermore, a trilateral (United States-Japan-EU) program was designed for the authorization of veterinary products and their marketing with technical and safety requirements (VICH). This program aimed to register veterinary products for their safe use [118].

In Europe, the European Medicines Agency (UMA) is formulated to evaluate and regulate pharmaceutical products used for animals and human health [119]. Despite the several international policies and regulating agencies, the regulation for monitoring various emerging water pollutants has not been completely formulated. Therefore, the concerned regulatory authorities and stakeholders must develop a strategic plan based on available scientific evidence and database. The developing countries face severe threats due to the lack of facilities and skilled personnel. Thus, there is a dire need for strict legislation at a regional level to formulate policies to overcome these alarming issues.

CONCLUSION

Emerging pollutants in the ecosystem have been recognized as an essential environmental issue. Many studies have reported the fate, occurrence, and ecotoxicity of numerous subgroups of EPs in river water, underground water, sediments, and effluent of wastewater treatment plants, in which trace concentrations range from nanograms per liter to milligram per liter. However, many EPs' fate and transport mechanisms are still not completely analyzed in the ecosystem. Moreover, the ecotoxicity of these EPs on different components of the environment and biological tissues is also not evaluated. Thus, there is a dire need for further exploration of various groups of undetected emerging pollutants in the environment along with their potential adverse impacts on the ecosystem.

Furthermore, a vast majority of legislative policies are designed and implemented in developed regions of the world. However, developing countries' condition is worse due to the lack of facilities, skilled personnel, and international waste disposal alongside freshwater resources. Hence, better policy regulations must be formulated at regional levels to regulate the occurrence of these pollutants in the environment.

CONSENT FOR PUBLICATION

Not applicable.

CONFLICT OF INTEREST

The author declares no conflict of interest, financial or otherwise.

ACKNOWLEDGEMENTS

Declared none.

REFERENCES

[1]   H. Salazar, P.M. Martins, B. Santos, M.M. Fernandes, A. Reizabal, V. Sebastián, G. Botelho, C.J. Tavares, J.L. Vilas-Vilela, and S. Lanceros-Mendez, "Photocatalytic and antimicrobial multifunctional nanocomposite membranes for emerging pollutants water treatment applications", *Chemosphere,* vol. 250, p. 126299, 2020.
[http://dx.doi.org/10.1016/j.chemosphere.2020.126299] [PMID: 32113095]

[2]   C. Peng, H. Tan, Y. Guo, Y. Wu, and D. Chen, "Emerging and legacy flame retardants in indoor dust from East China", *Chemosphere,* vol. 186, pp. 635-643, 2017.
[http://dx.doi.org/10.1016/j.chemosphere.2017.08.038] [PMID: 28818590]

[3]   A. Gogoi, P. Mazumder, V.K. Tyagi, G.G. Tushara Chaminda, A.K. An, and M. Kumar, "Occurrence and fate of emerging contaminants in water environment: A review", *Groundw. Sustain. Dev.,* vol. 6, no. January, pp. 169-180, 2018.
[http://dx.doi.org/10.1016/j.gsd.2017.12.009]

[4]   V. Geissen, H. Mol, E. Klumpp, G. Umlauf, M. Nadal, M. van der Ploeg, S.E.A.T.M. van de Zee, and C.J. Ritsema, "Emerging pollutants in the environment : A challenge for water resource management", *Int. Soil Water Conserv. Res.,* vol. 3, no. 1, pp. 1-9, 2015.
[http://dx.doi.org/10.1016/j.iswcr.2015.03.002]

[5]   M. Gavrilescu, K. Demnerová, J. Aamand, S. Agathos, and F. Fava, "Emerging pollutants in the environment: present and future challenges in biomonitoring, ecological risks and bioremediation", *N. Biotechnol.,* vol. 32, no. 1, pp. 147-156, 2015.
[http://dx.doi.org/10.1016/j.nbt.2014.01.001] [PMID: 24462777]

[6]   N.A. Khan, S.U. Khan, S. Ahmed, I.H. Farooqi, M. Yousefi, A.A. Mohammadi, and F. Changani, ""Recent trends in disposal and treatment technologies of emerging-pollutants- A critical review," TrAC -", *Trends Analyt. Chem.,* vol. 122, p. 115744, 2020.
[http://dx.doi.org/10.1016/j.trac.2019.115744]

[7]   C. Postigo, and D. Barceló, "Synthetic organic compounds and their transformation products in groundwater: occurrence, fate and mitigation", *Sci. Total Environ.,* vol. 503-504, pp. 32-47, 2015.
[http://dx.doi.org/10.1016/j.scitotenv.2014.06.019] [PMID: 24974362]

[8]   C. Peña-Guzmán, "Emerging pollutants in the urban water cycle in Latin America: A review of the current literature", *J. Environ. Manage,* vol. 237, 2019.
[http://dx.doi.org/10.1016/j.jenvman.2019.02.100]

[9]   A.B. Patel, S. Shaikh, K.R. Jain, C. Desai, and D. Madamwar, "Polycyclic Aromatic Hydrocarbons: Sources, Toxicity, and Remediation Approaches", *Front. Microbiol.,* vol. 11, no. November, p. 562813, 2020.
[http://dx.doi.org/10.3389/fmicb.2020.562813] [PMID: 33224110]

[10]   K. Gautam, and S. Anbumani, *Ecotoxicological effects of organic micro-pollutants on the environment.* Elsevier B.V., 2020.
[http://dx.doi.org/10.1016/B978-0-12-819594-9.00019-X]

[11]   C.G. Daughton, *Illicit Drugs in Municipal Sewage.,* 2001, pp. 348-364.
[http://dx.doi.org/10.1021/bk-2001-0791.ch020]

[12]   M.J. García-Galán, T. Garrido, J. Fraile, A. Ginebreda, M.S. Díaz-Cruz, and D. Barceló, "Simultaneous occurrence of nitrates and sulfonamide antibiotics in two ground water bodies of Catalonia (Spain)", *J. Hydrol. (Amst.),* vol. 383, no. 1–2, pp. 93-101, 2010.
[http://dx.doi.org/10.1016/j.jhydrol.2009.06.042]

[13] W. Brack, R. Altenburger, G. Schüürmann, M. Krauss, D. López Herráez, J. van Gils, J. Slobodnik, J. Munthe, B.M. Gawlik, A. van Wezel, M. Schriks, J. Hollender, K.E. Tollefsen, O. Mekenyan, S. Dimitrov, D. Bunke, I. Cousins, L. Posthuma, P.J. van den Brink, M. López de Alda, D. Barceló, M. Faust, A. Kortenkamp, M. Scrimshaw, S. Ignatova, G. Engelen, G. Massmann, G. Lemkine, I. Teodorovic, K.H. Walz, V. Dulio, M.T. Jonker, F. Jäger, K. Chipman, F. Falciani, I. Liska, D. Rooke, X. Zhang, H. Hollert, B. Vrana, K. Hilscherova, K. Kramer, S. Neumann, R. Hammerbacher, T. Backhaus, J. Mack, H. Segner, B. Escher, and G. de Aragão Umbuzeiro, "The SOLUTIONS project: challenges and responses for present and future emerging pollutants in land and water resources management", *Sci. Total Environ.,* vol. 503-504, pp. 22-31, 2015.
[http://dx.doi.org/10.1016/j.scitotenv.2014.05.143] [PMID: 24951181]

[14] O. Ojajuni, D. Saroj, and G. Cavalli, "Removal of organic micropollutants using membrane-assisted processes: a review of recent progress", *Environ. Technol. Rev.,* vol. 4, no. 1, pp. 17-37, 2015.
[http://dx.doi.org/10.1080/21622515.2015.1036788]

[15] F.D.L. Leusch, S.J. Khan, S. Laingam, E. Prochazka, S. Froscio, T. Trinh, H.F. Chapman, and A. Humpage, "Assessment of the application of bioanalytical tools as surrogate measure of chemical contaminants in recycled water", *Water Res.,* vol. 49, pp. 300-315, 2014.
[http://dx.doi.org/10.1016/j.watres.2013.11.030] [PMID: 24355290]

[16] O.R. Price, G.O. Hughes, N.L. Roche, and P.J. Mason, "Improving emissions estimates of home and personal care products ingredients for use in EU risk assessments", *Integr. Environ. Assess. Manag.,* vol. 6, no. 4, pp. 677-684, 2010.
[http://dx.doi.org/10.1002/ieam.88] [PMID: 20872648]

[17] A.B.A. Boxall, M.A. Rudd, B.W. Brooks, D.J. Caldwell, K. Choi, S. Hickmann, E. Innes, K. Ostapyk, J.P. Staveley, T. Verslycke, G.T. Ankley, K.F. Beazley, S.E. Belanger, J.P. Berninger, P. Carriquiriborde, A. Coors, P.C. Deleo, S.D. Dyer, J.F. Ericson, F. Gagné, J.P. Giesy, T. Gouin, L. Hallstrom, M.V. Karlsson, D.G. Larsson, J.M. Lazorchak, F. Mastrocco, A. McLaughlin, M.E. McMaster, R.D. Meyerhoff, R. Moore, J.L. Parrott, J.R. Snape, R. Murray-Smith, M.R. Servos, P.K. Sibley, J.O. Straub, N.D. Szabo, E. Topp, G.R. Tetreault, V.L. Trudeau, and G. Van Der Kraak, "Pharmaceuticals and personal care products in the environment: what are the big questions?", *Environ. Health Perspect.,* vol. 120, no. 9, pp. 1221-1229, 2012.
[http://dx.doi.org/10.1289/ehp.1104477] [PMID: 22647657]

[18] S.I. Mulla, A. Hu, Q. Sun, J. Li, F. Suanon, M. Ashfaq, and C.P. Yu, "Biodegradation of sulfamethoxazole in bacteria from three different origins", *J. Environ. Manage.,* vol. 206, pp. 93-102, 2018.
[http://dx.doi.org/10.1016/j.jenvman.2017.10.029] [PMID: 29059576]

[19] J.M. Brausch, and G.M. Rand, *A review of personal care products in the aquatic environment: Environmental concentrations and toxicity* Elsevier Ltd, 2011.

[20] Y. Yang, Y.S. Ok, K.H. Kim, E.E. Kwon, and Y.F. Tsang, "Occurrences and removal of pharmaceuticals and personal care products (PPCPs) in drinking water and water/sewage treatment plants: A review", *Sci. Total Environ.,* vol. 596-597, pp. 303-320, 2017.
[http://dx.doi.org/10.1016/j.scitotenv.2017.04.102] [PMID: 28437649]

[21] O.M. Rodriguez-Narvaez, J.M. Peralta-Hernandez, A. Goonetilleke, and E.R. Bandala, "Treatment technologies for emerging contaminants in water: A review", *Chem. Eng. J.,* vol. 323, pp. 361-380, 2017.
[http://dx.doi.org/10.1016/j.cej.2017.04.106]

[22] H.Q. Anh, "Antibiotics in surface water of East and Southeast Asian countries: A focused review on contamination status, pollution sources, potential risks, and future perspectives", *Sci. Total Environ.,* p. 142865, 2020.
[http://dx.doi.org/10.1016/j.scitotenv.2020.142865]

[23] C.L.S. Vilela, J.P. Bassin, and R.S. Peixoto, "Water contamination by endocrine disruptors: Impacts, microbiological aspects and trends for environmental protection", *Environ. Pollut.,* vol. 235, pp. 546-

559, 2018.
[http://dx.doi.org/10.1016/j.envpol.2017.12.098] [PMID: 29329096]

[24] B. Mathon, M. Coquery, Z. Liu, Y. Penru, A. Guillon, M. Esperanza, C. Miège, and J.M. Choubert, "Ozonation of 47 organic micropollutants in secondary treated municipal effluents: Direct and indirect kinetic reaction rates and modelling", *Chemosphere,* vol. 262, p. 127969, 2021.
[http://dx.doi.org/10.1016/j.chemosphere.2020.127969] [PMID: 33182096]

[25] A. Pal, Y. He, M. Jekel, M. Reinhard, and K.Y.H. Gin, "Emerging contaminants of public health significance as water quality indicator compounds in the urban water cycle", *Environ. Int.,* vol. 71, pp. 46-62, 2014.
[http://dx.doi.org/10.1016/j.envint.2014.05.025] [PMID: 24972248]

[26] W.T. Vieira, M.B. de Farias, M.P. Spaolonzi, M.G.C. da Silva, and M.G.A. Vieira, *Removal of endocrine disruptors in waters by adsorption, membrane filtration and biodegradation. A review* Springer International Publishing, 2020.
[http://dx.doi.org/10.1007/s10311-020-01000-1]

[27] A. Tursi, A. Beneduci, F. Chidichimo, N. De Vietro, and G. Chidichimo, "Remediation of hydrocarbons polluted water by hydrophobic functionalized cellulose", *Chemosphere,* vol. 201, pp. 530-539, 2018.
[http://dx.doi.org/10.1016/j.chemosphere.2018.03.044] [PMID: 29533802]

[28] M.C.V.M. Starling, C.C. Amorim, and M.M.D. Leão, "Occurrence, control and fate of contaminants of emerging concern in environmental compartments in Brazil", *J. Hazard. Mater.,* vol. 372, pp. 17-36, 2019.
[http://dx.doi.org/10.1016/j.jhazmat.2018.04.043] [PMID: 29728279]

[29] N.K. Khanzada, M.U. Farid, J.A. Kharraz, J. Choi, C.Y. Tang, L.D. Nghiem, A. Jang, and A.K. An, "Removal of organic micropollutants using advanced membrane-based water and wastewater treatment: A review", *J. Membr. Sci.,* vol. 598, p. 117672, 2020.
[http://dx.doi.org/10.1016/j.memsci.2019.117672]

[30] S.D. Richardson, and T.A. Ternes, "Water Analysis: Emerging Contaminants and Current Issues", *Anal. Chem.,* vol. 90, no. 1, pp. 398-428, 2018.
[http://dx.doi.org/10.1021/acs.analchem.7b04577] [PMID: 29112806]

[31] Y. Luo, W. Guo, H.H. Ngo, L.D. Nghiem, F.I. Hai, J. Zhang, S. Liang, and X.C. Wang, "A review on the occurrence of micropollutants in the aquatic environment and their fate and removal during wastewater treatment", *Sci. Total Environ.,* vol. 473-474, pp. 619-641, 2014.
[http://dx.doi.org/10.1016/j.scitotenv.2013.12.065] [PMID: 24394371]

[32] S.D. Richardson, and S.Y. Kimura, "Emerging environmental contaminants: Challenges facing our next generation and potential engineering solutions", *Environ. Technol. Innov.,* vol. 8, pp. 40-56, 2017.
[http://dx.doi.org/10.1016/j.eti.2017.04.002]

[33] C.D. Metcalfe, S. Chu, C. Judt, H. Li, K.D. Oakes, M.R. Servos, and D.M. Andrews, "Antidepressants and their metabolites in municipal wastewater, and downstream exposure in an urban watershed", *Environ. Toxicol. Chem.,* vol. 29, no. 1, pp. 79-89, 2010.
[http://dx.doi.org/10.1002/*etc.*27] [PMID: 20821422]

[34] G.A. Loraine, and M.E. Pettigrove, "Seasonal variations in concentrations of pharmaceuticals and personal care products in drinking water and reclaimed wastewater in southern California", *Environ. Sci. Technol.,* vol. 40, no. 3, pp. 687-695, 2006.
[http://dx.doi.org/10.1021/es051380x] [PMID: 16509304]

[35] S. Kleywegt, V. Pileggi, P. Yang, C. Hao, X. Zhao, C. Rocks, S. Thach, P. Cheung, and B. Whitehead, "Pharmaceuticals, hormones and bisphenol A in untreated source and finished drinking water in Ontario, Canada--occurrence and treatment efficiency", *Sci. Total Environ.,* vol. 409, no. 8, pp. 1481-1488, 2011.
[http://dx.doi.org/10.1016/j.scitotenv.2011.01.010] [PMID: 21315426]

[36]   S.D. Richardson, *Disinfection By-Products: Formation and Occurrence in Drinking Water.* Encycl. Environ. Heal, 2011, pp. 110-136.
[http://dx.doi.org/10.1016/B978-0-444-52272-6.00276-2]

[37]   H. Ilyas, I. Masih, and E.D. van Hullebusch, "Pharmaceuticals' removal by constructed wetlands: a critical evaluation and meta-analysis on performance, risk reduction, and role of physicochemical properties on removal mechanisms", *J. Water Health ,* vol. 18, no. 3, pp. 253-291, 2020.
[http://dx.doi.org/10.2166/wh.2020.213] [PMID: 32589615]

[38]   C.G. Daughton, *Pharmaceutical ingredients in drinking water.* Contam. Emerg. Concern Environ. Ecol. Hum. Heal. Considerations, 2010, pp. 9-68.

[39]   J.L. Oaks, "Diclofenac residues as the cause of vulture population decline in Pakistan", *Nature,* vol. 427, pp. 630-633, 2004.
[http://dx.doi.org/10.1038/nature02317]

[40]   K.A. Kidd, P.J. Blanchfield, K.H. Mills, V.P. Palace, R.E. Evans, J.M. Lazorchak, and R.W. Flick, "Collapse of a fish population after exposure to a synthetic estrogen", *Proc. Natl. Acad. Sci. USA,* vol. 104, no. 21, pp. 8897-8901, 2007.
[http://dx.doi.org/10.1073/pnas.0609568104] [PMID: 17517636]

[41]   S. Kim, K.H. Chu, Y.A.J. Al-Hamadani, C.M. Park, M. Jang, D-H. Kim, M. Yu, J. Heo, and Y. Yoon, "Removal of contaminants of emerging concern by membranes in water and wastewater: A review", *Chem. Eng. J.,* vol. 335, pp. 896-914, 2018.
[http://dx.doi.org/10.1016/j.cej.2017.11.044]

[42]   S. Tanwar, M. Di Carro, C. Ianni, and E. Magi, "Occurrence of PCPs in natural waters from Europe", *Handb. Environ. Chem.,* vol. 36, pp. 37-72, 2015.
[http://dx.doi.org/10.1007/698_2014_276]

[43]   F.O. Ehiguese, M.R. Alam, M.G. Pintado-Herrera, C.V.M. Araújo, and M.L. Martin-Diaz, "Potential of environmental concentrations of the musks galaxolide and tonalide to induce oxidative stress and genotoxicity in the marine environment", *Mar. Environ. Res.,* vol. 160, no. May, p. 105019, 2020.
[http://dx.doi.org/10.1016/j.marenvres.2020.105019] [PMID: 32907733]

[44]   L.M. Weatherly, and J.A. Gosse, "Triclosan exposure, transformation, and human health effects", *J. Toxicol. Environ. Health B Crit. Rev.,* vol. 20, no. 8, pp. 447-469, 2017.
[http://dx.doi.org/10.1080/10937404.2017.1399306] [PMID: 29182464]

[45]   D.W. Kolpin, E.T. Furlong, M.T. Meyer, E.M. Thurman, S.D. Zaugg, L.B. Barber, and H.T. Buxton, "Pharmaceuticals, hormones, and other organic wastewater contaminants in U.S. streams, 1999-2000: a national reconnaissance", *Environ. Sci. Technol.,* vol. 36, no. 6, pp. 1202-1211, 2002.
[http://dx.doi.org/10.1021/es011055j] [PMID: 11944670]

[46]   S. Das Sarkar, S.K. Nag, K. Kumari, K. Saha, S. Bandyopadhyay, M. Aftabuddin, and B.K. Das, "Occurrence and safety evaluation of antimicrobial compounds triclosan and triclocarban in water and fishes of the multitrophic niche of river Torsa, India", *Arch. Environ. Contam. Toxicol.,* vol. 79, no. 4, pp. 488-499, 2020.
[http://dx.doi.org/10.1007/s00244-020-00785-0] [PMID: 33215293]

[47]   E.K. Martin, S. Kemal, N.R. Henrik, E. Alexander, C. Natalia, J.C. Henrik, B. Thomas, B. Hans, and K. Erik, "Triclosan changes community composition and selects for specific bacterial taxa in marine periphyton biofilms in low nanomolar concentrations", *Ecotoxicology,* vol. 29, no. 7, pp. 1083-1094, 2020.
[http://dx.doi.org/10.1007/s10646-020-02246-9] [PMID: 32661899]

[48]   M. Lei, L. Zhang, J. Lei, L. Zong, J. Li, Z. Wu, and Z. Wang, "Overview of emerging contaminants and associated human health effects", *BioMed Res. Int.,* vol. 2015, p. 404796, 2015.
[http://dx.doi.org/10.1155/2015/404796] [PMID: 26713315]

[49]   L. Rani, "An extensive review on the consequences of chemical pesticides on human health and

environment", *J. Clean. Prod,* p. 124657, 2020.
[http://dx.doi.org/10.1016/j.jclepro.2020.124657]

[50]　R. Wang, Y. Yuan, H. Yen, M. Grieneisen, J. Arnold, D. Wang, C. Wang, and M. Zhang, "A review of pesticide fate and transport simulation at watershed level using SWAT: Current status and research concerns", *Sci. Total Environ.,* vol. 669, pp. 512-526, 2019.
[http://dx.doi.org/10.1016/j.scitotenv.2019.03.141] [PMID: 30884273]

[51]　F.P. Carvalho, "Pesticides, environment, and food safety", *Food Energy Secur.,* vol. 6, no. 2, pp. 48-60, 2017.
[http://dx.doi.org/10.1002/fes3.108]

[52]　Y. Leung, *Low-dose bisphenol a in a rat model of endometrial cancer : a clarity-BPA study,* 2020, pp. 1-12.

[53]　H. Cheng, H. Luo, Y. Hu, and S. Tao, "Release kinetics as a key linkage between the occurrence of flame retardants in microplastics and their risk to the environment and ecosystem: A critical review", *Water Res.,* vol. 185, p. 116253, 2020.
[http://dx.doi.org/10.1016/j.watres.2020.116253] [PMID: 32768659]

[54]　W. Choi, S. Lee, H.K. Lee, and H.B. Moon, "Organophosphate flame retardants and plasticizers in sediment and bivalves along the Korean coast: Occurrence, geographical distribution, and a potential for bioaccumulation", *Mar. Pollut. Bull.,* vol. 156, no. May, p. 111275, 2020.
[http://dx.doi.org/10.1016/j.marpolbul.2020.111275] [PMID: 32510414]

[55]　S. Graetz, M. Ji, S. Hunter, P.K. Sibley, and R.S. Prosser, "Deterministic risk assessment of firefighting water additives to aquatic organisms", *Ecotoxicology,* vol. 29, no. 9, pp. 1377-1389, 2020.
[http://dx.doi.org/10.1007/s10646-020-02274-5] [PMID: 32869175]

[56]　G. Daniel, A.R.R. Silva, D.M. de Souza Abessa, and S. Loureiro, "Fire Suppression Agents Combined with Gasoline in Aquatic Ecosystems: A Mixture Approach", In: *Environ. Toxicol. Chem,* 2020, pp. 0-3.
[http://dx.doi.org/10.1002/*etc.*4889]

[57]　P. Pathak, and K. Sushil, "Electronic waste: an emerging contaminant in the geo-environment", In: *Emerging Contaminants in the Environment.* Elsevier, 2022, pp. 275-286.
[http://dx.doi.org/10.1016/B978-0-323-85160-2.00018-4]

[58]　P. Wild, E. Bourgkard, and C. Paris, "Lung cancer and exposure to metals: the epidemiological evidence", *Methods Mol. Biol.,* vol. 472, pp. 139-167, 2009.
[http://dx.doi.org/10.1007/978-1-60327-492-0_6] [PMID: 19107432]

[59]　P.B. Tchounwou, C.G. Yedjou, A.K. Patlolla, and D.J. Sutton, "Heavy metal toxicity and the environment", *Mol. Clin. Environ. Toxicol,* pp. 133-164, 2012.

[60]　L.B. Barber, *Emerging Contaminants.* vol. 1. Elsevier Ltd., 2014.

[61]　C.D. Kassotis, L.N. Vandenberg, B.A. Demeneix, M. Porta, R. Slama, and L. Trasande, "Endocrine-disrupting chemicals: economic, regulatory, and policy implications", *Lancet Diabetes Endocrinol.,* vol. 8, no. 8, pp. 719-730, 2020.
[http://dx.doi.org/10.1016/S2213-8587(20)30128-5] [PMID: 32707119]

[62]　C.I. Kosma, D.A. Lambropoulou, and T.A. Albanis, "Occurrence and removal of PPCPs in municipal and hospital wastewaters in Greece", *J. Hazard. Mater.,* vol. 179, no. 1-3, pp. 804-817, 2010.
[http://dx.doi.org/10.1016/j.jhazmat.2010.03.075] [PMID: 20395039]

[63]　S. Esplugas, D.M. Bila, L.G.T. Krause, and M. Dezotti, "Ozonation and advanced oxidation technologies to remove endocrine disrupting chemicals (EDCs) and pharmaceuticals and personal care products (PPCPs) in water effluents", *J. Hazard. Mater.,* vol. 149, no. 3, pp. 631-642, 2007.
[http://dx.doi.org/10.1016/j.jhazmat.2007.07.073] [PMID: 17826898]

[64]　J. Wilkinson, P.S. Hooda, J. Barker, S. Barton, and J. Swinden, "Occurrence, fate and transformation of emerging contaminants in water: An overarching review of the field", *Environ. Pollut.,* vol. 231, no.

Pt 1, pp. 954-970, 2017.
[http://dx.doi.org/10.1016/j.envpol.2017.08.032] [PMID: 28888213]

[65] C.V.T. Rigueto, M.T. Nazari, C.F. De Souza, J.S. Cadore, V.B. Brião, and J.S. Piccin, "Alternative techniques for caffeine removal from wastewater: An overview of opportunities and challenges", *J. Water Process Eng.,* vol. 35, no. December, p. 2019, 2020.
[http://dx.doi.org/10.1016/j.jwpe.2020.101231]

[66] Y. Wang, J. Ma, J. Zhu, N. Ye, X. Zhang, and H. Huang, "Multi-walled carbon nanotubes with selected properties for dynamic filtration of pharmaceuticals and personal care products", *Water Res.,* vol. 92, pp. 104-112, 2016.
[http://dx.doi.org/10.1016/j.watres.2016.01.038] [PMID: 26845455]

[67] J.H. Li, B.X. Zhou, and W.M. Cai, "The solubility behavior of bisphenol A in the presence of surfactants", *J. Chem. Eng. Data,* vol. 52, no. 6, pp. 2511-2513, 2007.
[http://dx.doi.org/10.1021/je700235x]

[68] Z. Liu, X. Yan, M. Drikas, D. Zhou, D. Wang, M. Yang, and J. Qu, "Removal of bentazone from micro-polluted water using MIEX resin: kinetics, equilibrium, and mechanism", *J. Environ. Sci. (China),* vol. 23, no. 3, pp. 381-387, 2011.
[http://dx.doi.org/10.1016/S1001-0742(10)60441-X] [PMID: 21520806]

[69] S. Net, R. Sempéré, A. Delmont, A. Paluselli, and B. Ouddane, "Occurrence, fate, behavior and ecotoxicological state of phthalates in different environmental matrices", *Environ. Sci. Technol.,* vol. 49, no. 7, pp. 4019-4035, 2015.
[http://dx.doi.org/10.1021/es505233b] [PMID: 25730609]

[70] U.S. Environmental Protection Agency, *Technical Fact Sheet - Perfluorooctane Sulfonate (PFOS) and Perfluorooctanoic Acid (PFOA). EPA 505-F-17-001, United States Environmental Protection Agency,* 2017.

[71] P. Yan, J. Shen, Y. Zhou, L. Yuan, J. Kang, S. Wang, and Z. Chen, "Interface mechanism of catalytic ozonation in an α-Fe0.9Mn0.1OOH aqueous suspension for the removal of iohexol", *Appl. Catal. B,* vol. 277, no. May, p. 119055, 2020.
[http://dx.doi.org/10.1016/j.apcatb.2020.119055]

[72] N.- Deet, *Assessment Report i (Repellents and attractants),* 2010.

[73] A. D. Bernstein, A. Kieffer, and S. Francisco, "Lopamidol and Metrizamide for Myelography : Prospective CH₂OH".*, AJR,* vol. 143, pp. 869-873, 1984.

[74] M. I. Gamhewage, "Pesticides in the Water-Column and Bottom Sediments of Four Manitoba Rivers", *AGR online.,* 2020.

[75] J. Ao, L. Gao, T. Yuan, and G. Jiang, "Interaction mechanisms between organic UV filters and bovine serum albumin as determined by comprehensive spectroscopy exploration and molecular docking", *Chemosphere,* vol. 119, no. 800, pp. 590-600, 2015.
[http://dx.doi.org/10.1016/j.chemosphere.2014.07.019] [PMID: 25128891]

[76] M. Greenberg, M. Dodds, and M. Tian, "Naturally occurring phenolic antibacterial compounds show effectiveness against oral bacteria by a quantitative structure-activity relationship study", *J. Agric. Food Chem.,* vol. 56, no. 23, pp. 11151-11156, 2008.
[http://dx.doi.org/10.1021/jf8020859] [PMID: 19007234]

[77] P. R. Summary, A. Pesticides, and V. M. Authority, "Evaluation of the new active pyraclostrobin in the product cabrio fungicide". 2003.

[78] W. Fan, T. Yanase, H. Morinaga, S. Gondo, T. Okabe, M. Nomura, T. Komatsu, K. Morohashi, T.B. Hayes, R. Takayanagi, and H. Nawata, "Atrazine-induced aromatase expression is SF-1 dependent: implications for endocrine disruption in wildlife and reproductive cancers in humans", *Environ. Health Perspect.,* vol. 115, no. 5, pp. 720-727, 2007.
[http://dx.doi.org/10.1289/ehp.9758] [PMID: 17520059]

[79]   H. Ishak, I. Mokbel, J. Saab, J. Stephan, P. Paricaud, J. Jose, and C. Goutaudier, "Experimental measurements and thermodynamic modelling of aqueous solubilities, octanol-water partition coefficients and vapor pressures of dimethyl phthalate and butyl benzyl phthalate", *J. Chem. Thermodyn.,* vol. 131, pp. 286-293, 2019.
[http://dx.doi.org/10.1016/j.jct.2018.11.015]

[80]   D. Brooke, A. Footitt, and T.A. Nwaogu, *Environmental risk evaluation report: perfluorooctanesulphonate (PFOS).* Environ. Agency, 2004, p. 96.

[81]   I. Rodea-Palomares, F. Leganés, R. Rosal, and F. Fernández-Piñas, "Toxicological interactions of perfluorooctane sulfonic acid (PFOS) and perfluorooctanoic acid (PFOA) with selected pollutants", *J. Hazard. Mater.,* vol. 201-202, pp. 209-218, 2012.
[http://dx.doi.org/10.1016/j.jhazmat.2011.11.061] [PMID: 22177019]

[82]   C. Teodosiu, A.F. Gilca, G. Barjoveanu, and S. Fiore, "Emerging pollutants removal through advanced drinking water treatment: A review on processes and environmental performances assessment", *J. Clean. Prod.,* vol. 197, pp. 1210-1221, 2018.
[http://dx.doi.org/10.1016/j.jclepro.2018.06.247]

[83]   C. Di Poi, K. Costil, V. Bouchart, and M.P. Halm-Lemeille, "Toxicity assessment of five emerging pollutants, alone and in binary or ternary mixtures, towards three aquatic organisms", *Environ. Sci. Pollut. Res. Int.,* vol. 25, no. 7, pp. 6122-6134, 2018.
[http://dx.doi.org/10.1007/s11356-017-9306-9] [PMID: 28620858]

[84]   F. Prestinaci, P. Pezzotti, and A. Pantosti, "Antimicrobial resistance: a global multifaceted phenomenon", *Pathog. Glob. Health,* vol. 109, no. 7, pp. 309-318, 2015.
[http://dx.doi.org/10.1179/2047773215Y.0000000030] [PMID: 26343252]

[85]   J. Bryan-Wilson, "No time to wait",

[86]   E. Zuccato, S. Castiglioni, R. Bagnati, M. Melis, and R. Fanelli, "Source, occurrence and fate of antibiotics in the Italian aquatic environment", *J. Hazard. Mater.,* vol. 179, no. 1-3, pp. 1042-1048, 2010.
[http://dx.doi.org/10.1016/j.jhazmat.2010.03.110] [PMID: 20456861]

[87]   H. Hamid, and C. Eskicioglu, "Fate of estrogenic hormones in wastewater and sludge treatment: A review of properties and analytical detection techniques in sludge matrix", *Water Res.,* vol. 46, no. 18, pp. 5813-5833, 2012.
[http://dx.doi.org/10.1016/j.watres.2012.08.002] [PMID: 22939851]

[88]   M. Adeel, X. Song, Y. Wang, D. Francis, and Y. Yang, "Environmental impact of estrogens on human, animal and plant life: A critical review", *Environ. Int.,* vol. 99, pp. 107-119, 2017.
[http://dx.doi.org/10.1016/j.envint.2016.12.010] [PMID: 28040262]

[89]   Y. Zhang, W. Meng, C. Guo, J. Xu, T. Yu, W. Fan, and L. Li, "Determination and partitioning behavior of perfluoroalkyl carboxylic acids and perfluorooctanesulfonate in water and sediment from Dianchi Lake, China", *Chemosphere,* vol. 88, no. 11, pp. 1292-1299, 2012.
[http://dx.doi.org/10.1016/j.chemosphere.2012.03.103] [PMID: 22579454]

[90]   D. Zhang, B. Pan, H. Zhang, P. Ning, and B. Xing, "Contribution of different sulfamethoxazole species to their overall adsorption on functionalized carbon nanotubes", *Environ. Sci. Technol.,* vol. 44, no. 10, pp. 3806-3811, 2010.
[http://dx.doi.org/10.1021/es903851q] [PMID: 20394427]

[91]   C. Yang, G. Song, and W. Lim, "A review of the toxicity in fish exposed to antibiotics", *Comp. Biochem. Physiol. C Toxicol. Pharmacol.,* vol. 237, no. July, p. 108840, 2020.
[http://dx.doi.org/10.1016/j.cbpc.2020.108840] [PMID: 32640291]

[92]   E. Adikwu, and N. Brambaifa, "Ciprofloxacin Cardiotoxicity and Hepatotoxicity in Humans and Animals", *Pharmacol. Pharm.,* vol. 2012, no. April, pp. 207-213, 2012.
[http://dx.doi.org/10.4236/pp.2012.32028]

[93] G.H. Guo, F.C. Wu, H.P. He, R.Q. Zhang, and H.X. Li, "Screening level ecological risk assessment for synthetic musks in surface water of Lake Taihu, China", *Stochastic Environ. Res. Risk Assess.,* vol. 27, no. 1, pp. 111-119, 2013.
[http://dx.doi.org/10.1007/s00477-012-0581-1]

[94] I. Hussain, M.Z. Khan, A. Khan, I. Javed, and M.K. Saleemi, "Toxicological effects of diclofenac in four avian species", *Avian Pathol.,* vol. 37, no. 3, pp. 315-321, 2008.
[http://dx.doi.org/10.1080/03079450802056439] [PMID: 18568659]

[95] M. la Farré, S. Pérez, L. Kantiani, and D. Barceló, ""Fate and toxicity of emerging pollutants, their metabolites and transformation products in the aquatic environment," TrAC -", *Trends Analyt. Chem.,* vol. 27, no. 11, pp. 991-1007, 2008.
[http://dx.doi.org/10.1016/j.trac.2008.09.010]

[96] D. Wojcieszyńska, and U. Guzik, "Naproxen in the environment: its occurrence, toxicity to nontarget organisms and biodegradation", *Appl. Microbiol. Biotechnol.,* vol. 104, no. 5, pp. 1849-1857, 2020.
[http://dx.doi.org/10.1007/s00253-019-10343-x] [PMID: 31925484]

[97] K.A. Reimer, M.M. Rasmussen, and R.B. Jennings, "Reduction by propranolol of myocardial necrosis following temporary coronary artery occlusion in dogs", *Circ. Res.,* vol. 33, no. 3, pp. 353-363, 1973.
[http://dx.doi.org/10.1161/01.RES.33.3.353] [PMID: 4746723]

[98] J. Finn, M. Hui, V. Li, V. Lorenzi, N. de la Paz, S.H. Cheng, L. Lai-Chan, and D. Schlenk, "Effects of propranolol on heart rate and development in Japanese medaka (Oryzias latipes) and zebrafish (Danio rerio)", *Aquat. Toxicol.,* vol. 122-123, pp. 214-221, 2012.
[http://dx.doi.org/10.1016/j.aquatox.2012.06.013] [PMID: 22832281]

[99] S.E. Duirk, C. Lindell, C.C. Cornelison, J. Kormos, T.A. Ternes, M. Attene-Ramos, J. Osiol, E.D. Wagner, M.J. Plewa, and S.D. Richardson, "Formation of toxic iodinated disinfection by-products from compounds used in medical imaging", *Environ. Sci. Technol.,* vol. 45, no. 16, pp. 6845-6854, 2011.
[http://dx.doi.org/10.1021/es200983f] [PMID: 21761849]

[100] L.A. Feyk, and J.P. Giesy, "Development of a caffeine breath test to measure cytochrome P450-1A activity in birds", *Environ. Toxicol. Pharmacol.,* vol. 1, no. 1, pp. 51-61, 1996.
[http://dx.doi.org/10.1016/1382-6689(95)00009-7] [PMID: 21781663]

[101] M. Krause, A. Klit, M. Blomberg Jensen, T. Søeborg, H. Frederiksen, M. Schlumpf, W. Lichtensteiger, N.E. Skakkebaek, and K.T. Drzewiecki, "Sunscreens: are they beneficial for health? An overview of endocrine disrupting properties of UV-filters", *Int. J. Androl.,* vol. 35, no. 3, pp. 424-436, 2012.
[http://dx.doi.org/10.1111/j.1365-2605.2012.01280.x] [PMID: 22612478]

[102] C.P. Waring, and A. Moore, "The effect of atrazine on Atlantic salmon (Salmo salar) smolts in fresh water and after sea water transfer", *Aquat. Toxicol.,* vol. 66, no. 1, pp. 93-104, 2004.
[http://dx.doi.org/10.1016/j.aquatox.2003.09.001] [PMID: 14687982]

[103] K.W. Wilhelms, S.A. Cutler, J.A. Proudman, L.L. Anderson, and C.G. Scanes, "Effects of atrazine on sexual maturation in female Japanese quail induced by photostimulation or exogenous gonadotropin", *Environ. Toxicol. Chem.,* vol. 25, no. 1, pp. 233-240, 2006.
[http://dx.doi.org/10.1897/05-039R.1] [PMID: 16494247]

[104] "W. H. O. World Health Organization", *Antimony in Drinking-water. Background document for development of WHO Guidelines for Drinking-water Quality,* vol. 2, pp. 1-9, 2003.

[105] J. Ding, X. Shen, W. Liu, A. Covaci, and F. Yang, "Occurrence and risk assessment of organophosphate esters in drinking water from Eastern China", *Sci. Total Environ.,* vol. 538, pp. 959-965, 2015.
[http://dx.doi.org/10.1016/j.scitotenv.2015.08.101] [PMID: 26363608]

[106] M. Baril, "Environmental Health Criteria 209: Flame retardants: Tris(chloropropyl) phosphate and

tris(2-chloroethyl) phosphate", *Environ. Heal. Criteria,* 1998.

[107] R.L. Joshi, "Environmental Science", *JNMA J. Nepal Med. Assoc.,* vol. 10, no. 3, pp. 147-154, 2003.
[http://dx.doi.org/10.31729/jnma.1240]

[108] A.F. Lehner, F. Samsing, and W.K. Rumbeiha, "Organophosphate ester flame retardant-induced acute intoxications in dogs", *J. Med. Toxicol.,* vol. 6, no. 4, pp. 448-458, 2010.
[http://dx.doi.org/10.1007/s13181-010-0105-7] [PMID: 20717764]

[109] K. Kanda, S. Ito, D.H. Koh, E.Y. Kim, and H. Iwata, "Effects of tris(2-chloroethyl) phosphate exposure on chicken embryos in a shell-less incubation system", *Ecotoxicol. Environ. Saf,* vol. 207, 2021.
[http://dx.doi.org/10.1016/j.ecoenv.2020.111263]

[110] G.B. Post, P.D. Cohn, and K.R. Cooper, "Perfluorooctanoic acid (PFOA), an emerging drinking water contaminant: a critical review of recent literature", *Environ. Res.,* vol. 116, pp. 93-117, 2012.
[http://dx.doi.org/10.1016/j.envres.2012.03.007] [PMID: 22560884]

[111] M. Llorca, M. Farré, E. Eljarrat, S. Díaz-Cruz, S. Rodríguez-Mozaz, D. Wunderlin, and D. Barcelo, "Review of emerging contaminants in aquatic biota from Latin America: 2002-2016", *Environ. Toxicol. Chem.,* vol. 36, no. 7, pp. 1716-1727, 2017.
[http://dx.doi.org/10.1002/etc.3626] [PMID: 27666732]

[112] Y. Vystavna, Z. Frkova, H. Celle-Jeanton, D. Diadin, F. Huneau, M. Steinmann, N. Crini, and C. Loup, "Priority substances and emerging pollutants in urban rivers in Ukraine: Occurrence, fluxes and loading to transboundary European Union watersheds", *Sci. Total Environ.,* vol. 637-638, pp. 1358-1362, 2018.
[http://dx.doi.org/10.1016/j.scitotenv.2018.05.095] [PMID: 29801228]

[113] A. C. Lorrain, "France: google/monitoring obligation", *Directive 2000/31 IEC of the European Parliament and of the Council of 8 June 2000 on Certain Legal Aspects of Information Society Services, in Particular Electronic Commerce, In the Internal Market.,* 2013.
[http://dx.doi.org/10.1007/s40319-013-0039-0]

[114] S.S. Miraj, N. Parveen, and H.S. Zedan, "Plastic microbeads: small yet mighty concerning", *Int. J. Environ. Health Res.,* vol. 00, no. 00, pp. 1-17, 2019.
[http://dx.doi.org/10.1080/09603123.2019.1689233] [PMID: 31709817]

[115] S.S. Miraj, N. Parveen, and H.S. Zedan, "Plastic microbeads: small yet mighty concerning", *Int. J. Environ. Health Res.,* vol. 00, no. 00, pp. 1-17, 2019.
[http://dx.doi.org/10.1080/09603123.2019.1689233] [PMID: 31709817]

[116] S.S. Miraj, N. Parveen, and H.S. Zedan, "Plastic microbeads: small yet mighty concerning", *Int. J. Environ. Health Res.,* vol. 00, no. 00, pp. 1-17, 2019.
[http://dx.doi.org/10.1080/09603123.2019.1689233] [PMID: 31709817]

[117] S.A. Narayana, R.K. Pati, and P. Vrat, "Rpharmaceutical industry: A literature reviewesearch on management issues in the", *Int. J. Pharm. Healthc. Mark.,* vol. 6, no. 4, pp. 351-375, 2012.
[http://dx.doi.org/10.1108/17506121211283235]

[118] U. States, "For the adoption of ICH Q7: good manufacturing practice for active pharmaceutical ingredients", 2020.

[119] M.O. Barbosa, N.F.F. Moreira, A.R. Ribeiro, M.F.R. Pereira, and A.M.T. Silva, "Occurrence and removal of organic micropollutants: An overview of the watch list of EU Decision 2015/495", *Water Res.,* vol. 94, pp. 257-279, 2016.
[http://dx.doi.org/10.1016/j.watres.2016.02.047] [PMID: 26967909]

CHAPTER 2

Emerging Water Pollutants from Pharmaceuticals and Personal Care Products

Pranta Barua[1,*], **Joydeb Babu Pranta**[2] and **Nazia Hossain**[3,*]

[1] *Department of Electronic Materials Engineering, Kwangwoon University, Seoul 01897, South Korea*

[2] *Department of Civil Engineering, Stamford University, Bangladesh*

[3] *School of Engineering, RMIT University, Melbourne VIC 3001, Australia*

Abstract: Pharmaceuticals and personal care products (PCPs) are chemically modified products mostly used for beauty, cleaning, or health, such as disinfectants, fragrances, insect repellents, ultraviolet (UV) filters, and others. Due to the contamination caused by these chemically amended PCPs, water pollution has appeared as a hazardous condition for the water treatment and supply sector. A recent concern is that when these chemical compounds combine with water, they act as water pollutants and harm aquatic lives and the survival of human lives. Nowadays, the concern of water pollution by these chemicals is confined to water treatment complexities this contamination is leaving adverse effects on the environment. Most of these water pollutants borne by sewage effluents through wastewater plants develop because of the insufficient removal from treatment plants. Therefore, the emerging water pollutants caused by PCPs are responsible for environmental pollution. Hence, this chapter emphasized the state-of-the-art global application of PCPs, the mechanism of water pollution by PCPs, possible biohazards, and negative impacts on the environment. Besides that, various types of PCPs, along with the most applied chemical compounds in PCPs, have been discussed in this chapter. To minimize the contamination, suitable removal methods to enhance the removal efficacy have been discussed. The chapter also presents a future perspective for reducing the PCP contamination of surface water with cutting-edge technologies and wastewater treatment.

Keywords: Water pollutants, Pharmaceutical products, Personal care products, Surface water, Wastewater treatment technologies.

INTRODUCTION

The latest water testing strategies allow researchers to distinguish the small sums of chemicals in our water supplies. As a result, modern considerations uncover the

* **Corresponding author Pranta Barua and Nazia Hossain:** Department of Electronic Materials Engineering, Kwangwoon University, Seoul 01897, South Korea and School of Engineering, RMIT University, Australia; Tel: +61 480 123 691; E-mails: pranta.barua74@gmail.com and bristy808.nh@gmail.com

Shaukat Ali Mazari, Nabisab Mujawar Mubarak & Nizamuddin Sabzoi (Eds.)

nearness of drugs, individual care items, and other substances we utilize at workplaces, homes and even on ranches. These substances are commonly alluded to as "emerging pathogens" or "emerging contaminants." Dynamic fixings and additives of beauty care products, toiletries, and scents have been found in water. Nitro musk compounds utilized as scents in individual care items may have unfavorable natural impacts. These potential water quality contaminants are pharmaceuticals and personal care products (PCPs). Concurring to the Natural Assurance Office, PCPs are alluded to as "any items utilized by people for individual wellbeing or restorative reasons or utilized by agribusiness to improve development or wellbeing of livestock." Potential water quality contaminants are flushed into groundwater from an assortment of sources. Wastewater from sewage treatment facilities, ranoff from agrarian arrival employments, primarily from mechanical scale animal offices, and discharged from individual septic frameworks are the main prevalent occurrences. The ability of conventional sewage treatment to destroy medication or personal care item buildups varies greatly [1].

Water pollution is caused by a variety of contaminants, some of which can be seen in Fig. (**1**). PCPs are one of these pollutants.

The biological & chemical properties of various PCPs suggest that many properties are not helpfully taken out by ordinary water treatment measures, as displayed by pith in drinking water. In green development, they have been perceived as containing the best variation of woody biomass in maritime conditions [2]. The lipid substance of green growth gives a part in introducing the trophic exchange of lipophilic trademark adulterants. An assessment [3] isolated the closeness of utilized antimicrobial administrators triclocarban (TCC), triclosan (TCS) comparatively as its metabolite methyl-triclosan (M-TCS) in green advancement tests amassed around a treatment plant of wastewater in Texas. Centralizations of target PCPs in-water tests were low, going from 50 to 200mg/L. On the other hand, more vital degrees of 50-400 mg/g of new weight seen in the events' green turn. The resulting bio-aggregation factors ran at 700-1500 for MTCS, 900-2100 for TCS, and 1600-2700 for TCC, respectively. A noteworthy fact raised by many research that the PCPs in water-bodies are able to interconnect with the secretory system to make disproportionate effects/unsettling influence of homeostasis [3]. The World Health Organization described endocrine disruptors (ED) as an 'extranet substance or mix that modifies the secretory system's function(s). It is due to adverse health effects on a creature, its offspring, or a subpopulation.

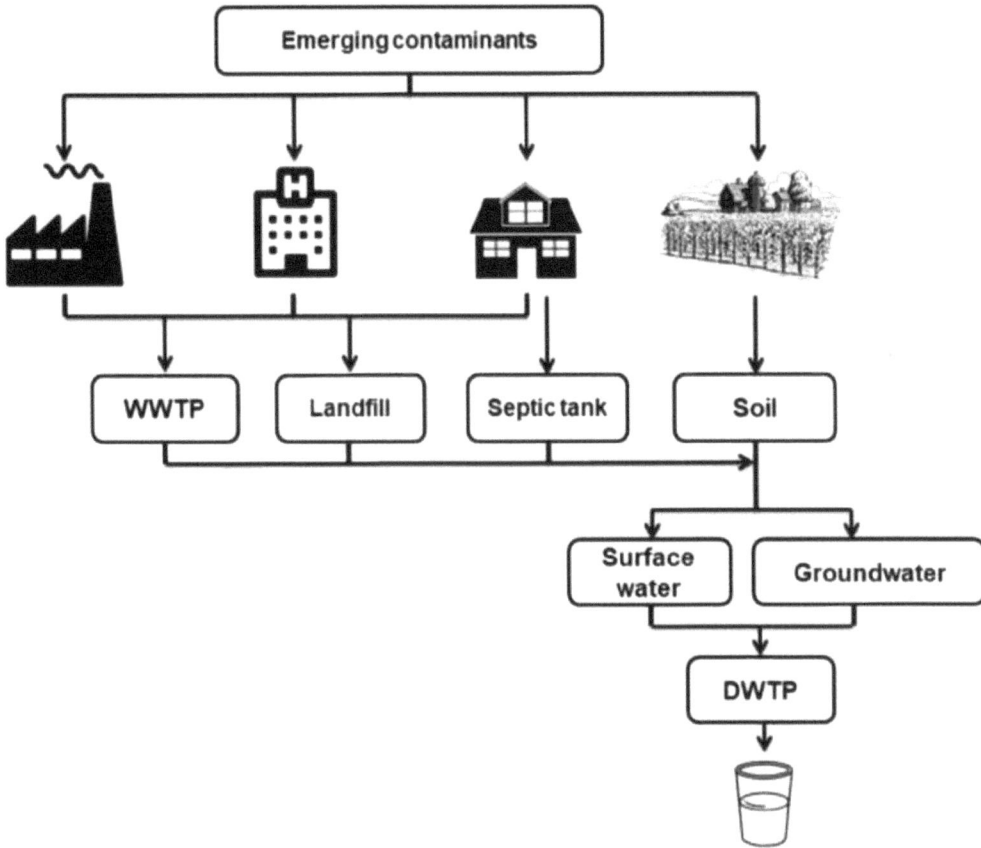

Fig. (1). Some pathways of water contamination [1].

EDs incorporated a tremendous gathering of synthetic concoctions from the standard (for example, mycotoxins and phytoestrogens). They engineered beginning (for instance, diethylstilbesterol (DES) and Bisphenol) in assortments of customer items (for example, PCPs, cleaning items, antimicrobials, food additives, and phthalates. Different endocrine-disrupting personal care products (PCPs) are now present that hamper our daily life. They have subdivided into sections: i) Disinfectants, ii) Conservation agents, iii) Fragrance, and iv) UV screens. The fragrance comes from different chemical compounds: musk xylol, musk ketone, galaxolide, tonalide, celestolied, and others. Also, UV screen is a mixture of varying contaminant chemicals, and they are benzophenone-3, homosalate, 4-methyl-benzylidyne camphor, octyl-methoxynniamine, octyl-dimethyl-PABA, and others [4].

PCPs incorporate countless engineered synthetic compounds utilized in regular items such as cleansers, creams, toothpaste, aroma, beautifying agents, and sunscreens. Their broad utilization, inappropriate removal, and wasteful treatment of urban wastewater add to the pollution of water bodies by PCPs and their metabolites. The pollution of the water repositories by PCPs is enthusiasm because of their likely harmfulness to oceanic biological systems and people. The same number has been accounted for as ecological relentless, bioactive, bioaccumulative, and endocrine disturbing mixes [5 - 7]. As per Rutgers researchers, medication and individual consideration items from human discharge may prompt turning out to contaminate water. The researchers tried the capacity of microscopic organisms in slop from a sewage treatment plant to separate two broadly utilized drug items: naproxen, a non-steroidal mitigating drug, and guaifenesin, an expectorant in many hack and cold medicines. They likewise tried two essential mixes in close-to-home consideration items: oxybenzone, a fundamental fixing in numerous sunscreens, and methylparaben, additive-in-like beautifiers. As indicated, microorganisms that don't expect oxygen develop in the emission-separated methylparaben. However, the organisms mostly diverged the three different synthetic concoctions – and made new contaminants [8].

PCPs in nature are often found in sea-going situations because PCPs break up effectively and don't dissipate at ordinary temperatures and weights. Practices, for example, utilization of sewage slop ("bio-solids") and recycled water for water systems carries PCPs into contact with the dirt. Notwithstanding anti-toxins and steroids, more than 100 individual PCPs have been distinguished (2007) in ecological examples and drinking water [9]. Recent studies show that around 5000 ingredients are available in markets that are depicted as PCPs. Most of the chemicals synthesizing PCPs are mentioned as emerging contaminants. A recent database from the US showed that a person is using six products daily. From these six products, the person is exposed to 85 chemicals. For a woman, around 12 ingredients are used daily, and it is assumed that the woman is exposed to 168 different chemicals from PCPs in daily routine [10].

A previous study surveyed articles containing fixations in water lattices of synthetic compounds. A place with individual consideration items, reported as developing contaminations concurring the US-EPA definition, was incorporated as appropriate for investigation [11]. Different studies and experiments are being performed to determine the contamination level of water caused by PCPs. The chapter aims to review the current prospect of PCP as a contaminant for water pollution, and how these chemicals derived from PCPs are being treated around the world will also be discussed and outlined. In addition, future perspectives and challenges have been elaborated.

POTENTIAL PCPS AS EMERGING CONTAMINANTS FOR WATER POLLUTION

Categories of water pollution are comprised of four classes: infectious agent, inorganic material, organic material, and minute pollutants. Significant concerns emerge from the discovery of synthetic substances for which there is proof that they may unfavorably influence sea-going life. The accompanying segments sum up a portion of the significant worries about the aquatic environment.

Beauty Products

Harmful synthetic substances in beautifiers washed down channels winds up in seas, harming the eco-framework and making demise sea-going species. Animals influenced by poisons that end up in soil can endure conceptive, hereditary, and formative changes just as numerous sorts of diseases. Brands including a couple of characteristic fixings veiling other poisonous fixings are quickly draining regular assets. For instance, beautifiers with palm oil add to rapid deforestation and environmental change. Popular brands getting on board with the common fleeting trend by merely adding a couple of regular fixings to beauty care products containing a bunch of other dangerous fixings are duping clients and harming the earth with the destruction of ordinary assets impractical cultivating rehearses [12]. Different beauty products (sunscreen, lotions, nail polish, powder, body wash, shampoo) contain toxic elements that are being washed and mixed with water from lack, rivers, ponds, and others. They are a direct threat to the aquatic environment: Butylated hydroxyanisole (BHA) and butylated hydroxytoluene (BHT) cause fish and shellfish demise. BHT causes hereditary transformations in amphibians. Sodium Laureth sulfate modifies fish conduct and expands mortality, Whereas dioxane makes passing insects. Diethanolamine (DEA) develops in the earth and responds with nitrates to frame nitrosamines, profoundly cancer-causing to human and creature life. DEA is lethally poisonous to land and water creatures, scavengers, fish, nematodes, flatworms, and creature plankton. Livestock presented to synthetic compounds in the soil can experience the ill effects of conceptive issues and disease [12].

Along with these conventional toxic elements, a new harmful component of beauty products is being introduced that will be a future threat to the freshwater environment. UV filters, plastic microbeads, and others are now being treated as prospective concerns for the aquatic environment. Literature reports the event of natural sunscreen mixes likewise in the water of pools. The proximity and gathering of UV direct in these workplaces depend in a general sense upon their washing off from bathers' skin during swimming and on the bather load. In pools, UV channels occur at higher centers diverged from other water bodies (streams, lakes, seas), where substance pollutions experience high debilitating, while in pools, the water is reused [13, 14]. Most of the little plastic particles in the oceans originate from the fracture of more important things; nonetheless, there is proof that micro-plastics enter the marine condition legitimately from various sources, including makeup, attire, and current cycles. Plastic microbeads are available as grating scrubbers in different individual considerations and corrective items, for example, hand chemicals, toothpaste, face scours, bubble baths, shampoos, and cleansers, where they supplant characteristic shedding materials recently used (pumice, oats, apricot husks).

Notwithstanding the capacity of dead skin expulsion and profound purging, plastic microbeads can likewise assume an embellishing function close to home consideration items. Around 93% of the microbeads are made of polyethylene; however, in some cases, the material they are produced is polypropylene, nylon, polyethylene terephthalate, or polymethyl methacrylate [15 - 18].

Cleaning Products

The ecological effects of cleaning things include the results of blends in purging things. Cleaning administrators can be organically active with outcomes from delicate to severe. These cleaning things can contain hazardous engineered materials that detrimentally affect nature. Developmental and endocrine disruptors have been associated with cleaning administrators. Distinctive poisonous synthetic compounds are being utilized in blending cleaning items, which can harshly jeopardize the amphibian condition. A portion of the substances are recorded, for example, alkylphenol ethoxylates (APEs), triclosan, sodium hypochlorite (family unit fade), ammonium hydroxide (ammonia), propellant gas, phosphates, and others [19, 20].

APEs are extensively used in family things such as chemicals and generally convenient cleaning things. These precisely manufactured products are found in 55% of the nuclear family, which are the main reasons for market cleaning [21]. They are powerless to bacterial or biochemical mortification to alkylphenols: lipophilic, internal secretion imitating mixes [20]. APEs have been genuinely associated with endocrine interference. Other assessments found that hormone-mimicking alkylphenols part with the oestradiol receptor and avoid the right official limit of oestradiol. Male trout in alkylphenol polluted streams exhibited diminished testicular turn of events and arranged on different occasions more vitellogenin than male trout's control. The tremendous measure of vitellogenin, a predecessor of lipo-and phosphoproteins that make up egg-vitellus protein, in the male trout people from River Lea of England outperformed females not well before ovulation [20, 21].

Triclosan (TCS) is an adversary of microbial engineers commonly used in nuclear families as a foe of bacterial and unfriendly to parasitic experts found in chemicals and different disinfectants. 96% of family things that contain TCS come out of the channel. Therefore, TCS is commonly present in the land, water conditions, and levels have been attempted all through the US. Noteworthy, TCS distinctive degrees were found in wastewater (max 26.2 µg obsession) and the maximum amount in biosolids obtained in sewage garbage (max 35,000 µg center). TCS is found in 57.6% of light of everything and streams attempted all through the USA.

Additionally, levels of TCS are traveled through the water into marine life. Diverse manufactured creations exposed because of TCS are known as debasement items. During wastewater treatment, methyltriclosan (MTCS) is made based on TCS's methylation, which isn't biodegradable and unimaginably productive all through the earth [21, 22]. Different synthetic compounds are likewise liable for defiling water.

Phosphates have generally been named a chemical in an extensive show of disinfecting items. Phosphates are common in cleaners of various families is pentasodium triphosphate (PTSP). PTSP and various phosphates can't be taken entirely out during the drain water treatment or sewerage system. It is associated with eutrophication that includes over-the-top improvement of green development, which ingests the oxygen in the water. Because of the nonappearance of oxygen, all maritime living things going from plants to aqua lives will pass on. Eutrophication is a considerable danger that can pound sea life's natural frameworks fast, making it inconceivable for marine life to get by further [23]. The Ecological Protection Agency has tabulated phosphorus, nitrogen, antacid, and engineered substances accumulated under the articulation "Temperamental Biotic Compounds" as the most terrible regular dangers in family cleaners. According to the Canadian Labor Environmental Alliance Society, chemicals in the dishwasher are 30 to 40% phosphorus. A soluble base is a multipurpose nuclear family cleaner in many cleaning things, which clean anything from defatting to antisepticising and dispensing with sensitivity. Volatile organic compounds (VOCs) are used in many cleaning specialties. The chemicals light up pieces of clothing, eradicate oil from dishes, and act as a disinfectant for washroom cleaners in our daily lifestyle. Nitrogen is present in glass and surface cleaning things; likewise, the invention is discovered in floor cleaners [24].

Health Products

Developing contaminants comprises a broad gathering of engineered and common substances, including drugs, individual consideration items, steroid hormones, and agrochemicals. The observation of leftover medicines on earth has featured many substances in wastewater treatment plants' effluents and surface waters extending from ng/L to µg/L. Most of these mixes are released into nature persistently through homegrown sewage treatment frameworks [25]. One of the viable secret experiments for PCPs was performed in Germany during the 1990s. It figured out stores having a point with 32 various restorative sections (antiphlogistics, lipid regulators, mental meds, antiepileptic drugs, beta-blockers, and β2-sympathomimetics similarly as five metabolites) in German metropolitan sewage treatment plant deliveries, stream, and stream waters. Because of their inescapable closeness in maritime conditions, tremendous quantities of these drugs must be named appropriate normal poisons [26]. In 2011, drug blends were recognized at low obsessions in 2.3% of 1231 instances of groundwater (center significance wells=61m) used for open drinking water deftly in California. Of 14 medication blends examined, 7 were perceived at obsessions more significant than or comparable to strategy acknowledgment limits: acetaminophen (torment soothing, most extraordinary center 1.89 kg/L), caffeine (energizer, 0.29 kg/L), carbamazepine (outlook stabilizer, 0.42 kg/L), codeine (opiate torment diminishing, 0.214 kg/L), p-xanthine (caffeine metabolite, 0.12 kg/L), sulfamethoxazole (antidote poison, 0.17 kg/L), and trimethoprim (against disease, 0.018 kg/L). Center recognized centralization of medication blends resembled those of capricious regular blends and was higher than pesticides [27].

Various synthetic compounds are divided into analgesics and mitigating drugs (codeine, ibuprofen, acetaminophen, acetylsalicylic corrosive, diclofenac), mental prescriptions (diazepam), b-blockers (Metoprolol, Timolol), steroids (estradiol, estrone), lipid controllers, hormones, sulphonamides (human and veterinary medications), antibiotic medications (human and veterinary medications). Another assessment was reached in two South Wales (UK) streams. The proximity and fate of 56 medications, singular thought things, endocrine disruptors, and illicit drugs were inspected in stream Taff and stream Ely, which were checked for 10 months. The more significant part of the centralizations of micro-pollutants was in the extent of 1-9 kg/L, and their levels were subject to the level of water debilitating (precipitation). The arrival of treated wastewater spouting into the stream course was the essential driver of water-polluting with PCPs. These medications were overwhelmingly antibacterial medicines (trimethoprim, erythromycin-H_2O, and amoxicillin) against inflammatories and analgesics drugs 11(paracetamol, tramadol, codeine, naproxen, ibuprofen, and diclofenac) and antiepileptic drugs (carbamazepine and gabapentin) [28].

Checking and consistent assessment of medication developments and PCPs in sewerage treatment plants, streams, 13 ground, and drinking water has been passed on in the latest decade in Greece. Different acidic medication developments (ibuprofen, naproxen, ketoprofen), business drugs (clofibric, diclofenac, phenazone, polyphenazone, 2,4-dichlorobenzoin destructive), singular thought materials, and endocrine upsetting manufactured mixes have been directed by gas or liquid chromatography-mass spectrometry. The picked areas of study were city and clinical facility wastewater treatment plants, sewage overflow, streams, ground, and drinking water. Spotlights on the influent and effluents varied, and the ejection capability went from 15-80% [29 - 32].

GLOBAL SCENARIO OF PCPS APPLICATION AND WATER POLLUTION

Different countries are now concerned about how applications of PCPs are contaminating the water sources. Many research, experiments, and studies are being accomplished globally to ensure the contamination level and remove the contaminant from water. Many new regulations are being launched for the corresponding authorities to reduce the usage of poisonous chemicals in PCPs and find an alternative to the chemicals that can be harnessed naturally from the environment.

From statistics of former years, the information says that Spain and the USA possess the most necessary number of PCPs not covered as an emerging pollutant in water with 42 and 36 blends. At the same time, the United Kingdom and Germany have 22 and 20, respectively [28, 33 - 38]. A whole of 43 PCPs were represented as EPs on surface water from a couple of countries around the world, for instance, Australia, Germany, China, Denmark, France, Czech Republic, India, Romania, Japan, Singapore, Antarctic, South Korea, Spain, Taiwan, Switzerland, United Kingdom, and United States [22, 35, 38 - 44]. Another study perceived musk xylene and musk ketone (built musk) in 100% and 80% independently of 74 models from Tama River and Tokyo Bay in Japan. The levels and scattering of 12 antimicrobials were inspected in water from the Mekong Delta, Vietnam, and differentiated and those in the Tamagawa Waterway, Japan. While a few blends, for instance, sulfamethoxazole, sulfamethazine, trimethoprim, and erythromycin-H_2O were recognized in Vietnam at concentrates someplace in the scope of 7 and 360 ng/L. Moreover, more antimicrobials existed in the Japanese metropolitan stream, tallying sulfamethoxazole, sulfapyridine, trimethoprim, erythromycin-H_2O, azithromycin, clarithromycin, and roxithromycin at centers going from 4 to 448 ng/L [45, 46].

Table 1 shows how PCPs are affecting wildlife very badly. Another assessment drove in the metropolitan riverine water of the Pearl River Delta at Guangzhou, South China, revealed the closeness of the estrogenic hormone, estrone, at the best centralization of 65 ng/L. At the same time, acidic medications, such as salicylic destructive, clofibric destructive, and ibuprofen, were distinguished in most water tests with the most noteworthy centralizations of 2098, 248, and 1417 ng/L independently. A later report in Taihu Lake, China, recognized eight artificially unique blends, to be specific: roxithromycin, erythromycin, ibuprofen, diclofenac, propranolol, carbamazepine, E2, and EE2 in surface water and residue tests with most noteworthy concentrations in the extent of 8.74-118 ng/L and 0.78-42.5 ng/L dry weight independently. Four medication developments in wastewater STP and seawater around the spouting delivery zone are developed in Northern Taiwan. The medication obsessions assessed in influent were: clofibric destructive (104-109 ng/L), diclofenac (152-185 ng/L), ibuprofen (724-2200 ng/L), and ketoprofen (128-184 ng/L). Contrasting concentrations in radiating were: 95-102 ng/L, 100-131 ng/L, 552-1600 ng/L, and 68-128 ng/L exclusively [36, 47, 48]. PCPs are recognized as EPs in water lattices, and their sources are introduced in Table **1**. These were found in boundless spots and overall landmasses [22, 28, 33 - 40]. Pakistan and Malaysia also advanced their research on water pollutants for groundwater. They found some latest outcomes that will help experts ensure a sustainable solution to water contamination from PCPs [49, 50].

Table 1. Different personal care products [22, 28, 33 - 40].

Compound	Source	Matrix
MNT	Toothpaste	Wastewater,
CMP	Cosmetic	Wastewater
BHT	Cosmetic	Sewerage, surface water,
BPB	Cosmetic	Sewerage, surface water,
CP	Cosmetic	Sewerage, surface water,
OMC	Sunscreen	Sewerage
MPB	Cosmetic	Sewerage, surface water, groundwater
OPP	Cosmetic	Sewerage, surface water, groundwater
NP1EO	Fumigant stick, soap	Sewerage, surface water, groundwater
3,4,5,6-Tetrabromo-o-cresol	Fumigant stick, soap	Sewerage
2-NP	Cleansing agent	Surface water
4-NP	Cleansing agent	Sewerage, surface water, sludge
TCC	Cleaning agent	Sewerage, surface water, groundwater, sludge
TCS	Cleaning agent	Sewerage, surface water, groundwater, sludge
DEET	Cleansing agent	Sewerage, surface water, groundwater, sludge
ACN	Aroma	Groundwater

ENVIRONMENTAL IMPACT OF WATER POLLUTION BY PCPS

The way toward distinguishing the presence of substance mixes in any ecological grid doesn't imply that it is of tension or may hurt. In any case, huge concerns ascend from the location of synthetic substances mean that they may unfavorably influence sea-going life. The accompanying segments sum up a portion of critical worries about the presence of PCPs in freshwater and marine conditions.

Toxicity

A detailed view of the extraordinary and wearisome toxic substance levels of PCP blends in water has been coordinated [51]. Among the compound explored, Parabens (alkyl-p-hydroxybenzoate), Triclosan (TCS) (Irgasan DP 300, a chlorinated diphenyl ether: 2,4,4'-trichloro-2'-hydroxy-diphenyl ether), triclocarban (TCC), designed musk, six sunscreen masters (SSAs), and N, N-diethyl-m-toluamide (DEET) were highlighted as having the immense potential for a serious threat to water.

Parabens (alkyl-p-hydroxybenzoates) are enthusiastically deployed set-ups of antimicrobial added substances in cosmetics (skin creams, tanning lotions, *etc.*), toiletries, medicates, and also goods (up to 0.1% wt/wt). Regardless of how the unprecedented destructiveness of these blends is amazingly low, it is reported [51] that blends (methyl through butyl homologs) show low estrogenic activity in a couple of tests. Butyl-paraben revealed the greatest genuine authority to the rat estrogen receptor at obsessions one to two noteworthy degrees higher than nonylphenol and exhibited estrogenic activity in a yeast estrogen screen at 10–6 M [51]. The assessment relies upon confined biological concentration and toxicity data; benzyl-, butyl-and propyl-paraben may have contrasting effects on land and water-bodies. Daughton and Ternes [51] executed parabens to a confined danger to marine animals; nevertheless, parabens, expressly benzyl-, butyl-and propyl-paraben, can move weak-level estrogenic reactions. *In vitro* examinations drove with fish MCF-7 cell lines and yeast, estrogenic screening analyses displayed parabens could rouse estrogenic reply at weak levels. Dussault *et al.* [51] found that parabens can cause vitellogenin (VTG) to blend in fish when introduced to low obsessions. The low-level introduction to parabens may cause estrogenic effects at normally significant core interests. Additional assessments have been coordinated to investigate the impact of parabens on sexual endpoints, recollecting spermatogenesis, and serum testosterone for male rodents [52]. Both butyl- and propyl-paraben generally quelled spermatogenesis anyway didn't impact serum testosterone [52]. These results exhibit normal effects in maritime living things diligently introduced to parabens. The data on biological obsessions propose a solitary irrelevant risk to marine animals as effect centers are regularly 1000x upper than what has been found in surface water.

Triclosan (Irgasan DP 300, a chlorinated diphenyl ether: 2, 4, 4'-trichloro-2 '-hydroxy-diphenyl ether) has been commonly used for directly around 30 years in a gigantic scope of purchaser things. It is merged at < 1% in Colgate's Total toothpaste, the essential toothpaste supported by the US Food and Drug Administration (FDA) to battle gum malady. While triclosan is enrolled with the US EPA as a pesticide, it is an uninhibitedly common over-the-counter (OTC) drug. Triclosan's is widely used in footwear (in hosiery and insoles of shoes called Odor-Eaters), hand soap, skin irritation creams (*e.g.*, Clearasil), in a wide variety of plastic stuff from children's toys to kitchen utensils, for instance, cutting sheets. High usages can achieve the quick arrival of triclosan to sewage structures. Likewise, this compound can find its way into tolerating waters, dependent on microbial degradation insurance. Okamura *et al.* [53] found traces of triclosan stretching from 0.05 to 0.15 µg/L in water. In some cases, triclosan has been seen as a biocide, a toxin having a wide-broadening, unknown mechanism(s) of action for this circumstance. McMurry *et al.* [54] mentioned that triclosan is an antibacterial agent with enzymatic targets (lipid blend). In that limit, minute living beings could make assurance from triclosan. Additionally, with all enemies of microbes on the earth, this could provoke the progression of resistance and change in microbial organization structure (grouped assortment).

In PCPs from 1957, Triclosan (TCC) was utilized and seen in surface water at 6.75 ug/L [55]. It is accepted that TCC occurs in WWTP gushing and surface water as triclocarban (TCS); notwithstanding, until 2004, TCC couldn't be distinguished at base levels [55]. TCC has been identified at higher focuses and more now and again in WWTP emanating and surface water than TCS or M-TCS throughout the most recent five years [3]. Furthermore, TCC has exhibited an inclination to bioaccumulate greater than TCS or M-TCS in amphibian life forms. Different disinfectants (phenol, 4-methyl phenol, and biphenylol) are regularly used in family units and can be delivered into sea-going conditions. These mixes have been recognized in surface water or WWTP profluent [56], with phenol discovered over regularly than 4-methyl phenol and biphenylol, just as in more noteworthy fixations (as high as 1.3 ug/L) [56].

Fabricated musk is aromas applied to widespread things together with antiperspirants and chemicals. The musk is either nitro musk, which was initialed in the last 1800s, or polycyclic musk found in the 1950s [51]. The largely applied nitro musks are musk xylene (MX) and musk ketone (MK), while musk ambrette (MA), musk moskene (MM), and musk tibetene (MT) are used less routinely [51]. Nitro musk is continuously disposed of involving their environmental eagerness and anticipated that their disposal would cause a threat to marine species [51]. Polycyclic musk is recently applied in higher sums than nitro musk with celestolide (ABDI), galaxolide (HHCB), and toxalide (AHTN) applied most consistently, and traseloide (ATII), phantolide (AHMI), and cashmeran (DPMI) used less frequently [51]. HHCB and AHTN fabrication alone have consistently been surveyed at 1 million pounds and have been determined to the High Production Volume List by the USEPA [57]. Various studies first perceived nitro musk on earth and drove the essential immense watching examination on MX and MK [45, 58]. MX and MK were obtained more significantly than 80% of stream water tests, WWTP profluent, surface water fish, and shellfish in Japan. Obsessions were generally critical in WWTP profluent going from 25 to 36 ng/L and 140–410 ng/L for MX and MK, respectively. Different researchers recognized severe groupings of musk in lipids from new and saltwater fish and mollus [53]. Like nitro musk, polycyclic musk is non-harmful to animals of land and water. Developmental paces of cowardly animals and advancement and improvement for Eels fish are the highest sensitive endpoints read to date for polycyclic musk.

The solidified centralizations of six sunscreen administrators (SSAs) perceived in perch (Perca fluvi-atilis) in the pre-summer of 1991 was as strong as 2.0 mg/kg lipid and in bug (rutilus L) in the mid-year of 1993, as high as 0.50 mg/kg lipid. Methyl benzylidene camphor (MBC) was recognized in bugs from three other German lakes. These lipophilic SSAs indicate to happen comprehensively in fish from little lakes applied for recreational swimming. Both fish species had body loads of SSA similar to PCBs.

DEET is the most ordinarily utilized dynamic fixing in bug anti-agents and is industrious in sea-going conditions. Even though DEET has been recognized worldwide in sewage treatment plants (STPs), its fixations in influents and effluents are generally weak. The degree of DEET is essentially diminished in winter because of decreased utilization [59]. It is described in [52] that DEET was obtained in 95% of broken-down examples, with a middle grouping of around 0.2 µg/L. Nonetheless, DEET cannot aggregate in oceanic life forms, as shown by its low bioconcentration factor [57]. Generally, the poisonousness of PCPs in the sea-going condition stretches out past the intense impacts seen when helpful levels are reached or surpassed. Late investigations have indicated PCP poisonousness to shift contingent upon the uncovered creature, introduction term, contaminant focus, and formative window at which presentation happens.

Additionally, the constant follow-the-level presentation impacts, particularly at certain touchy phases of improvement, are bound to clarify watched anomalies inside uncovered non-target life forms than intense high portion exposure. As numerous drug contaminants are naturally presented after human or veterinary use, metabolite focuses might be huger than that of parent mixes. For instance, some acetylated metabolites of anti-infection agents (N4-acetylsulfapyridine) were more harmful than the parent compound (sulfapyridine) in green growth. Moreover, under unfortunate conditions in the oceanic environment, dynamic drug operators' presence can change their toxicological ingredients. To represent, the photodegradation results of naproxen accounted for more harmful impacts than the parent compound on green growth, rotifers, and micro-crustaceans [52]. Acidic drug mixes could inspire diverse toxicological reactions at various pH levels in uncovered non-target creatures [58], and metals that appeared to amass in waterway biofilms have appeared to expand the harmfulness of explicit anti-toxin pollutants (fluoroquinolones and antibiotic medications) in an added substance way.

Bioaccumulation

Whether PCPs are perceived in the freshwater condition at low concentration, massive amounts of them and their metabolites are regularly ground-breaking and impact non-target land and water, capable creatures. A few assessments have explored the presence of PCPs on non-target animals, mainly fish. The introduction of goldfish (*Carassius auratus*) to waterborne gemfibrozil at an ordinarily noteworthy concentrate surpassing 14 days accomplished a plasma bio-focus factor of 113 [60, 61]. Another study [62] demonstrated bioaccumulation of antiepileptic drug carbamazepine (CBZ) by the green turn of events - Pseudokirchneriella sub-capitata and the shellfish - *Thamnocephalus platyurus* with bioaccumulation portions of 2.2 and 12.6. The bioaccumulation factors for caffeine, diphenhydramine, diltrazem, carbamazepine, and ibuprofen were 2.0, 16, 16, 1.4, and 28. Oxazepam was recognized as a high focus in Eurasian roost fish with a bioaccumulation factor of 12 [63]. This assessment [63] showed the assortment of fluoxetine in snails with the bioaccumulation factor of 3000. Around 145 PCPs were found in wild and bound mussels from the Grand River, Ontario, and 43 drugs from various classes were seen in mussel tissues, with bioaccumulation factors running from 0.66 for metformin 32,022 for sertraline. As explicit from drugs, PCPs have been found in green turn of events, which recall the most dazzling abundance of plant biomass for the oceanic condition. The lipid substance of green improvement gives a portion highlighting trophic exchange of ordinary lipophilic contaminants.

Persistence

The physicochemical characteristics of different PCPs interprets that these PCPs are not discarded by ordinary water treatment measures as discovered by their core in drinking water [64]. The failure to influence the total expulsion of PCPs from the waste treatment plant addresses a possible danger to oceanic living animals and general thriving. Surprisingly, watching appraisals shows that PCPs have discovered their way into the aquatic condition and are comprehensive [65]. The extensive idea of PCPs' utilization has gotten together with the inspiring partner of new prescriptions with the market [51]. Pseudo-inventive meds are proposed to have more basic potential for organic imagination than other standard contaminants like pesticides, considering how their source dependably restores when followed up by regular cycles such as biodegradation photodegradation, and particulate sorption. Thus, sedates may adequately carry on as predictable mixes in light of their consistent movement into the earth [66]. Loffler organized ten drugs and medicine metabolites into low, moderate, and high affirmation mixes as indicated by their dispersal time (DT50) in water/development tests. Paracetamol, Ibuprofen, 2-hyroxyibuprofen, and CBZ-diol demonstrated low steadfast (DT50 ¼ 3.1-7 days), Oxazepam, Iopromide, and Ivermectin were viewed as acceptably consistent (DT50 ¼ 15-54 days). Simultaneously, Clofibric, Diazepam, Carbamazepine was evaluated unflinching (DT50 ¼ 119-328 days) [67]. A report showed that anxiolytic solution (Oxazepam) expanded life in freshwater lakes due to past information and making metropolitan individuals [68].

POTENTIAL TECHNIQUES FOR PCPS REMOVAL FROM POLLUTED WATER

Wastewater and Sewage Water Treatment Methods

The launch or defilement of typical contaminants present in the water relies on a few segments, including source water quality, treatment cycles, objectives, and brand name compound properties of toxic substances, for example, atomic weight, relative hydrophobicity, sweet-smelling carbon content, and practical social affair strategy [69]. Various single and hybrid technologies exist to remove PCPs, which partially or almost fully remove these pollutants. These processes vary from coagulation to sedimentation, flotation, activated carbon adsorption, advanced oxidation processes, membrane separation and biological treatment. Various studies indicate that it is impossible to remove these pollutants with a single technology fully. However, hybrid technologies can reach a maximum removal capacity of PCPs. For example, the application of conventional treatment methods, membrane reactors, and advanced posttreatment methods resulting in a hybrid wastewater treatment technology appears to be the best.

The coagulation, sedimentation, and filtration have been found to kill around 25% of most PCPs and are ordinarily deficient for shedding common isolated contaminants [69]. Layer filtration measures (nanofiltration (NF) and RO) are good choices for the expulsion of PCPs from wastewater. UF and microfiltration

(MF) have appeared to treat PCPs well. Their takeoff introductions are low since layer pore sizes are more prominent than PCP particles. For evaluation, pressure-driven layer cycles, NF, and RO were applied to the drinking water treatment [69]. These cycles show crucial PCP flight efficiencies; these layers are still scarcely penetrable to some toxic substances [64]. AOPs, such as ozonation, UV, photocatalysis, and Fenton response, have been utilized for drinking water treatment (*e.g.*, smell/taste control and refining) and wastewater sanitization [70]. A previous study [71] found that the explanation cycle with ferric chloride ($FeCl_3$) as coagulant addressed the discharge of just 15% of the ordinary social occasion of basic contaminants during wastewater creation at a US drinking water treatment. The expulsion of diclofenac, ibuprofen, benzafibrate, carbamazepine, and sulfamethoxazole by ferric sulfate-helped coagulation in both Milli-Q water [72]. Free chlorine and chloramines can treat a degree of PCPs, by hydrophilic mixes [71]. Free chlorine is more skilled than chloramines at discarding PCPs [73]. For instance, advanced water treatment developments, ozonation, granular started carbon (GAC) adsorption, and brilliant (UV) light have ensured treating PCPs. For UV treatment, a normal purifying part (5–50 mJ/cm^2) was a couple of critical degrees lower than the measurements required for clearing most engineered mixes [69, 73].

Surface Water Cleansing Methods

A PCP and atrazine removal plant for surface water treatment in Windsor, Canada, observed that coagulation/flocculation/sedimentation and twofold media filtration treatment commonly lacked ozonation followed by conventional treatment updated the removal of carbamazepine, cotinine, caffeine, and atrazine [74]. Diminishing by customary centralization of five organophosphorus fire retardants (TBEP, TDIP, TCEP, TPP, and TBP), the musk smells AHTN and HHCB, the bug repellent DEET, and the pesticide compound metolachlor was under 25% through oxidation with free chlorine [69]. Snyder *et al.* [2] considered the discharge of 36 EDCs and PCPs in both surface water and wastewater utilizing O_3 and O_3/H_2O_2. Ozone estimations of 1.25 mg/l or more were satisfactory in accomplishing over 80% freeing from 22 mixes in surface water. The appraisal results show that O_3/H_2O_2 could decrease the contact time required by O_3 alone. The expansion of H_2O_2 to O_3 indicated a decrease for most mixes yet express mixes: for example, androstenedione, progesterone, testosterone, caffeine, metolachlor, and pentoxifylline [2]. Moreover, cleansing by OH revolutionaries is inadequate, differentiated and O_3, and there is an opportunity of confining additional disinfection reactions (Snyder *et al.* 2006; Wert *et al.* 2007). To date, less effort is pondered on the turn of events, fate, acknowledgment, and harmfulness of oxidation symptoms of EDCs and PCPs [69].

Bromate, which doesn't use defilement in natural procedures, is the fundamental consequence of ozonation coordinated in drinking water treatment. Previous studies observed that regular ozone tends to dispose of PCPS and would not make enormous proportions of bromate [3, 78].

Biota

Around the globe, it has been demonstrated that preface to WWTP effluents containing PCPs is associated with a degree of destructive impacts on sea creatures' growth [75]. A study discovered bioaccumulation of a blend of estrogenic contaminants in fish tissues, thus determining vitellogenin and perhaps adding to the feminization of wild fish staying in the UK streams [75]. A previous study [72] reported endocrine-upsetting fabricated manifestations (EDCs) in water, development, and biota of Venice's coastline front tidal pond. Their evaluation showed that most picked mixes were found in water and silt at a middle running from 2.8 to 211 ng/L and 3.1-289 mg/kg dry weight by a wide margin. The combinations perceived in the Mediterranean mussel (*Mytilis galloprovincialis*) were 17a-ethinylestradiol, and nonylphenol at fixation goes 7.2e240 ng/g in dry weight [41]. In another study [76], rainbow trout were acquainted with sedate sewage effluents, levonorgestrel was assembled in fish blood at centralizations of 8.5-12 ng/ml. A previous study [77] evaluated that galaxolide and tonalide in tilapia and bream fish tests were collected from Rhine River, Germany, at 81 and 5.5 ng/g wet weights. Alvarez-Munoz *et al.* [78] investigated the presence of prescriptions in shellfish, mollusk, and mussel tests amassed from the Ebro delta, Spain. The outcomes revealed that unavoidable mixes apparent were the mental medication venlafaxine and the counter defilement azithromycin, with the most significant focuses on mussels (2.7 ng/g) and shellfish (3.0 ng/g). A Spanish evaluation inspected holding up prescriptions in pork, veal, sheep and chicken muscle, liver, and kidney correspondingly as salmon, ocean bass, and sole substance bought at a near to advertise.

FUTURE PERSPECTIVE AND CHALLENGES

The soiling of water by PCPs has raised concerns over perils to general prosperity and natural frameworks. Such polluting contaminants can challenge land cutoff points and contaminate fresher locales. An aspect of the organochlorine pesticides, which are presently restricted in industrialized nations, is being used to battle pollution such as intestinal contamination in making nations. Just two essential exercises are dealing with prescriptions on the earth. One outcome from the evaluation has happened in different European/Scandinavian nations, winding up at ground zero in rules from the EU.

It is essential to see, in any case, that responsibility in the United States for watching prescriptions on the earth doesn't exist at present rest with either the FDA or the EPA. Generally, any other elective framework for contemplating the ecological danger presented by drugs has been proposed. PCPs are rising contaminants with physicochemical characteristics from different toxins (*e.g.*, driving forward trademark contaminations). It is incorporated that oceanic living things are acquainted continuously with PCPs for an astounding range cycle. There is a mounting check that these contaminants' obligatory presence in the sea condition may negatively affect land water capable life. The significant point wellspring of PCPs into land and water capable conditions is WWTPs, close to different sources, for example, natural floods, PCP making protests, and hydroponics. It is recommended that WWTPs are not fit for taking out PCPs during treatment measures. Accordingly, treated effluents conveyed into enduring water bodies may contain colossal PCP advancements. Regardless of the constant advances in interpretive techniques that award the delicate multi-improvement evaluation of several PCPs in various natural cross segments and show the unintended regular presence of such created materials. It reported that more assessments of PCPs are relied upon portraying their regular presence in making nations, as there is less information for Africa, Asia, and South America showed up contrastingly corresponding to Europe and North America [79].

Besides, while sorption to buildup particles was proposed to expect a limit in picking the destiny of PCPs in the freshwater land and water capable condition, no individual evaluations are watching out for the direct segments of PCPs in freshwater frameworks or how silt may go about as a sink for these contaminants or wellspring of PCPs for base managing oceanic biota. It is important to energize the current energy about the toxicological repercussions of the reliable prologue to complex blends of PCPs at sub-steady levels in both objective and non-target creatures. More outstanding evaluation is expected to portray the impact of such presentation on the status of general thriving in debased zones (*e.g.*, impacts on malarial obstruction in Africa or luxury/fecundability in districts defiled with estrogenic hormones).

Further evaluation is required to provide importance to rare groupings of PCPs in different standard compartments. Another primary area is bioaccumulation of PCPs in land and water capable living things, for example, green turn of events, shellfish, and fish. Further assessment of the possible bioaccumulation by oceanic biota is required, including its suggestions for human presence by using polluted fish/shellfish.

CONCLUSION

Water pollutants are the prime concern throughout the world. Different health hazards are now endangering human lives because of water pollution. PCPs are now a threat to the aquatic environment and freshwater globally, among the various contaminants. Moreover, a sharp rise in daily product use predicts that the poisonous chemicals will be the leading cause of different chronic diseases like a failure of kidneys, lungs, and liver and asthma, bronchial problems, and others. Therefore, this chapter emphasized various sources of PCPs such as beauty products, cleaning products, and health products as emerging contaminants for water pollution; the global scenario of PCPs application and water pollution, the hazardous impact of PCPs on human life and wildlife; potential techniques for PCPs removal from wastewater and future perspectives and challenges. This chapter also outlined different facts, reasons, effects, and prospective solutions to prevent contamination caused by PCPs. Based on the previous experimental studies, it can be strongly recommended that decisive actions should be taken by the corresponding authority to reduce the contamination level and to find alternative natural or green chemicals that will be friendly to the aquatic environment. Moreover, it is necessary to ensure a safe and healthy water environment to secure the lives of animals of the lake, sea, ocean, and other water resources.

CONSENT FOR PUBLICATION

Not applicable.

CONFLICT OF INTEREST

The author declares no conflict of interest, financial or otherwise.

ACKNOWLEDGEMENTS

Declared none.

REFERENCES

[1]   PCPs and Drinking Water, www.watersystemscouncil.org

[2]   H.B. Quesada, A.T.A. Baptista, L.F. Cusioli, D. Seibert, C. de Oliveira Bezerra, and R. Bergamasco, "Surface water pollution by pharmaceuticals and an alternative of removal by low-cost adsorbents: A review", *Chemosphere,* vol. 222, pp. 766-780, 2019.
[http://dx.doi.org/10.1016/j.chemosphere.2019.02.009] [PMID: 30738319]

[3]   S.A. Snyder, "Occurrence, treatment, and toxicological relevance of EDCs and pharmaceuticals in water", *Ozone Sci. Eng.,* vol. 30, no. 1, pp. 65-69, 2008.
[http://dx.doi.org/10.1080/01919510701799278]

[4]   M.A. Coogan, R.E. Edziyie, T.W. La Point, and B.J. Venables, "Algal bioaccumulation of

triclocarban, triclosan, and methyl-triclosan in a North Texas wastewater treatment plant receiving stream", *Chemosphere,* vol. 67, no. 10, pp. 1911-1918, 2007.
[http://dx.doi.org/10.1016/j.chemosphere.2006.12.027] [PMID: 17275881]

[5]   E. Wielogórska, C.T. Elliott, M. Danaher, O. Chevallier, and L. Connolly, "Validation of an ultra high performance liquid chromatography–tandem mass spectrometry method for detection and quantitation of 19 endocrine disruptors in milk", *Food Control,* vol. 48, pp. 48-55, 2015.
[http://dx.doi.org/10.1016/j.foodcont.2014.06.001]

[6]   B.D. Blair, J.P. Crago, C.J. Hedman, and R.D. Klaper, "Pharmaceuticals and personal care products found in the Great Lakes above concentrations of environmental concern", *Chemosphere,* vol. 93, no. 9, pp. 2116-2123, 2013.
[http://dx.doi.org/10.1016/j.chemosphere.2013.07.057] [PMID: 23973285]

[7]   B.D. Blair, J.P. Crago, C.J. Hedman, R.J.F. Treguer, C. Magruder, L.S. Royer, and R.D. Klaper, "Evaluation of a model for the removal of pharmaceuticals, personal care products, and hormones from wastewater", *Sci. Total Environ.,* vol. 444, pp. 515-521, 2013.
[http://dx.doi.org/10.1016/j.scitotenv.2012.11.103] [PMID: 23295178]

[8]   B. Blair, A. Nikolaus, C. Hedman, R. Klaper, and T. Grundl, "Evaluating the degradation, sorption, and negative mass balances of pharmaceuticals and personal care products during wastewater treatment", *Chemosphere,* vol. 134, pp. 395-401, 2015.
[http://dx.doi.org/10.1016/j.chemosphere.2015.04.078] [PMID: 25985097]

[9]   Health Europa, https://www.healtheuropa.eu/personal-care-products-water-pollutants/90895/

[10]   MedicineNet, https://www.medicinenet.com/script/main/art.asp?articlekey=87749#tocf

[11]   P.C. Products, *Everyday Pollutants on Your Body.*.https://mytapscore.com/blogs/tips-fo--taps/personal-care-products-everyday-pollutants-on-your-body

[12]   M.E. Cooley, "Symptoms in adults with lung cancer. A systematic research review", *J. Pain Symptom Manage.,* vol. 19, no. 2, pp. 137-153, 2000.
[http://dx.doi.org/10.1016/S0885-3924(99)00150-5] [PMID: 10699541]

[13]   Real Cost of Cosmetics, *Environmental Impact.*.https://www.adorncosmetics.com.au/blog/post/environmental-impact/

[14]   A. Sánchez Rodríguez, M. Rodrigo Sanz, and J.R. Betancort Rodríguez, "Occurrence of eight UV filters in beaches of Gran Canaria (Canary Islands). An approach to environmental risk assessment", *Chemosphere,* vol. 131, pp. 85-90, 2015.
[http://dx.doi.org/10.1016/j.chemosphere.2015.02.054] [PMID: 25792520]

[15]   T.L.L. Teo, H.M. Coleman, and S.J. Khan, "Chemical contaminants in swimming pools: Occurrence, implications and control", *Environ. Int.,* vol. 76, pp. 16-31, 2015.
[http://dx.doi.org/10.1016/j.envint.2014.11.012] [PMID: 25497109]

[16]   L.S. Fendall, and M.A. Sewell, "Contributing to marine pollution by washing your face: Microplastics in facial cleansers", *Mar. Pollut. Bull.,* vol. 58, no. 8, pp. 1225-1228, 2009.
[http://dx.doi.org/10.1016/j.marpolbul.2009.04.025] [PMID: 19481226]

[17]   M. Cole, P. Lindeque, C. Halsband, and T.S. Galloway, "Microplastics as contaminants in the marine environment: A review", *Mar. Pollut. Bull.,* vol. 62, no. 12, pp. 2588-2597, 2011.
[http://dx.doi.org/10.1016/j.marpolbul.2011.09.025] [PMID: 22001295]

[18]   T. Gouin, J. Avalos, I. Brunning, K. Brzuska, J. de Graaf, J. Kaumanns, T. Konong, M. Meyberg, K. Rettinger, and H. Schlatter, "Use of micro-plastic beads in cosmetic products in Europe and their estimated emissions to the North Sea environment", *SOFW J,* vol. 3, pp. 1-33, 2015.

[19]   I.E. Napper, A. Bakir, S.J. Rowland, and R.C. Thompson, "Characterisation, quantity and sorptive properties of microplastics extracted from cosmetics", *Mar. Pollut. Bull.,* vol. 99, no. 1-2, pp. 178-185, 2015.
[http://dx.doi.org/10.1016/j.marpolbul.2015.07.029] [PMID: 26234612]

[20]   S.H. Swan, K.M. Main, F. Liu, S.L. Stewart, R.L. Kruse, A.M. Calafat, C.S. Mao, J.B. Redmon, C.L. Ternand, S. Sullivan, and J.L. Teague, Study for Future Families Research Team, "Decrease in anogenital distance among male infants with prenatal phthalate exposure", *Environ. Health Perspect.,* vol. 113, no. 8, pp. 1056-1061, 2005.
[http://dx.doi.org/10.1289/ehp.8100] [PMID: 16079079]

[21]   T.M. Leatherland, and A.M. Warhurst, "An Environmental Assessment of Alkylphenol Ethoxylates and Alkylphenols", *J. Appl. Ecol.,* vol. 32, no. 4, p. 890, 1995.
[http://dx.doi.org/10.2307/2404838]

[22]   C.A. Staples, J. Weeks, J.F. Hall, and C.G. Naylor, "Evaluation of aquatic toxicity and bioaccumulation of C8- and C9-alkylphenol ethoxylates", *Environ. Toxicol. Chem.,* vol. 17, no. 12, pp. 2470-2480, 1998.
[http://dx.doi.org/10.1002/etc.5620171213]

[23]   https://www.govinfo.gov/content/pkg/FR-2020-03-09/.../FR-2020-03-09.pdf

[24]   https://acp.copernicus.org/articles/20/4275/2020/acp-20-4272020.pdf

[25]   S. Marketplace, *Greener Products and Services.*.https://www.epa.gov/greenerproducts

[26]   T.A. Ternes, "Occurrence of drugs in German sewage treatment plants and rivers", *Water Res.,* vol. 32, no. 11, pp. 3245-3260, 1998.
[http://dx.doi.org/10.1016/S0043-1354(98)00099-2]

[27]   M.S. Fram, and K. Belitz, "Occurrence and concentrations of pharmaceutical compounds in groundwater used for public drinking-water supply in California", *Sci. Total Environ.,* vol. 409, no. 18, pp. 3409-3417, 2011.
[http://dx.doi.org/10.1016/j.scitotenv.2011.05.053] [PMID: 21684580]

[28]   B Kasprzyk-Hordern, RM Dinsdale, and AJ Guwy, "The occurrence of pharmaceuticals, personal care products, endocrine disruptors and illicit drugs in surface water in South Wales, UK", *Water Res,* vol. 42, no. 13, pp. 3498-3518, 2008.

[29]   Th. Heberer, B. Fuhrmann, K. Schmidt-Baumler, D. Tsipi, V. Koutsouba, and A. Hiskia, "Occurence of pharmaceutical residues in sewage, river, ground, and drinking water in Greece and Berlin (Germany)", In: *ACS Symposium Series,* CG Daughton, TL Jones-Lepp, Eds., vol. 791. ACS publications: Washington DC, 2001, pp. 70-83.
[http://dx.doi.org/10.1021/bk-2001-0791.ch004]

[30]   V. Koutsouba, T. Heberer, B. Fuhrmann, K. Schmidt-Baumler, D. Tsipi, and A. Hiskia, "Determination of polar pharmaceuticals in sewage water of Greece by gas chromatography–mass spectrometry", *Chemosphere,* vol. 51, no. 2, pp. 69-75, 2003.
[http://dx.doi.org/10.1016/S0045-6535(02)00819-6] [PMID: 12586139]

[31]   C.I. Kosma, D.A. Lambropoulou, and T.A. Albanis, "Occurrence and removal of PPCPs in municipal and hospital wastewaters in Greece", *J. Hazard. Mater.,* vol. 179, no. 1-3, pp. 804-817, 2010.
[http://dx.doi.org/10.1016/j.jhazmat.2010.03.075] [PMID: 20395039]

[32]   V.G. Samaras, N.S. Thomaidis, A.S. Stasinakis, and T.D. Lekkas, "An analytical method for the simultaneous trace determination of acidic pharmaceuticals and phenolic endocrine disrupting chemicals in wastewater and sewage sludge by gas chromatography-mass spectrometry", *Anal. Bioanal. Chem.,* vol. 399, no. 7, pp. 2549-2561, 2011.
[http://dx.doi.org/10.1007/s00216-010-4607-6] [PMID: 21197532]

[33]   M. Beretta, V. Britto, T.M. Tavares, S.M.T. da Silva, and A.L. Pletsch, "Occurrence of pharmaceutical and personal care products (PPCPs) in marine sediments in the Todos os Santos Bay and the north coast of Salvador, Bahia, Brazil", *J. Soils Sediments,* vol. 14, no. 7, pp. 1278-1286, 2014.
[http://dx.doi.org/10.1007/s11368-014-0884-6]

[34]   P. Emnet, S. Gaw, G. Northcott, B. Storey, and L. Graham, "Personal care products and steroid hormones in the Antarctic coastal environment associated with two Antarctic research stations,

McMurdo Station and Scott Base", *Environ. Res.,* vol. 136, pp. 331-342, 2015.
[http://dx.doi.org/10.1016/j.envres.2014.10.019] [PMID: 25460654]

[35] K. Osenbrück, H.R. Gläser, K. Knöller, S.M. Weise, M. Möder, R. Wennrich, M. Schirmer, F. Reinstorf, W. Busch, and G. Strauch, "Sources and transport of selected organic micropollutants in urban groundwater underlying the city of Halle (Saale), Germany", *Water Res.,* vol. 41, no. 15, pp. 3259-3270, 2007.
[http://dx.doi.org/10.1016/j.watres.2007.05.014] [PMID: 17575997]

[36] L.P. Padhye, H. Yao, F.T. Kung'u, and C.H. Huang, "Year-long evaluation on the occurrence and fate of pharmaceuticals, personal care products, and endocrine disrupting chemicals in an urban drinking water treatment plant", *Water Res.,* vol. 51, pp. 266-276, 2014.
[http://dx.doi.org/10.1016/j.watres.2013.10.070] [PMID: 24262763]

[37] X. Peng, Y. Yu, C. Tang, J. Tan, Q. Huang, and Z. Wang, "Occurrence of steroid estrogens, endocrine-disrupting phenols, and acid pharmaceutical residues in urban riverine water of the Pearl River Delta, South China", *Sci. Total Environ.,* vol. 397, no. 1-3, pp. 158-166, 2008.
[http://dx.doi.org/10.1016/j.scitotenv.2008.02.059] [PMID: 18407320]

[38] J.P.R. Sorensen, D.J. Lapworth, D.C.W. Nkhuwa, M.E. Stuart, D.C. Gooddy, R.A. Bell, M. Chirwa, J. Kabika, M. Liemisa, M. Chibesa, and S. Pedley, "Emerging contaminants in urban groundwater sources in Africa", *Water Res.,* vol. 72, pp. 51-63, 2015.
[http://dx.doi.org/10.1016/j.watres.2014.08.002] [PMID: 25172215]

[39] G.-G. Ying, and R.S. Kookana, "Triclosan in wastewaters and biosolids from Australian wastewater treatment plants", *Environ. Int.,* vol. 33, pp. 199-205, 2007.
[http://dx.doi.org/10.1016/j.envint.2006.09.008]

[40] I.J. Buerge, H.R. Buser, M.D. Müller, and T. Poiger, "Behavior of the polycyclic musks HHCB and AHTN in lakes, two potential anthropogenic markers for domestic wastewater in surface waters", *Environ. Sci. Technol.,* vol. 37, no. 24, pp. 5636-5644, 2003.
[http://dx.doi.org/10.1021/es0300721] [PMID: 14717174]

[41] E. Carmona, V. Andreu, and Y. Picó, "Occurrence of acidic pharmaceuticals and personal care products in Turia River Basin: From waste to drinking water", *Sci. Total Environ.,* vol. 484, pp. 53-63, 2014.
[http://dx.doi.org/10.1016/j.scitotenv.2014.02.085] [PMID: 24686145]

[42] J. Cavalheiro, A. Prieto, M. Monperrus, N. Etxebarria, and O. Zuloaga, "Determination of polycyclic and nitro musks in environmental water samples by means of microextraction by packed sorbents coupled to large volume injection-gas chromatography–mass spectrometry analysis", *Anal. Chim. Acta,* vol. 773, pp. 68-75, 2013.
[http://dx.doi.org/10.1016/j.aca.2013.02.036] [PMID: 23561908]

[43] Y. Chen, J. Vymazal, T. Březinová, M. Koželuh, L. Kule, J. Huang, and Z. Chen, "Occurrence, removal and environmental risk assessment of pharmaceuticals and personal care products in rural wastewater treatment wetlands", *Sci. Total Environ.,* vol. 566-567, pp. 1660-1669, 2016.
[http://dx.doi.org/10.1016/j.scitotenv.2016.06.069] [PMID: 27342641]

[44] J.J. Jiang, C.L. Lee, and M.D. Fang, "Emerging organic contaminants in coastal waters: Anthropogenic impact, environmental release and ecological risk", *Mar. Pollut. Bull.,* vol. 85, no. 2, pp. 391-399, 2014.
[http://dx.doi.org/10.1016/j.marpolbul.2013.12.045] [PMID: 24439316]

[45] B. Kasprzyk-Hordern, R.M. Dinsdale, and A.J. Guwy, "Multiresidue methods for the analysis of pharmaceuticals, personal care products and illicit drugs in surface water and wastewater by solid-phase extraction and ultra performance liquid chromatography–electrospray tandem mass spectrometry", *Anal. Bioanal. Chem.,* vol. 391, no. 4, pp. 1293-1308, 2008.
[http://dx.doi.org/10.1007/s00216-008-1854-x] [PMID: 18253724]

[46] Z. Moldovan, "Occurrences of pharmaceutical and personal care products as micropollutants in rivers

from Romania", *Chemosphere,* vol. 64, no. 11, pp. 1808-1817, 2006.
[http://dx.doi.org/10.1016/j.chemosphere.2006.02.003] [PMID: 16540150]

[47]   T. Yamagishi, T. Miyazaki, S. Horii, and K. Akiyama, "Synthetic musk residues in biota and water from Tama River and Tokyo Bay (Japan)", *Arch. Environ. Contam. Toxicol.,* vol. 12, no. 1, pp. 83-89, 1983.
[http://dx.doi.org/10.1007/BF01055006] [PMID: 6830312]

[48]   S. Managaki, A. Murata, H. Takada, B.C. Tuyen, and N.H. Chiem, "Distribution of macrolides, sulfonamides, and trimethoprim in tropical waters: ubiquitous occurrence of veterinary antibiotics in the Mekong Delta", *Environ. Sci. Technol.,* vol. 41, no. 23, pp. 8004-8010, 2007.
[http://dx.doi.org/10.1021/es0709021] [PMID: 18186329]

[49]   Z. Xie, G. Lu, J. Liu, Z. Yan, B. Ma, Z. Zhang, and W. Chen, "Occurrence, bioaccumulation, and trophic magnification of pharmaceutically active compounds in Taihu Lake, China", *Chemosphere,* vol. 138, pp. 140-147, 2015.
[http://dx.doi.org/10.1016/j.chemosphere.2015.05.086] [PMID: 26070079]

[50]   M.J. Hilton, and K.V. Thomas, "Determination of selected human pharmaceutical compounds in effluent and surface water samples by high-performance liquid chromatography–electrospray tandem mass spectrometry", *J. Chromatogr. A,* vol. 1015, no. 1-2, pp. 129-141, 2003.
[http://dx.doi.org/10.1016/S0021-9673(03)01213-5] [PMID: 14570326]

[51]   N. Hossain, and L. Mahmud, "Experimental Investigation of Water Quality and Inorganic Solids in Malaysian Urban Lake, Taman Tasik Medan Idaman", *Lakes Reservoirs: Res. Manage.,* vol. 24, no. 2, pp. 107-114, 2019.
[http://dx.doi.org/10.1111/lre.12259]

[52]   S.M.T. Qadri, M.A. Islam, A. Raza, and N. Hossain, "Physico-Chemical Analysis, Classification of Ground Water, and Impact of Water Quality on the Health of People in Khushab City, Pakistan", *European Journal of Geosciences,* vol. 1, no. 1, pp. 57-74, 2019.
[http://dx.doi.org/10.34154/2019-EJGS-0101-57-74/euraass]

[53]   C.G. Daughton, and T.A. Ternes, "Pharmaceuticals and personal care products in the environment: agents of subtle change", *Environ. Health Perspect.,* vol. 107, p. 907e938, 1999.
[http://dx.doi.org/10.1289/ehp.99107s6907]

[54]   È.B. Dussault, V.K. Balakrishnan, E. Sverko, K.R. Solomon, and P.K. Sibley, "Toxicity of human pharmaceuticals and personal care products to benthic invertebrates", *Environ. Toxicol. Chem.,* vol. 27, no. 2, pp. 425-432, 2008.
[http://dx.doi.org/10.1897/07-354R.1] [PMID: 18348646]

[55]   J.W. Kim, H. Ishibashi, R. Yamauchi, N. Ichikawa, Y. Takao, M. Hirano, M. Koga, and K. Arizono, "Acute toxicity of pharmaceutical and personal care products on freshwater crustacean (Thamnocephalus platyurus) and fish (Oryzias latipes)", *J. Toxicol. Sci.,* vol. 34, no. 2, pp. 227-232, 2009.
[http://dx.doi.org/10.2131/jts.34.227] [PMID: 19336980]

[56]   T. Okumura, and Y. Nishikawa, "Gas chromatography-mass spectrometry determination of triclosans in water, sediment and fish samples *via* methylation with diazomethane", *Anal. Chim. Acta,* vol. 325, no. 3, pp. 175-184, 1996.
[http://dx.doi.org/10.1016/0003-2670(96)00027-X]

[57]   L.M. McMurry, M. Oethinger, and S.B. Levy, "Triclosan targets lipid synthesis", *Nature,* vol. 394, no. 6693, pp. 531-532, 1998.
[http://dx.doi.org/10.1038/28970] [PMID: 9707111]

[58]   R.U. Halden, and D.H. Paull, "Analysis of triclocarban in aquatic samples by liquid chromatography electrospray ionization mass spectrometry", *Environ. Sci. Technol.,* vol. 38, no. 18, pp. 4849-4855, 2004.
[http://dx.doi.org/10.1021/es049524f] [PMID: 15487795]

[59]   D.W. Kolpin, E.T. Furlong, M.T. Meyer, E.M. Thurman, S.D. Zaugg, L.B. Barber, and H.T. Buxton, "Pharmaceuticals, hormones, and other organic wastewater contaminants in U.S. streams, 1999-2000: a national reconnaissance", *Environ. Sci. Technol.,* vol. 36, no. 6, pp. 1202-1211, 2002.
[http://dx.doi.org/10.1021/es011055j] [PMID: 11944670]

[60]   S.T. Glassmeyer, E.T. Furlong, D.W. Kolpin, J.D. Cahill, S.D. Zaugg, S.L. Werner, M.T. Meyer, and D.D. Kryak, "Transport of chemical and microbial compounds from known wastewater discharges: potential for use as indicators of human fecal contamination", *Environ. Sci. Technol.,* vol. 39, no. 14, pp. 5157-5169, 2005.
[http://dx.doi.org/10.1021/es048120k] [PMID: 16082943]

[61]   T. Yamagishi, T. Miyazaki, S. Horii, and S. Kaneko, "Identification of musk xylene and musk ketone in freshwater fish collected from the Tama River, Tokyo", *Bull. Environ. Contam. Toxicol.,* vol. 26, no. 1, pp. 656-662, 1981.
[http://dx.doi.org/10.1007/BF01622152] [PMID: 7260436]

[62]   Q. Sui, J. Huang, S. Deng, G. Yu, and Q. Fan, "Occurrence and removal of pharmaceuticals, caffeine and DEET in wastewater treatment plants of Beijing, China", *Water Res,* vol. 44, no. 2, pp. 417-426, 2010.

[63]   D.R. Dietrich, and Y-J. Chou, "Ecotoxicology of musk", In: *Pharmaceuticals and personal care products in the environment: scientific and regulatory issues.,* C.G. Daughton, T.L. Jones-Lepp, Eds., American Chemical Society, 2001, pp. 156-167.
[http://dx.doi.org/10.1021/bk-2001-0791.ch009]

[64]   C. Mimeault, A.J. Woodhouse, X.S. Miao, C.D. Metcalfe, T.W. Moon, and V.L. Trudeau, "The human lipid regulator, gemfibrozil bioconcentrates and reduces testosterone in the goldfish, Carassius auratus", *Aquat. Toxicol.,* vol. 73, no. 1, pp. 44-54, 2005.
[http://dx.doi.org/10.1016/j.aquatox.2005.01.009] [PMID: 15892991]

[65]   G. Vernouillet, P. Eullaffroy, A. Lajeunesse, C. Blaise, F. Gagné, and P. Juneau, "Toxic effects and bioaccumulation of carbamazepine evaluated by biomarkers measured in organisms of different trophic levels", *Chemosphere,* vol. 80, no. 9, pp. 1062-1068, 2010.
[http://dx.doi.org/10.1016/j.chemosphere.2010.05.010] [PMID: 20557923]

[66]   T. Brodin, S. Piovano, J. Fick, J. Klaminder, M. Heynen, and M. Jonsson, "Ecological effects of pharmaceuticals in aquatic systems—impacts through behavioural alterations", *Philos. Trans. R. Soc. Lond. B Biol. Sci.,* vol. 369, no. 1656, p. 20130580, 2014.
[http://dx.doi.org/10.1098/rstb.2013.0580] [PMID: 25405968]

[67]   S.A. Snyder, P. Westerhoff, Y. Yoon, and D.L. Sedlak, "Pharmaceuticals, personal care products, and endocrine disruptors in water: implications for the water industry", *Environ. Eng. Sci.,* vol. 20, no. 5, pp. 449-469, 2003.
[http://dx.doi.org/10.1089/109287503768335931]

[68]   Q. Bu, B. Wang, J. Huang, S. Deng, and G. Yu, "Pharmaceuticals and personal care products in the aquatic environment in China: A review", *J. Hazard. Mater.,* vol. 262, pp. 189-211, 2013.
[http://dx.doi.org/10.1016/j.jhazmat.2013.08.040] [PMID: 24036145]

[69]   C.J. Houtman, A.M. van Oostveen, A. Brouwer, M.H. Lamoree, and J. Legler, "Identification of estrogenic compounds in fish bile using bioassay-directed fractionation", *Environ. Sci. Technol.,* vol. 38, no. 23, pp. 6415-6423, 2004.
[http://dx.doi.org/10.1021/es049750p] [PMID: 15597899]

[70]   D. Löffler, J. Römbke, M. Meller, and T.A. Ternes, "Environmental fate of pharmaceuticals in water/sediment systems", *Environ. Sci. Technol.,* vol. 39, no. 14, pp. 5209-5218, 2005.
[http://dx.doi.org/10.1021/es0484146] [PMID: 16082949]

[71]   J. Klaminder, T. Brodin, A. Sundelin, N.J. Anderson, J. Fahlman, M. Jonsson, and J. Fick, "F. J. Long-term persistence of an anxiolytic drug (Oxazepam) in a large fresh- water lake", *Environ. Sci. Technol.,* vol. 49, no. 17, pp. 10406-10412, 2015.

[http://dx.doi.org/10.1021/acs.est.5b01968] [PMID: 26196259]

[72] P. Westerhoff, "Removal of endocrine disruptors, pharmaceuticals and personal care products during water treatment", *Southwest Hydrol.,* vol. 2, no. 6, pp. 18-19, 2003.

[73] M.M. Huber, A. GÖbel, A. Joss, N. Hermann, D. LÖffler, C.S. McArdell, A. Ried, H. Siegrist, T.A. Ternes, and U. von Gunten, "Oxidation of pharmaceuticals during ozonation of municipal wastewater effluents: a pilot study", *Environ. Sci. Technol.,* vol. 39, no. 11, pp. 4290-4299, 2005.
[http://dx.doi.org/10.1021/es048396s] [PMID: 15984812]

[74] P.E. Stackelberg, E.T. Furlong, M.T. Meyer, S.D. Zaugg, A.K. Henderson, and D.B. Reissman, "Persistence of pharmaceutical compounds and other organic wastewater contaminants in a conventional drinking-water-treatment plant", *Sci. Total Environ.,* vol. 329, no. 1-3, pp. 99-113, 2004.
[http://dx.doi.org/10.1016/j.scitotenv.2004.03.015] [PMID: 15262161]

[75] N. Vieno, T. Tuhkanen, and L. Kronberg, "Removal of pharmaceuticals in drinking water treatment: effect of chemical coagulation", *Environ. Technol.,* vol. 27, no. 2, pp. 183-192, 2006.
[http://dx.doi.org/10.1080/09593332708618632] [PMID: 16506514]

[76] D. Khiari, *Endocrine disruptors, pharmaceuticals and personal care products in drinking water: an overview of AwwaRF research to date.,* 2007. http:// www.cwwa.ca/pdf_files/AwwaRF_EDC%20 article1.pdf

[77] W. Hua, E.R. Bennett, and R.J. Letcher, "Ozone treatment and the depletion of detectable pharmaceuticals and atrazine herbicide in drinking water sourced from the upper Detroit River, Ontario, Canada", *Water Res.,* vol. 40, no. 12, pp. 2259-2266, 2006.
[http://dx.doi.org/10.1016/j.watres.2006.04.033] [PMID: 16777173]

[78] E. Wert, F. Rosarioortiz, D. Drury, and S. Snyder, "Formation of oxidation byproducts from ozonation of wastewater", *Water Res.,* vol. 41, no. 7, pp. 1481-1490, 2007.
[http://dx.doi.org/10.1016/j.watres.2007.01.020] [PMID: 17335867]

[79] R. Gibson, M.D. Smith, C.J. Spary, C.R. Tyler, and E.M. Hill, "Mixtures of estrogenic contaminants in bile of fish exposed to wastewater treatment works effluents", *Environ. Sci. Technol.,* vol. 39, no. 8, pp. 2461-2471, 2005.
[http://dx.doi.org/10.1021/es048892g] [PMID: 15884336]

CHAPTER 3

Emerging Water Pollutants from Food and Packaging Industry

Muhammad Ashar Ayub[1], Muhammad Zia ur Rehman[2,*], Muhammad Umair[2], Sana Rana[1], Zahoor Ahmad[3], Nimra Khalid[1], Hafiza Salma[1] and **Husnain Zia[2]**

[1] *Institute of Agro-Industry and Environment, The Islamia University of Bahawalpur, Punjab, Pakistan*

[2] *Institute of Soil & Environmental Sciences, University of Agriculture, Faisalabad 38000, Pakistan*

[3] *Department of Botany, University of Central Punjab, Bahawalpur Campus, Pakistan*

Abstract: Rapid industrialization and continuous mechanization of the food industry have increased waste production which is a source of various kinds of contaminants especially emerging water pollutants (EWPs). The industry of food (agriculture and processed food) and packaging industry are major sources of these emerging pollutants. Agrochemicals are also a source of pollutants which are contaminating the food chain and underground water. The dairy and meat industries are source of excess antibiotics, lactating hormones, medicines (*via* exudates and effluents), animal fats, acids, sludge, organic compounds and persisting chemicals in meat and milk. The food processing industry can also contribute various pollutants, like dyes, preservatives, sanitizing and disinfecting agents, as well as alcoholic and phenolic residues from the beverage and wine industries.The food packaging industry is also a major source of food preservatives, dyes, glue, and non-recyclable/one-time-use materials (plastic and polystyrene), which can be broken down into micro/nano plastics leading their way to water reserves and ultimately to the human food chain. Effluents of the food and packaging industry are rich in organic material and can support the growth of various pathogenic bacteria and fungi which can become a source of EWPs and can compromise human health that's why comprehensive information about these pollutants is needed. Keeping in mind all of these aspects present draft is compiled. This chapter covers various aspects of emerging water pollutants released from food-relevant industries.

Keywords: Emerging pollutants, Water pollution, Food industry waste, Packaging industry waste, Plastic waste, Wastewater treatment, Environmental remediation.

---

* **Corresponding author Muhammad Zia ur Rehman:** Institute of Soil & Environmental Sciences, University of Agriculture, Faisalabad, 38000, Pakistan; E-mail: ziasindhu1399@gmail.com

INTRODUCTION

The emerging water pollutants (EWPs) are not new contaminants, but these have been recognized as a threat recently. With an ever-increasing global population, food production and consumption are expected to be rising tremendously in the following years. With increasing demand and production of food, the release of pollutants relevant to the industry is also getting intensified, and multi-pollutant intrusion in water resources is expected to rise [1 - 5]. The main pollutants pertinent to the food industry are pesticide residues [3], heavy metals, pathogens [5], and excessive nutrients [6 - 8], which eventually get dumped into water resources [1]. Livestock is another integral part of food production and is also responsible for the discharge of nutrients [7], pathogens [5], and pharmaceutical products, like antibiotics in water bodies [1]. The emerging pollutants of these kinds have become a concern for global communities and a threat to all freshwater reserves [9]. A key result of a population growth is excessive food production, thus releasing more contaminants into water bodies and very soon most of water bodies are expected to be contaminated with EWPs coming from food production and processing [10, 11]. A similar kind of multi-pollutant contamination can be observed worldwide due to the food production industry. Depleting freshwater resources worldwide and aquatic pollution from excessive fertilizer and manure application (nutrient release *via* runoff) are also major concerns nowadays [2].

Besides the food production industry, the food packaging/processing industry is another major contributor to emerging water pollutants. Industrial effluents (produced over courses of industrial processes) are not properly treated before being discharged into water bodies, and thus are another major source of aquatic pollution. Among the major food processing industries, breweries, soft-drink production companies, and cooking (vegetable) oil-producing industries are prominent. The brewery industry releases wastewater containing various emerging pollutants, like spent cooling water, grain liquor, hop liquor, yeast recovery system liquor, *etc.* These must be processed properly before discharge into water bodies [12, 13]. Food processing waste (sludge and water) is not usually subjected to proper treatment and drained as such in water bodies in developing countries [14]. Waste produced from the food processing industry commonly has a large number of organic materials, suspended solids, high biological oxygen demand (BOD) or chemical oxygen demand (COD), suspended oils/lipids, and most have over permissible levels of other pollutants [15]. Processed food needs to be packed, which gives rise to another industry (food packaging industry) that acts as a leading contributor of pollutants to the aquatic system, like volatile suspended solids and organic materials. Materials being used for packaging come in contact with various synthetic materials like adhesives, coating, and inks which can be washed away and act as a source of hydrocarbon pollution in water [16]. Once reaching landfills, these packages also act as a slow-releasing source of pollutants (*via* landfill leachates), which can contain numerous pollutants along with food remains. Old packaging constituents can be a source of plasticizers and heavy metals like lead (Pb) and cadmium (Cd) being released from inks and pigments [17 - 20].

Henceforward, this contaminant inundation from food production, processing, and packaging industries will get intensified, triggering a massive threat to the aquatic biota, which may impact back society (*via* pathogen contamination) and destroy aquatic nature (*via* algal blooms due to nutrient excess) [21].

TYPES AND CLASSIFICATION OF EMERGING WATER POLLUTANTS FROM THE FOOD INDUSTRY

Agriculture

To meet the needs of the ever-increasing global population, recent decades have seen an immense expansion of agriculture practices, which is putting a lot of pressure on natural resources like water. The area which was meant to be irrigated for agricultural purposes has doubled in the recent century (from 0.139 billion hectares in 1961 to 0.320 billion hectares in 2012) as reported by [22 - 26]. Another major consumer of water in agriculture is aquaculture adoption, which has increased 20-fold since the 1980s, especially in Asia [27 - 30]. This increased water use resulted in untreated/wastewater irrigation as an alternate but accessible approach being opted widely [31 - 34]. The wastewater being applied in agriculture is the source of various pollutants, leaching down to groundwater and moving to freshwater bodies *via* runoff.

Moreover, intensive agriculture has led to excessive application of pesticides and chemical fertilizers, which have become a source of emerging water pollutants. Aquaculture may also add remains of food in the water streams, which can cause the eutrophication of freshwater lakes. Water pollution from agriculture has many adverse effects on human health, like the development of the blue-baby syndrome and methemoglobinemia (a fatal illness in infants caused due to nitrate pollution in water). Pesticide accumulation in water may lead their way into our food chain, which has shown many adverse effects on human and have seen a ban on main broad-spectrum pesticides like dichlorodiphenyltrichloroethane (DDT) and organo-phosphates. However, some of these are still being and used in developing countries [35 - 38]. The aquatic system is on the verge of pollution with emerging pollutants (pesticides, nutrients, and fertilizer residues) responsible for eutrophication, disturbing the aquatic flora and fauna. The agriculture-sourced environment and water pollution have cost billions of dollars annually only in Economic Co-operation and Development (OECD) countries alone [39 - 41].

Livestock related effluents contain pollutants like antibiotics, vaccines, growth promoters, and hormones. Leaching or movement of these contaminants with runoff from livestock or aquaculture farms and manure/slurry-fed agriculture soils is a potent threat to water bodies [39 - 41]. Another potent toxic class of EWPs are heavy metals pollutants which are being applied in agriculture (indirectly as a part of agrochemicals), and they can also lead their way to freshwater resources. More than 700 emerging pollutants, their metabolites, and products have been listed to be present in the aquatic environment [42]. The potential contribution of the agriculture sector in the spread of these emerging pollutants in the environment and water needs thorough management [42 - 44].

Livestock

Most of the water used by the dairy and poultry industry returns to the environment in various forms, some of which are reusable. In contrast, others are contaminated with a wide range of contaminants [43 - 48]. Dairy arming and poultry has emerg as a major industry in the past few decades owing to increased food demand. A major contributor of emerging pollutants from dairyindustr is livestock waste effluent, which is being discharged unattended into the environment. Excreta of dairy animals contains a significant amount of nutrients (N, P, and K), antibodies (residues), heavy metals, and pathogens, especially E. Coli, which can contaminate soil and water bodies if discharged without formal processing [33, 49]. Various mechanisms can be involved in transferring and translocating dairy wastewater-sourced contaminants to reach soil and water bodies. Malpracticing in waste discharge and its direct introduction into agricultural soils, dumping in canals and water channels, and runoff from flooded soils can lead to their massive deposition in water bodies and groundwater.

The main emerging water contaminants present in dairy water discharge are nutrients, as livestock consumes a lot of nutrients over the growing year. For example, a milk-producing cow consumes 163.7 kg N and 22.6 kg P over one growing year [50, 51]. Livestock animal waste is reported to have significant amounts of these nutrients as hog manure (76.2 g N/kg dry weight), turkey waste (59.6 g N/kg dry weight), poultry layer manure (49 g N/kg dry weight), sheep/goat farm manure (44.4 g N/kg dry weight), broilers waste (40 g N/kg dry weight), and milk-producing cattle manure (32-39 g N/kg dry weight). For phosphorus, same animals' systems release P (g/kg dry weight) in waste as 20.8 (poultry layer), 17.6 (hogs), 16.5 (turkey), poultry broilers (16.9) and dairy cattle (6-10) [35, 50, 52]. These nutrients can get into water bodies and cause eutrophication leading to odour and an unchecked population of microbes. Besides nutrient incorporation in water bodies, livestock effluents also add organic matter to water bodies once discharged unchecked, which ultimately increases the biological oxygen demand of the system [53].

In addition to nutrient input, livestock (dairy, poultry) and slaughterhouse effluents contribute to many microbial communities in water bodies. A wide range of pathogenic contaminants is present in their faeces, which can last there for a very long period and multiply due to the surplus presence of nutrients and organic matter to feed on. Wastewater from livestock farms (effluents, manure, and sludge) has a significant amount of pathogenic bacteria species responsible for various infections in humans. It is responsible for many human diseases, especially diarrhoea [54, 55]. It is reported that *Campylobacter spp.* can survive in livestock dung for a considerably long time. If safe disposal of waste effluents is not administered, then it can be a health hazard for water and humans ultimately [48, 56, 57]. *Escherichia Coli* is another important human pathogen found in contaminated waters and has a rigorous growth in livestock waste. It is responsible for various diseases like cholecystitis, cholangitis, urinary tract infection, and diarrhoea [58, 59]. Another class of microbial pollutants excessively found in livestock waste/effluents belongs to *Salmonella spp.* *Salmonella ublin* is a commonly associated human pathogen found in cattle dung among various species members. The contaminated effluent containing this species can make its way to water bodies *via* unchecked dumping and raise concerns for human health [60]. Waste effluents of livestock farms also has some more pathogens like *Clostridium Botulinum* and some viral pathogens which raise concern for drinking water as they are responsible for various diseases like foot and mouth disease, talfan disease and swine vesicular disease [35, 50].

Another class of emerging pollutants found in livestock effluents (especially poultry and aquaculture) and waste are drug/antibiotics residues, mainly including antimicrobial substances and hormones for growth regulation. Anti-microbial drugs are primarily given to treat some illnesses. They are also being administered as a prophylactic (giving medicines to healthy animals) approach which is mainly responsible for an excessive amount of these drug remnants in wastewater [35, 50, 61 - 63]. The major categories of antibiotics found in livestock waste is fluoroquinolones, sulphonamides, and tetracyclines, which are critical and responsible for antibiotic resistance in humans if they contact the humanoid food chain [64]. Likewise, hormones fed to livestock and fish population for their growth and productivity enhancement are transported to their end products (milk, meat, eggs) which, upon consumption by the human, can be very dangerous [65, 66].

Processed Food Industry

Production of processed food has become a worldwide need and emerging industry with a significant contribution to emerging water pollutant load in freshwater. The composition of effluents being released from the food processing industry is primarily controlled by the type of food being processed as well as involved processes as summarized below:

- Beverages: Most of the water in this industry is used for washing and rinsing cans, bottles, cleaning equipment, containers, and floor and effluents usually have high organic matter (sweeteners), suspended solids, and BOD. Even after the treatment of wastewater, organic constituents can be expected to be present.
- Breweries: Besides being the main ingredient of processed food, this industry uses water mostly for washing, cleaning, and processing purposes. The water has high BOD, yeast, and leftover product containing waste effluent, which can be a source of pollution if discharged untreated.
- Fats and Oil: This industry converts various plant oils into edible vegetable oil and ghee. The vegetable oil is produced from rapeseed, corn, and soybean. The refining process is done in which dust, saccharides, proteins, gummy substances, fatty acids, pigments-smelling substances, and other items are removed. These substances can be found in effluents of the oil industry and are of emerging concern for water bodies.
- Milk and Dairy Products: In the dairy and dairy products manufacturing units, water is used for washing, cooling, air conditioning, boilers, sanitation, *etc*. The wastewater released from these industries can contain remains of machines and floor washings, accidental leakages or raw material and products, and dumping off some specific contaminated/spoiled raw material or expired products. These all constituents lead to high BOD and pathogen load in wastewater, which can be a concern for the health of water bodies.
- Agriculture Product Processing: In these factories, agricultural products are processed to produce chips, ketchup, wheat starch, potato starch, sweet potato starch, and sugar. The wastewater effluents of these industries can have very high organic matter contents due to spills, excessive preservers, and washing materials.
- Takeout Dishes: These dishes are mostly made as disposable (one-time use), directly impacting soil and landfill pollution, affecting nearby water reserve quality. During the dish-making process, a wide range of chemicals (inks, glue, detergents, microplastic) can make their way to effluents.

Food Packaging Industry

The food packaging industry has become enormous in the past few decades, and now packaging material is being made from a variety of materials like ceramics, glass, metal, paper (paperboards), cardboard, wax (natural and artificial), and plastic (recyclable as well as non-recyclable). Most food packaging comprises of plastic, paper, rigid plastic, and glass packaging labeled with inks and pigments. Plastic is a major packaging material worldwide which can be biodegradable (made from plant extracts) or non-degradable (derived from petroleum and hydrocarbon derivatives and polymers). Most of the packaging uses a coating of wax and also comes with pigment (inks) used for printing [67, 68]. Most food packaging is only used once and discarded (either buried in landfills or added to litter carried into water bodies). Non-degradable plastic packaging, once it reaches landfills, sits there for years without degradation, but chemicals and dyes present in/on them often leach (*via* landfill leachates) and contaminate the groundwater and soil [69]. Due to excessive and explorative production of plastic packaging, scientists are using this type of pollution as a key geologic indicator of our time (as around 8.3 billion tons of plastic have been produced and most of which have been dumped since 150) [70, 71]. Plastic pollution is affecting water bodies too badly that it has become so intense that the United Nations chief of oceans has termed it a planetary crisis. The issue of one-time plastic packaging is too immense that about one-third of plastic ends up either in soil or in freshwater. Though not biodegradable, once the plastic is broken down into small plastic grains (microplastic), it becomes more drastic as it is invisible and can penetrate the human food chain *via* aquatic foods. Microplastic generating form landfills can disrupt soil fauna as well [69, 70, 72].

The food packaging industry uses a wide range of chemicals few of which are classified below:

Intentionally Added

Per- and Poly-fluorinated Alkyl Substances (PFAS)

These are grease-proofing agents used for paper packaging and are often highly persisting and bioaccumulating *via* direct exposure to dumped food packaging [73]. Massive exposure to PFAS in humans can lead to various disorders, including endocrine gland function disruption [74].

Ortho-phthalates

These compounds are primarily added to plastic packaging and present in printing inks and pigments. Food packaging like cellophane, paper, and plastic can be a big source of these compounds, which can either directly enter the foodstuff or make their way to water bodies along landfill leachates or runoff. They are currently under review for safety assessment by FDA as they are associated with numerous health-hazardous [75 - 79].

Perchlorate

This class of contaminants is used in plastic for dry food packaging as it acts as an anti-static agent [80, 81]. These compounds are radially deposited in food and freshwater bodies with the main source of dumped food packaging (water pollution) and direct absorption in food. Excessive exposure to perchlorate is responsible for disruptive brain growth in children and disturbs the normal functioning of thyroid glands [82].

Benzophenone

Another important and concerning pollutant belong to the benzophenone class, which is prominently being released into water bodies *via* landfill leachates and primarily being sourced from cardboard food packaging. It's also used as a plasticizer in rubber articles and is primarily not recycled. It has been reported to have carcinogenic properties if long-term exposure to the contaminant is observed [83].

Residual

Ethyl and Methyl Glycol, Toluene, and n-methyl-pyrrolidone (NMP)

Some food packaging ingredients are responsible for the residual release of contaminants like glycols, toluene, and pyrrolidones. These are mostly solvents used in inks. They are left on food packaging for a long time and can make their way to either food being covered in contaminated packaging or *via* contaminated water. First-class (methyl/ethyl glycol) is a prominently large and universal group of ink solvents used worldwide on various packages. On the other hand, toluene and NMPs are also an integral part of packaging printing and are left unattended once packaging ends up in dumps. These solvents can get adsorbed in food directly or absorbed in the soil, transport *via* runoff or leach down *via* landfill leachates into water reserves, posing deVere threat to humans [84, 85].

Bisphenol A, B, F, S

This class of compounds is used to make the epoxy lining of plastic/tin cans, making polycarbonate plastic and some forms of inks.e Similar to NMPs, is also highly persistent and poses a severe threat to the environment if not managed properly. Exposure of these pollutants to humans *via* either direct (contaminated packed food) or indirect (contaminated soil/water bodies) ways can lead to various abnormalities linked to endocrine gland abnormalities. These are banned from being used in food packaging stuff such as baby milk bottles or infant formula packaging. However, their multi-purpose use still causes their huge upsurge in inclusion in the environment [86 - 89].

Along with the direct metal body packaging issue, adulterating other toxic HMs in food may also come with food packaging. Although these metals are not intentionally added, food packaging may contain many of them.

Table **1** shows summarized information regarding the contribution of various industries to EWPs.

Table 1. A Review of emerging pollutants being released from various food-related industries.

Industry	Pollutant Source	Pollutant Type and Remarks	References
Agriculture (Crop Farming	Pesticides	organophosphates (OPPs) and organochlorines (OClPs) found in surface and groundwater	[90]
		In many regions of India, drinking water found contaminated with organophosphate and organochlorine	[91, 92]
		Many pesticides residues are even found in treated drinking water, including p,p'-DDE, lindane, and endosulfan	[93]
	Fertilizer (nutrients)	Leaching of nutrients from farm activities is the cause of eutrophication in the Baltic sea	[94, 95]
	Heavy Metals	Heavy metals in surface water are coming from agricultural soils by leaching	[96]
		Fertilization of soil with organic materials increased the leaching of Cd, Pb, and Ni in groundwater to unsatisfactory levels	[17]
Agriculture (Live Stock)	Antibiotics	The high concentration of chlortetracycline and enrofloxacin found in cow, pig, and chicken manure residues	[97]
		A considerable amount of ciprofloxacin, enrofloxacin, and florfenicol found in dairy farm effluent, a nearby river, and pond water samples	[98]
		tetracycline deposits found in soil supplied with liquid manures	[99]
		Residues of sulfamethazine, oxytetracycline, tetracycline, sulfadiazine, and sulfamethoxazole at high levels are found in surface water around the animal farms	[100]
	Nutrients	Dairy wastewater contains a large amount of phosphorus and ammonium-N	[101]
		Manure contains phosphate and nitrate lead to surface water contamination *via* runoff	[103]

(Table 1) cont.....

Industry	Pollutant Source	Pollutant Type and Remarks	References
-	Organic pollutants	Oil & Grease found 4 times higher than the safe limit in dairy effluents	[102]
	Organic and chemical pollutants	BOD and COD of dairy effluent were found higher than the permissible limit	
	Pathogens	Dairy wastewater contains a large number of Escherichia coli	[101]
		E. coli, *Campylobacter*, and *Salmonella* were found to survive in the soil after application for about one month, leading to environmental contamination	[104]
		Clostridium botulinum, Erysipelothrix rhusiopathiae, Clostridium perfringens, Bacillus anthracis, Leptospira interrogans, and *Brucella abortus* found in poultry, swine, goat, and cattle manures	[44]
Food Processing Industry	Chemicals	Wastewater of edible oil industries has high chemical oxygen demand, total dissolved solids, biological oxygen demand, grease, total suspended solids, fats, sulfate, oil, and phosphate	[105]
		Effluents from edible oil industries have a high amount of lipids	[106]
		Oil industries wastewater contains a large number of nutrients and chemicals like oil, phosphate, grease, and sulfate	[107]
		Sodium salts produced from fatty acids soap during neutralization in oil processing insert sulfate and phosphate in wastewater	[108]
		a complex mixture of phenolic and carbohydrate compounds also found in wastewater released from oil factories	[109]
		Wastewater from the food industry is highly polluted with organic matter, oil, dissolved solids, grease, and suspended solids	[110]
		Wastewaters from the dairy industry have high Chemical oxygen demand, and organic matter	[111]
		Wastewaters from the cheese whey industry have high organic matter	[112]
		Wastewaters from the starch industry have high organic matter	[113]
		Wastewaters from slaughterhouse have high Chemical oxygen demand	[114]
		Wastewaters from the pharmaceutical industry have high Chemical oxygen demand	[115]
		Wastewaters from the cassava starch industry have high Chemical oxygen demand	[116]

(Table 1) cont.....

Industry	Pollutant Source	Pollutant Type and Remarks	References
Food Packaging Industry	epoxy resins	bisphenol A, bisphenol A diglycidyl ether	[117]
	plasticizers	Epoxidized soybean oil (ESBO)	[118]
	Paper adhesives or plasticizers	printing inks or mineral oils	[119]
	Polymers	polystyrene (PS), poly-ethylene (PE), polyethylene terephthalate (PET), polycarbonate (PC), high-density polyethylene (HDPE), and polyvinyl chloride (PVC)	[120]
	Adhesives	acrylic, rubber, hot-melt, or polyurethane adhesives	[121]

SOURCE-SINK RELATION AND MONITORING OF EWPS

Water pollution with emerging contaminants has become an alarming issue worldwide, and proper plans are being asked for aiming at clear investigation and detection of EWPs. Every day, around 2 million tonnes of industrial, agricultural, and sewage waste is being dumped directly into the world's waters which is equivalent to the net weight of entire humans. According to keen estimates, the average production of wastewater is expected to hit 1500 km^3 which is even more than real freshwater flowing in rivers [122 - 124], and it's ultimately going to be dumped into world waters. Lack of adequate pre-treatment of wastewater and unchecked discharge makes alarming conditions, especially in third-world countries [122, 125 - 127]. The EPs coming into wastewater can be from the point and non-point or diffused sources. EPs' transport later depends upon EP properties [128 - 130], including solubility, adsorption, and persistence. While EPs from city effluents are directly discharged into rivers, they are of concern as their fate of transportation and accumulation is unknown. If organic, they can degrade, but still, their byproduct can be toxic to the environment, and thus the need for proper detection remains necessary.

EPs can spread through the air, agricultural runoff, and erosion in rural environments and make their way to water bodies. Charged EPs (pharmaceutical products used in the livestock industry) have a very diverse fate in water and soil due to their varying adsorption trends [131]. The presence of manure and sludge (derived from livestock waste is a source of a substantial number of EPs) can affect EPs sorption behavior in soil [132].

Once degraded, organic EPs create humic acid and other ions that can affect the remaining EPs present in soil [133, 134]. Another class of EPs, nanoparticles (NPs), has become another concern for the world's waters and their fate in water is still unknown. Research is mandatory to work on this class of EPs for better understanding. The same goes for inorganic EPs [130].

Management of EWPs

The chief source of emerging water pollutants is unattended wastewater released into water bodies without proper treatment. Treatment and processing of wastewater is the most appropriate approach to tackle the influx of EWPs into water bodies.

WASTEWATER TREATMENT

There are many ways to stop these EPs from reaching water bodies: one method is wastewater treatment. As discussed earlier, wastewater from the dairy and livestock industry is immensely polluted with EWPs (organic and inorganic), which must be taken out of effluents before flushing into water bodies. To manage dairy industry effluent-derived EWPs influx into water bodies, pre-treatment of effluents (especially whey treatment at the start) is considered an appropriate approach. The dairy and livestock effluents contain many EWPs, including organic matter and nutrients, which can severely deteriorate water quality. That's why worldwide regulations are being imposed for the proper disposal of livestock waste and effluents [135 - 137, 138]. Among the various stages of wastewater treatment, a few important steps are discussed below:

Pre-treatment

Pre-treatment of industrial effluents is needed, which mostly involves dampening flow, pH, and organic load in dairy industry effluent management. Now industries install pre-treatment systems that reduce waste loadings and enable industries to dump effluents directly into the municipal system. The pre-treatment implant industries are encouraged to maximize solid material using air flotation, neutralization, and flocculant/polyelectrolytes. Though pre-treatment of dairy effluents can significantly remove solid waste but is usually less efficient in removing organic load (the main reason for higher BOD of wastewater). The processed water is again a source of worry for the environment that must go through more treatment processes. The organic constituents of wastewater can be transformed into fertilizer by composting them properly.

Land Treatment

In this step of dairy wastewater treatment, effluents are discharged rightly into the soil (which acts as an adsorbent for harmful contents of water). That is how the water becomes fit for disposal.

Organic Loading

Organic-matter-rich wastewater can be converted to food for microorganisms if applied properly and disposed of as non-harmful by-products, including slimes and beneficial microbes. Usually, dairy and livestock industry waste effluents have high BOD due to lactose, fats, and proteins, which can act as a food constituent for microbial growth. The dairy industry uses the aerobic or anaerobic treatment of waste (rich in organic constituents), with the prior one being most effective for moderate strength wastewaters while later is used for concentrated wastewater treatments [139, 140].

Nutrient Removal

Industrial wastewaters contain a significant amount of nutrients (primarily N and P) which can be a source of eutrophication if it reaches water bodies. To effectively manage N and P rich waste, an activated sludge system is used where these nutrients become food for microbes with few modifications needed to match environmental needs [141].

CONCLUSION

Emerging water pollutants can be sourced from various industrial and agricultural food processes and have become an alarming concern due to their unchecked presence. The food and packaging industries are major sources of various emerging pollutants. The dairy and meat industries are a source of excess antibiotics, lactating hormones, medicines (*via* exudates and effluents), animal fats, acids, sludge, organic compounds, and persisting chemicals in meat and milk. The food processing industry produces various pollutants like dyes, preservatives, sanitizing, disinfecting agents, *etc*. The food packaging industry is a major source of food preservatives, dyes, glue, and non-recyclable one-time-use materials (plastic and polystyrene), which can be broken down into micro/nano plastics leading their way to water reserves and ultimately to the human food chain. These EWPs must be addressed properly to ensure their minimum encounter with the human food chain.

CONSENT FOR PUBLICATION

Not applicable.

CONFLICT OF INTEREST

The author declares no conflict of interest, financial or otherwise.

ACKNOWLEDGEMENTS

Declared none.

REFERENCES

[1]   G.P. Robertson, and P.M. Vitousek, "Nitrogen in Agriculture: Balancing the Cost of an Essential Resource", *Annu. Rev. Environ. Resour.*, vol. 34, no. 1, pp. 97-125, 2009.
       [http://dx.doi.org/10.1146/annurev.environ.032108.105046]

[2]   M. Springmann, M. Clark, D. Mason-D'Croz, K. Wiebe, B.L. Bodirsky, L. Lassaletta, W. de Vries, S.J. Vermeulen, M. Herrero, K.M. Carlson, M. Jonell, M. Troell, F. DeClerck, L.J. Gordon, R. Zurayk, P. Scarborough, M. Rayner, B. Loken, J. Fanzo, H.C.J. Godfray, D. Tilman, J. Rockström, and W. Willett, "Options for keeping the food system within environmental limits", *Nature,* vol. 562, no. 7728, pp. 519-525, 2018.
       [http://dx.doi.org/10.1038/s41586-018-0594-0] [PMID: 30305731]

[3]   A. Ippolito, M. Kattwinkel, J.J. Rasmussen, R.B. Schäfer, R. Fornaroli, and M. Liess, "Modeling global distribution of agricultural insecticides in surface waters", *Environ. Pollut.,* vol. 198, pp. 54-60, 2015.
       [http://dx.doi.org/10.1016/j.envpol.2014.12.016] [PMID: 25555206]

[4]   M. Strokal, C. Kroeze, M. Wang, Z. Bai, and L. Ma, "The MARINA model (Model to Assess River Inputs of Nutrients to seAs): Model description and results for China", *Sci. Total Environ.,* vol. 562, pp. 869-888, 2016.
       [http://dx.doi.org/10.1016/j.scitotenv.2016.04.071] [PMID: 27115624]

[5]   L.C. Vermeulen, J. Benders, G. Medema, and N. Hofstra, "Global Cryptosporidium Loads from Livestock Manure", *Environ. Sci. Technol.,* vol. 51, no. 15, pp. 8663-8671, 2017.
       [http://dx.doi.org/10.1021/acs.est.7b00452] [PMID: 28654242]

[6]   B. Gu, A.M. Leach, L. Ma, J.N. Galloway, S.X. Chang, Y. Ge, and J. Chang, "Nitrogen footprint in China: food, energy, and nonfood goods", *Environ. Sci. Technol.,* vol. 47, no. 16, pp. 9217-9224, 2013.
       [http://dx.doi.org/10.1021/es401344h] [PMID: 23883136]

[7]   L. Ma, F. Wang, W. Zhang, W. Ma, G. Velthof, W. Qin, O. Oenema, and F. Zhang, "Environmental assessment of management options for nutrient flows in the food chain in China", *Environ. Sci. Technol.,* vol. 47, no. 13, pp. 7260-7268, 2013.
       [http://dx.doi.org/10.1021/es400456u] [PMID: 23656482]

[8]   M. Strokal, C. Kroeze, M. Wang, and L. Ma, "Reducing future river export of nutrients to coastal waters of China in optimistic scenarios", *Sci. Total Environ.,* vol. 579, pp. 517-528, 2017.
       [http://dx.doi.org/10.1016/j.scitotenv.2016.11.065] [PMID: 27884528]

[9]   R.J. Diaz, and R. Rosenberg, *Spreading Dead Zones and Consequences for Marine Ecosystems*, 2008.
       [http://dx.doi.org/10.1126/science.1156401]

[10]   A.A. Li, M.M. Strokal, Z.Z.H. Bai, C.C. Kroeze, L.L. Ma, and F.F.S. Zhang, "Modelling reduced coastal eutrophication with increased crop yields in Chinese agriculture", *Soil Res.,* vol. 55, no. 6, p. 506, 2017.
       [http://dx.doi.org/10.1071/SR17035]

[11]   M. Wang, L. Ma, M. Strokal, Y. Chu, and C. Kroeze, "Exploring nutrient management options to increase nitrogen and phosphorus use efficiencies in food production of China", *Agric. Syst.,* vol. 163, pp. 58-72, 2018.
       [http://dx.doi.org/10.1016/j.agsy.2017.01.001]

[12]   J. Lu, J. Wu, C. Zhang, and Y. Zhang, "Possible effect of submarine groundwater discharge on the pollution of coastal water: Occurrence, source, and risks of endocrine disrupting chemicals in coastal

groundwater and adjacent seawater influenced by reclaimed water irrigation", *Chemosphere,* vol. 250, p. 126323, 2020.
[http://dx.doi.org/10.1016/j.chemosphere.2020.126323] [PMID: 32126332]

[13]   J. Corrales, L.A. Kristofco, W.B. Steele, B.S. Yates, C.S. Breed, E.S. Williams, and B.W. Brooks, "Global Assessment of Bisphenol A in the Environment", *Dose Response,* vol. 13, no. 3, 2015.
[http://dx.doi.org/10.1177/1559325815598308] [PMID: 26674671]

[14]   T.G. Amabye, "Effect of Food Processing Industries Effluents on the Environment: A Case Study of MOHA Mekelle Bottling Company, Tigray, Ethiopia", *Industrial Chemistry,* vol. 1, no. 2, 2015.
[http://dx.doi.org/10.4172/2469-9764.1000110]

[15]   G.T. Kroyer, "Impact of food processing on the environment—an overview", *Lebensm. Wiss. Technol.,* vol. 28, no. 6, pp. 547-552, 1995.
[http://dx.doi.org/10.1016/0023-6438(95)90000-4]

[16]   E. Pongrácz, "The environmental impacts of packaging", In: *Environmentally Conscious Materials and Chemicals Processing.* John Wiley & Sons, Inc., 2007, pp. 237-278.
[http://dx.doi.org/10.1002/9780470168219.ch9]

[17]   J. Wierzbowska, S. Sienkiewicz, S. Krzebietke, and T. Bowszys, "Heavy Metals in Water Percolating Through Soil Fertilized with Biodegradable Waste Materials", *Water Air Soil Pollut.,* vol. 227, no. 12, p. 456, 2016.
[http://dx.doi.org/10.1007/s11270-016-3147-x] [PMID: 27942079]

[18]   R.C. Kroner, D.G. Ballinger, and H.P. Kramer, "Evaluation of Laboratory Methods for Analysis of Heavy Metals in Water", *J. Am. Water Works Assoc.,* vol. 52, no. 1, pp. 117-124, 1960.
[http://dx.doi.org/10.1002/j.1551-8833.1960.tb00456.x]

[19]   J. Wen, Z. Li, N. Luo, M. Huang, R. Yang, and G. Zeng, "Investigating organic matter properties affecting the binding behavior of heavy metals in the rhizosphere of wetlands", *Ecotoxicol. Environ. Saf.,* vol. 162, pp. 184-191, 2018.
[http://dx.doi.org/10.1016/j.ecoenv.2018.06.083] [PMID: 29990730]

[20]   S.E.M. Selke, "Packaging and the environment: Alternatives, trends and solutions", *Packag. Technol. Sci.,* vol. 4, no. 1, pp. 49-50, 1991.
[http://dx.doi.org/10.1002/pts.2770040109]

[21]   A. Li, C. Kroeze, T. Kahil, L. Ma, and M. Strokal, "Water pollution from food production: lessons for optimistic and optimal solutions", *Curr. Opin. Environ. Sustain.,* vol. 40, pp. 88-94, 2019.
[http://dx.doi.org/10.1016/j.cosust.2019.09.007]

[22]   AQUASTAT website, "AQUASTAT - FAO's Information System on Water and Agriculture", *Food and Agriculture Organization of the United Nations (FAO).,* 2016.

[23]   W. Busscher, "Spending our water and soils for food security", *J. Soil Water Conserv.,* vol. 67, no. 3, pp. 228-234, 2012.
[http://dx.doi.org/10.2489/jswc.67.3.228]

[24]   T. Sternberg, "Water megaprojects in deserts and drylands", *Int. J. Water Resour. Dev.,* vol. 32, no. 2, pp. 301-320, 2016.
[http://dx.doi.org/10.1080/07900627.2015.1012660]

[25]   A.B.M. Moore, "A review of sawfishes (Pristidae) in the Arabian region: diversity, distribution, and functional extinction of large and historically abundant marine vertebrates", *Aquat. Conserv.,* vol. 25, no. 5, pp. 656-677, 2015.
[http://dx.doi.org/10.1002/aqc.2441]

[26]   J. Ben-Iwo, V. Manovic, and P. Longhurst, "Biomass resources and biofuels potential for the production of transportation fuels in Nigeria", *Renew. Sustain. Energy Rev.,* vol. 63, pp. 172-192, 2016.
[http://dx.doi.org/10.1016/j.rser.2016.05.050]

[27]   Food and Agriculture Organization of the United Nations (FAO), *The State of World Fisheries and Aquaculture, Opportunities and Challenges 2014.* FAO Fisheries and Aquaculture Department: Rome, 2014.

[28]   FAO (Food and Agriculture Organization of the United Nations), "The State of Food Insecurity in the World Meeting the 2015 interation hunger targets: taking stock of uneven progress", *State Food Insecurity World,* 2015.

[29]   "The State of World Fisheries and Aquaculture. Food and Agriculture Organization of the United Nations", *Aquaculture,* 2010.

[30]   *International Fund for Agricultural Development (IFAD), United Nations Children's Fund (UNICEF), World Food Program (WFP), and World Health Organization (WHO).* The State of Food Security and Nutrition in the World. Transforming Food Systems For Affordable Healthy Diets, 2020.

[31]   A.L. Thebo, P. Drechsel, E.F. Lambin, and K.L. Nelson, "A global, spatially-explicit assessment of irrigated croplands influenced by urban wastewater flows", *Environ. Res. Lett.,* vol. 12, no. 7, p. 074008, 2017.
[http://dx.doi.org/10.1088/1748-9326/aa75d1]

[32]   M. Azhar, M. Zia ur Rehman, S. Ali, M.F. Qayyum, A. Naeem, M.A. Ayub, M. Anwar ul Haq, A. Iqbal, and M. Rizwan, "Comparative effectiveness of different biochars and conventional organic materials on growth, photosynthesis and cadmium accumulation in cereals", *Chemosphere,* vol. 227, pp. 72-81, 2019.
[http://dx.doi.org/10.1016/j.chemosphere.2019.04.041] [PMID: 30981972]

[33]   H.R. Ahmad, *"Wastewater Irrigation-Sourced Plant Nutrition: Concerns and Prospects," Plant Micronutrients.* Springer International Publishing, 2020, pp. 417-434.
[http://dx.doi.org/10.1007/978-3-030-49856-6_18]

[34]   M.A. Ayub, "Restoration of Degraded Soil for Sustainable Agriculture", In: *Soil Health Restoration and Management.* Springer, 2020, pp. 31-81.
[http://dx.doi.org/10.1007/978-981-13-8570-4_2]

[35]   J. Mateo-Sagasta, S. M. Zadeh, and H. Turral, "Water pollution from agriculture: A global review", *Exec. Summ.,* 2017.

[36]   *Water pollution from agriculture: a global review Executive summary.* FAO IWMI, 2017.

[37]   D. Hope, M.F. Billett, and M.S. Cresser, "A review of the export of carbon in river water: Fluxes and processes", *Environ. Pollut.,* vol. 84, no. 3, pp. 301-324, 1994.
[http://dx.doi.org/10.1016/0269-7491(94)90142-2] [PMID: 15091702]

[38]   S. Bolisetty, M. Peydayesh, and R. Mezzenga, "Sustainable technologies for water purification from heavy metals: review and analysis", *Chem. Soc. Rev.,* vol. 48, no. 2, pp. 463-487, 2019.
[http://dx.doi.org/10.1039/C8CS00493E] [PMID: 30603760]

[39]   K. Parris, "Impact of agriculture on water pollution in OECD countries: Recent trends and future prospects", *Int. J. Water Resour. Dev.,* vol. 27, no. 1, pp. 33-52, 2011.
[http://dx.doi.org/10.1080/07900627.2010.531898]

[40]   C. Ground, W. Board, M. Of, and W. Resources, "Ground water quality in shallow aquifers of india", *Ground Water,* 2010.

[41]   D. Quality, "Glyphosate and AMPA in Drinking-water", *America (NY),* 2005.

[42]   W. Brack, V. Dulio, and J. Slobodnik, "The Norman Network and its activities on emerging environmental substances with a focus on effect-directed analysis of complex environmental contamination", *Environ. Sci. Eur.,* vol. 24, no. 1, p. 29, 2012.
[http://dx.doi.org/10.1186/2190-4715-24-29]

[43]   "Manipulation of animal diets to affect manure production, composition and odors: state of the science", *Animal Agriculture and the Environment, National Center for Manure & Animal Waste*

Management White Papers American Society of Agricultural and Biological Engineers, 2006. [http://dx.doi.org/10.13031/2013.20259]

[44]   P.S. Sobsey, L.A. Khatib, VR. Hill, and E. Alocilja, "Animal Agriculture and the Environment, National Center for Manure & Animal Waste Management White Papers", 2006. [http://dx.doi.org/10.13031/aae2006.2013]

[45]   D. Parker, "Controlling agricultural nonpoint water pollution: costs of implementing the Maryland Water Quality Improvement Act of 1998", *Agric. Econ.,* vol. 24, no. 1, pp. 23-31, 2000. [http://dx.doi.org/10.1111/j.1574-0862.2000.tb00090.x]

[46]   S. Kappes, "US FOOD ANIMAL AGRICULTURE: DIRECTIONS AND IMPLICATIONS", In: *Livestock Environment VII.* American Society of Agricultural and Biological Engineers: Beijing, China, 2005. [http://dx.doi.org/10.13031/2013.19112]

[47]   P. Gerber, N. Key, F. Portet, and H. Steinfeld, "Policy options in addressing livestock's contribution to climate change", *Animal,* vol. 4, no. 3, pp. 393-406, 2010. [http://dx.doi.org/10.1017/S1751731110000133] [PMID: 22443943]

[48]   B. Hald, M.N. Skov, E.M. Nielsen, C. Rahbek, J.J. Madsen, M. Wainø, M. Chriél, S. Nordentoft, D.L. Baggesen, and M. Madsen, "Campylobacter jejuni and Campylobacter coli in wild birds on Danish livestock farms", *Acta Vet. Scand.,* vol. 58, no. 1, p. 11, 2015. [http://dx.doi.org/10.1186/s13028-016-0192-9] [PMID: 26842400]

[49]   W. Janczukowicz, M. Zieliński, and M. Dębowski, "Biodegradability evaluation of dairy effluents originated in selected sections of dairy production", *Bioresour. Technol.,* vol. 99, no. 10, pp. 4199-4205, 2008. [http://dx.doi.org/10.1016/j.biortech.2007.08.077] [PMID: 17976980]

[50]   K.T. Osman, *Soil degradation, conservation and remediation.* vol. 9789400775. Springer, 2014. [http://dx.doi.org/10.1007/978-94-007-7590-9]

[51]   H. Arriaga, M. Pinto, S. Calsamiglia, and P. Merino, "Nutritional and management strategies on nitrogen and phosphorus use efficiency of lactating dairy cattle on commercial farms: An environmental perspective", *J. Dairy Sci.,* vol. 92, no. 1, pp. 204-215, 2009. [http://dx.doi.org/10.3168/jds.2008-1304] [PMID: 19109280]

[52]   D.N. Miller, and V.H. Varel, "In vitro study of the biochemical origin and production limits of odorous compounds in cattle feedlots", *J. Anim. Sci.,* vol. 79, no. 12, pp. 2949-2956, 2001. [http://dx.doi.org/10.2527/2001.79122949x] [PMID: 11811446]

[53]   A. Pal, K.Y.H. Gin, A.Y.C. Lin, and M. Reinhard, "Impacts of emerging organic contaminants on freshwater resources: Review of recent occurrences, sources, fate and effects", *Sci. Total Environ.,* vol. 408, no. 24, pp. 6062-6069, 2010. [http://dx.doi.org/10.1016/j.scitotenv.2010.09.026] [PMID: 20934204]

[54]   F. Liu, R. Ma, Y. Wang, and L. Zhang, "The Clinical Importance of Campylobacter concisus and Other Human Hosted Campylobacter Species", *Front. Cell. Infect. Microbiol.,* vol. 8, p. 243, 2018. [http://dx.doi.org/10.3389/fcimb.2018.00243] [PMID: 30087857]

[55]   C. Pedati, S. Koirala, T. Safranek, B.F. Buss, and A.V. Carlson, "Campylobacteriosis Outbreak Associated with Contaminated Municipal Water Supply — Nebraska, 2017", *MMWR Morb. Mortal. Wkly. Rep.,* vol. 68, no. 7, pp. 169-173, 2019. [http://dx.doi.org/10.15585/mmwr.mm6807a1] [PMID: 30789878]

[56]   R.R. Kumar, B.J. Park, and J.Y. Cho, "Application and environmental risks of livestock manure", *J. Korean Soc. Appl. Biol. Chem.,* vol. 56, no. 5, pp. 497-503, 2013. [http://dx.doi.org/10.1007/s13765-013-3184-8]

[57]   G.D. Inglis, T.A. McAllister, F.J. Larney, and E. Topp, "Prolonged survival of Campylobacter species in bovine manure compost", *Appl. Environ. Microbiol.,* vol. 76, no. 4, pp. 1110-1119, 2010.

[http://dx.doi.org/10.1128/AEM.01902-09] [PMID: 20023098]

[58]   T.T.Y. Guan, and R.A. Holley, *Hog Manure Management, the Environment and Human Health.*
       Springer US, 2003.
       [http://dx.doi.org/10.1007/978-1-4615-0031-5]

[59]   E.D. Berry, and D.N. Miller, "Cattle feedlot soil moisture and manure content: II. Impact on
       Escherichia coli O157", *J. Environ. Qual.,* vol. 34, no. 2, pp. 656-663, 2005.
       [http://dx.doi.org/10.2134/jeq2005.0656] [PMID: 15758118]

[60]   J.L. Mawdsley, R.D. Bardgett, R.J. Merry, B.F. Pain, and M.K. Theodorou, "Pathogens in livestock
       waste, their potential for movement through soil and environmental pollution", *Appl. Soil Ecol.,* vol. 2,
       no. 1, pp. 1-15, 1995.
       [http://dx.doi.org/10.1016/0929-1393(94)00039-A] [PMID: 32288277]

[61]   J.C. Chee-Sanford, R.I. Mackie, S. Koike, I.G. Krapac, Y.F. Lin, A.C. Yannarell, S. Maxwell, and R.I.
       Aminov, "Fate and transport of antibiotic residues and antibiotic resistance genes following land
       application of manure waste", *J. Environ. Qual.,* vol. 38, no. 3, pp. 1086-1108, 2009.
       [http://dx.doi.org/10.2134/jeq2008.0128] [PMID: 19398507]

[62]   K.R. Kim, G. Owens, S.I. Kwon, K.H. So, D.B. Lee, and Y.S. Ok, "Occurrence and Environmental
       Fate of Veterinary Antibiotics in the Terrestrial Environment", *Water Air Soil Pollut.,* vol. 214, no. 1-
       4, pp. 163-174, 2011.
       [http://dx.doi.org/10.1007/s11270-010-0412-2]

[63]   Z. He, X. Cheng, G.Z. Kyzas, and J. Fu, "Pharmaceuticals pollution of aquaculture and its
       management in China", *J. Mol. Liq.,* vol. 223, pp. 781-789, 2016.
       [http://dx.doi.org/10.1016/j.molliq.2016.09.005]

[64]   A. Van Epps, and L. Blaney, "Antibiotic Residues in Animal Waste: Occurrence and Degradation in
       Conventional Agricultural Waste Management Practices", *Curr. Pollut. Rep.,* vol. 2, no. 3, pp. 135-
       155, 2016.
       [http://dx.doi.org/10.1007/s40726-016-0037-1]

[65]   I.G. Lange, A. Daxenberger, B. Schiffer, H. Witters, D. Ibarreta, and H.H.D. Meyer, "Sex hormones
       originating from different livestock production systems: fate and potential disrupting activity in the
       environment", *Anal. Chim. Acta,* vol. 473, no. 1-2, pp. 27-37, 2002.
       [http://dx.doi.org/10.1016/S0003-2670(02)00748-1]

[66]   S.H. Jeong, D.J. Kang, M.W. Lim, C.S. Kang, and H.J. Sung, "Risk assessment of growth hormones
       and antimicrobial residues in meat", *Toxicol. Res.,* vol. 26, no. 4, pp. 301-313, 2010.
       [http://dx.doi.org/10.5487/TR.2010.26.4.301] [PMID: 24278538]

[67]   B.A. Zeeb, J.S. Amphlett, A. Rutter, and K.J. Reimer, "Potential for phytoremediation of
       polychlorinated biphenyl-(PCB-)contaminated soil", *Int. J. Phytoremediation,* vol. 8, no. 3, pp. 199-
       221, 2006.
       [http://dx.doi.org/10.1080/15226510600846749] [PMID: 17120525]

[68]   J. Muncke, "Hazards of Food Contact Material: Food Packaging Contaminants", In: *Encyclopedia of
       Food Safety.* Elsevier, 2014, pp. 430-437.
       [http://dx.doi.org/10.1016/B978-0-12-378612-8.00218-3]

[69]   R.C. Thompson, C.J. Moore, F.S. vom Saal, and S.H. Swan, "Plastics, the environment and human
       health: current consensus and future trends", *Philos. Trans. R. Soc. Lond. B Biol. Sci.,* vol. 364, no.
       1526, pp. 2153-2166, 2009.
       [http://dx.doi.org/10.1098/rstb.2009.0053] [PMID: 19528062]

[70]   R. Geyer, J.R. Jambeck, and K.L. Law, "Production, use, and fate of all plastics ever made", *Sci. Adv.,*
       vol. 3, no. 7, pp. e1700782-e1700782, 2017.
       [http://dx.doi.org/10.1126/sciadv.1700782] [PMID: 28776036]

[71]   J. Zalasiewicz, C.N. Waters, J.A. Ivar do Sul, P.L. Corcoran, A.D. Barnosky, A. Cearreta, M.

Edgeworth, A. Gałuszka, C. Jeandel, R. Leinfelder, J.R. McNeill, W. Steffen, C. Summerhayes, M. Wagreich, M. Williams, A.P. Wolfe, and Y. Yonan, "The geological cycle of plastics and their use as a stratigraphic indicator of the Anthropocene", *Anthropocene,* vol. 13, pp. 4-17, 2016.
[http://dx.doi.org/10.1016/j.ancene.2016.01.002]

[72]   R. A. Rudel, and L.J. Perovich, "Endocrine disrupting chemicals in indoor and outdoor air", *Atmos. Environ.,* vol. 43, no. 1, pp. 170-181, 1994.
[http://dx.doi.org/10.1016/j.atmosenv.2008.09.025]

[73]   T. Stoiber, S. Evans, and O.V. Naidenko, "Disposal of products and materials containing per- and polyfluoroalkyl substances (PFAS): A cyclical problem", *Chemosphere,* vol. 260, p. 127659, 2020.
[http://dx.doi.org/10.1016/j.chemosphere.2020.127659] [PMID: 32698118]

[74]   E.M. Andersson, K. Scott, Y. Xu, Y. Li, D.S. Olsson, T. Fletcher, and K. Jakobsson, "High exposure to perfluorinated compounds in drinking water and thyroid disease. A cohort study from Ronneby, Sweden", *Environ. Res.,* vol. 176, p. 108540, 2019.
[http://dx.doi.org/10.1016/j.envres.2019.108540] [PMID: 31252203]

[75]   J. Li, H. Liu, and J. Paul Chen, "Microplastics in freshwater systems: A review on occurrence, environmental effects, and methods for microplastics detection", *Water Res.,* vol. 137, pp. 362-374, 2018.
[http://dx.doi.org/10.1016/j.watres.2017.12.056] [PMID: 29580559]

[76]   S. Net, A. Delmont, R. Sempéré, A. Paluselli, and B. Ouddane, "Reliable quantification of phthalates in environmental matrices (air, water, sludge, sediment and soil): A review", *Sci. Total Environ.,* vol. 515-516, pp. 162-180, 2015.
[http://dx.doi.org/10.1016/j.scitotenv.2015.02.013] [PMID: 25723871]

[77]   D. Eerkes-Medrano, R.C. Thompson, and D.C. Aldridge, "Microplastics in freshwater systems: A review of the emerging threats, identification of knowledge gaps and prioritisation of research needs", *Water Res.,* vol. 75, pp. 63-82, 2015.
[http://dx.doi.org/10.1016/j.watres.2015.02.012] [PMID: 25746963]

[78]   P. Otero, S. K. Saha, S. Moane, J. Barron, G. Clancy, and P. Murray, "Improved method for rapid detection of phthalates in bottled water by gas chromatography-mass spectrometry", *J Chromatography B.,* vol. 997, pp. 229-235, 2015.
[http://dx.doi.org/10.1016/j.jchromb.2015.05.036]

[79]   D. Gao, Z. Li, H. Wang, and H. Liang, "An overview of phthalate acid ester pollution in China over the last decade: Environmental occurrence and human exposure", *Sci. Total Environ.,* vol. 645, pp. 1400-1409, 2018.
[http://dx.doi.org/10.1016/j.scitotenv.2018.07.093] [PMID: 30248862]

[80]   R.J. Kendall, "SETAC globalization", *Environ. Toxicol. Chem.,* vol. 11, no. 11, pp. 1511-1512, 1992.
[http://dx.doi.org/10.1002/etc.5620111101]

[81]   M.V. Maffini, L. Trasande, and T.G. Neltner, "Perchlorate and Diet: Human Exposures, Risks, and Mitigation Strategies", *Curr. Environ. Health Rep.,* vol. 3, no. 2, pp. 107-117, 2016.
[http://dx.doi.org/10.1007/s40572-016-0090-3] [PMID: 27029550]

[82]   C.W. Trumpolt, M. Crain, G.D. Cullison, S.J.P. Flanagan, L. Siegel, and S. Lathrop, "Perchlorate: Sources, uses, and occurrences in the environment", *Rem. J.,* vol. 16, no. 1, pp. 65-89, 2005.
[http://dx.doi.org/10.1002/rem.20071]

[83]   W.A.C. Anderson, and L. Castle, "Benzophenone in cartonboard packaging materials and the factors that influence its migration into food", *Food Addit. Contam.,* vol. 20, no. 6, pp. 607-618, 2003.
[http://dx.doi.org/10.1080/0265203031000109486] [PMID: 12881135]

[84]   O.W. Lau, and S.K. Wong, "Contamination in food from packaging material", *J. Chromatogr. A,* vol. 882, no. 1-2, pp. 255-270, 2000.
[http://dx.doi.org/10.1016/S0021-9673(00)00356-3] [PMID: 10895950]

[85]   K. Marsh, and B. Bugusu, "Food packaging--roles, materials, and environmental issues", *J. Food Sci.,* vol. 72, no. 3, pp. R39-R55, 2007.
[http://dx.doi.org/10.1111/j.1750-3841.2007.00301.x] [PMID: 17995809]

[86]   N. Rastkari, R. Ahmadkhaniha, M. Yunesian, L.J. Baleh, and A. Mesdaghinia, "Sensitive determination of bisphenol A and bisphenol F in canned food using a solid-phase microextraction fibre coated with single-walled carbon nanotubes before GC/MS", *Food Addit. Contam. Part A Chem. Anal. Control Expo. Risk Assess.,* vol. 27, no. 10, pp. 1460-1468, 2010.
[http://dx.doi.org/10.1080/19440049.2010.495730] [PMID: 20658403]

[87]   S.C. Cunha, C. Almeida, E. Mendes, and J.O. Fernandes, "Simultaneous determination of bisphenol A and bisphenol B in beverages and powdered infant formula by dispersive liquid–liquid micro-extraction and heart-cutting multidimensional gas chromatography-mass spectrometry", *Food Addit. Contam. Part A Chem. Anal. Control Expo. Risk Assess.,* vol. 28, no. 4, pp. 513-526, 2011.
[http://dx.doi.org/10.1080/19440049.2010.542551] [PMID: 21240700]

[88]   F. Vilarinho, R. Sendón, A. van der Kellen, M.F. Vaz, and A.S. Silva, "Bisphenol A in food as a result of its migration from food packaging", *Trends Food Sci. Technol.,* vol. 91, pp. 33-65, 2019.
[http://dx.doi.org/10.1016/j.tifs.2019.06.012]

[89]   J. Zhou, X.H. Chen, S.D. Pan, J.L. Wang, Y.B. Zheng, J.J. Xu, Y.G. Zhao, Z.X. Cai, and M.C. Jin, "Contamination status of bisphenol A and its analogues (bisphenol S, F and B) in foodstuffs and the implications for dietary exposure on adult residents in Zhejiang Province", *Food Chem.,* vol. 294, pp. 160-170, 2019.
[http://dx.doi.org/10.1016/j.foodchem.2019.05.022] [PMID: 31126448]

[90]   S.Z. Lari, N.A. Khan, K.N. Gandhi, T.S. Meshram, and N.P. Thacker, "Comparison of pesticide residues in surface water and ground water of agriculture intensive areas", *J. Environ. Health Sci. Eng.,* vol. 12, no. 1, p. 11, 2014.
[http://dx.doi.org/10.1186/2052-336X-12-11] [PMID: 24398360]

[91]   C.P. Kaushik, H.R. Sharma, and A. Kaushik, "Organochlorine pesticide residues in drinking water in the rural areas of Haryana, India", *Environ. Monit. Assess.,* vol. 184, no. 1, pp. 103-112, 2012.
[http://dx.doi.org/10.1007/s10661-011-1950-9] [PMID: 21409364]

[92]   M.H. Dehghani, "Corrigendum to 'Optimizing the removal of organophosphorus pesticide malathion from water using multi-walled carbon nanotubes", *Chem. Eng. J.,* vol. 321, p. 669, 2017.
[http://dx.doi.org/10.1016/j.cej.2017.04.009]

[93]   N. Thacker, J. Bassin, V. Deshpande, and S. Devotta, "Trends of organochlorine pesticides in drinking water supplies", *Environ. Monit. Assess.,* vol. 137, no. 1-3, pp. 295-299, 2008.
[http://dx.doi.org/10.1007/s10661-007-9764-5] [PMID: 17530432]

[94]   W. Ning, A.B. Nielsen, L.N. Ivarsson, T. Jilbert, C.M. Åkesson, C.P. Slomp, E. Andrén, A. Broström, and H.L. Filipsson, "Anthropogenic and climatic impacts on a coastal environment in the Baltic Sea over the last 1000 years", *Anthropocene,* vol. 21, pp. 66-79, 2018.
[http://dx.doi.org/10.1016/j.ancene.2018.02.003]

[95]   O.P. Savchuk, "Large-Scale Nutrient Dynamics in the Baltic Sea, 1970–2016", *Front. Mar. Sci.,* vol. 5, p. 95, 2018.
[http://dx.doi.org/10.3389/fmars.2018.00095]

[96]   L.T.C. Bonten, P.F.A.M. Römkens, and D.J. Brus, "Contribution of Heavy Metal Leaching from Agricultural Soils to Surface Water Loads", *Environ. Forensics,* vol. 9, no. 2-3, pp. 252-257, 2008.
[http://dx.doi.org/10.1080/15275920802122981]

[97]   L. Zhao, Y.H. Dong, and H. Wang, "Residues of veterinary antibiotics in manures from feedlot livestock in eight provinces of China", *Sci. Total Environ.,* vol. 408, no. 5, pp. 1069-1075, 2010.
[http://dx.doi.org/10.1016/j.scitotenv.2009.11.014] [PMID: 19954821]

[98]   R. Wei, F. Ge, M. Chen, and R. Wang, "Occurrence of ciprofloxacin, enrofloxacin, and florfenicol in

animal wastewater and water resources", *J. Environ. Qual.,* vol. 41, no. 5, pp. 1481-1486, 2012.
[http://dx.doi.org/10.2134/jeq2012.0014] [PMID: 23099939]

[99] G. Hamscher, S. Sczesny, H. Höper, and H. Nau, "Determination of persistent tetracycline residues in soil fertilized with liquid manure by high-performance liquid chromatography with electrospray ionization tandem mass spectrometry", *Anal. Chem.,* vol. 74, no. 7, pp. 1509-1518, 2002.
[http://dx.doi.org/10.1021/ac015588m] [PMID: 12033238]

[100] R. Wei, F. Ge, S. Huang, M. Chen, and R. Wang, "Occurrence of veterinary antibiotics in animal wastewater and surface water around farms in Jiangsu Province, China", *Chemosphere,* vol. 82, no. 10, pp. 1408-1414, 2011.
[http://dx.doi.org/10.1016/j.chemosphere.2010.11.067] [PMID: 21159362]

[101] R.M. Monaghan, and L.C. Smith, "Minimising surface water pollution resulting from farm-dairy effluent application to mole-pipe drained soils. II. The contribution of preferential flow of effluent to whole-farm pollutant losses in subsurface drainage from a West Otago dairy farm", *N. Z. J. Agric. Res.,* vol. 47, no. 4, pp. 417-428, 2004.
[http://dx.doi.org/10.1080/00288233.2004.9513610]

[102] Deepak Kumar, Kushal Desai, and Dharmendra Gupta, "Pollution Abatement in Milk Dairy Industry", *Journal of Current Pharma Research,* vol. 1, no. 2, pp. 145-152, 2011.
[http://dx.doi.org/10.33786/JCPR.2011.v01i02.010]

[103] D.R. Smith, P.R. Owens, A.B. Leytem, and E.A. Warnemuende, "Nutrient losses from manure and fertilizer applications as impacted by time to first runoff event", *Environ. Pollut.,* vol. 147, no. 1, pp. 131-137, 2007.
[http://dx.doi.org/10.1016/j.envpol.2006.08.021] [PMID: 17029684]

[104] F.A. Nicholson, S.J. Groves, and B.J. Chambers, "Pathogen survival during livestock manure storage and following land application", *Bioresour. Technol.,* vol. 96, no. 2, pp. 135-143, 2005.
[http://dx.doi.org/10.1016/j.biortech.2004.02.030] [PMID: 15381209]

[105] C. Nweke, P. Igbokwe, and J. Nwabanne, *"Kinetics of Batch Anaerobic Digestion of Vegetable Oil Wastewater" Open Journal of Water Pollution and Treatment,* vol. 2014, no. 2, pp. 1-10, 2014.
[http://dx.doi.org/10.15764/WPT.2014.02001]

[106] P. Saranya, K. Ramani, and G. Sekaran, "Biocatalytic approach on the treatment of edible oil refinery wastewater", *RSC Advances,* vol. 4, no. 21, p. 10680, 2014.
[http://dx.doi.org/10.1039/c3ra43668c]

[107] C.C. Mar, Y. Fan, F.L. Li, and G.R. Hu, "Bioremediation of wastewater from edible oil refinery factory using oleaginous microalga Desmodesmus sp. S1", *Int. J. Phytoremediation,* vol. 18, no. 12, pp. 1195-1201, 2016.
[http://dx.doi.org/10.1080/15226514.2016.1193466] [PMID: 27260474]

[108] S. Sharma, and H. Simsek, "Treatment of canola-oil refinery effluent using electrochemical methods: A comparison between combined electrocoagulation + electrooxidation and electrochemical peroxidation methods", *Chemosphere,* vol. 221, pp. 630-639, 2019.
[http://dx.doi.org/10.1016/j.chemosphere.2019.01.066] [PMID: 30665092]

[109] G. Louhıchı, L. Bousselmı, A. Ghrabı, and I. Khounı, "Process optimization via response surface methodology in the physico-chemical treatment of vegetable oil refinery wastewater", *Environ. Sci. Pollut. Res. Int.,* vol. 26, no. 19, pp. 18993-19011, 2019.
[http://dx.doi.org/10.1007/s11356-018-2657-z] [PMID: 29987464]

[110] L. Guerrero, F. Omil, R. Méndez, and J.M. Lema, "Anaerobic hydrolysis and acidogenesis of wastewaters from food industries with high content of organic solids and protein", *Water Res.,* vol. 33, no. 15, pp. 3281-3290, 1999.
[http://dx.doi.org/10.1016/S0043-1354(99)00041-X]

[111] E.V. Ramasamy, S. Gajalakshmi, R. Sanjeevi, M.N. Jithesh, and S.A. Abbasi, "Feasibility studies on the treatment of dairy wastewaters with upflow anaerobic sludge blanket reactors", *Bioresour.*

Technol., vol. 93, no. 2, pp. 209-212, 2004.
[http://dx.doi.org/10.1016/j.biortech.2003.11.001] [PMID: 15051084]

[112] S.V. Kalyuzhnyi, E.P. Martinez, and J.R. Martinez, "Anaerobic treatment of high-strength cheese-whey wastewaters in laboratory and pilot UASB-reactors", *Bioresour. Technol.,* vol. 60, no. 1, pp. 59-65, 1997.
[http://dx.doi.org/10.1016/S0960-8524(96)00176-9]

[113] B. Rajbhandari, and A.P. Annachhatre, "Anaerobic ponds treatment of starch wastewater: case study in Thailand", *Bioresour. Technol.,* vol. 95, no. 2, pp. 135-143, 2004.
[http://dx.doi.org/10.1016/j.biortech.2004.01.017] [PMID: 15246437]

[114] A. Torkian, A. Eqbali, and S.J. Hashemian, "The effect of organic loading rate on the performance of UASB reactor treating slaughterhouse effluent", *Resour. Conserv. Recycling,* vol. 40, no. 1, pp. 1-11, 2003.
[http://dx.doi.org/10.1016/S0921-3449(03)00021-1]

[115] Y.A. Oktem, O. Ince, P. Sallis, T. Donnelly, and B.K. Ince, "Anaerobic treatment of a chemical synthesis-based pharmaceutical wastewater in a hybrid upflow anaerobic sludge blanket reactor", *Bioresour. Technol.,* vol. 99, no. 5, pp. 1089-1096, 2008.
[http://dx.doi.org/10.1016/j.biortech.2007.02.036] [PMID: 17449241]

[116] X. Colin, J.L. Farinet, O. Rojas, and D. Alazard, "Anaerobic treatment of cassava starch extraction wastewater using a horizontal flow filter with bamboo as support", *Bioresour. Technol.,* vol. 98, no. 8, pp. 1602-1607, 2007.
[http://dx.doi.org/10.1016/j.biortech.2006.06.020] [PMID: 16973355]

[117] A.G. Cabado, S. Aldea, C. Porro, G. Ojea, J. Lago, C. Sobrado, and J.M. Vieites, "Migration of BADGE (bisphenol A diglycidyl-ether) and BFDGE (bisphenol F diglycidyl-ether) in canned seafood", *Food Chem. Toxicol.,* vol. 46, no. 5, pp. 1674-1680, 2008.
[http://dx.doi.org/10.1016/j.fct.2008.01.006] [PMID: 18289761]

[118] G.A. Pedersen, L.K. Jensen, A. Fankhauser, S. Biedermann, J.H. Petersen, and B. Fabech, "Migration of epoxidized soybean oil (ESBO) and phthalates from twist closures into food and enforcement of the overall migration limit", *Food Addit. Contam. Part A Chem. Anal. Control Expo. Risk Assess.,* vol. 25, no. 4, pp. 503-510, 2008.
[http://dx.doi.org/10.1080/02652030701519088] [PMID: 18348048]

[119] C. Nerín, E. Contín, and E. Asensio, "Kinetic migration studies using Porapak as solid-food simulant to assess the safety of paper and board as food-packaging materials", *Anal. Bioanal. Chem.,* vol. 387, no. 6, pp. 2283-2288, 2007.
[http://dx.doi.org/10.1007/s00216-006-1080-3] [PMID: 17237924]

[120] R. Paseiro-Cerrato, A.R.B. De Quirós, R. Sendón, J. Bustos, M.I. Santillana, J.M. Cruz, and P. Paseiro-Losada, "Chromatographic Methods for the Determination of Polyfunctional Amines and Related Compounds Used as Monomers and Additives in Food Packaging Materials: A State-of-t-e-Art Review", *Compr. Rev. Food Sci. Food Saf.,* vol. 9, no. 6, pp. 676-694, 2010.
[http://dx.doi.org/10.1111/j.1541-4337.2010.00133.x] [PMID: 33467824]

[121] M. Aznar, P. Vera, E. Canellas, C. Nerín, P. Mercea, and A. Störmer, "Composition of the adhesives used in food packaging multilayer materials and migration studies from packaging to food", *J. Mater. Chem.,* vol. 21, no. 12, p. 4358, 2011.
[http://dx.doi.org/10.1039/c0jm04136j]

[122] U.N. WWAP, "Water for People Water for Life", 2003.

[123] WWAP (World Water Assessment Programme), *Managing Water under Uncertainty and Risk.,* 2012.

[124] D. Molden, *Water for food water for life: A Comprehensive assessment of water management in agriculture.,* 2013.

[125] C.L. Moe, and R.D. Rheingans, "Global challenges in water, sanitation and health", *J. Water Health,*

vol. 4, no. S1, suppl. Suppl. 1, pp. 41-57, 2006.
[http://dx.doi.org/10.2166/wh.2006.0043] [PMID: 16493899]

[126] R.B. Jackson, S.R. Carpenter, C.N. Dahm, D.M. McKnight, R.J. Naiman, S.L. Postel, and S.W. Running, "Water in a changing world", *Ecol. Appl.*, vol. 11, no. 4, pp. 1027-1045, 2001.
[http://dx.doi.org/10.1890/1051-0761(2001)011[1027:WIACW]2.0.CO;2]

[127] U.N. Water, *Human Rights | UN-Water*. UN Water - Water Facts and Human Rights, 2019.

[128] M. Farré, S. Pérez, K. Gajda-Schrantz, V. Osorio, L. Kantiani, A. Ginebreda, and D. Barceló, "First determination of C60 and C70 fullerenes and N-methylfulleropyrrolidine C60 on the suspended material of wastewater effluents by liquid chromatography hybrid quadrupole linear ion trap tandem mass spectrometry", *J. Hydrol. (Amst.)*, vol. 383, no. 1-2, pp. 44-51, 2010.
[http://dx.doi.org/10.1016/j.jhydrol.2009.08.016]

[129] V. Geissen, F.Q. Ramos, P. de J Bastidas-Bastidas, G. Díaz-González, R. Bello-Mendoza, E. Huerta-Lwanga, and L.E. Ruiz-Suárez, "Soil and water pollution in a banana production region in tropical Mexico", *Bull. Environ. Contam. Toxicol.*, vol. 85, no. 4, pp. 407-413, 2010.
[http://dx.doi.org/10.1007/s00128-010-0077-y] [PMID: 20734023]

[130] V. Geissen, H. Mol, E. Klumpp, G. Umlauf, M. Nadal, M. van der Ploeg, S.E.A.T.M. van de Zee, and C.J. Ritsema, "Emerging pollutants in the environment: A challenge for water resource management", *Int. Soil Water Conserv. Res.*, vol. 3, no. 1, pp. 57-65, 2015.
[http://dx.doi.org/10.1016/j.iswcr.2015.03.002]

[131] T.L. ter Laak, S.O. Agbo, A. Barendregt, and J.L.M. Hermens, "Freely dissolved concentrations of PAHs in soil pore water: measurements via solid-phase extraction and consequences for soil tests", *Environ. Sci. Technol.*, vol. 40, no. 4, pp. 1307-1313, 2006.
[http://dx.doi.org/10.1021/es0514803] [PMID: 16572790]

[132] S.C. Monteiro, and A.B.A. Boxall, "Factors affecting the degradation of pharmaceuticals in agricultural soils", *Environ. Toxicol. Chem.*, vol. 28, no. 12, pp. 2546-2554, 2009.
[http://dx.doi.org/10.1897/08-657.1] [PMID: 19580336]

[133] D. Kasel, S.A. Bradford, J. Šimůnek, M. Heggen, H. Vereecken, and E. Klumpp, "Transport and retention of multi-walled carbon nanotubes in saturated porous media: Effects of input concentration and grain size", *Water Res.*, vol. 47, no. 2, pp. 933-944, 2013.
[http://dx.doi.org/10.1016/j.watres.2012.11.019] [PMID: 23228890]

[134] D. Kasel, S.A. Bradford, J. Šimůnek, T. Pütz, H. Vereecken, and E. Klumpp, "Limited transport of functionalized multi-walled carbon nanotubes in two natural soils", *Environ. Pollut.*, vol. 180, pp. 152-158, 2013.
[http://dx.doi.org/10.1016/j.envpol.2013.05.031] [PMID: 23770315]

[135] American Public Health Association (APHA), *Standard Methods for the Examination of Water and Wastewater*. 21st ed. APHA: Washington, DC, USA, 2005.

[136] American Public Health Association, *Addressing law enforcement violence as a public health issue*, 2016.

[137] American Public Health Association (APHA), *Standard Methods for the Examination of Water and Wastewater*. 21st ed. Am. Water Work. Assoc. Water Environ. Fed: Washington, DC, 2005.

[138] J.P. Kushwaha, V.C. Srivastava, and I.D. Mall, "An overview of various technologies for the treatment of dairy wastewaters", *Crit. Rev. Food Sci. Nutr.*, vol. 51, no. 5, pp. 442-452, 2011.
[http://dx.doi.org/10.1080/10408391003663879] [PMID: 21491269]

[139] M. Baroudi, R. Kabbout, H. Bakkour, F. Dabboussi, S. Taha, and J. Halwani, "Characterization, physicochemical and biological treatment of sweet whey (major pollutant in dairy effluent)", *Asian J. Water Environ. Pollut.*, 2012.

[140] A. Nakbi, M. Bouzid, F. Ayachi, N. Bouaziz, and A. Ben Lamine, "Quantitative characterization of sucrose taste by statistical physics modeling parameters using an analogy between an experimental

physicochemical isotherm of sucrose adsorption on β-cyclodextrin and a putative biological sucrose adsorption from sucrose dose-taste response curve (psychophysics and electrophysiology)", *J. Mol. Liq.,* vol. 298, p. 111950, 2020.
[http://dx.doi.org/10.1016/j.molliq.2019.111950]

[141] A. Tawfik, M. Sobhey, and M. Badawy, "Treatment of a combined dairy and domestic wastewater in an up-flow anaerobic sludge blanket (UASB) reactor followed by activated sludge (AS system)", *Desalination,* vol. 227, no. 1-3, pp. 167-177, 2008.
[http://dx.doi.org/10.1016/j.desal.2007.06.023]

CHAPTER 4

Engineered Nanomaterials as Emerging Water Pollutants

Abdul Sattar Jatoi[1,*], **Shuakat Ali Mazari**[1], **Zubair Hashmi**[1], **Shoaib Ahmed**[1], **Nabisab Mujawar Mubarak**[2], **Rama Rao Karri**[2], **Nizamuddin Sabzoi**[3], **Rashid Abro**[1], **Asif Shah**[4], **Abdul Qayoom Memon**[1], **Abdul Karim Shah**[1] and **Atta Muhammad**[1]

[1] *Department of Chemical Engineering Dawood University of Engineering and Technology Karachi 74800, Pakistan*

[2] *Petroleum and Chemical Engineering, Faculty of Engineering, Universiti Teknologi Brunei, Bandar Seri Begawan BE1410, Brunei Darussalam*

[3] *School of Engineering, RMIT University, Melbourne 3000, Australia*

[4] *Department of Metallurgy and Materials Engineering Dawood, University of Engineering and Technology Karachi*

Abstract: Nanotechnology has many advantages, and its applications are spread to every field, from engineering to medicine and space to agriculture. Owing to the immense advantages of nano-size particles, nano-based materials are widely applied in wastewater treatment. These nanomaterials are developed and utilized in different sizes, shapes, and chemical compositions. These nanomaterials are characterized by their unique physical, chemical and biological properties. Besides the immense benefits of nanomaterials, they also have few environmental implications. This chapter presents the pros and cons of nanomaterials and their implications on the environment. Further, the effect of various nanomaterials on the aquatic environment, including the behaviour and toxicity on the aquatic ecosystem, is discussed. Finally, future directions to minimize the toxic effect of nanomaterials on the aquatic ecosystem and the need for improvement in the nanomaterials are presented.

Keywords: Engineered nanomaterials, Metal oxide nanomaterials, Carbon-based nanomaterials, Toxicity, Emerging water pollutants, Environmental impact, Environmental assessment, Pollutant monitoring, Nanoelectronic devices, Nanotextile materials, Remediation technologies, Wastewater treatment plants, Sewage treatment plants, Transport and environmental fate of nanomaterials, Nanomaterial waste handling.

---

* **Corresponding author Abdul Sattar Jatoi:** Department of Chemical Engineering Dawood University of Engineering and technology Karachi, 74800, Pakistan; E-mail: abdul.sattar@duet.edu.pk

Shaukat Ali Mazari, Nabisab Mujawar Mubarak & Nizamuddin Sabzoi (Eds.)

INTRODUCTION

Nanotechnology is a rapidly expanding research field and has found numerous applications in various fields. Currently, nanomaterials can be used in widely diverse applications such as wastewater treatment, catalysis, separation, and purification processes, bio-medical applications, electronic and optoelectronic accessories (such as light-emitting diodes (LEDs), solar cells, *etc.* Over the past decade, the production of nanomaterials has grown exponentially, and further research has now focused on improving their performance and applicability [1 - 3]. No doubt, nanotechnology has been effectively utilized in many areas to improve human efforts [4, 5]. Nanomaterials (NM) are mostly divided into conducive and non-conducive materials depending on their sources and applications [6, 7]. Although NMs available naturally or intentionally produced from organic activities, such as the combustion of fossil fuels or volcanic explosions, construction works, *etc* [8]. However, after the usage of these nanomaterials in various applications, they are left over as sludge or waste material, which eventually is dumped as solid waste. Inappropriate dumping of these NMs poses a severe threat to the environment and water bodies [9, 10]. Therefore, the immense NM applications poses risks to the environment and human health due to their severe ill effects such as toxicity, mutagenicity, *etc.*

Earlier research findings have observed that certain NMs are poisonous to marine creatures, mammals, and birds [11, 12]. Synthesized nanomaterials can be metal-based or carbon-based. Metal-based NMs incorporate metal oxides, magnetic nanoparticles (mNPs), noble metal NPs, silica NPs, and zeolites [13]. Carbon-based NMs are presented as fullerene (C60), graphene oxide (GO), and carbon nanotube (CNT) [14]. Due to their exceptional chemical, mechanical, photosensitive, physical, and thermal characteristics, NMs have attracted great consideration in science and engineering. Further, due to their small size and higher significant area and aspect ratio, they present unique characteristics [15]. CNTs are one-atom-layer sheets of carbon panes into a prolonged, echoing cylindrical structure [16]. Common techniques for synthesizing CNT are arc discharge and graphite CVD. Due to the endless chain of C-C covalent bonds, CNTs have greater mechanical characteristics than other materials. Graphene is a two-dimensional hexagonal frame carbon coating with a thickness of one atom. They have significant thermal, electrical, mechanical, ophthalmic characteristics and can combine with other elements [17, 18].

Several NMs have been utilized for water, effluent, and intake water treatment *via* adsorption, decontamination, membrane filtration, and photocatalysis. NMs incorporate magnetic nanoparticles, dye-doped silica NPs, noble metal NPs, quantum dots, CNTs, metal oxide NPs, nanofibers, zeolites, and graphene [19].

Due to their extremely permeable shape, exceptional geomorphology and geometry, and functionally dynamic surface, nano sorbents have the ability to eliminate toxins with unique hydrophobicity, and enhance the effectiveness of adsorption. Nano sorbents have higher kinetics due to available active sites and free surface, elevated elimination capabilities, and can be chemically stimulated for numerous applications [20]. The use of NMs in larger-scale water treatment raises many fundamental technological issues, including accumulation, difficulty in separating NMs after treatment, leakage into the water, and potential biological and human health consequences [21]. Earlier investigations have described that NMs influence their surrounding; they contaminate the atmosphere, accumulate in the ecosystem, and manipulate the life progression of inhabiting creatures. Although NMs are distributed in marine ecosystems, assessing their future harmful effects remains vague. Thus, there is a pressing need for more research to generate data that can be employed for probability evaluation, checking, and drafting regulatory guidelines.

NMs endure compound revolutions by intermingling with different media and creatures upon penetrating the ecosystem [22]. This results in various processes, such as accumulation, dissolution, alluviation, and translocation of the foodstuff [23]. Ecological experience with NMs can be evaluated by deciding the quantity of NMs circulated throughout the ending-of-life cycle. Though existing risk assessment techniques may help determine the prospective consequences of NMs, a new set of rules should be developed to tackle their exceptional characteristics and impacts. These techniques should consider the risk and dangers presented by NMs. From the numerous accessible methods, life cycle assessment (LCA) procedures for assessing the sustainability of nanotechnology have been encouraged. Moreover, there are few statistics on the impacts of long-term exposure at smaller concentrations.

Furthermore, few surveys about the collaborative effects among NMs, pure organic matter, and micro-toxins. To have a comprehensive analytic appraisal structure, it is essential to utilize a consistent approach and parameters for expecting the toxic effects of NMs. This chapter intends to assess the effect of nanomaterials on water pollution and their impacts on the ecosystem, ascertain the resources of nanomaterials in the marine ecosystem and assess the haulage and future nanomaterials.

SOURCES OF NANOMATERIALS

Engineered and Non-engineered Nanomaterials

Nanomaterials can be roughly divided into engineered and non-engineered nanomaterials . The engineered nanomaterials (ENMs) are synthetic and devised for task-specific applications. Consequently, the production and user requirement of ENMs has increased throughout the years. At the same time, unmanufactured nanomaterials exist naturally in natural surroundings or are accidentally produced during volcanic eruptions, combustion, construction works, *etc* [24, 25].

Carbon- and Metal-based Nanomaterials

Carbon source nanomaterials are very useful resources due to their great content and low manufacturing cost compared with other nanomaterials. CNTs are microtubes of graphitic carbon with an external dimension of 30 nm and 1 μm [26]. They are divided into 3 categories: single-wall carbon nanotubes (SWCNTs), double-wall carbon nanotubes (DWCNTs), and multi-wall carbon nanotubes (MWCNTs).

Different carbon materials, like graphene, carbon onion, nano-diamonds, and diamond-like carbons, have been developed as carbon nanomaterials. Due to their new physiochemical and thermo-mechanical characteristics, CNTs have aroused great interest in many science, technology, medicine, and engineering disciplines. C-based materials can be generated from a variety of traditional carbon resources, like coal [27], biomass [28, 29], activated carbon [30, 31], carbon black [32], and graphite [33]. The comprehensive approaches are known as top to bottom and bottom to top. The top-down approach uses various technologies like high-energy ball milling, electrospinning, nanolithography, pulsed laser ablation, laser deposition, pulse laser, arid inscription, and anodizing to lessen the extent of the raw material from large size to nanoscale [34]. The bottom-up methodology uses small particles or material species like atoms or molecules to create objects whose final dimensions are in the nanometer zone. These technologies encompass sol-gel handling, CVD, self-assembly and bio-assisted fusion, laser pyrolysis, electrochemical/ electroplating, atomic or molecular condensation, and supercritical fluid synthesis [35].

For example, nanomaterials like zinc oxide use bottom-up methods (*i.e.*, sol-gel method) to synthesize silver nanoparticles and nZVI. At the same time, polymer or ceramic nanofibers can be prepared using electrospinning technology through a top-down approach. C-based NMs applications include electronics, ecological contamination control, effluent treatment, civil engineering, agricultural, energy sector, textiles, biomedical, directed drug delivery, sensors, *etc* [36]. For example, MgO nanoparticles have been employed to restrict heavy metal ions [37], and nano-zero-valent iron (nZVI) as a reactive absorbent fence to purify groundwater [38] and zinc from pollutants such as organic pesticides. TiO_2 and nanoparticles have been incorporated in sunscreens to reduce ultraviolet emissions [39].

TRANSPORT AND ENVIRONMENTAL FATE OF NANOMATERIALS

The effect of NM in the ecosystem differs depending on their physical and chemical characteristics and their interactions with other pollutants [40]. After being released into the atmosphere, NMs will accumulate in mud, water, air, and sediments. To enrich the basic comprehension of NMs and mitigation measures to minimize the NMs toxicity, it is vital to identify their fate in the ecosystem [41]. The carbons in carbon nanomaterials (CNMs) remains unaffected or crumbles. Various microorganisms can use CNMs as a carbon source to release CO_2 into the atmosphere. The decomposition ratio of CNMs is usually very slow, but due to the mass production of CNMs globally, the carbon produced is very limited. Since CNMs can also be correlated with ecological data, it is tricky to pursue the relocation of CNM-related carbons in the ecosystem. CNM carbon transfer path and its inclusion in the carbon cycle need further research. In addition, CNMs directly affect the carbon cycle due to their harmful impacts on fungi and bacteria [42].

Nanoparticles can enter the environment through wastewater treatment plants, wastewater from industrial sources, and runoff from soil and urban roads [43, 44]. After entering the aquatic environment, their behaviour and influence are related by the following processes: aerobic and anaerobic, biodegradation- diffusion combination, reaction with other pollutants, decaying of natural organic matter (NOM), and biological degradation. Fig. (**1**) shows an influx of different nanomaterials to the environment, including water bodies. Various physical and chemical intrications depend on the properties of water, such as ionic strength, pH, and the strength of electromagnetic energy. In aquatic systems, the major transformation of NPs is homogeneous or heterogeneous aggregation, which depends on the interaction of attraction and repulsion and the frequency of surface collisions of NPs. Reduced combination, the specific surface area, and interface free energy of NPs reduce their responsiveness. Prior studies have confirmed the harmful impacts of aquatic creatures. Although the creatures in the real ecosystem are anticipated to be exposed to multiple NMs for a long time, the acute exposure of single type of NMs is still at naïve study. Direct observation and quantitative analysis of silver oxide released by AgNPs of different sizes in water show that AgNPs are unevenly distributed in the ecosystem [42].

Nanoparticles can also participate through fish like tilapia, a popular cultivated fish in Zimbabwe and the main regional fish eaten by ordinary people. Additional research on luminescence imaging technology for Fauna shows that silver oxide is mainly scattered in cell fragments, phagosomes and metal-charged particles [45]. A survey stated the harmful effects of nanoparticles on the liver and heart of African catfish. In an additional report, an escalation in intracellular reactive oxygen species (ROS) in zebrafish (Danio rerio) embryos exposed to ZnO-NP has been observed. These consequences show that compared to other metal salts, the ZnO-NPs can pass through the chorionic level, reach embryos or larvae, and prove to be more poisonous than the equivalent at these stages. High levels of animals and humans may be unprotected from NMs through skin interaction, breathing, nutrition contamination, and imbibing water contamination.

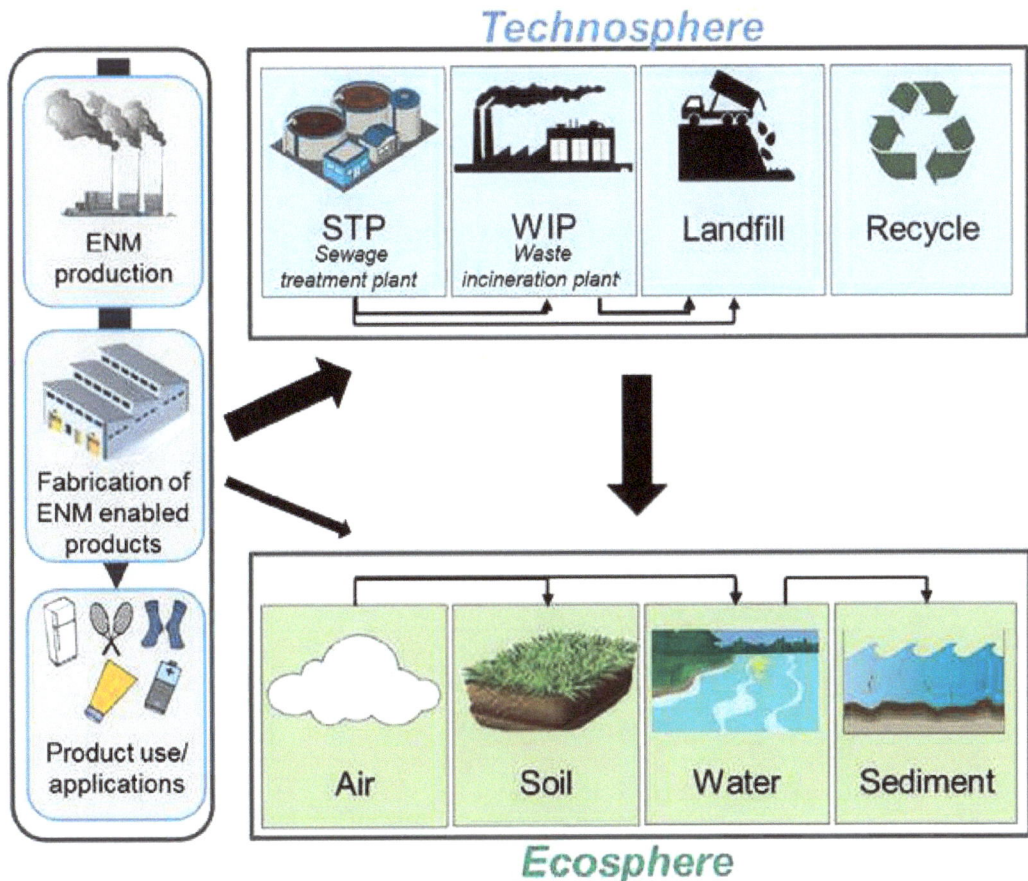

Fig. (1). Transport and Environmental fate of nanomaterials [46].

Table 1. Provides a summary of various routes of nanomaterials various and their impacts.

Nanoparticles	Fungi/Plant Association	Applied Concentration	Effect	Ref.
ZnO-NPs	Aquatic microcosm experiment	30, 300, 3000 ng/L	Negative impact on microbial enzymes activities and fungal community structure	[47]

(Table 1) cont.....

Nanoparticles	Fungi/Plant Association	Applied Concentration	Effect	Ref.
ZnO-NPs	*Glomus versiforme/ caledonium*-Maize	800–3200 mg/kg loamy soil	ZnO-NPs caused toxicity to AMF symbiosis	[48, 49]
Nano diamonds	*Phanerochaete chrysosporium*	0.01–1 mg/ml	Destruction of fungal mycelia, loss of cytoplasm, cell wall damage, inhibition of laccase and manganese peroxidase activity through oxidative stress	[50]
ZnO-NPs	*Funneliformis mosseae*- Maize	500 mg /kg soil	Negative impact on AMF association	[51]
ZnO-NPs	*Mycena citricolor* and *Colletotrichum* sp.	6, 9, and 12 mmol /L	Damage to fungal cell wall, vegetative and reproductive structures, ~93% inhibition in growth of *Colletotrichum* sp. and ~97% for *M. citricolor*	[52]
TiO$_2$-NPs	aAMF-Rice consortium	8, 16, and 33 mg/kg sandy soil	Inhibition of AMF symbiosis in rice roots	[53]
MoO$_3$	*Aspergillus flavus* and *A. niger*	10 and 200 mg/L	Significant growth inhibition, NPs induced nuclear condensation, change in hyphae morphology, metabolic changes leading to apoptosis,	[54]
Ag-NPs	aAMF-Tomato	12–36 mg/kg soil	Reduction in AMF colonization by Ag-NPs in a dose-dependent manner	[55]
Ag-NPs and Fe$_2$O$_3$-NPs	Mycorrhizal clover (*Trifolium repens*)	0.032–3.2 mg/ kg Fe2O3- NPs and 0.01–1 mg/kg Ag-NPs	Drastic reduction in biomass of mycorrhizal clover, root nutrient acquisition of AMF, and glomalin content	[56]
Ag-NPs	*Glomus aggregatum*-Faba bean	800 µg/kg sandy soil-loam mixture	Reduced mycorrhizal colonization, glomalin content, and mycorrhizal responsiveness	[57]

The Toxicity of Nanomaterials in Aquatic Systems

The rapid utilization of NMs in consumer and industrialized products has increased attention towards harmful impacts on the environment and human health (Fig. **2**). exhibits toxicity of nanomaterials in aquatic systems. The aquatic nature is the main zone of convergence of nanomaterials and finally enters the aquatic ecosystem through numerous direct and indirect channels. Various NMs like nano-valent zero copper, Fe$_2$O$_3$-biocarbon nanocomposite, and nano-MgO composite has been used to eradicate contaminants from the aqueous environment. Quantum dots (QDs) are a type of NM with a dimension of 2-10 nm, which have also been widely used in industrialization and consumer goods. However, there are significant data on their concentration in the marine environment. The environment toxicological impacts of quantum dots are of major concern. Among other applications, NMs have also been used in high-performance separation membrane technology [58]. These NM-based membranes can have anti-fouling, anti-bacterial, photodegradation characteristics and excellent permeability, thus providing novel benefits for instant water treatment processes. Therefore, the existence of NMs in distinct ecological systems can lead to long term toxicologic problems. Apart from the nature of the nanoparticle, the properties and impact of nanomaterials depend on their size, shape, and structure. The ratio of absorption of nanomaterials into various compartments differs in their characteristics. Routes of contact between humans and mammals *via* various means like food stuff, died clothes and paints, fitness goods, or therapeutic applications [59 - 62].

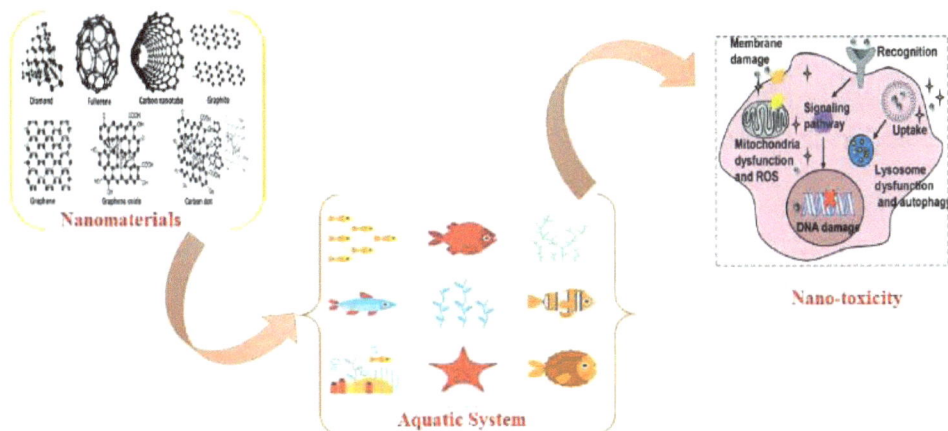

Fig. (2). Toxicity of nanomaterials in aquatic system.

Toxicity in Plants

The degree of toxicity of NMs in plants hinges on the source of the precursor materials utilized to develop nanomaterials and the final characteristics of NMs as highlighted in Fig. (**3**). For example, carbon nanotubes impede the progress of particular plant life by meddling with cell activity and provoking tubocurarine [63]. More specifically, certain C-based NMs often disrupt the photosynthesis of plants, reduce seed germination, and cause chromosomal aberrations in rice or corn. However, C-based NMs have also been observed to upsurge the water holding capacity of plants and fruit production. Nano-oxide (Gd_2O_3) and nano-silica (SiO_2), and other technical nanomaterials inhibit the root elongation of cucumber [64]. Although it has a negative effect on Arabidopsis, nano-SiO_2 can increase the death of biomass and quality of rice. Silver nanoparticles (AgNP) are among the most frequently handled nanomaterials, which change plants in many aspects, including germination rate, shoot length, cell structure integrity and chromosome system [65]. It is reported that AgNPs with a size of 20 nm reduce the germination rate and germination length of barley but have no consequence on the growth rate of flax. A proportional study of silver oxide and AgNPs on barley stated that silver oxide had added significant effects on propagation and seedling growth than AgNPs. These two types of silver oxides also lead to chlorophyllation, manifested by a decline in chlorophyll [66]. Photosynthetic colourants and more organelles are the major edifices influenced by silver noxiousness. Furthermore, when contact with AgNPs, malondialdehyde (an index of membrane damage) was greater in sunken macrophytes than in Juncus emission, indicating that artificial nanoparticles' pressure on the human body is greater than in submerged large plants. Although they are used in many engineering functions like textiles, paints, agrochemicals and microelectronics, copper nanoparticles (CuNP) and CuO nanoparticles (CuONP) cause multiple noxious belongings in cucumber plants, containing oxidation tension and genotoxicity. The size of CuNP is between 10 and 30 nm, resulting in a decline in overall biomass and chlorophyll substance. Nano-toxicity has been demonstrated by the accumulation of CuNP in plant tissues of cultivated Clostridium, and the concentration in root tissues is very high [67]. In another study, AgNPs and CuNPs did not show phytotoxic impacts on growing considerations Quercus in the nursery. However, the changes in ultrastructure were noted in chloroplasts but not in stem and cell skins [63, 68].

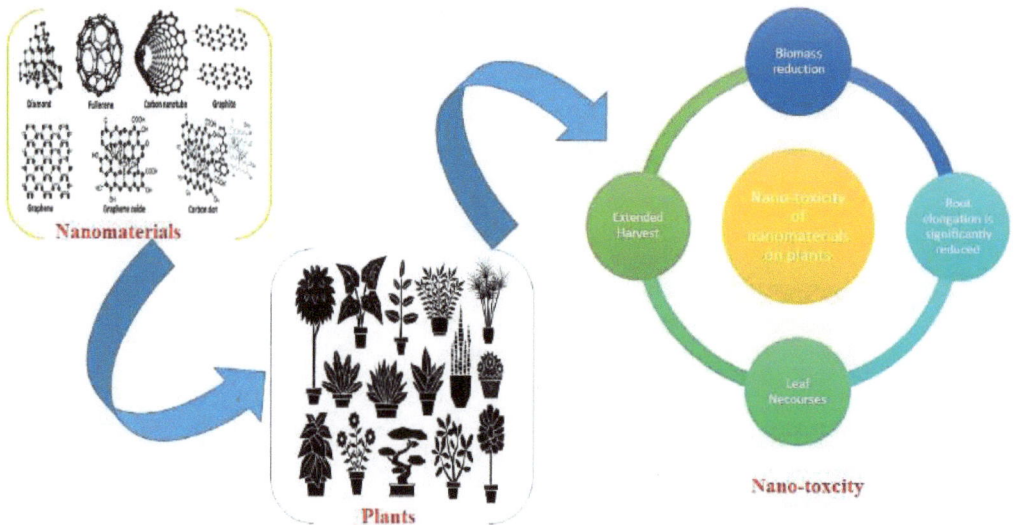

Fig. (3). Toxicity of nanomaterials in plants.

Toxicity in Animals

Nanoparticles have a high-level surface-to-volume proportion, which improves absorption and interaction with aquatic organisms. Due to the nanometer size, nanomaterials can span most bodily disorders in mammals and humans, like the blood-brain fence, blood-placental fence, and accumulation in essential tissues (like brain, liver, lung, spleen), inducing persistent or severe injury at the cellular point *etc* [69]. Fig. (**4**) shows the toxicity of nanomaterials to animals. The poisonous impacts of NMs on animals have gotten more study consideration than on plants. Compared with terrestrial organisms and humans, these impacts are detected in lesser concentrations of marine creatures. The NM and C-based NMs have attracted significant investigation awareness and countless commercial endeavours [70]. The fatality of carbon nanotubes alters on factors such as dimension, concentration, and contact. SWCNTs are more contaminated than MWCTs [71]. For example, single-walled carbon nanotubes are more cytotoxic and fibrogenic than multi-walled carbon nanotubes. It has been reported that prolonged occupational exposure to SWCNTs can cause skin cancer to irretrievable malevolent transformation in humans. Nevertheless, in several surveys, the toxicity of certain CNMs is not obvious. For example, no impermanence was noticed in the severe harmfulness assessment (96 hours of experience) using SWCNTs on Longhair Or (fish). Yet, at the greatest concentration of SWCNTs, most fish will swim in the vicinity of the water surface, and at the end of the disclosure phase, several fish will also excrete murky faeces, which might be a gauge of CNT intake. Analogous findings were detected in studies utilizing creek microcrustaceans (water fleas). In antithesis to the behaviour of paradigm animals, SWCNTs persuaded severe harmfulness in freshwater algae because of the reduced growth rate [72, 73]. Some toxicity studies have pointed out that carboxyl groups reduce the venomousness of MWCNTs due to the functionalization of CNM. Nevertheless, the precise structure of this alleviation impact has not been properly revealed.

Silver nanoparticles have received extensive study and business awareness. They are still among the most studied NMs; they have been used in water treatment, medical dressings, microelectronic trips, agronomy, nanomedicine, and textiles. This widespread use requires in-depth research on the harmfulness of AgNPs. *In vivo* toxicology experiments on land invertebrates, vertebrates, marine organisms, and higher plants have shown that AgNPs can trigger spinal cord, heart, and eye abnormalities. Experiments on the toxicity of AgNPs in humans have demonstrated that AgNPs can persuade oxidative stress in macrophages. The findings show that compared to larger NPs, smaller particles (<20 nm) have greater cytotoxicity. However, when the dosage is based on the surface area, AgNP with a size of 20 nm and 40 nm appears to have an equivalent cytotoxic impact. Curiously, gold nanoparticles of comparable size do not affect macrophages, indicating that gold nanoparticles may be organically immobile.

Further research on the harmfulness of AgNPs to mammalian cells eliminated the previous conclusions that silver is non-toxic to colloidal silver; apart from that, shellac is triggered by silver colloids, and when mammalian cells are visible to toxic doses of AgNPs, there may be differences in the mechanism of cell death [74]. Histopathological studies on rats show that the organs (liver, kidney, thymus, and spleen) where AgNPs precipitated had greater damage than silver ions. Numerous surveys have shown that functionalized AgNPs can decrease their harmfulness. For example, AgNPs summarized with polyethene glycol spinoffs show a relationship between the harmfulness of AgNPs and ion kind in solution or a specific situation. Harmfulness of AgNP to zebrafish embryos was dose-reliant, with different impacts, such as reduced heart frequency, extreme transience, and stalled formulating. The risk of AgNPs is intensity-reliant on marine genera [75]. During manufacturing, workers may inhale CuNPs used in paints and pesticide formulations, which may cause lung soreness and a powerful impervious system retort. In aquatic systems, CuNPs are toxic to the louse Enchytraeus crypticus, thereby affecting their survival and reproduction yield. However, the toxicity level depends on the form of copper, and the order is as follows: copper nitrate ($CuNO_3$>Cu-NPs) > copper ground land. Nano-zero-valent iron (nZVI) is an alternative NM whose applications are currently commercialized, particularly in the ecological redress of earth and groundwater. A research report stated that nZVI is non-toxic to marine creatures at a concentration of up to 0.1 g/L, the cut-off value of the EU's dangerous label [76 - 79].

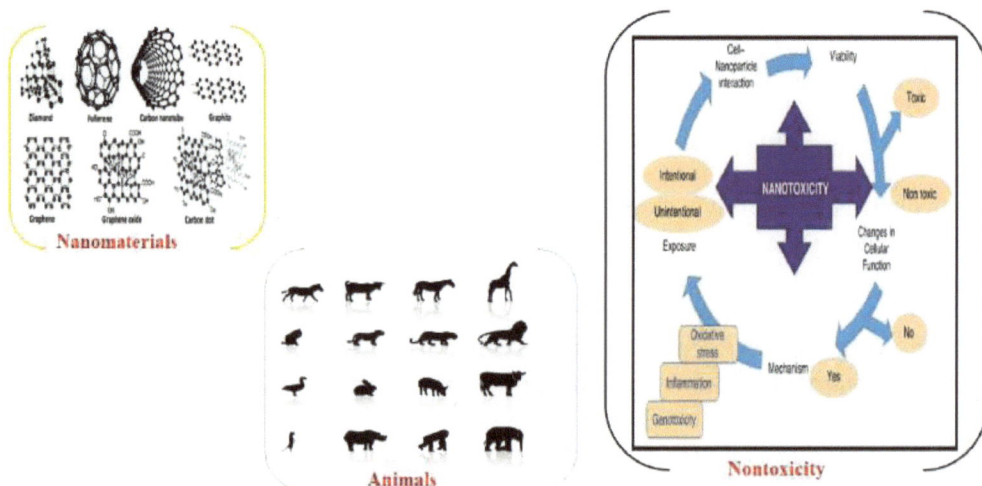

Fig. (4). Toxicity of nanomaterials in animals.

Toxicity Mechanisms

NPs generate noxiousness in living things, including humans, and trigger a broad range of damage. At the cellular stage, the absorption of NMs occurs by cytological impact, *i.e.*, phagocytosis, macropinocytosis, and clathrin-mediated endocytosis. Purposefully, NMs endure as particles in the aquatic hierarchy and can infiltrate biological integuments due to their tiny size. Fig. (**5**) presents various mechanisms of toxicities of nanomaterials. In the aqua hierarchy, NPs form a mixture that can interact with amasses, heaps, and other tinctures [80]. This happens easily in marine ecologies due to these schemes' widespread limy pH, which elevates the ionic potency and a rich fount of colloidal interruptions and NOM. Shoreline zones with a high intensity of extracts, natural materials, and discharges or drips loaded with NPs from activity display. Nevertheless, in freshwater structures, NP gathers have a greater prospect of progressively dropping to the base and amassing in sediments where they influence the benthal genus and are taken up by Fauna.

Sometimes, the NPs are more likely to be reprocessed into the biota by bioaccumulation and biotransformation. Polysaccharides have been shown to improve the alluviation of NPs and reduce their agility. NPs can absorb carbon-based contaminants, which raises the impact of toxicity. For instance, the toxicity of phenanthrene to water flea in the existence of fullerene in water presented by an edict of enormity, and the accumulation of various nanoparticles in grouse bodies increase the existence of TiO_2 NPs. In adding to the synthetic variation of NMs, the existence of organic stuff can additionally enhance their liposome in marine structures. Perhaps, deferments of nanotubes and fullerenes grow into extra balance in the existence of humic and fulvic acids, which are enclosed in residues; in this manner, accumulates are not constructed or are minor. CNP can be taken up by protozoon with nourishment, as shown for amoeba, ciliophora, water fleas, and copepods. The capability of CNP to enter the body of marine protozoon will differ on their composition and variation. For instance, immaculate C60 fullerenes entered zebrafish embryos using the amniote while hydroxylated CNT [81]. Nano noxiousness experimentations concentrate on sociable creatures like water flea and omit model species of invertebrates and vertebrates. These varieties are crucial because of their existence in human nutrition. Comprehensively, while statistics on the performance of NMs in numerous analeptic forms suggest the presence of bioaccumulation at different analeptic degrees, the precise structures are yet mysterious [82 - 84].

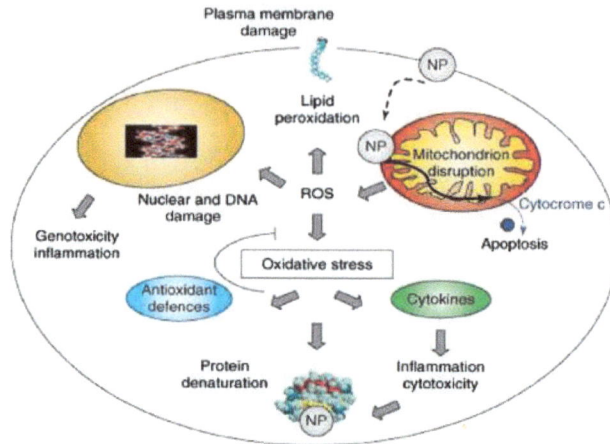

Fig. (5). Mechanisms of toxicity related to nanomaterials.

Methods for the Evaluation of Nanotoxicity

Extensive nanomaterials must be carefully evaluated, with specific awareness of the route of experience, dose expression, biocompatibility, pharmacokinetics, toxicology, and pathological sequelae. The distinctive little size of NMs is aimed at the invulnerable approach of the organism. Many techniques can be used to assess the toxicologic impacts of nanomaterials [85]. These include *In vivo* and *in vitro* toxicities, physiochemical evaluation practices, and nano information processing. Even though existing injuriousness assay techniques can be utilized to ascertain the possible dangers of NMs. Modern procedures still need to be studied to examine their unique properties and effects [86]. Therefore, it is not enough to use physicochemical characteristics as inaccessible attributes for management functions. These assessments should also consider the hazards and risks of exposure. Therefore, other methods need to provide useful classifications.

Based on the different methods available, the European Commission recommends using life cycle analysis (LCA) ways to assess the longevity of nanomaterials. Life cycle assessment is important for analyzing, evaluating, and managing the ecological impact NMs. This information and the economics of LCA can provide information for policies establishing nanotechnology [87]. There is still a lack of ecotoxicological data on NMs under exposure conditions in the real environment and long-term ecosystem noxiousness as analysis is inadequate. To precisely understand deadliness statistics and assess various investigations, it is important to study the exposure conditions carefully. A method is needed to distinguish between the toxicity caused by NM and the impacts of adsorbed contaminants. The lack of data on the enduring consequences of long-term exposure to low concentration limits the inference from lab-scale findings to real-life functions.

In addition, there are few findings on the interaction among NM, NOM, and micropollutants. Table **2** shows various methods used for *in vitro* assessment. It is crucial to build harmonized methods and models to improve the analysis and assessment technique to predict the toxic consequences of various distinct exposures *In vivo* and *in vitro*. Due to their reactivity or the number of hypothetical parameters in a single contact, several paradigm impacts have been explored. In general, the investigational outcomes are hard to explain as the experimental design and possible parameters are very wide, the actual exposure is lacking, and the data-related effects of related experiments are rare. There is a lack of data on model invertebrates and vertebrate species. Therefore, the process of bioconcentration and bioconcentration should be considered, and the toxicity of NMs to these species should be further studied. Invertebrates as laboratory animals are not restricted by law [88, 89].

Table 2. Few examples of conventional tests used for *in vitro* toxicity assessment [90, 91].

Test	Evaluation Endpoint
Dichlorodihydrofluorescein diacetate (DCFH-DA)	Oxidative stress (direct method)
Inductively-coupled plasma mass spectrometry (ICP-MS)	Evaluation of cell internalization of nanomaterials
Nitroblue tetrazolium (NBT) assays	Oxidative stress (direct method)
Confocal microscopy	Evaluation of cell internalization of nanomaterials
Glutathione (GSH) content	Oxidative stress (indirect method)
Transmission electron microscopy (TEM) and scanning electron microscopy (SEM)	Evaluation of cell internalization of nanomaterials
Malondialdehyde (MDA) formation	Oxidative stress (indirect method)
Neutral red test	Assessment of membrane integrity and cell necrosis (accidental cell death)
Superoxide dismutase (SOD) activity	Oxidative stress (indirect method)
Cytokine release	Inflammatory reaction and phagocytosis
Flow cytometry with annexin-V assay	Apoptosis (programmed cell death) induced by ENMs
Vital dyes (*e.g.*, trypan blue and propidium iodide	Assessment of membrane integrity and cell necrosis (accidental cell death)
Measurement of the mitochondrial membrane potential	Apoptosis induced by ENMs

Test	Evaluation Endpoint
Lactate dehydrogenase (LDH)	Assessment of membrane integrity and cell necrosis (accidental cell death)
Protein markers of apoptosis, including Bax, Bcl2, p53 and caspases	Apoptosis induced by ENMs
3-(4, 5-dimethyl-2-thiazolyl)-2,5-diphenyl-2H-tetrazolium bromide (MTT) assay	Measurement of cellular metabolic activity
Adenosine triphosphate (ATP) luminescence assays	Measurement of cellular metabolic activity

FUTURE RESEARCH DIRECTIONS

Nanomaterials have attracted several industrial as well as consumer applications. Their use varies from various sophisticated instruments to personal care products. With their diverse applications, the quantitative consumption of nanomaterials is on the rise, and the same is for waste generation. Their exposure has endangered humans, the environment, and the ecosystem. The fate of nanomaterials can be analyzed by utilizing physicochemical properties, examining their aggregation and constancy alongside accumulation ratios. The growing and subsequent ecological noxious impacts of NMs on plants and other species in the marine ecological system have been extensively reported. The ecological noxious of NMs vary on their physicochemical properties, characteristics of the marine atmosphere, and the extent of exposure. There is necessary to (1) investigative methods and protocols to sense and evaluate NMs in numerous conservational eco-systems, predominantly in effluent and marine situations, and (2) establishing tentative strategy and emergent novel bio signs for the ecological impression valuation of NMs.

Investigation of modern practices is essential to tackle the exceptional characteristics and impacts of NMs. The usefulness of physicochemical characteristics as an inaccessible trait for governing objectives is deficient. LCA can be employed to assess, appraise, and supervise the management impacts of the NMs. This evidence, jointly with the monetary factors, can enlighten strategy and ecological noxiousness relief methodologies centered on the life cycle. A comprehensive examination of exposure circumstances is crucial to construe toxicity data and assess various findings. It is essential to build a consistent approach and recommendations for expecting the harmful consequences from different individuals to groups both *In vivo* and *in vitro*. Data on vertebrate species are deficient and therefore, the further investigation should be conducted to inspect NMs toxity in these species. The environmental toxicology assessment scheme of NMs is essential for development of policies on the advancement and utilization of NMs. Treated NMs should be biocompatible with additional ingredients and wield the least possible unfavourable impacts on human health.

CONCLUSION

The enhanced use of nanomaterials inevitably exposes the ecology and anthropological ecosystem. While the impacts of NMs are tricky to detect, attaining presumptions around the eco-friendly influences of NMs is a challenge. The presently employed risk evaluation methodologies in nanotoxicology, which concentrates on the malignancy of immaculate NMs, involve adjustment to encompass the estimation of fragments circulated across the whole life expectancy of NMs. Toxicological findings are essential in standardizing and devising less toxic NMs. Various ways around which NMs intrude into animals during breathing, makes the exposure more concerned. Though many experiments have been conducted to identify such relations, most of these investigate specific species. But data on the performance and outcome of NMs not very much available, especially from a real-life perspective. A general examination into environment is important to interpret toxicity data and precisely equate several findings. Techniques are needed to differentiate between NM-stimulated fatality and the consequences of adsorbed contaminants. Furthermore, experimental data on the lingering impacts from enduring exposures at minimal intensities is unavailable, limiting the extrapolation of test laboratory results.

CONSENT FOR PUBLICATION

Not applicable.

CONFLICT OF INTEREST

The author declares no conflict of interest, financial or otherwise.

ACKNOWLEDGEMENTS

Declared none.

REFERENCES

[1]   J.S. Duhan, R. Kumar, N. Kumar, P. Kaur, K. Nehra, and S. Duhan, "Nanotechnology: The new perspective in precision agriculture", *Biotechnol. Rep. (Amst.),* vol. 15, pp. 11-23, 2017.
[http://dx.doi.org/10.1016/j.btre.2017.03.002] [PMID: 28603692]

[2]   A. Mehmood, F.S.A. Khan, N.M. Mubarak, Y.H. Tan, R.R. Karri, M. Khalid, R. Walvekar, E.C. Abdullah, S. Nizamuddin, and S.A. Mazari, "Magnetic nanocomposites for sustainable water purification—a comprehensive review", *Environ. Sci. Pollut. Res. Int.,* vol. 28, no. 16, pp. 19563-19588, 2021.
[http://dx.doi.org/10.1007/s11356-021-12589-3] [PMID: 33651297]

[3]   A.S. Jatoi, F.S. Ahmed Khan, S.A. Mazari, N.M. Mubarak, R. Abro, J. Ahmed, M. Ahmed, H. Baloch,

and N. Sabzoi, "Current applications of smart nanotextiles and future trends", In: *Nanosensors and Nanodevices for Smart Multifunctional Textiles.*, A. Ehrmann, T.A. Nguyen, P. Nguyen Tri, Eds., Elsevier, 2021, pp. 343-365.
[http://dx.doi.org/10.1016/B978-0-12-820777-2.00019-4]

[4]   J.N. Sahu, H. Zabed, R.R. Karri, S. Shams, and X. Qi, "Applications of nano-biotechnology for sustainable water purification", In: *Industrial Applications of Nanomaterials.*, 2019, pp. 313-340.
[http://dx.doi.org/10.1016/B978-0-12-815749-7.00011-6]

[5]   R.R. Karri, S. Shams, and J.N. Sahu, "Overview of Potential Applications of Nano-Biotechnology in Wastewater and Effluent Treatment", In: *Nanotechnology in Water and Waste water Treatment.* Elsevier, 2019, pp. 87-100.
[http://dx.doi.org/10.1016/B978-0-12-813902-8.00004-6]

[6]   R.P. Ollier, M.E. Villanueva, G.J. Copello, V.A. Alvarez, and L.M. Sanchez, "Engineered nanomaterials for emerging contaminant removal from wastewater",

[7]   M. Zeng, M. Chen, D. Huang, S. Lei, X. Zhang, L. Wang, and Z. Cheng, *Engineered two-dimensional nanomaterials: an emerging paradigm for water purification and monitoring,* 2020.Mater. Horiz.,

[8]   P.J.J. Alvarez, C.K. Chan, M. Elimelech, N.J. Halas, and D. Villagrán, "Emerging opportunities for nanotechnology to enhance water security", *Nat. Nanotechnol.,* vol. 13, no. 8, pp. 634-641, 2018.
[http://dx.doi.org/10.1038/s41565-018-0203-2] [PMID: 30082804]

[9]   T. Singh, S. Shukla, P. Kumar, V. Wahla, V.K. Bajpai, and I.A. Rather, "Application of nanotechnology in food science: perception and overview", *Front. Microbiol.,* vol. 8, p. 1501, 2017.
[http://dx.doi.org/10.3389/fmicb.2017.01501] [PMID: 28824605]

[10]   S.A. Mazari, N.M. Mubarak, A.S. Jatoi, R. Abro, A. Shah, A.K. Shah, N. Sabzoi, H. Baloch, V. Kumar, and Z. Lghari, "Environmental impact of using nanomaterials in textiles", In: *Nanosensors and Nanodevices for Smart Multifunctional Textiles.*, A. Ehrmann, T.A. Nguyen, P. Nguyen Tri, Eds., Elsevier, 2021, pp. 321-342.
[http://dx.doi.org/10.1016/B978-0-12-820777-2.00018-2]

[11]   W. Elmer, and J.C. White, "The future of nanotechnology in plant pathology", *Annu. Rev. Phytopathol.,* vol. 56, no. 1, pp. 111-133, 2018.
[http://dx.doi.org/10.1146/annurev-phyto-080417-050108] [PMID: 30149792]

[12]   M.S. Mauter, I. Zucker, F. Perreault, J.R. Werber, J.H. Kim, and M. Elimelech, "The role of nanotechnology in tackling global water challenges", *Nat. Sustain.,* vol. 1, no. 4, pp. 166-175, 2018.
[http://dx.doi.org/10.1038/s41893-018-0046-8]

[13]   D. Jassby, T.Y. Cath, and H. Buisson, "The role of nanotechnology in industrial water treatment", *Nat. Nanotechnol.,* vol. 13, no. 8, pp. 670-672, 2018.
[http://dx.doi.org/10.1038/s41565-018-0234-8] [PMID: 30082807]

[14]   F.S.A. Khan, N.M. Mubarak, Y.H. Tan, M. Khalid, R.R. Karri, R. Walvekar, E.C. Abdullah, S. Nizamuddin, and S.A. Mazari, "A comprehensive review on magnetic carbon nanotubes and carbon nanotube-based buckypaper for removal of heavy metals and dyes", *J. Hazard. Mater.,* vol. 413, p. 125375, 2021.
[http://dx.doi.org/10.1016/j.jhazmat.2021.125375] [PMID: 33930951]

[15]   Y. Zare, K.Y. Rhee, and S.J. Park, "A developed equation for electrical conductivity of polymer carbon nanotubes (CNT) nanocomposites based on Halpin-Tsai model", *Results Phys.,* vol. 14, p. 102406, 2019.
[http://dx.doi.org/10.1016/j.rinp.2019.102406]

[16]   S. Kim, Y. Zare, H. Garmabi, and K.Y. Rhee, "Variations of tunneling properties in poly (lactic acid) (PLA)/poly (ethylene oxide) (PEO)/carbon nanotubes (CNT) nanocomposites during hydrolytic degradation", *Sens. Actuators A Phys.,* vol. 274, pp. 28-36, 2018.
[http://dx.doi.org/10.1016/j.sna.2018.03.004]

[17] A.G. Olabi, M.A. Abdelkareem, T. Wilberforce, and E.T. Sayed, "Application of graphene in energy storage device – A review", *Renew. Sustain. Energy Rev.,* vol. 135, p. 110026, 2021.
[http://dx.doi.org/10.1016/j.rser.2020.110026]

[18] B. Bonelli, F.S. Freyria, I. Rossetti, and R. Sethi, *Nanomaterials for the Detection and Removal of Wastewater Pollutants.* Elsevier, 2020.

[19] X. Yu, H. Cheng, M. Zhang, Y. Zhao, L. Qu, and G. Shi, "Graphene-based smart materials", *Nat. Rev. Mater.,* vol. 2, no. 9, p. 17046, 2017.
[http://dx.doi.org/10.1038/natrevmats.2017.46]

[20] A.G. Olabi, M.A. Abdelkareem, T. Wilberforce, and E.T. Sayed, "Application of graphene in energy storage device – A review", *Renew. Sustain. Energy Rev.,* vol. 135, p. 110026, 2021.
[http://dx.doi.org/10.1016/j.rser.2020.110026]

[21] J. Wang, X. Jin, C. Li, W. Wang, H. Wu, and S. Guo, "Graphene and graphene derivatives toughening polymers: Toward high toughness and strength", *Chem. Eng. J.,* vol. 370, pp. 831-854, 2019.
[http://dx.doi.org/10.1016/j.cej.2019.03.229]

[22] S. Shukla, R. Khan, and C.M. Hussain, *Nanoremediation, in The Handbook of Environmental Remediation.,* pp. 443-467, 2020.

[23] B.R. Smith, and S.S. Gambhir, "Nanomaterials for In vivo imaging", *Chem. Rev.,* vol. 117, no. 3, pp. 901-986, 2017.
[http://dx.doi.org/10.1021/acs.chemrev.6b00073] [PMID: 28045253]

[24] T.A. Saleh, *Trends in the sample preparation and analysis of nanomaterials as environmental contaminants.* Trends in Environmental Analytical Chemistry, 2020, p. e00101.

[25] S. Zeb, N. Ali, Z. Ali, M. Bilal, B. Adalat, S. Hussain, S. Gul, F. Ali, R. Ahmad, and H.M. Iqbal, "Silica-based nanomaterials as designer adsorbents to mitigate emerging organic contaminants from water matrices", *J. Water Process Eng.,* vol. 38, p. 101675, 2020.

[26] X. Zhang, G. Li, D. Wu, J. Liu, and Y. Wu, "Recent advances on emerging nanomaterials for controlling the mycotoxin contamination: From detection to elimination", *Food Front.,* vol. 1, no. 4, pp. 360-381, 2020.
[http://dx.doi.org/10.1002/fft2.42]

[27] X. Wang, C. Liu, S. Chen, L. Chen, K. Li, and N. Liu, "Impact of coal sector's de-capacity policy on coal price", *Appl. Energy,* vol. 265, p. 114802, 2020.

[28] M. Kaur, N.M. Mubarak, B.L.F. Chin, M. Khalid, R.R. Karri, R. Walvekar, E.C. Abdullah, and F.A. Tanjung, "Extraction of reinforced epoxy nanocomposite using agricultural waste biomass", In: *IOP Conference Series.* Materials Science and Engineering, 2020.
[http://dx.doi.org/10.1088/1757-899X/943/1/012021]

[29] R.R. Karri, J.N. Sahu, and B.C. Meikap, "Improving efficacy of Cr (VI) adsorption process on sustainable adsorbent derived from waste biomass (sugarcane bagasse) with help of ant colony optimization", *Ind. Crops Prod.,* vol. 143, p. 111927, 2020.
[http://dx.doi.org/10.1016/j.indcrop.2019.111927]

[30] M.H. Dehghani, R.R. Karri, Z.T. Yeganeh, A.H. Mahvi, H. Nourmoradi, M. Salari, A. Zarei, and M. Sillanpää, "Statistical modelling of endocrine disrupting compounds adsorption onto activated carbon prepared from wood using CCD-RSM and DE hybrid evolutionary optimization framework: Comparison of linear vs non-linear isotherm and kinetic parameters", *J. Mol. Liq.,* vol. 302, p. 112526, 2020.
[http://dx.doi.org/10.1016/j.molliq.2020.112526]

[31] R.R. Karri, and J.N. Sahu, "Process optimization and adsorption modeling using activated carbon derived from palm oil kernel shell for Zn (II) disposal from the aqueous environment using differential evolution embedded neural network", *J. Mol. Liq.,* vol. 265, pp. 592-602, 2018.

[32]   S. Khodabakhshi, P.F. Fulvio, and E. Andreoli, "Carbon black reborn: Structure and chemistry for renewable energy harnessing", *Carbon,* vol. 162, pp. 604-649, 2020.
[http://dx.doi.org/10.1016/j.carbon.2020.02.058]

[33]   D.P. Finegan, A. Quinn, D.S. Wragg, A.M. Colclasure, X. Lu, C. Tan, T.M.M. Heenan, R. Jervis, D.J.L. Brett, S. Das, T. Gao, D.A. Cogswell, M.Z. Bazant, M. Di Michiel, S. Checchia, P.R. Shearing, and K. Smith, "Spatial dynamics of lithiation and lithium plating during high-rate operation of graphite electrodes", *Energy Environ. Sci.,* vol. 13, no. 8, pp. 2570-2584, 2020.
[http://dx.doi.org/10.1039/D0EE01191F]

[34]   L. Kumar, V. Ragunathan, M. Chugh, and N. Bharadvaja, "Nanomaterials for remediation of contaminants: a review", *Environ. Chem. Lett.,* vol. 19, no. 4, pp. 3139-3163, 2021.
[http://dx.doi.org/10.1007/s10311-021-01212-z]

[35]   J. Scaria, A. Gopinath, and P. Nidheesh, "A versatile strategy to eliminate emerging contaminants from the aqueous environment: Heterogeneous Fenton process", *J. Clean. Prod.,* p. 124014, 2020.

[36]   B.S. Rathi, P.S. Kumar, and P.L. Show, "A review on effective removal of emerging contaminants from aquatic systems: Current trends and scope for further research", *J. Hazard. Mater.,* vol. 409, p. 124413, 2021.
[http://dx.doi.org/10.1016/j.jhazmat.2020.124413] [PMID: 33183841]

[37]   T. Williams, C. Walsh, K. Murray, and M. Subir, "Interactions of emerging contaminants with model colloidal microplastics, C 60 fullerene, and natural organic matter – effect of surface functional group and adsorbate properties", *Environ. Sci. Process. Impacts,* vol. 22, no. 5, pp. 1190-1200, 2020.
[http://dx.doi.org/10.1039/D0EM00026D] [PMID: 32250376]

[38]   A.B. Zeidman, O.M. Rodriguez-Narvaez, J. Moon, and E.R. Bandala, *Removal of antibiotics in aqueous phase using silica-based immobilized nanomaterials: A review.* Environmental Technology & Innovation, 2020, p. 101030.

[39]   S. Alipoori, H. Rouhi, E. Linn, H. Stumpfl, H. Mokarizadeh, M.R. Esfahani, A. Koh, S.T. Weinman, and E.K. Wujcik, "Polymer-Based Devices and Remediation Strategies for Emerging Contaminants in Water", *ACS Appl. Polym. Mater.,* vol. 3, no. 2, pp. 549-577, 2021.
[http://dx.doi.org/10.1021/acsapm.0c01171]

[40]   E. Suhendra, C.H. Chang, W.C. Hou, and Y.C. Hsieh, "A review on the environmental fate models for predicting the distribution of engineered nanomaterials in surface waters", *Int. J. Mol. Sci.,* vol. 21, no. 12, p. 4554, 2020.
[http://dx.doi.org/10.3390/ijms21124554] [PMID: 32604975]

[41]   A. Avellan, M. Simonin, S.M. Anderson, N.K. Geitner, N. Bossa, E. Spielman-Sun, E.S. Bernhardt, B.T. Castellon, B.P. Colman, J.L. Cooper, M. Ho, M.F. Hochella Jr, H. Hsu-Kim, S. Inoue, R.S. King, S. Laughton, C.W. Matson, B.G. Perrotta, C.J. Richardson, J.M. Unrine, M.R. Wiesner, and G.V. Lowry, "Differential reactivity of copper-and gold-based nanomaterials controls their seasonal biogeochemical cycling and fate in a freshwater wetland mesocosm", *Environ. Sci. Technol.,* vol. 54, no. 3, pp. 1533-1544, 2020.
[http://dx.doi.org/10.1021/acs.est.9b05097] [PMID: 31951397]

[42]   U.M. Graham, A.K. Dozier, G. Oberdörster, R.A. Yokel, R. Molina, J.D. Brain, J.M. Pinto, J. Weuve, and D.A. Bennett, "Tissue Specific Fate of Nanomaterials by Advanced Analytical Imaging Techniques - A Review", *Chem. Res. Toxicol.,* vol. 33, no. 5, pp. 1145-1162, 2020.
[http://dx.doi.org/10.1021/acs.chemrestox.0c00072] [PMID: 32349469]

[43]   R.R. Karri, G. Ravindran, and M.H. Dehghani, *Dehghani, Wastewater—Sources, Toxicity, and Their Consequences to Human Health, in Soft Computing Techniques in Solid Waste and Wastewater Management.,* 2021, pp. 3-33.

[44]   M.H. Dehghani, G.A. Omrani, and R.R. Karri, *Solid Waste—Sources, Toxicity, and Their Consequences to Human Health, in Soft Computing Techniques in Solid Waste and Wastewater Management.,* pp. 205-213, 2021.

[45] T. Piplai, T. Parsai, A. Kumar, and B.J. Alappat, "Understanding Interactions of Nanomaterials with Soil: Issues and Challenges Ahead", *Environmental Chemistry for a Sustainable World,* vol. 27, pp. 117-141, 2020.
[http://dx.doi.org/10.1007/978-3-030-26672-1_4]

[46] J.R. Lead, G.E. Batley, P.J.J. Alvarez, M.N. Croteau, R.D. Handy, M.J. McLaughlin, J.D. Judy, and K. Schirmer, "Nanomaterials in the environment: Behavior, fate, bioavailability, and effects-An updated review", *Environ. Toxicol. Chem.,* vol. 37, no. 8, pp. 2029-2063, 2018.
[http://dx.doi.org/10.1002/etc.4147] [PMID: 29633323]

[47] S. Adhikari, A. Adhikari, S. Ghosh, D. Roy, I. Azahar, D. Basuli, and Z. Hossain, "Assessment of ZnO-NPs toxicity in maize: An integrative microRNAomic approach", *Chemosphere,* vol. 249, p. 126197, 2020.
[http://dx.doi.org/10.1016/j.chemosphere.2020.126197] [PMID: 32087455]

[48] Z. Huang, X. Zheng, D. Yan, G. Yin, X. Liao, Y. Kang, Y. Yao, D. Huang, and B. Hao, "Toxicological effect of ZnO nanoparticles based on bacteria", *Langmuir,* vol. 24, no. 8, pp. 4140-4144, 2008.
[PMID: 18341364]

[49] P.A. Arciniegas-Grijalba, M.C. Patiño-Portela, L.P. Mosquera-Sánchez, J.A. Guerrero-Vargas, and J.E. Rodríguez-Páez, "ZnO nanoparticles (ZnO-NPs) and their antifungal activity against coffee fungus Erythricium salmonicolor", *Appl. Nanosci.,* vol. 7, no. 5, pp. 225-241, 2017.
[http://dx.doi.org/10.1007/s13204-017-0561-3]

[50] V.N. Mochalin, O. Shenderova, D. Ho, and Y. Gogotsi, "The properties and applications of nanodiamonds", *Nat. Nanotechnol.,* vol. 7, no. 1, pp. 11-23, 2012.
[http://dx.doi.org/10.1038/nnano.2011.209] [PMID: 22179567]

[51] R. Vandebriel, and W. De Jong, "A review of mammalian toxicity of ZnO nanoparticles", *Nanotechnol. Sci. Appl.,* vol. 5, pp. 61-71, 2012.
[http://dx.doi.org/10.2147/NSA.S23932] [PMID: 24198497]

[52] K.M. Reddy, K. Feris, J. Bell, D.G. Wingett, C. Hanley, and A. Punnoose, "Selective toxicity of zinc oxide nanoparticles to prokaryotic and eukaryotic systems", *Appl. Phys. Lett.,* vol. 90, no. 21, p. 213902, 2007.
[http://dx.doi.org/10.1063/1.2742324] [PMID: 18160973]

[53] O. Kose, M. Tomatis, L. Leclerc, N.B. Belblidia, J.F. Hochepied, F. Turci, J. Pourchez, and V. Forest, "Impact of the physicochemical features of TiO2 nanoparticles on their in vitro toxicity", *Chem. Res. Toxicol.,* vol. 33, no. 9, pp. 2324-2337, 2020.
[http://dx.doi.org/10.1021/acs.chemrestox.0c00106] [PMID: 32786542]

[54] L. Zhang, Z. Liu, L. Jin, B. Zhang, H. Zhang, M. Zhu, and W. Yang, "Self-assembly gridding α-MoO3 nanobelts for highly toxic H2S gas sensors", *Sens. Actuators B Chem.,* vol. 237, pp. 350-357, 2016.
[http://dx.doi.org/10.1016/j.snb.2016.06.104]

[55] K. Kawata, M. Osawa, and S. Okabe, "in vitro toxicity of silver nanoparticles at noncytotoxic doses to HepG2 human hepatoma cells", *Environ. Sci. Technol.,* vol. 43, no. 15, pp. 6046-6051, 2009.
[http://dx.doi.org/10.1021/es900754q] [PMID: 19731716]

[56] A. Fouda, S.E.D. Hassan, E. Saied, and M.S. Azab, "An eco-friendly approach to textile and tannery wastewater treatment using maghemite nanoparticles (γ-Fe2O3-NPs) fabricated by Penicillium expansum strain (K-w)", *J. Environ. Chem. Eng.,* vol. 9, no. 1, p. 104693, 2021.
[http://dx.doi.org/10.1016/j.jece.2020.104693]

[57] Y. Budama-Kilinc, R. Cakir-Koc, T. Zorlu, B. Ozdemir, Z. Karavelioglu, A.C. Egil, and S. Kecel-Gunduz, *Assessment of nano-toxicity and safety profiles of silver nanoparticles.* IntechOpen, 2018.
[http://dx.doi.org/10.5772/intechopen.75645]

[58] K.T. Yong, W.C. Law, R. Hu, L. Ye, L. Liu, M.T. Swihart, and P.N. Prasad, "Nanotoxicity assessment

of quantum dots: from cellular to primate studies", *Chem. Soc. Rev.,* vol. 42, no. 3, pp. 1236-1250, 2013.
[http://dx.doi.org/10.1039/C2CS35392J] [PMID: 23175134]

[59]   M.B. Caixeta, P.S. Araújo, B.B. Gonçalves, L.D. Silva, M.I. Grano-Maldonado, and T.L. Rocha, "Toxicity of engineered nanomaterials to aquatic and land snails: A scientometric and systematic review", *Chemosphere,* vol. 260, p. 127654, 2020.
[http://dx.doi.org/10.1016/j.chemosphere.2020.127654] [PMID: 32758772]

[60]   N. Chaukura, T.C. Madzokere, N. Mugocheki, and T.M. Masilompane, *The Impact of Nanomaterials in Aquatic Systems.,* 2020.
[http://dx.doi.org/10.1002/9781119592990.ch10]

[61]   P. Saxena, V. Sangela, S. Ranjan, V. Dutta, N. Dasgupta, M. Phulwaria, D.S. Rathore, and Harish, "Aquatic nanotoxicology: impact of carbon nanomaterials on algal flora", *Energy Ecol. Environ.,* vol. 5, no. 4, pp. 240-252, 2020.
[http://dx.doi.org/10.1007/s40974-020-00151-9]

[62]   N. Malhotra, O.B. Villaflores, G. Audira, P. Siregar, J.S. Lee, T.R. Ger, and C.D. Hsiao, *"Toxicity Studies on Graphene-Based Nanomaterials in Aquatic Organisms: Current Understanding Molecules",* vol. 25, no. 16, p. 3618, 2020.
[http://dx.doi.org/10.3390/molecules25163618] [PMID: 32784859]

[63]   M. Ke, Y. Li, Q. Qu, Y. Ye, W.J.G.M. Peijnenburg, Z. Zhang, N. Xu, T. Lu, L. Sun, and H. Qian, "Offspring toxicity of silver nanoparticles to Arabidopsis thaliana flowering and floral development", *J. Hazard. Mater.,* vol. 386, p. 121975, 2020.
[http://dx.doi.org/10.1016/j.jhazmat.2019.121975] [PMID: 31884364]

[64]   C. Kole, D.S. Kumar, and M.V. Khodakovskaya, *Plant nanotechnology: Principles and practices.* Springer, 2016.
[http://dx.doi.org/10.1007/978-3-319-42154-4]

[65]   A. Yan, and Z. Chen, "Impacts of silver nanoparticles on plants: a focus on the phytotoxicity and underlying mechanism", *Int. J. Mol. Sci.,* vol. 20, no. 5, p. 1003, 2019.
[http://dx.doi.org/10.3390/ijms20051003] [PMID: 30813508]

[66]   L.R. Pokhrel, and B. Dubey, "Evaluation of developmental responses of two crop plants exposed to silver and zinc oxide nanoparticles", *Sci. Total Environ.,* vol. 452-453, pp. 321-332, 2013.
[http://dx.doi.org/10.1016/j.scitotenv.2013.02.059] [PMID: 23532040]

[67]   K.A. Mosa, M. El-Naggar, K. Ramamoorthy, H. Alawadhi, A. Elnaggar, S. Wartanian, E. Ibrahim, and H. Hani, "Copper Nanoparticles Induced Genotoxicty, Oxidative Stress, and Changes in Superoxide Dismutase (SOD) Gene Expression in Cucumber (Cucumis sativus) Plants", *Front. Plant Sci.,* vol. 9, no. 872, p. 872, 2018.
[http://dx.doi.org/10.3389/fpls.2018.00872] [PMID: 30061904]

[68]   P. Ahmad, M.N. Alyemeni, A.A. Al-Huqail, M.A. Alqahtani, L. Wijaya, M. Ashraf, C. Kaya, and A. Bajguz, "Zinc Oxide Nanoparticles Application Alleviates Arsenic (As) Toxicity in Soybean Plants by Restricting the Uptake of as and Modulating Key Biochemical Attributes, Antioxidant Enzymes, Ascorbate-Glutathione Cycle and Glyoxalase System", *Plants,* vol. 9, no. 7, p. 825, 2020.
[http://dx.doi.org/10.3390/plants9070825] [PMID: 32630094]

[69]   J. Jeevanandam, A. Barhoum, Y.S. Chan, A. Dufresne, and M.K. Danquah, "Review on nanoparticles and nanostructured materials: history, sources, toxicity and regulations", *Beilstein J. Nanotechnol.,* vol. 9, pp. 1050-1074, 2018.
[http://dx.doi.org/10.3762/bjnano.9.98] [PMID: 29719757]

[70]   R. Mohammadpour, D.L. Cheney, J.W. Grunberger, M. Yazdimamaghani, J. Jedrzkiewicz, K.J. Isaacson, M.A. Dobrovolskaia, and H. Ghandehari, "One-year chronic toxicity evaluation of single dose intravenously administered silica nanoparticles in mice and their Ex vivo human hemocompatibility", *J. Control. Release,* vol. 324, pp. 471-481, 2020.

[http://dx.doi.org/10.1016/j.jconrel.2020.05.027] [PMID: 32464151]

[71]   D. Li, J. Ji, Y. Yuan, and D. Wang, "Toxicity comparison of nanopolystyrene with three metal oxide nanoparticles in nematode Caenorhabditis elegans", *Chemosphere,* vol. 245, p. 125625, 2020. [http://dx.doi.org/10.1016/j.chemosphere.2019.125625] [PMID: 31855754]

[72]   Y. Liu, Y. Zhao, B. Sun, and C. Chen, "Understanding the toxicity of carbon nanotubes", *Acc. Chem. Res.,* vol. 46, no. 3, pp. 702-713, 2013. [http://dx.doi.org/10.1021/ar300028m] [PMID: 22999420]

[73]   K. Pikula, V. Chaika, A. Zakharenko, Z. Markina, A. Vedyagin, V. Kuznetsov, A. Gusev, S. Park, and K. Golokhvast, "Comparison of the level and mechanisms of toxicity of carbon nanotubes, carbon nanofibers, and silicon nanotubes in bioassay with four marine microalgae", *Nanomaterials (Basel),* vol. 10, no. 3, p. 485, 2020. [http://dx.doi.org/10.3390/nano10030485] [PMID: 32182662]

[74]   C. Liao, Y. Li, and S. Tjong, "Bactericidal and Cytotoxic Properties of Silver Nanoparticles", *Int. J. Mol. Sci.,* vol. 20, no. 2, p. 449, 2019. [http://dx.doi.org/10.3390/ijms20020449] [PMID: 30669621]

[75]   S.C. Jagdale, R.U. Hude, and A.R. Chabukswar, *Zebrafish: A Laboratory Model to Evaluate Nanoparticle Toxicity, in Model Organisms to Study Biological Activities and Toxicity of Nanoparticles.,* 2020.

[76]   M.P. Sutunkova, S.N. Solovyeva, I.N. Chernyshov, S.V. Klinova, V.B. Gurvich, V.Y. Shur, E.V. Shishkina, I.V. Zubarev, L.I. Privalova, and B.A. Katsnelson, "Manifestation of systemic toxicity in rats after a short-time inhalation of lead oxide nanoparticles", *Int. J. Mol. Sci.,* vol. 21, no. 3, p. 690, 2020. [http://dx.doi.org/10.3390/ijms21030690] [PMID: 31973040]

[77]   N. Hadrup, A.K. Sharma, K. Loeschner, and N.R. Jacobsen, "Pulmonary toxicity of silver vapours, nanoparticles and fine dusts: A review", *Regul. Toxicol. Pharmacol.,* vol. 115, p. 104690, 2020. [http://dx.doi.org/10.1016/j.yrtph.2020.104690] [PMID: 32474071]

[78]   L. Zhou, H. Xie, X. Chen, J. Wan, S. Xu, Y. Han, D. Chen, Y. Qiao, L. Zhou, S. Zheng, and H. Wang, "Dimerization-induced self-assembly of a redox-responsive prodrug into nanoparticles for improved therapeutic index", *Acta Biomater.,* vol. 113, pp. 464-477, 2020. [http://dx.doi.org/10.1016/j.actbio.2020.07.007] [PMID: 32652227]

[79]   D. Bobori, A. Dimitriadi, S. Karasiali, P. Tsoumaki-Tsouroufli, M. Mastora, G. Kastrinaki, K. Feidantsis, A. Printzi, G. Koumoundouros, and M. Kaloyianni, "Common mechanisms activated in the tissues of aquatic and terrestrial animal models after TiO2 nanoparticles exposure", *Environ. Int.,* vol. 138, p. 105611, 2020. [http://dx.doi.org/10.1016/j.envint.2020.105611] [PMID: 32126387]

[80]   J.T. Buchman, N.V. Hudson-Smith, K.M. Landy, and C.L. Haynes, "Understanding nanoparticle toxicity mechanisms to inform redesign strategies to reduce environmental impact", *Acc. Chem. Res.,* vol. 52, no. 6, pp. 1632-1642, 2019. [http://dx.doi.org/10.1021/acs.accounts.9b00053] [PMID: 31181913]

[81]   C.Y. Usenko, S.L. Harper, and R.L. Tanguay, "In vivo evaluation of carbon fullerene toxicity using embryonic zebrafish", *Carbon,* vol. 45, no. 9, pp. 1891-1898, 2007. [http://dx.doi.org/10.1016/j.carbon.2007.04.021] [PMID: 18670586]

[82]   Y. Zhang, Y. Hai, Y. Miao, X. Qi, W. Xue, Y. Luo, H. Fan, and T. Yue, "The toxicity mechanism of different sized iron nanoparticles on human breast cancer (MCF7) cells", *Food Chem.,* vol. 341, no. Pt 2, p. 128263, 2021. [http://dx.doi.org/10.1016/j.foodchem.2020.128263] [PMID: 33038805]

[83]   A. Rana, K. Yadav, and S. Jagadevan, "A comprehensive review on green synthesis of nature-inspired metal nanoparticles: Mechanism, application and toxicity", *J. Clean. Prod.,* vol. 272, p. 122880, 2020. [http://dx.doi.org/10.1016/j.jclepro.2020.122880]

[84]   X. Chen, S. Zhu, X. Hu, D. Sun, J. Yang, C. Yang, W. Wu, Y. Li, X. Gu, M. Li, B. Liu, L. Ge, Z. Gu, and H. Xu, "Toxicity and mechanism of mesoporous silica nanoparticles in eyes", *Nanoscale,* vol. 12, no. 25, pp. 13637-13653, 2020.
[http://dx.doi.org/10.1039/D0NR03208E] [PMID: 32567638]

[85]   R.A. Maheshwari, D.B. Sen, A.S. Zanwar, and A.K. Sen, "Evaluation of nanotoxicity using zebrafish: preclinical model", *Nanocarriers: Drug Delivery System.,* pp. 173-197, 2021.
[http://dx.doi.org/10.1007/978-981-33-4497-6_7]

[86]   H.N. Abdelhamid, "General methods for detection and evaluation of nanotoxicity", In: *Nanotoxicity.* Elsevier, 2020, pp. 195-214.
[http://dx.doi.org/10.1016/B978-0-12-819943-5.00009-9]

[87]   K. Ettrup, A. Kounina, S.F. Hansen, J.A.J. Meesters, E.B. Vea, and A. Laurent, "Development of comparative toxicity potentials of TiO2 nanoparticles for use in life cycle assessment", *Environ. Sci. Technol.,* vol. 51, no. 7, pp. 4027-4037, 2017.
[http://dx.doi.org/10.1021/acs.est.6b05049] [PMID: 28267926]

[88]   E. Gwag, L. Abelmann, A. Manz, Y.J. Kim, and B. Sung, *For In Vitro Nanotoxicity Evaluation.*

[89]   X. Liu, K. Shan, X. Shao, X. Shi, Y. He, Z. Liu, J.A. Jacob, and L. Deng, "Nanotoxic effects of silver nanoparticles on normal HEK-293 cells in comparison to cancerous HeLa cell line", *Int. J. Nanomedicine,* vol. 16, pp. 753-761, 2021.
[http://dx.doi.org/10.2147/IJN.S289008] [PMID: 33568905]

[90]   F. Bettazzi, and I. Palchetti, "Nanotoxicity assessment: A challenging application for cutting edge electroanalytical tools", *Anal. Chim. Acta,* vol. 1072, pp. 61-74, 2019.
[http://dx.doi.org/10.1016/j.aca.2019.04.035] [PMID: 31146866]

[91]   Y. Qian, X. Meng, H. Liu, X. Wang, and H. Wang, "Bifunctional Nanoparticles as a Recyclable Fluorescent Sensor for pH and Cu2+ Detection and Removal of Heavy Metal Ions", *Nano,* vol. 15, no. 4, p. 2050048, 2020.
[http://dx.doi.org/10.1142/S1793292020500484]

CHAPTER 5

Emerging Water Pollutants from Industrial Processes

N.W.C. Jusoh[1,*]**, N.F. Jaafar**[2]**, A. Masudi**[3] **and P.Y. Liew**[1]

[1] Department of Chemical and Environmental Engineering (ChEE), Malaysia-Japan International Institute of Technology (MJIIT), Universiti Teknologi Malaysia (UTM) Kuala Lumpur, Jalan Sultan Yahya Petra, 54100 Kuala Lumpur, Malaysia

[2] School of Chemical Sciences, Universiti Sains Malaysia, 11800 USM Penang, Malaysia

[3] Clean Energy Research Centre, Korea Institute of Science and Technology, P.O. Box 131, Cheongryang, Seoul 136-791, Republic of Korea

Abstract: The widespread problem of water pollution endangers human health. Every year unsafe water causes more deaths than war and integrates all other forms of violence. Industrial activities create a large source of emerging water pollutants that are highly harmful to humans and the environment. This leads to increasing concern for the possible ecological impact of these pollutants on the environment. This chapter identifies various emerging water pollutants produced from different industrial processes (*e.g.*, petrochemical production, textile, paper and pulp, semiconductor/ electroplating, and metal production). The possible route of pollution formation is discussed in this chapter. In addition, the impact of the produced water pollutants on the environment and health has been elucidated. Furthermore, the concerns of emerging contaminants remain a moving subject as the new pollutants continuously are being produced in industrial processes. In response, an insight into the challenges of minimizing water pollution is also focused on mutual benefit.

Keywords: Denim industry, Emerging water pollutants, Heavy industries, Industrial engineering, Industrial process, Manufacturing processes, Paper and pulp industry, Petrochemical industry, Semiconductor/electroplating.

INTRODUCTION

Industrial processes utilize water and release numerous compounds in wastewater, which may accumulate in drinking water bodies that disrupt everyday life. The discharge of industrially polluted water into clean water is a significant issue [1]. These practices have produced various contaminations that disrupted the water

---

* **Corresponding author N.W.C.Jusoh:** Department of Chemical and Environmental Engineering (ChEE), Malaysia-Japan International Institute of Technology (MJIIT), Universiti Teknologi Malaysia (UTM) Kuala Lumpur, Jalan Sultan Yahya Petra, 54100 Kuala Lumpur, Malaysia; E-mail:nurfatehah@utm.my

Shaukat Ali Mazari, Nabisab Mujawar Mubarak & Nizamuddin Sabzoi (Eds.)

cycle and caused great concern due to their possible effect on the environment and human health [2, 3]. Emerging water pollutants (EWPs) have received much attention from researchers and public administrations. EWPs are defined as 'any synthetic or naturally occurring chemical or microorganism that is not ordinarily monitored in the environment, however, has a great potential to influent the environment and cause known or suspected adverse ecological and human health effects [4]. EWPs come from organic species such as perfluorinated compounds, pesticides, pharmaceuticals, hormones, micro-pollutants, endocrine disruptors, disinfection by products, sunscreens, flame retardants, industrially related synthetic dyes, and dyes-containing hazardous pollutants, *etc* [5, 6]. Although these pollutants are present in low concentrations of wastewater, they still become a threat to water bodies and the ecosystem if not controlled.

In many developing countries, producing products or goods from the industrial process has become a major activity. These activities result in the discharge of industrial effluents that may contain pollutants at levels that could affect the quality of receiving waters and the aquatic ecosystem [7]. There are regulatory requirements and guidelines in all countries on the discharge of industrial effluents into water bodies. Due to that, most factories must meet the minimum standard requirements. However, not all the chemicals or contaminants are listed in the guidelines, especially the EWPs. The chemical industry may release many substances not considered by regulatory requirements and might be the EWPs. These compounds may be the final products, precursors, intermediates of the process, or impurities and by-products. The main aim of this chapter is to explain in detail the EWPs discharged from industrial processes, specifically in petrochemical production, textile, paper pulp, semiconductor/plating, and metal production industry.

PETROCHEMICAL INDUSTRY

Petrochemical products from the oil and gas industry could be converted into various goods, including polymers, pharmaceuticals, dyes, solvents, and other chemicals [8]. There are about 2500 types of products produced from crude oil. The crude petroleum is extracted from the ground. The crude oil goes through refining processes, separating the fuels and chemicals such as light hydrocarbon, naphtha, kerosene, diesel, gas oil, *etc*. According to the difference between boiling points, these products and intermediates undergo other processes to produce fuel gas, propane, gasoline, jet fuel, petrochemicals, kerosene, diesel, gasoline, and fuel oil precursors. These processes require industries like refineries and petrochemical complexes, and major methods may involve distillation, catalysis reaction, polymerization, purification, *etc*. [9]. The oil and gas industry contributes to a huge portion of the country's economy, such as 50% (Saudi

Arabia), 30% (UAE), 14.5% (Malaysia), 1.2% (United Kingdom), and 8% (United States of America) of their country's gross domestic product (GDP) value.

The wastewater generation in petroleum refineries and petrochemicals plants (PRPP) is higher than the crude oil production by 40-160% [10]. Some physical parameters measure water quality like other wastewater, including pH, hardness, turbidity, heat, taste, and odor. To carry out the bulk characterization of the water quality, various tests are performed for the determination of oil and grease (O&G), total hydrocarbon (TBC), biochemical oxygen demand (BOD), soluble BOD, chemical oxygen demand (COD), and soluble COD [11]. Metals are frequently found in the PRPP wastewater, such as Cd, Ni, Hg, Pb, and vanadium [12]. Inorganic materials, such as fluorides, phosphates, sulphides, and chlorides, are also found in the PRPP wastewater [12]. The other common contaminants are summarised in Table **1** [10].

Table 1. Common wastewater contaminants from petrochemical processes [10].

Petrochemical Processes	Wastewater Contaminants
Drilling, exploration, transportation	Polymers, ionic emulsifiers, non-ionic emulsifiers
Crude desalting	Free oil. Ammonia, sulfides and suspended solids
Crude oil distillation	Sulfides, ammonia, phenols, oil, chlorides
Thermal cracking	Phenols, hydrogen sulfate, ammonia
Catalytic cracking	Sulfides, ammonia, phenols, oil, cyanide
Hydrocracking	Sulfides
Polymerization	Sulfides, mercaptans, ammonia
Alkylation	Sulfides spent caustic oil
Isomerazation	Phenols
Reforming	Sulfides
Hydro-treating	Sulfides, ammonia, phenols

Besides the contaminants mentioned above, the composition of PRPP wastewater also consists of an aromatic organic compound, which includes polyaromatic hydrocarbons (PAHs), phenolic substances, and aliphatic hydrocarbon compounds. For example, phenols are highly soluble in water, which could be detected in refinery wastewater at a few milligrams per liter to 7000 mg/L [13]. These compounds in the wastewater streams are hardly degradable by nature, thus affecting marine life and the ecosystem.

The United States Environmental Protection Agency (USEPA) has identified about 16 polycyclic aromatic hydrocarbons (PAHs) as potentially hazardous to human health and the environment. The benzo[a]pyrene has the highest potential as carcinogenic material among these PAHs. The PAHs are recalcitrant, bioaccumulative, and semi-volatile lipophilic organic pollutants, which are widespread in the environment and enter into environmental matrices *via* natural and anthropogenic sources [14]. These components are well known for their carcinogenic, immunotoxic, mutagenic, and teratogenic properties [14, 15]. These pollutants are biodegradable with exposure to the microbial community [16]. Due to the nature of hydrophobics, the pollutants need to be dissolved in water for biodegradation. PAHs' remediation techniques must be developed due to their effects on human and environmental health.

TEXTILE INDUSTRY

Textile is one of the fastest-growing industries in modern societies. This industry contributes to more job creation and faster urbanization and fosters market wealth. The increasing gross industrial output in China reached up to 80% from 1990 to 2015. China also accounted for around one-third of apparel and textile export worldwide [17]. In Brazil, the textile industry created approximately 1.5 million jobs with annual yearly revenue of US $ 52 billion. Brazil is also one of the export leaders for denim, knitwear, and jeanswear [18]. Meanwhile, there were 2324 textile industries in India, from traditional to modern machinery, which consumed 56% of total dye production in the country. This industry requires 35 million employees with 1 trillion dollars in revenue worldwide [19].

Depending on the product type, the textile industry releases some pollutants into the environment. In general, the stages flow of the wet textile industry is presented in Fig. (**1**). Dyeing is one of the crucial stages in textile manufacture for the indulgence of dyes, which later get into wastewater streams. In this step, some dyes or colorants are added to the fibers with the simultaneous addition of several chemicals to increase dye adsorption to the cotton. In the finishing step, some remaining dyes and chemicals are discharged as textile effluents that harm the environment [20, 21].

The textile industry also discharges traces of hazardous metals into environments (air, water, and soil). This industry also emits around 2000-3000 m^3/j of synthetic dyes, sometimes persistent in water [22]. Additionally, the synthetic stain fibers produce polyester, polyamide, or acetate, non-biodegradable. Finally, these fibers end up as solid waste [23]. The problem associated with textile industry wastewater is alkaline due to excessive use of chemicals during bleaching. The existence of dyes on water surfaces reduces the amount of light penetration and limits the growth of organisms. At the same time, metals originated from metal complexes to dyes and additives during the process [24].

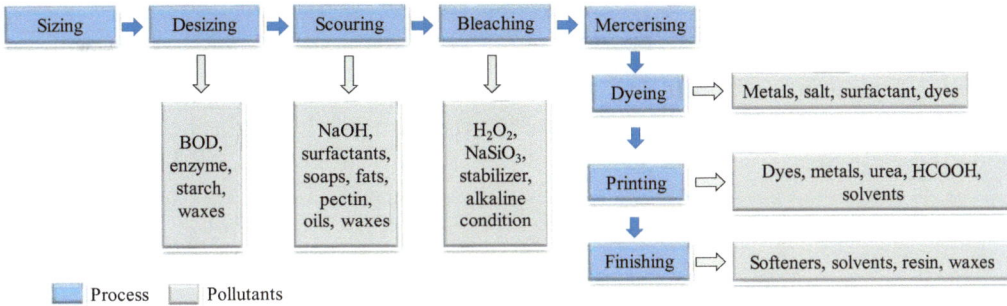

Fig. (1). Flow of wet textile industry.

Heavy metals such as Cr, Zn, Fe, Hg, and Pb are problematic to remove from the environment, while metal complexes in dyes involve Co, Cu, and Cr. These pollutants lead to several diseases through direct contact, inhalation, or ingestion. For example, inhalation or contact with an azo dye causes acute respiratory pathologies, allergic skin, or lung cancer. The survey of ninety workers of textile companies proved that daily exposure to hazardous chemicals induces several unhealthy symptoms [22]. Exposure to textile effluents induces mutation in small organisms and disturbs the hematopoietic and reproductive systems in rats and mice [25]. The concentration of dye discharge varies depending on several factors, such as type of industry, use of chemicals, region, cost of wastewater treatment, implementation of environmental standards and regulations, *etc*. The real textile effluent increased rapidly from 10-50 mg/L in 1991 [26] to 100-200 mg/L in 1994 [27], then became 600-800 mg/L in 1998 [28]. Hence, the dyes and chemical uses in the industries should receive serious concerns from responsible authorities.

In addition to dyes, textile industries also employed several chemical auxiliaries during textile manufacture. Alkylphenol ethoxylates (APEOs) are utilized in bleaching and scouring, which may be disruptive hormones and are non-biodegradable [29]. Hence, many countries in Europe limited the utilization of this chemical and searched for their potential replacement. Sodium sulphate/chloride is widely used as an exhausting agent, which is still challenging to remove in textile effluents. The only way to reduce this pollutant is achieved by diluting the water discharge. However, this is not a removal technique but rather a concentration lowering tactic. The alternative way is by impregnating with lower salt concentration to minimize the pollutant with continuous effort to develop a copper-free paint as a greener industry [11].

Formaldehyde is used as one of the important chemicals as a stabilizer in dye formulation or dispersant. However, direct contact with formaldehyde leads to respiratory problems or skin irritation. Candidly, the utilization of aldehydes is not preferable as dye reducers, and more suggested to employ sodium hydrosulphide. Meanwhile, urea was advised to substitute EDTA as a dye auxiliary with careful pH and temperature control to minimize the number of dye auxiliaries [30]. Another problem is that textile companies often use perfluoroalkyl chains as water and oil repellent.

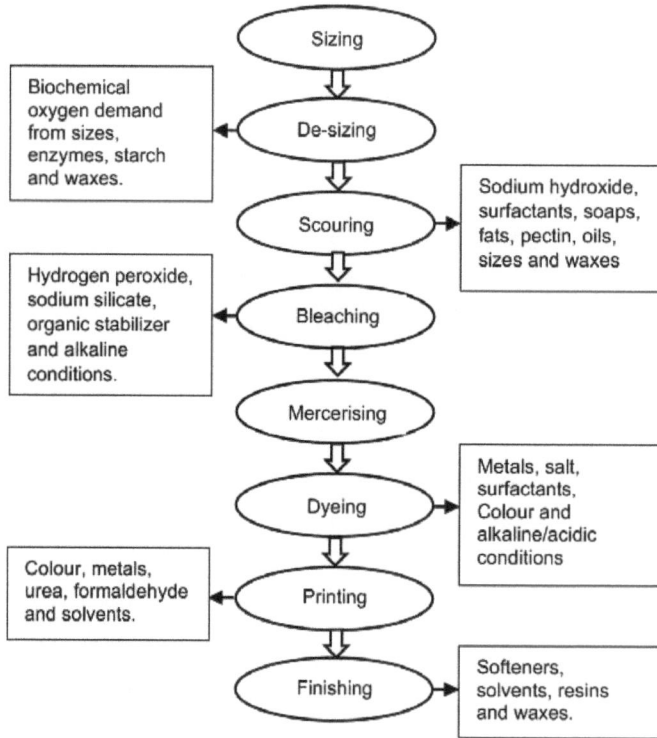

Fig. (2). Various point sources of textile water contamination [20].

PAPER AND PULP INDUSTRY

The paper and pulp industry consumes extensive water and energy. This industry is ranked 5th in energy consumption, while water depends on the industry capacity. The sector could utilize 60 m^3 of freshwater per ton of produced paper [31]. In 2020, the worldwide industry will have 500 million tons of paper and release 100 million kg of hazardous pollutants into the environment [32]. The other problems were high organic content (estimated from COD) and high organic halide with disposal of solid residue consisting of heavy metal trace, pathogens, and chlorinated compounds [33].

The paper industry could be classified as (1) pulping and (2) bleaching. In the first process, lignin is removed from wood chips, increasing the fibers' strength. A combination of mechanical and chemical treatments is employed with medium efficiency and better paper quality than standalone treatment [34]. Initially, the wood feedstock proceeds to high temperature and pressure in alkaline (NaOH and Na_2S) or acidic (H_2SO_3 and sulphide ions) conditions. However, 5-10 wt.% of lignin could not be removed in this stage and causing a dark colour. Then, the paper is whitened in the second stage using some chemicals. In short, the bleaching process is used in a sequential process from elemental chlorine treatment, alkali, chlorine dioxide, alkali, to chlorine dioxide. Along with chemical discharge to water surfaces, this industry also emits several hazardous gases such as nitrogen oxides (NO_x), sulphur oxides (SO_x), and volatile organic carbons (VOCs) [35].

The water discharge of pulp and paper industries contains wood chips, remaining cellulose, extractive chemicals, and the trace of an organic compound. Some organic halides may also accumulate in effluents because of the bleaching reaction between chlorine compounds and chemicals. In addition, the water discharge also contains high content of BOD and COD due to the excessive content of organic derivatives [36]. The paper industry's darkish river water could lead to anaerobic decomposition and produce hydrogen sulphide (H_2S), releasing it into the environment with limited photosynthetic activity. This would affect secondary and tertiary consumers. In addition, the evolution of organism reproduction may also occur and smaller gonads. The dissolved solids are also crucial for this industry since they contain sulphate and chlorides of Mg, Ca, and Na. These substances accumulate in soil and cause improper nutrients that seriously affect crop yield and quality and transfer to human beings as toxic compounds. Lastly, the amount of oxygen depletes the soil and decelerates plant growth [37].

The amount of generated waste could be minimized by recovering chemicals and recycling [38]. Chemical recovery should be conducted, especially in pulping stage, to increase economic return. Although chemical recovery has been applied in several industries, there are limited mechanisms or systems to recover the sulphate mills. In addition, it is also crucial to develop a baghouse scrubber or electrostatic precipitator to reduce air pollution. Secondly, recycling must be synchronized with the recent technologies. For example, the conventional pulping process could be substituted by either organic, acid, or bio-pulping [39]. Organic pulping is conducted using methanol or ethanol but requires extensive energy. It is feasible for chemical recovery in acid pulping, but the acidic discharge may problematically lead to the environment.

Meanwhile, bio-pulping is promising with less chemical and energy, lower pollutants, and better pulp quality. Bio-pulping depends on an enzyme that can imitate lignin degradation in nature. Then, the bleaching stage is employed to total chlorine-free approach using ozone, hydrogen peroxide, and caustic soda to reduce chlorinated pollutants. Meanwhile, the number of toxic pollutants could be reduced by careful planning to improve the process efficiency by exploring potential feedstocks [40].

The solid waste of the pulp industry also received increasing concern with escalating demand for paper in modern industries. Several methods such as anaerobic digestion, land application, and thermal process are employed to minimize the disposal of solid residue. The method selection may vary depending on the industry profile's characteristics. Another option is to convert the mill of wastewater to ethanol tested in the Pacific Northwest [41]. Finally, the solid waste from old, corrugated containers (OCC) could also be utilized in the paper board manufacturing industry.

SEMICONDUCTOR AND ELECTROPLATING INDUSTRY

The semiconductor industry plays an important role because semiconductors have been widely used in various industries as essential components where it becomes a primary component to enable the technology. The current interest of society requires up-to-date electronic devices to contribute to the rapid expansion of microelectronic features, which are used in various electronic industries. Other than electronic devices, semiconductors are also widely used to manufacture photovoltaic (PV) cells. The semiconductor industry's growth positively impacts the economy; however, this trend is causing concern about its effect on the environment. This industry can pollute the environment *via* its manufacturing plants and different process points, such as collecting electronic waste and depletion of raw materials [42].

A few indicators are currently accessible to monitor products' environmental risk and impacts, which could help characterize ecological impacts. Life Cycle Assessment (LCA) is a powerful method to show a global vision of the effects created by by-products [43]. LCA has become popular in the industry as one of the important tools due to its capability to identify product improvement, decision making, evaluation of a product, environmental performance, market claims, *etc* [44]. This method produces a profile containing the information about the baseline in managing the ecological state of a product besides proposing approaches to improve it. After considering all aspects, including biodiversity, eco-systems quality, climate change, *etc.*, the profile produces without lessening or escalating any effect. The good point about this method is that it considers all aspects of preventing impact transfer by improving one type of impact, which probably can weaken another one previously neglected or not considered important for the industry. Still, bear in mind each industry has their own concerns, so it is important to identify the most significant and priority impact during production of profile through this process.

There are various semiconductor associations globally, such as European Union (EU), Japan, and others, which are committed to lowering the waste produced by the industry. As an example, the EU initiated the restriction of the use of certain hazardous substances (RoHS) [45], waste electrical and electronic equipment (WEEE) [46], and eco-design of energy-using products (EuP) [47] to create guidelines for monitoring the eco-friendly use of semiconductors around the world. The use of various chemicals in the production of semiconductors cannot be irrefutable. Thus, it is not surprising to note that wastewater produced from this industry has now been considered an emerging pollutant; that must In the production of semiconductors, they use a lot of per-fluorinated chemicals (PFCs), which can cause the emissions of carbon tetrafluoride (CF_4), hexafluoroethane (C_2F_6), octafluoropropane (C_3F_8), nitrogen trifluoride (NF_3), fluoroform (CHF_3), and sulphur hexafluoride (SF_6). Additionally, the production processes include many acid solutions, organic solvents, and toxic gases where, consequently, most plants produce organic/inorganic wastewater and sludge with complicated pollution problems [48].

The hydrophobic and lipophobic properties of PFCs make them an important component in the semiconductor industry, leading to ubiquitous production and usage for more than half a century. The excessive use of PFCs has indirectly been causing PFCs to be released into the environment in large quantities, contributing to the high pollution of PFCs. The PFCs are very stable under environmental conditions, especially per-fluorooctanoic acid (PFOA) and per-flurooctane sulfonate (PFOS), have hydrolytic half-lives of more than 41 years, and they also can resist direct and indirect photolysis [49]. Due to their resistance to weakening in natural environments and being expected to appear in various environmental matrices, the PFCs can be found highly in waterbodies near this industry. Due to that situation, PFOS can accumulate in aquatic systems as a medium since water and sediment can be considered the final sinks of PFOS.

The PFOA and PFOS cause the greatest concern from an environmental and health perspective because of their demonstrated abundance, where PFOS can be derived from perfluorooctane sulfonamides (N-ethyl perfluorooctane sulfonaminoethanol and N-ethyl perfluorooctane sulfonamide) [50] and PFOA from fluorotelomer alcohols [51]. PFOS and PFOA can be derived from degradative processes such as biodegradation, biotransformation, and atmospheric oxidation. PFOS is more abundant in biota because of its higher-bioaccumulation potential [52].

The properties of PFOS not only have chemical stability with water and oil repellence, but they also have high surface activity, which is commonly used as surfactants in semiconductors and the metal plating industry. The emissions of PFOS for metal plating and semiconductor has been classified as direct and indirect sources, respectively [53]. According to Paul *et al.* [54], direct sources are released during the manufacture and application of PFOS, while indirect sources are emissions of PFOS impurities formed during the manufacture of per-flurooctane sulfonyl fluoride (POSF) derivatives or by a breakdown in the environment form PFOS precursors.

As the production of semiconductors involves enormous processes, which potentially generate toxic waste, it is very common for the industry to have proper waste management to fulfill the requirements of environmental policies and standards. In addition, nowadays, most industries have become aware of their responsibility toward the environment and increased public awareness and interest regarding industrial waste generation. Generally, the production of semiconductors involves ten steps, and most of the steps would generate pollutants for the water [55]. This wastewater includes various metals (*e.g.*, aluminum, copper, chromium, nickel, arsenic, silver), fluorides (*e.g.*, hydrofluoric acid, boron trifluoride), and spent solvents (*e.g.*, methanol, acetone, isopropyl alcohol). Thus, these major pollutants need a proper waste treatment to comply with the regulations for the disposal or recycling of wastewater. The selection of method for wastewater treatment involves various processes depending upon the type of pollutant present in the wastewater. The removal of fluorides can be accomplished using total suspended solid (TTS), which can be done through coagulation, flocculation, and clarification. The treatment process for aqueous waste usually involves pH treatment to cleaning and rinsing operations because the pH values may affect the wastewater treatment at publicly treatment works (POTWs). At the same time, spent solvents are commonly collected to store on-site and removed after every 90 days for purification and reuse. The solvent sludge can be used as a fuel in cement kilns.

It is reported that the wastewater of the electroplating industry contains a toxic and hazardous composition of up to 29%. The wastewater of the electroplating industry includes heavy metals such as arsenic, cobalt, copper, chromium, mercury, iron, nickel, zinc, lead, and cadmium. It also contains acids, alkalis, cyanide-complex, and significant amounts of BOD/COD, DS, SS, TS, and turbidity. The toxicity of electroplating industry wastewater can cause kidney failure, thyroid dysfunction, sleeplessness, tiredness, rheumatoid arthritis, affects the circulatory system and neural system, causes gastrointestinal mucosa irritation, and leads to lung cancer [56, 57].

METAL PRODUCTION INDUSTRY

The metal production industry plays an important role in processing metals such as nickel (Ni), aluminum (Al), copper (Cu), titanium (Ti), and zinc (Zn). Metal production includes mining, pyrometallurgical and hydrometallurgical consume large amounts of energy and water during the processes, which tend to produce large volumes of waste. Besides water and energy usage in metal production, these processes also involve chemicals such as acids and solvents, potentially polluting the environment.

For example, lead (Pb) and Zn are commonly mined together as both come from the same source, like sulphide ore (5.5% Pb, 8.6& Zn). Separate processes are often used for these mixed ores to produce refined Pb and Zn using the Pb blast furnace process and electrolytic Zn process. However, Pb and Zn are produced by the imperial smelting process. Commonly, all three processes can be combined as the main processing route for Pb and Zn production [58]. Besides, through the production of Zn refining, this process also produces cadmium (Cd) during the purification process through the Cd plant.

Compared with the production of Ni and Cu, both use pyrometallurgical and hydrometallurgical processes. Cu mainly uses major processes like hydrometallurgical, which includes three processes of leaching heap (ore), pressure (conc.), and atmospheric (cont.) [59]. While in the production of Ni, both major processes play an important role in the combination of flash smelting (pyrometallurgical) and acid pressure leaching (hydrometallurgical) [60]. The difference between these two is due to the different feed of Ni, which requires laterite ore.

The importance of steel and stainless steel in producing various goods is undeniable. Generally, the production of steel can be divided into two main production processes, which are integrated (from iron ore) and electric (from scrap) steelmaking [58]. There are different feeds to produce steel and stainless steel, where earlier requires only iron ore (64% Fe), while the former needs a mixture of pig iron (94% Fe), chromite ore (27% Cr, 17.4% Fe)) and laterite ore (2.4% Ni, 13.4% Fe).

Even though the production of each metal happens using different feedstocks, however, all ores are obtained through the mining process. Mining processes involve the extraction of metals and minerals from the earth, which exert pressure on the environment at many stages. During the mining activities, various environmental problems arise, such as air pollution, noise, land degradation, soil contamination, surface and groundwater pollution, and degradation of forests and biodiversity. The water resources are most affected during mining activities through erosion, sedimentation, acid mine drainage, spill/tailing, and destruction of the hydrological cycle and rainfall. Among them, acid mine drainage has a devastating long-term impact and serious threat to water streams and aquatic life and can be associated with the mobilization of contaminants [61]. It primarily depends on the type of rock material and accessibility of water and oxygen. It can occur at abandoned and active mine sites, which can dissolve various toxic metals like Cu, Cd, As, Hg and Pb. While for metal like Fe, it has the potential to coat the bottom stream with an orange-red colour.

Apart from being polluted mainly due to acid mine drainage during the mining, other side effects can indirectly pollute the water system: spills/tailing and erosion. There is rising concern about accidental spillage of chemicals from storage or processing facilities that can pollute the surface water due to leaching, potentially resulting in residuals after the collection of metals. Besides, mining processes also involve the combination of rinsing, physical isolation, detoxification of heap, leach, and washing coal, which requires a lot of industrial chemicals [62].

The impact on the mining site after completion of mining also contributes to undesired materials into streams, as illustrated in Fig. (**3**). The erosion usually happens to waste materials such as waste rock and tailing. The waste rock usually is disposed of near to mines location to reduce the cost; however, without proper management on site, it could contribute to metal bearing elements, which consequently erode into water streams [63]. The major sources of erosion at the mining site include dump leaches, open pit areas, piles and dams, waste rock and overburden piles, ore stockpiles, exploration areas, and reclamation areas [64].

Throughout these processes, directly (mining and process) and indirectly (consumption of raw materials) make the use of water around mining areas and intake various chemicals and unwanted solids. Thus, industries need to invest in treatment or adopt environment-friendly technologies to ensure the optimized use of water and lower its contamination. Moreover, processes need to be developed and employed to treat water coming from contaminated reservoirs, pipelines, or storage facilities. Currently, the common practice is to manage and conserve the resources scientifically, besides looking for a substitute for the mineral resources that are being widely used and proper recycling of used metals. LCA is also being applied in metal production because this method is reliable and manages to evaluate the potential environmental impacts during the show.

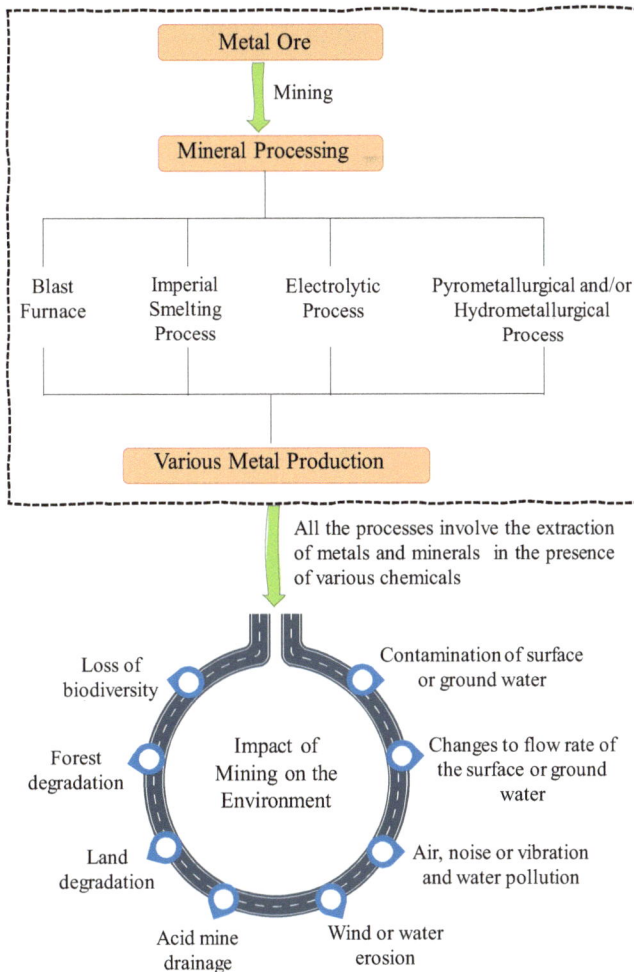

Fig. (3). Simple processing route for metal production and their impact on the environment.

CHALLENGES IN MINIMIZING WATER POLLUTANTS FROM INDUSTRIAL PROCESS

A successful approach to minimizing the emerging water pollutants from industrial processes needs to be highly focused. Many ways have been proposed to reduce the contaminants, especially in the liquid phase of the industrial process. It can be expressed as 'cleaner production, which covers raw materials to process and output with better efficiency and is safer for organisms, including humans [65]. There are three ways for cleaner production in the industry: minimizing waste and feedstock, recycling, and product development. In waste minimization, the industrial process may reduce or eliminate hazardous materials that enter the production process. Here, alternative materials may be used, which pose less hazardous ingredients. Good operating practices such as waste minimization programs, management and personnel practices, loss prevention, and waste segregation can help reduce waste at their source.

Additionally, the recycling technique can be applied to utilize the waste material as a raw material for another process. This can be done by discussing different industrial players who may benefit from each other. In some cases, certain chemicals' wastewater with high purity can be sold as a by-product [29]. The handy tools and methods like environmental impact assessment (EIA), environmental management system (EMS) and life cycle assessment (LCA) be used for further evaluation and assessment of emerging water pollutants released by the industrial practices. Furthermore, detailed guidelines on emerging industrial water pollutants may be determined, assessed, regulated, and monitored by the United Nations Environment Programme (UNEP), the United Nations Industrial Development Organization (UNIDO), The Organization For Economic Co-Operation And Development (OECD) and The World Business Council For Sustainable Development (WBCSD) [66].

Although several ways can be used to minimize the water pollutants, there still exist huge challenges in performing them due to technology, manpower, and finances. Some industries may care about profit without thinking about their discharged wastewater on health and the environment. However, others may take serious steps. Nonetheless, standard guidelines may streamline the release of emerging water pollutants. Lack of awareness of industrial players and the public on the emerging water pollutants has also become a major challenge due to the lack of important data in the public domain. Herein, toxicologists need to play an extremely important role in determining how these compounds affect our health and the environment. Besides, additional research must be conducted into how the production of EWPs can be prevented.

CONCLUSION

Mostly, industrial players have on-site wastewater treatment facilities for industrial effluents. However, the existing treatment facilities heavily focus on treating conventional regulated contaminants in the wastewater. The data and guidelines on the treatment of emerging water pollutants need more research, evaluation, and assessment. Industrial pollutants like polycyclic aromatic hydrocarbons (PAHs), perflurooctanoic acid (PFOA), perflurooctane sulfonate (PFOS), and several others need to be focused on. In the current understanding, the EWPs have a wide and flexible definition for different industries, which increases the difficulty of setting related control policies. These emerging pollutants have high impacts on human health and the environment. There is an urgent need to strengthen scientific knowledge and adopt appropriate technologies and policies to monitor and control EWPs in water resources. Then, the effect of these compounds can be prevented and controlled for a better future.

CONSENT FOR PUBLICATION

Not applicable.

CONFLICT OF INTEREST

The author declares no conflict of interest, financial or otherwise.

ACKNOWLEDGEMENTS

Declared none.

REFERENCES

[1]   K. Adesina Adegoke, O. Samuel Agboola, J. Ogunmodede, A. Oluyomi Araoye, and O. Solomon Bello, "Metal-organic frameworks as adsorbents for sequestering organic pollutants from wastewater", *Mater. Chem. Phys.,* vol. 253, p. 123246, 2020.
[http://dx.doi.org/10.1016/j.matchemphys.2020.123246]

[2]   D. Barceló, "Emerging pollutants in water analysis", *Trends Analyt. Chem.,* vol. 22, no. 10, pp. xiv-xvi, 2003.
[http://dx.doi.org/10.1016/S0165-9936(03)01106-3]

[3]   P.E. Rosenfeld, and L.G.H. Feng, "16 - Emerging Contaminants", In: *Risks of Hazardous Wastes,* P.E. Rosenfeld, L.G.H. Feng, Eds., William Andrew Publishing: Boston, 2011, pp. 215-222.

[4]   T. Smital, "Acute and chronic effects of emerging contaminants", In: *Handbook of Environmental Chemistry* vol. 5. Water Pollution, 2008, pp. 105-142.

[5]   T. Rasheed, M. Bilal, F. Nabeel, M. Adeel, and H.M.N. Iqbal, "Environmentally-related contaminants of high concern: Potential sources and analytical modalities for detection, quantification, and treatment", *Environ. Int.,* vol. 122, pp. 52-66, 2019.
[http://dx.doi.org/10.1016/j.envint.2018.11.038] [PMID: 30503315]

[6]   A.A. Khan, R. Ahmad, and I. Ahmad, "Density functional theory study of emerging pollutants removal from water by covalent triazine based framework", *J. Mol. Liq.,* vol. 309, p. 113008, 2020.
[http://dx.doi.org/10.1016/j.molliq.2020.113008]

[7]   R. Guerra, "Water Analysis | Industrial Effluents", In: *Encyclopedia of Analytical Science (Second Edition)* Oxford: Elsevier, 2005, pp. 289-299.P. Worsfold, C. Townshend, and C. Poole,
[http://dx.doi.org/10.1016/B0-12-369397-7/00111-4]

[8]   G.S. Cholakov, "Control of pollution in the petroleum industry", *Pollution Control Technologies,* vol. 3, pp. 1-10, 2010.

[9]   Y. Alqaheem, A. Alomair, M. Vinoba, and A. Perez, ""Polymeric gas-separation membranes for petroleum refining,"", *International Journal of Polymer Science,* vol. 2017, 2017.
[http://dx.doi.org/10.1155/2017/4250927]

[10]   M. Jain, A. Majumder, P.S. Ghosal, and A.K. Gupta, "A review on treatment of petroleum refinery and petrochemical plant wastewater: A special emphasis on constructed wetlands", *J. Environ. Manage.,* vol. 272, p. 111057, 2020.
[http://dx.doi.org/10.1016/j.jenvman.2020.111057] [PMID: 32854876]

[11]   M. Al Zarooni, and W. Elshorbagy, "Characterization and assessment of Al Ruwais refinery

wastewater", *J. Hazard. Mater.,* vol. 136, no. 3, pp. 398-405, 2006.
[http://dx.doi.org/10.1016/j.jhazmat.2005.09.060] [PMID: 16859828]

[12]   S. Jafarinejad, and S.C. Jiang, "Current technologies and future directions for treating petroleum refineries and petrochemical plants (PRPP) wastewaters", *J. Environ. Chem. Eng.,* vol. 7, no. 5, p. 103326, 2019.
[http://dx.doi.org/10.1016/j.jece.2019.103326]

[13]   P. Praveen, and K-C. Loh, "Simultaneous extraction and biodegradation of phenol in a hollow fiber supported liquid membrane bioreactor", *J. Membr. Sci.,* vol. 430, pp. 242-251, 2013.
[http://dx.doi.org/10.1016/j.memsci.2012.12.021]

[14]   S.J. Varjani, E. Gnansounou, and A. Pandey, "Comprehensive review on toxicity of persistent organic pollutants from petroleum refinery waste and their degradation by microorganisms", *Chemosphere,* vol. 188, pp. 280-291, 2017.
[http://dx.doi.org/10.1016/j.chemosphere.2017.09.005] [PMID: 28888116]

[15]   S-H. Seo, H-O. Kwon, M-K. Park, I-S. Lee, and S-D. Choi, "Contamination characteristics of polycyclic aromatic hydrocarbons in river and coastal sediments collected from the multi-industrial city of Ulsan, South Korea", *Mar. Pollut. Bull.,* vol. 160, p. 111666, 2020.
[http://dx.doi.org/10.1016/j.marpolbul.2020.111666] [PMID: 33181941]

[16]   S.J. Varjani, and V.N. Upasani, "A new look on factors affecting microbial degradation of petroleum hydrocarbon pollutants", *Int. Biodeterior. Biodegradation,* vol. 120, pp. 71-83, 2017.
[http://dx.doi.org/10.1016/j.ibiod.2017.02.006]

[17]   B. Lin, and H. Zhao, "Technological progress and energy rebound effect in China's textile industry: Evidence and policy implications", *Renew. Sustain. Energy Rev.,* vol. 60, pp. 173-181, 2016.
[http://dx.doi.org/10.1016/j.rser.2016.01.069]

[18]   G.C. de Oliveira Neto, J.M. Ferreira Correia, P.C. Silva, A.G. de Oliveira Sanches, and W.C. Lucato, "Cleaner Production in the textile industry and its relationship to sustainable development goals", *J. Clean. Prod.,* vol. 228, pp. 1514-1525, 2019.
[http://dx.doi.org/10.1016/j.jclepro.2019.04.334]

[19]   A. Desore, and S.A. Narula, "An overview on corporate response towards sustainability issues in textile industry", *Environ. Dev. Sustain.,* vol. 20, no. 4, pp. 1439-1459, 2018.
[http://dx.doi.org/10.1007/s10668-017-9949-1]

[20]   D.A. Yaseen, and M. Scholz, "Textile dye wastewater characteristics and constituents of synthetic effluents: a critical review", *Int. J. Environ. Sci. Technol.,* vol. 16, no. 2, pp. 1193-1226, 2019.
[http://dx.doi.org/10.1007/s13762-018-2130-z]

[21]   V.B. Leon, B.A.F. Negreiros, C.Z. Brusamarello, G. Petroli, M. Di Domenico, and F.B. Souza, "Artificial neural network for prediction of color adsorption from an industrial textile effluent using modified sugarcane bagasse: Characterization, kinetics and isotherm studies", *Environ. Nanotechnol. Monit. Manag.,* vol. 14, p. 100387, 2020.
[http://dx.doi.org/10.1016/j.enmm.2020.100387]

[22]   H. Tounsadi, Y. Metarfi, M. Taleb, K. El Rhazi, and Z. Rais, "Impact of chemical substances used in textile industry on the employee's health: Epidemiological study", *Ecotoxicol. Environ. Saf.,* vol. 197, p. 110594, 2020.
[http://dx.doi.org/10.1016/j.ecoenv.2020.110594] [PMID: 32335392]

[23]   E. Ozturk, H. Koseoglu, M. Karaboyaci, N.O. Yigit, U. Yetis, and M. Kitis, "Sustainable textile production: cleaner production assessment/eco-efficiency analysis study in a textile mill", *J. Clean. Prod.,* vol. 138, pp. 248-263, 2016.
[http://dx.doi.org/10.1016/j.jclepro.2016.02.071]

[24]   D.A. Yaseen, and M. Scholz, "Comparison of experimental ponds for the treatment of dye wastewater under controlled and semi-natural conditions", *Environ. Sci. Pollut. Res. Int.,* vol. 24, no. 19, pp. 16031-16040, 2017.

[http://dx.doi.org/10.1007/s11356-017-9245-5] [PMID: 28537020]

[25]   V. Suryavathi, S. Sharma, S. Sharma, P. Saxena, S. Pandey, R. Grover, S. Kumar, and K.P. Sharma, "Acute toxicity of textile dye wastewaters (untreated and treated) of Sanganer on male reproductive systems of albino rats and mice", *Reprod. Toxicol.,* vol. 19, no. 4, pp. 547-556, 2005.
[http://dx.doi.org/10.1016/j.reprotox.2004.09.011] [PMID: 15749270]

[26]   I.G. Laing, "The impact of effluent regulations on the dyeing industry", *Review of Progress in Coloration and Related Topics,* vol. 21, no. 1, pp. 56-71, 1991.
[http://dx.doi.org/10.1111/j.1478-4408.1991.tb00081.x]

[27]   F. Gähr, F. Hermanutz, and W. Oppermann, "Ozonation- An important technique to comply with new German laws for textile wastewater treatment", *Water Sci. Technol.,* vol. 30, no. 3, pp. 255-263, 1994.
[http://dx.doi.org/10.2166/wst.1994.0115]

[28]   P.C. Vandevivere, R. Bianchi, and W. Verstraete, "Review: Treatment and reuse of wastewater from the textile wet-processing industry: Review of emerging technologies", *J. Chem. Technol. Biotechnol.,* vol. 72, no. 4, pp. 289-302, 1998.
[http://dx.doi.org/10.1002/(SICI)1097-4660(199808)72:4<289::AID-JCTB905>3.0.CO;2-#]

[29]   T. Toprak, and P. Anis, "Textile industry's environmental effects and approaching cleaner production and sustainability, an overview", *J. of Textile Eng & Fashion Techn,* vol. 2, no. 4, pp. 429-442, 2017.
[http://dx.doi.org/10.15406/jteft.2017.02.00066]

[30]   P. Kumari, S. Singh, and N. Rose, "Eco-Textiles: For Sustainable Development", *Int. J. Sci. Eng. Res.,* vol. 4, pp. 1379-1390, 2013.

[31]   G. Thompson, J. Swain, M. Kay, and C.F. Forster, "The treatment of pulp and paper mill effluent: a review", *Bioresour. Technol.,* vol. 77, no. 3, pp. 275-286, 2001.
[http://dx.doi.org/10.1016/S0960-8524(00)00060-2] [PMID: 11272013]

[32]   N. Ferronato, and V. Torretta, "Waste Mismanagement in Developing Countries: A Review of Global Issues", *Int. J. Environ. Res. Public Health,* vol. 16, no. 6, p. 1060, 2019.
[http://dx.doi.org/10.3390/ijerph16061060] [PMID: 30909625]

[33]   M.C. Monte, E. Fuente, A. Blanco, and C. Negro, "Waste management from pulp and paper production in the European Union", *Waste Manag.,* vol. 29, no. 1, pp. 293-308, 2009.
[http://dx.doi.org/10.1016/j.wasman.2008.02.002] [PMID: 18406123]

[34]   D. Pokhrel, and T. Viraraghavan, "Treatment of pulp and paper mill wastewater--a review", *Sci. Total Environ.,* vol. 333, no. 1-3, pp. 37-58, 2004.
[http://dx.doi.org/10.1016/j.scitotenv.2004.05.017] [PMID: 15364518]

[35]   K. Bahar, Z. Cetecioglu, and O. Ince, "Pollution prevention in the pulp and paper industries", In: *Environmental Management in Practice.,* E. Broniewisz, Ed., , 2011.

[36]   J.A. Rintala, and J.A. Puhakka, "Anaerobic treatment in pulp- and paper-mill waste management: A review", *Bioresour. Technol.,* vol. 47, no. 1, pp. 1-18, 1994.
[http://dx.doi.org/10.1016/0960-8524(94)90022-1]

[37]   G.K. Mandeep, G.K. Gupta, H. Liu, and P. Shukla, "Gupta, H. Liu, and P. Shukla, "Pulp and paper industry–based pollutants, their health hazards and environmental risks", *Curr. Opin. Environ. Sci. Health,* vol. 12, pp. 48-56, 2019.
[http://dx.doi.org/10.1016/j.coesh.2019.09.010]

[38]   G. Mandeep, G. Kumar Gupta, and P. Shukla, "Insights into the resources generation from pulp and paper industry wastes: Challenges, perspectives and innovations", *Bioresour. Technol.,* vol. 297, p. 122496, 2020.
[http://dx.doi.org/10.1016/j.biortech.2019.122496] [PMID: 31831257]

[39]   H. Vashi, O.T. Iorhemen, and J.H. Tay, "Aerobic granulation: A recent development on the biological treatment of pulp and paper wastewater", *Envi. Techn. & Innov.,* vol. 9, pp. 265-274, 2018.
[http://dx.doi.org/10.1016/j.eti.2017.12.006]

[40]   P. Bajpai, "Chapter 19 - Pulp Bleaching", In: *Biermann's Handbook of Pulp and Paper (Third Edition)*, P. Bajpai, Ed., Elsevier, 2018, pp. 465-491.

[41]   T.K. Das, and A.K. Jain, "Pollution prevention advances in pulp and paper processing", *Environ. Prog.*, vol. 20, no. 2, pp. 87-92, 2001.
[http://dx.doi.org/10.1002/ep.670200211]

[42]   H. Paudyal, and K. Inoue, "Adsorptive removal of trace concentration of fluoride from water using cerium loaded dried orange juice residue", *J. of Inst. of Sci. and Techn.*, vol. 23, no. 1, pp. 43-48, 2018.
[http://dx.doi.org/10.3126/jist.v23i1.22160]

[43]   A. Villard, A. Lelah, and D. Brissaud, "Drawing a chip environmental profile: environmental indicators for the semiconductor industry", *J. Clean. Prod.*, vol. 86, pp. 98-109, 2015.
[http://dx.doi.org/10.1016/j.jclepro.2014.08.061]

[44]   L-T. Lu, I.K. Wernick, T-Y. Hsiao, Y-H. Yu, Y-M. Yang, and H-W. Ma, "Balancing the life cycle impacts of notebook computers: Taiwan's experience", *Resour. Conserv. Recycling*, vol. 48, no. 1, pp. 13-25, 2006.
[http://dx.doi.org/10.1016/j.resconrec.2005.12.010]

[45]   D. E. European Commission, *Restriction of the Use of Certain Hazardous Substances in Electrical and Electronic Equipment.*Brussels, Belgium, 2002.

[46]   *Waste Electrical and Electronic Equipment.*Brussels, Belgium, 2002.

[47]   *D. E. European Commission.* Energy End-use Efficiency and Energy Services: Brussels, Belgium, 2005.

[48]   H. Chein, T.M. Chen, S.G. Aggarwal, C-J. Tsai, and C-C. Huang, "Inorganic acid emission factors of semiconductor manufacturing processes", *J. Air Waste Manag. Assoc.*, vol. 54, no. 2, pp. 218-228, 2004.
[http://dx.doi.org/10.1080/10473289.2004.10470898] [PMID: 14977323]

[49]   J.P. Giesy, S.A. Mabury, J.W. Martin, K. Kannan, P.D. Jones, and J.L. Newsted, "Perfluorinated Compounds in the Great Lakes", In: *Persistent Organic Pollutants in the Great Lakes,* R.A. Hites, Ed., Springer: Berlin Heidelberg, 2006, pp. 391-438.
[http://dx.doi.org/10.1007/698_5_046]

[50]   B. Boulanger, J.D. Vargo, J.L. Schnoor, and K.C. Hornbuckle, "Evaluation of perfluorooctane surfactants in a wastewater treatment system and in a commercial surface protection product", *Environ. Sci. Technol.*, vol. 39, no. 15, pp. 5524-5530, 2005.
[http://dx.doi.org/10.1021/es050213u] [PMID: 16124283]

[51]   K. Prevedouros, I.T. Cousins, R.C. Buck, and S.H. Korzeniowski, "Sources, fate and transport of perfluorocarboxylates", *Environ. Sci. Technol.*, vol. 40, no. 1, pp. 32-44, 2006.
[http://dx.doi.org/10.1021/es0512475] [PMID: 16433330]

[52]   A.A. Jensen, and H. Leffers, "Emerging endocrine disrupters: perfluoroalkylated substances", *Int. J. Androl.*, vol. 31, no. 2, pp. 161-169, 2008.
[http://dx.doi.org/10.1111/j.1365-2605.2008.00870.x] [PMID: 18315716]

[53]   S. Xie, T. Wang, S. Liu, K.C. Jones, A.J. Sweetman, and Y. Lu, "Industrial source identification and emission estimation of perfluorooctane sulfonate in China", *Environ. Int.*, vol. 52, pp. 1-8, 2013.
[http://dx.doi.org/10.1016/j.envint.2012.11.004] [PMID: 23266910]

[54]   A.G. Paul, K.C. Jones, and A.J. Sweetman, "A first global production, emission, and environmental inventory for perfluorooctane sulfonate", *Environ. Sci. Technol.*, vol. 43, no. 2, pp. 386-392, 2009.
[http://dx.doi.org/10.1021/es802216n] [PMID: 19238969]

[55]   F.G.A. Vagliasindi, and S.R. Poulsom, "Waste Generation and Management in the Semiconductor Industry: A Case Study", *Water Sci. Technol.*, vol. 29, no. 9, pp. 331-341, 1994.
[http://dx.doi.org/10.2166/wst.1994.0501]

[56]   S. Rajoria, M. Vashishtha, and V.K. Sangal, "Treatment of electroplating industry wastewater: a review on the various techniques", *Environ. Sci. Pollut. Res. Int.,* 2022.
[http://dx.doi.org/10.1007/s11356-022-18643-y] [PMID: 35084684]

[57]   S. Rajoria, M. Vashishtha, and V.K. Sangal, "Review on the treatment of electroplating industry wastewater by electrochemical methods", *Mater. Today Proc.,* vol. 47, pp. 1472-1479, 2021.
[http://dx.doi.org/10.1016/j.matpr.2021.04.165]

[58]   *Handbook of extractive metallurgy.* Wiley-VCH, 1997.

[59]   *B. A and D. W, Extractive metallurgy of copper vol.* 3rd ed. Pergamon, 1994.

[60]   J. Kyle, *Pressure Acid Leaching of Australian Nickel-Cobalt Laterites,* 1996.

[61]   D. Jhariya, R. Khan, and G. S. Thakur, *Impact of Mining Activity on Water Resource: An Overview study.,* 2016.

[62]   H. T, *Living with Earth, An Introduction to Environmental Geology.* PHI Learning Private Limited, 2012.

[63]   "S. K L N, and P. S. Dhinwa, "Impact of mining on rural environment and economy–a case study, Kota district, Rajasthan", *Int. J. Rem. Sens. Geosci,* vol. 2, pp. 21-26, 2012.

[64]   S. Pattanayak, S. Saha, P. Sahu, E. Sills, A. Singha, and J-C. Yang, "Mine over matter? Health, wealth and forests in a mining area of Orissa", *Indian Growth Dev. Rev.,* vol. 3, no. 2, pp. 166-185, 2010.
[http://dx.doi.org/10.1108/17538251011084473]

[65]   B.F. Giannetti, F. Agostinho, J.J.C. Eras, Z. Yang, and C.M.V.B. Almeida, "Cleaner production for achieving the sustainable development goals", *J. Clean. Prod.,* vol. 271, p. 122127, 2020.
[http://dx.doi.org/10.1016/j.jclepro.2020.122127]

[66]   P. Glavič, and R. Lukman, "Review of sustainability terms and their definitions", *J. Clean. Prod.,* vol. 15, no. 18, pp. 1875-1885, 2007.
[http://dx.doi.org/10.1016/j.jclepro.2006.12.006]

CHAPTER 6

Risk Assessment of Emerging Water Pollutants

Zahra Zahra[1,*], Zunaira Habib[2] and Brian Moon[3]

[1] *Department of Civil & Environmental Engineering, University of California-Irvine, Irvine, CA 92697, USA*

[2] *Institute of Environmental Sciences and Engineering (IESE), School of Civil and Environmental Engineering (SCEE), National University of Sciences and Technology, Islamabad, Pakistan*

[3] *Plamica Labs, Batten Hall, 125 Western Ave, Allston, MA 02163, USA*

Abstract: A complex mixture of pollutants in wastewater runs down from different sources into the aquatic environment, with potential hazards to aquatic organisms, human health, and the environment. Among these water pollutants, F^-, NO_3^-, and heavy metals (Cd, Pb Hg, Zn, Cr, Ni, As, *etc.*) are considered conventional pollutants, whereas nanomaterials, pharmaceutical compounds, personal care products, pesticides, endocrine disrupting compounds (EDCs), artificial sweeteners, surfactants, *etc.* are known as emerging water pollutants. This cocktail of water pollutants in the aquatic ecosystem is a real danger, leading to detrimental effects. This chapter discussed the environmental risk assessment (ERA) of the emerging water pollutants, especially the nanomaterials. The ERA of emerging pollutants will help indicate potential risks associated with these substances, highlighting the importance of their hazard identification, dose-response and exposure assessment, and risk characterization. This information will give insights into the recent findings related to the pollutants' effects and their assessment approach.

Keywords: Agrochemicals, Emerging water pollutants, Nanomaterials, Personnel care products, Pharmaceuticals, Risk assessment.

INTRODUCTION

Water is the key component of life. Globally, increased wastewater discharge occurs due to intensive urbanization, industrialization, and several other factors. About 400 billion m^3 of wastewater is released from different sectors annually, contaminating about 5500 billion m^3 of freshwater annually [1]. About 80–90% of the total wastewater contaminated the Asian and Pacific region's surface water sources/environment [2]. According to estimates, the industrial discharge of wast-

---

* **Corresponding author Zahra Zahra:** Department of Civil & Environmental Engineering, University of California-Irvine, Irvine, CA 92697, USA; E-mail:nzahra@uci.edu

Shaukat Ali Mazari, Nabisab Mujawar Mubarak & Nizamuddin Sabzoi (Eds.)

water into the environment will be doubled by 2025 [3], which will lead to further deterioration of surface water reservoirs.

Wastewater is a complex mixture of pollutants discharged from different sources reaching the aquatic environment, with potential hazards to aquatic organisms, human health, and the environment. Among these water pollutants, F^-, NO_3^- and heavy metals (Cd, Cr, Pb Hg, Zn, Ni, As, *etc.*) are considered conventional pollutants [4], whereas nanomaterials, pharmaceuticals & personal care products, endocrine disrupting compounds (EDCs), artificial sweeteners, surfactants, *etc.* are known as emerging water pollutants. Emerging water pollutants are chemical substances that have the potential to threaten human health or the environment and lack published health criteria. An "emerging pollutant" can be determined from an unknown cause, a novel exposure to humans, or *via* a new detection method [5]. Emerging pollutants are mainly the materials that make their way into surface water bodies through various anthropogenic activities [6]. In another study, emerging pollutants are defined as natural or chemical substances synthesized by man, exist in numerous environmental compartments, and have the potential to induce toxicity that can alter the metabolism of living organisms [7]. The compounds containing these emerging pollutants can be divided into three main categories: (i) Compounds that enter the environment by progressive industrialization. (ii) Compounds that have been detected recently, however, entered our environment years ago and (iii) the compounds that are already examined and recognized for having potential adverse impacts on humans or the ecosystem [8]. The modern lifestyle has increased the demand for new products in each sector such as industry, transport, agriculture, pharmaceutics, *etc.* As chemicals are used, they end up in the environment in the form of nonbiodegradable or hazardous wastes. Due to inappropriate wastewater treatment facilities, these chemicals run down from wastewater streams into surface/fresh water and consequently pollute our environment. Therefore, several concerns are rising about the harmful effects of these emerging pollutants, especially on humans and other organisms.

This book chapter discussed several kinds of emerging water pollutants and their sources. These pollutants include a wide range of chemical substances that we use regularly. They are being added to our environment, such as pharmaceutical compounds, engineered nanomaterials, personal care products, pesticides, perfluorinated compounds, *etc.* This chapter mainly discussed the environmental risk assessment (ERA) of these emerging water pollutants, especially the nanomaterials, which will help identify the potential risks associated with these substances.

EMERGING WATER POLLUTANTS

Nanotechnology had left no domain untouched by its scientific novelties. Although nanotechnology is early, it appears to affect different areas significantly. Nanotechnology has great potential and can serve in various sectors, including agriculture, health care, communication, and information technologies. In doing so, the negative effects of nanomaterials must not be ignored, such as ecotoxicity. The term 'nanomaterials' is used for all materials with a nano range size in at least one of their dimensions [9]. The increased use of nanomaterials in various fields has raised a worldwide concern about their release and influence on surroundings and the environment. Nanomaterials are already interacting with our environment since many scientists and engineers are handling the matter at the nanoscale. These nanomaterials are produced at a massive scale and are used directly in numerous commercial products and indirectly released into the environment [10]. As far as the manufacturing of nanomaterials is concerned, it has increased due to the increased demand for these nanomaterials in industries [11], recording a major boost for personal coating and care products, including anti-wetting products, cosmetics, and textiles. Other applications include electronics, engineering, energy, sports, construction, automotive, food and drink, medicines, *etc.* According to the Global Nanomaterials Market Size Report, 2010-2027, the market size of nanomaterials was approximately 8.52 billion USD in 2019, which is estimated to increase up to 9.58 billion USD in 2020. The compound annual growth rate of the nanomaterials market is calculated to increase by 13% from 2019 to 2027 [12]. Likewise, these nanomaterials' prompt synthesis and use have increased their release in the environment and, ultimately, their interaction with abiotic or biotic components. Although Engineered Nanoparticles (ENPs) are noteworthy, their long-lasting exposure and existence in the environment are not bereft of toxic and hazardous biological effects. Therefore, it is crucial to understand the complete interaction of these nanomaterials within the ecosystem, their likely dangerous effects, and their consequences on human health as well as identify and assess nanotechnology for the safe application of these nanomaterials [13].

Among global concerns, the environmental impacts caused by pharmaceutical compounds need serious consideration as their uses cannot be avoided. Therefore, an appropriate risk assessment of their presence in the environment is of key importance. The potential risks of pharmaceutical compounds to humans and other organisms in the aquatic ecosystem are a matter of concern even at sublethal concentrations since the first case was brought to the horizon in 1985 [14]. However, the ecotoxicological risks associated with the increased concentrations of pharmaceutical compounds in aquatic ecosystems are still not known completely [15]. Among other emerging pollutants, personal care products and agrochemicals released in aquatic bodies need special attention due to their significant impacts. The cosmetic industry is the fastest-growing entity with a substantial share in the global market and is predicted to garner $429.8 billion by 2022.

Similarly, the active ingredients used in cosmetics are called "The beast of beauty" due to their adverse impacts on consumers or aquatic bodies and require strict regulations based on their environmental impact assessment data [16]. In contrast, as a vibrant part of modern farming, agrochemicals have an estimated consumption of 1-2.5 billion tons of active ingredients annually on a global scale. According to ecological risk assessment, the active ingredients in agrochemicals pose a high risk to aquatic biota [17]. Table **1** summarizes the sources and concentrations of these emerging water pollutants found in different water bodies.

Table 1. Summary table having concentrations and sources of emerging water pollutants.

Type of Pollutants	Source in Water	Water Body	Concentration	References
Nanomaterials				
TiO_2	Pharmaceutical compounds, household and personal care products, food coloring, industrial and domestic wastewater discharge	Surface water	3 ng/L-10 µg/L	[18]
SiO_2	Printer toner, drugs, cosmetics, varnishes, additives for food, biosensors	Surface water	0.7 ng/L	[19]
Fullerenes	Energy systems, medical and bio-nanotechnology, environmental remediation, and drug transport	Surface water	98 ng/L	[20]
Au	Consumption of commercial cosmetics	Surface water	140 ng/L	[19]
Ag	Medical appliances, building materials, industrial and domestic wastewater discharge	Surface water	10-100 ng/L	[21]
CeO_2	Automobile catalyzer, ceramics, cosmetics, coatings and paint industries	Surface water	0.01 ng/L	[22]
Pharmaceuticals				
Ibuprofen	Pharmaceutical industries, human consumption, hospitals	Surface water	1-2370 ng/L	[23]
Paracetamol	Direct disposal in sewage, feces/urine discharge, hospitals, inappropriate handling of industrial effluents	Surface water	110-10000 ng/L	[24]
17a-ethynylestradiol	Feces/urine discharge, hospitals, release of treated effluent in environment	Surface water	< 0.98-10.2 ng/L	[24]

(Table 1) cont.....

Type of Pollutants	Source in Water	Water Body	Concentration	References
Nanomaterials				
Amoxicillin	Feces/urine discharge, hospitals, release of treated effluent in environment	Surface water	< 2.5-245 ng/L	[23]
Erythromycin	Hospitals, sewage leakages, effluent from wastewater treatment plants	Surface water	< 0.5-159 ng/L	[23]
Atenolol	Hospitals, human consumption, release of treated effluent in environment	Surface water	< 1-487 ng/L	[23]
Personal Care Product				
Triclosan	Human consumption at household level	Surface water	140-2300 ng/L	[24]
Sulfamethoxazole	Effluent discharged from industries, healthcare facilities and wastewater treatment plants	Surface water	< 1-46 ng/L	[25]
Benzophenone-4	Directly from recreational activities and industrial effluent, indirectly from wastewater treatment plants	Surface water	< 1-600 ng/L	[26]
Aroflorone	Consumption at the household level, effluent from industries and wastewater treatment pants	Surface water	< 0.17-0.48 ng/L	[27]
Isobornyl acetate	Consumption at the household level, effluent discharge from wastewater treatment plants	Surface water	< 0.18-0.65 ng/L	[27]
Linalool	Anthropogenic activities, food industries	Surface water	< 0.5-0.6 ng/L	[27]
Agrochemicals				
Atrazine	Runoff from fields	Sea water	8.7–64.8 ng/L	[28]
Simazine	Crop and non-crop areas	Sea water	1.4–5.3 ng/L	[28]
Triadimenol	Control fungal disease and enter water bodies *via* seepage	Seawater	4.3–20.8 ng/L	[28]
Acetochlor	Run off from corn and soybean fields	Surface water	20–6300 ng/L	[29]
Diazinon	Effluent discharge from residential and non-food buildings	Surface water	10–900 ng/L	[29]
Lindane	Human consumption at the household level and from forest land	Surface water	5–15 ng/L	[29]

Here, concern for the negative effects of emerging water pollutants on human health and the environment is mainly focused on the environmental risk assessment (ERA). This book chapter aims to demonstrate the ERA of the emerging water pollutants, focusing on nanomaterials. We have briefly discussed the risks associated with pharmaceutical compounds, personal care products, and pesticides, giving insights into the potential risks associated with these emerging pollutants.

RISK ASSESSMENT OF EMERGING WATER POLLUTANTS

Environmental risk assessment of pollutants has been considered the most applicable approach to identifying and quantifying their potential hazards and associated risks [30]. We have briefly discussed the dangers of emerging water pollutants on human health and the environment, as shown in Fig. (**1**) Table **2** presents various assessment levels for emerging water pollutants under EU guidelines and protocols.

Table 2. Risk Assessment methods for emerging water pollutants under EU Registration, Evaluation and Authorization of Chemicals framework [31].

Groups of substances	Risk assessment methods
Pharmaceuticals	Phase I: Prescreening predicted environmental concentration restricted to aquatic compartment If Partition Coefficient of substance >4.5, screen in a stepwise procedure for bioaccumulative and toxic substances profile Action limit: if predicted environmental concentration > 0.01 ug/L, Phase II Tier A: screening-environmental fate and effect analysis. Predicted No-Effect Concentration (No Observed Effect Concentration) Initial prediction of risk- predicted environmental concentration. Predicted No-Effect Concentration (water, groundwater, microorganisms) >1, Tier octanol-water partitioning coefficient >1000, The bioconcentration factor, Tier organic carbon partition coefficient > 10000, L/Kg (terrestrial compartment) If the substance is not ready biodegradable from the water-sediment study, effects on sediment, Tier B.Phase II Tier B: Extended substance and compartment-specific refinement and risk assessment.
Industrial chemicals	Tier I: Initial effect and exposure evaluation. Worst-case scenario Default assumptions and no site-specific information Tier II: Refined effect and exposure valuation. Site-specific data. Conventional scenarios Tier III: Higher tier effect assessment. Site-specific data on environment and societies and exposure of possible ecosystems. Introduction of Derived No-Effect Level for human health risk assessment

Groups of substances	Risk assessment methods
Plant protection products	Tier I: No Predicted No-Effect Concentration. The basis for calculation: LC50/EC50 or No Observed Effect Concentration / no observable effect level Individual information-information always required-additional requirements lied down in the Directives and guidelines -intrinsic properties-exposition (extended use)-the result of the risk assessment (higher tier studies) Tier II: No emission scenarios. Simple predicted environmental concentration calculation-water: drift values and 30 cm deep water body-soil: application rate in 5 cm soil-Birds and mammals: residues in the grass, fruits Tier III: Risk assessment EC50 or Predicted No-Effect Concentration compared with acute toxicity/exposure ratio; short term dietary toxicity/exposure ratio; long term dietary toxicity/exposure should be > 1, 5, 10, or 100 hazard quotients for oral and contact exposure must be reported
Biocides	Tier I: Effect assessment (Predicted No-Effect Concentration) Illustrative for compartment-core information-further requirements according to the different products types-additional data for refinement Tier II: Emission scenarios for every product type (predicted environmental concentration) Tier III: Risk assessment
Nanomaterials	Provisions according to the Registration, Evaluation, Authorization and Restriction of Chemicals framework

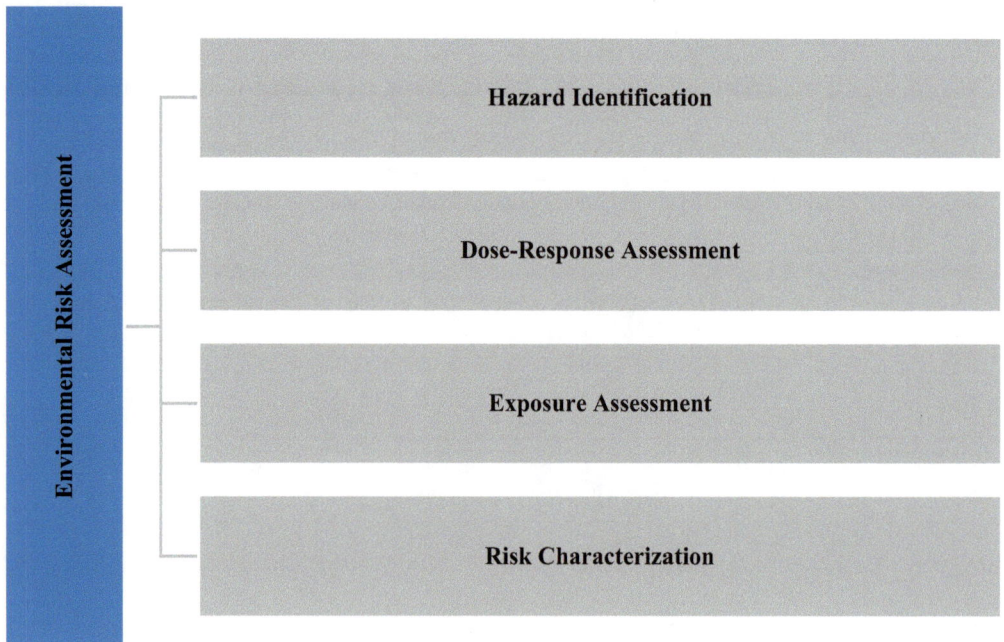

Fig. (1). Framework for environmental risk assessment.

Nanomaterials

Hazard Identification of Nanomaterials

Hazard identification is described as identifying adverse impacts produced by the nanomaterials. This step includes collecting nanotoxicity information for which risk assessment can be performed. Nanomaterials released from different sources such as industries, municipalities, hospitals, and agricultural run-off contaminate fresh and marine water. As a result, aquatic organisms are exposed to different types of Nanomaterials and get affected [32]. Although Nanomaterials possess beneficial properties, they are generally toxic whenever their concentration exceeds the threshold level [33]. The risk assessment of these Nanomaterials is important as various toxicity studies have shown their effects on humans and various species of the aquatic food chain at various trophic levels [34, 35]. For example, TiO_2 nanoparticles (NPs) have been shown to cause oxidative damage and apoptosis to human liver cells [36]. TiO_2 NPs in the anatase phase have been reported to cause enough damage to human dermal fibroblast (HDF) cells at LC_{50} of 3.6 µg·mL^{-1} [37]. Another study reported the reduction in *via*bility of human lymphoblastoid cells upon exposure to TiO_2 NPs (0–130 µg·mL^{-1} for 6–48 h) [38]. CuO NPs have been reported to cause oxidative stress revealing toxic effects on humans [39]. ZnO has been reported to induce cytotoxic effects and poses higher toxicity than other metallic NPs [40]. Studies on toxicity, pro-inflammatory and genotoxicity of ZnO NPs in human nasal mucous have been reported. DNA damage along with cytotoxic effects was observed at NPs concentration of 10 ($p < 0.05$) and 50 µg/mL ($p < 0.01$), respectively [41]. Thus, precautionary measurements are strictly requested in handling ZnO NPs in industries.

Nanosilver is commercially available in the form of superfine silver in the suspension of de-ionized water, having a size ranging from 5-50 nm. They are widely used in consumer products, water treatment, textile engineering, and strong antibacterial agents [42]. According to the USEPA secondary drinking-water regulations, a value of 0.1 mg/L is assigned for silver [43]. However, the maximum acute concentration for silver in freshwater is 3.2 µg/L and for saltwater is 1.9 µg/L depending on the acute toxicity of silver to macroinvertebrates and fish [44]. At higher concentrations, silver is toxic to all aquatic organisms such as fish, crustaceans, various plants, algae, and bacteria [45]. Besides their enormous usage, they also cause different ecological and toxicological effects from 10–100 ng/L or lower concentrations [46]. Prolonged exposure to silver can cause permanent bluish-gray discoloration of the skin (argyria) and eyes (argyrosis). Moreover, exposure to other soluble silver compounds can also induce toxic effects, including damage to the liver, kidney, respiratory and intestinal tract, eyes and skin irritation, and alterations in human blood cells [44, 45]

After the hazard identification, the potential of NPs to cause harm is evaluated in the hazard assessment. Many considerations should be taken into account while characterizing hazards, *e.g.*, the extent of biological uptake and metabolism, which type of toxicological studies have been implemented (*e.g.*, ecotoxicity, *in-vitro* or *in-vivo*), what kind of effects have been studied (*e.g.*, irritation, inflammation, stress, reproductive effects, mutation, growth reduction or even mortality, *etc.*), what the other concerns are regarding end-points, what about the data from experiments, either they are linked to realistic exposure mechanisms, *etc.* But unfortunately, in the case of NPs, many of these (and other) concerns have not been considered properly, resulting in inaccessible facts due to a lack of investigations. Until recently, conclusive studies regarding the environmental and health risks of NPs are still inadequate. However, in recent years, several experimental studies (*in vivo* and *in vitro*) have focused on the health effects of living organisms upon NPs exposure.

Dose-Response Assessment of Nanomaterials

Dose-response assessments are based on the facts and findings gained from laboratory-based experimentation or computational or mathematical modelling. In the case of NPs, the dose-response relationships involve various factors that need to be optimized, and no reference guidelines have been developed yet. For example, the fate and transport of NPs at the organism level involve various processes such as uptake, bioaccumulation, bioassimilation, dissolution, and elimination [47]. NPs undergo various transformations in the aquatic ecosystem. Their fate and behavior depend on other factors, *i.e.*, NPs type, size, properties, aggregation, dissolution, oxidation, toxicity, and water quality [48]. These transformations of NPs involve kinetic parameters based on nanoparticle properties. Functionalization of nanoparticles performed through surface coating or capping agents such as organic materials, natural organic matter, *etc.*, can affect their accumulation factor [49] and, consequently, the exposure. Many studies are available on optimizing NPs exposure concentration in different conditions. There exists a gap in calculating actual doses of NPs exposure to the human system for the risk assessment stage, which should be considered. Due to the lack of data about NPs reference doses for humans, NPs in their respective ionic forms can be used as reference guidelines to represent the reference doses of NPs [50]. Some of the studies related to the dose-response assessment of nanomaterials are cited here. For example, ecotoxicity of TiO_2 against *Daphnia magna* was observed to be LC_{50} at 5.5 mg/L after 48 h exposure [51]. TiO_2 NPs had been reported to kill human dermal fibroblast (HDF) cells at LC_{50} of 3.6 μgmL^{-1} [37]. Another study observed that the *via*bility of human lymphoblastoid cells may decrease due to exposure to TiO_2 NPs at 0–130 μgmL^{-1} for 6–48 h [38]. Silver NPs at concentrations >40 μgmL^{-1} have been reported to cause cell necrosis, leading to permanent rupture of the cell membrane in the human body [45]. Similarly, these NPs have also been reported to cause toxicity at a concentration as low as 40 ug/L in *Daphnia pulex* adults [52].

Exposure Assessment

Exposure assessment estimates the concentrations of chemical substances to which human populations and environmental systems are exposed [53]. The main routes of human exposure to these emerging water pollutants can either be ingestion/drinking or penetration through skin pores. In addition to this, the level or duration of exposure also plays a significant role in determining the extent of toxicity.

Ingestion/ drinking

In the case of exposure assessment of NPs from surface water, the concentration range of NPs was selected from 1ng/L to 1mg/L. This concentration range was selected to identify an environmentally relevant concentration between nano and micro range up to a maximum concentration level of 1mg/L [54, 55]. Humans can directly uptake these NPs through ingestion/ drinking of water polluted with NPs leading towards the targeted organs in the human body. From another perspective, the route of NPs uptake by humans involves the consumption of fish or other aquatic organisms. For this purpose, the bioaccumulation and bioassimilation factors need to be considered both at the individual and trophic levels [50]. Bioaccumulation of NPs in species can be determined through different factors such as NPs type, size, properties, and functionalization [47]. NPs characteristics can be modified by changing these factors, subsequently affecting the uptake of NPs. For example, ZnO NPs having different functionalized surfaces showed varied bioavailability and NPs uptake by aquatic organisms from crustaceans to zebrafish. In this regard, ZnO-octyl-NPs showed a significant uptake of 37230 mg Zn/Kg dry weight, whereas ZnO-OH-NPs showed no significant uptake with 287mg Zn/body Kg dry weight. In another study, CeO_2 NPs (positively charged) have been reported to induce more toxicity and bioavailability to *C. elegans* compared to neutral or negatively charged species [49].

Moreover, humans can be exposed to various NPs *via* various media like water, soil, and air. Plants also offered a prospective route for the transfer of NPs to the environment and ultimately paved the way for their bioaccumulation into the food chain and consequently enter the human body [56]. The use of NPs has also gained much attention in the agriculture and food processing industries to eliminate the food shortage caused by global environmental challenges [57]. These NPs are used for various purposes, such as enhancing shelf life, nutrient adsorbent capacity, and improved taste and texture. Ingestion is a major exposure route for these NPs. NPs, either directly or indirectly in drugs and food products, get absorbed *via* the digestive system, followed by accumulation in lymphatic tissues [58]. Depending on their nature, these NPs may block gastrointestinal tracks, leading to death [59].

Similarly, a study was conducted on the cytotoxic effect of different ENPs being used in packaging (ZnO and MgO) and as additives (TiO_2 and SiO_2) [59]. Also, they can damage DNA. But unfortunately, there is a lack of data that completely describes the ingestion process for assessing the severe toxic effects of ENPs in food-based products and goods [60].

Penetration by skin

The exposure of NPs to the skin, either by excessive use of cosmetics containing these NPs or handling at a lab scale, ultimately results in deep penetration of NPs into the blood circulation system, causing serious health issues [61]. Unintentional exposure of these NPs to the human body occurs through various recreational activities like swimming. For penetration to the skin, these NPs either follow the transcellular pathway or follicular pathway *via* hair follicles [62]. Although the skin is an active barrier that does not allow toxic contaminants to enter the body, this barrier's vulnerability is influenced by sweat glands and hair follicles, which facilitate these NPs to enter the body *via* skin [63]. Rapid distribution, poor nanoparticles, and the bioaccumulation nature of these NPs are crucial factors responsible for cellular toxicity.

Risk Characterization

Risk characterization is performed based on all the information gathered during the hazard identification, dose-response, and exposure assessment. It helps determine and communicate the actual likelihood of risks to exposed populations. Risk characterization can be assessed quantitatively by comparing the predicted environmental concentration (PEC) or measured environmental concentration (MEC) value of a chemical substance with its predicted no-effect concentration (PNEC). The PNEC is the concentration level below which the exposure to that chemical substance is not expected to induce harm. The PEC/PNEC or MEC/PNEC ratio is called the risk quotient (RQ), which is used to characterize the risk level. For example, a study reported the RQ value as less than 0.001, suggesting there was little or no risk to aquatic organisms. However, the RQ value for TiO_2 has been reported to be in the range of 0.7 to 16, which is suggested to pose risks [64].

In surface water, a recent study assessed 3 NPs, including ZnO, TiO2, and CuO. ZnO NPs were reported to pose a higher risk relative to CuO and TiO_2 NPs. Due to the lack of a reference dose for ZnO NPs, the Zn ion reference dose was used to estimate Zn nanotoxicity. According to the estimates, the maximum allowable concentration for Zn ions (*i.e.*, 1.66 mg/L Zn ions) was higher than ZnO NPs (*i.e.*, 0.092mg/L Zn ions), suggesting that both the exposure to Zn ions and ZnO NPs should be avoided to maintain the water quality and protecting human health. A similar trend was also observed in the case of CuO NPs and Cu ions exposure [50]. Some studies focused on the risk assessment of nanoparticle exposure to aquatic organisms. For example, a study that estimated the PEC value for NPs exposure using a probabilistic approach suggested that NPs exposure poses a low risk to freshwater organisms [65].

Similarly, in another study, Ag NPs, ZnO NPs, TiO_2 NPs, CNTs and fullerene were estimated to induce low or no risk to aquatic organisms [66]. This estimation-based report usually focuses on the ecological risks caused by NPs and lacks the assessment procedure for human health risks [50]. Therefore, based on the outcomes gained from the literature, ENPs can cause unusual environmental problems, impose hazards to human health, or affect these two simultaneously. However, at this point, with this state of knowledge, it is quite difficult to make any conclusive statement about the definite risks of ENPs exposure.

Pharmaceuticals

With the increasing usage of pharmaceutical compounds, their presence in various environmental matrices has augmented the need to evaluate their risks. This approach assesses the dose of a pharmaceutical compound detrimental to specific specie present in the polluted aquatic ecosystem. The risk assessment is based on a risk quotient (RQ) that is the ratio between the Predicted or Measured Environmental Concentration (MEC) to the Predicted No-Effect Concentration (PNEC) of the pollutant [67]. According to the literature [68, 69], RQ < 0.1 represents the minimum risk to the organisms, $0.1 \leq RQ \leq 1$ represents the medium level risk, whereas RQ=1 indicates high-level risk [70, 71]. Several studies have been conducted to assess pharmaceutical compounds released in waterbodies against different aquatic species. Risk characterization was based on the RQ values and are classified as: low risk ($0.01 \leq RQ < 0.1$), medium risk ($0.1 \leq RQ < 1$), or high risk ($RQ \geq 1$) [72, 73]. We have briefly summarized the risk characterization for some of the PCs in Table **3**.

Table 3. Summary of the risk characterization of pharmaceutical contaminants exposed to different aquatic organisms.

Type of pharmaceutical contaminants	Species	RQ value	Risk characterization	References
Paracetamol	*D. magna*	0.04	Low risk	[67]
Paracetamol	*S. proboscideu*	2.3	High risk	[74]
Acetaminophen	Crustaceans	0.321	Moderate risk	[75]
Acetaminophen	Zebrafish	1.6	High risk	[76]
Ibuprofen	*O. latipes*	1.9	High risk	[67]
Ibuprofen	*T. rubrum*	1.8	High risk	[74]
Diclofenac	*O. mykiss*	0.016	No risk	[67]
Diclofenac	Fish	1.02	High risk	[77]
Diazepam	*D. magna*	0.013	Low risk	[74]
Fluoxetine	Fish	9.06	High risk	[77]
Fluoxetine	*Daphnids*	1.06	High risk	[77]
Ofloxacin	Algae	3.137	High risk	[75]
Oxazepam	*D. magna*	0.3	Moderate risk	[67]
Clarithromycin	Algae	0.41	Moderate risk	[78]

Among medicines consumption, paracetamol (also known as acetaminophen) is a commonly used medicine that has been reported to induce side effects at a reproductive or embryonic level in *Daphnia magna* or fish even at low concentrations of 0.5 µg/L [76, 79, 80]. The RQ values for paracetamol have been found for different aquatic organisms such as *Daphnia* (64) [70, 71], *D. magna* (1.6) [67], *S. proboscideus* (9.2) [81], and green algae (0.11) [82], respectively. The RQ values indicate that *Daphnia, D. magna*, and *S. proboscideus* were at high risk, whereas green algae with a 0.11 RQ value showed low risk posed by paracetamol. In the case of ciprofloxacin, RQ values for green algae and *Microcystis aeruginosa* were observed to be 220 [83] and 440 [84], which indicated a high level of risk. However, for fish and daphnia species, RQ values have been reported to be low (<0.1), indicating no risk.

In the case of ofloxacin, the RQ values against *Daphnia* (56), fish (153) [83], *Vibrio fisheri* (4050) [81], Green algae (4050) [83], *Pseudokirchneriella subcapitata* (*P. subcapitata*) (7364) [84], and *Pseudomonas putida* (81,000) [85], were found to be very high and can cause increased risk. RQ values were also calculated for moxifloxacin against fish, *daphnia* and green algae and reported that the RQ values were less than 1, indicating no effect on these species in the receiving environment [83]. RQ values in case of exposure to naproxen have been reported to be 177 and 70 for fish and *Ceriodaphnia dubia* (*C. dubia*), respectively [70, 71, 81]. These results suggested that these species were at high risk.

RQ values for ibuprofen were calculated against green algae (253) [82], fish (1,013) [70, 71], and *O. latipes* (167, 300) [67], and the values were found to be very high against the investigated species, especially for the *O.latipes* RQ >>> 1 value indicated that ibuprofen could cause the highest level of risk to this species. In the case of diclofenac, RQ values were reported against fish (26) [70, 71], green algae (76) [82], and *Oncorhynchus mykiss* (O. mykiss) (12,600) [81]. These values warrant a high risk (RQ>>1) of diclofenac for the investigated species. All the PCs except moxifloxacin have been reported to induce high risk (RQ > 1) against most investigated species. Based on the following facts, it is estimated that the presence of pharmaceutical waste in the water bodies is highly detrimental to the aquatic organisms and ecosystem.

Household and Personal Care Products (HPCPs)

The widespread production and consumption of household personal care products (HPCPs) in human life have made them ubiquitous within commerce. These products are divided into three main categories depending on their use in daily life, *i.e.*, leave-on products (sunscreen, hand and face cream, lotion and deodorant), rinse-off products (shampoo, liquid soap, shower gel, toothpaste, conditioner, and shaving foam) and make-up products (nail polish, foundation, lip balm, lipstick). People are exposed to various other HPCPs on a typical day. For example, a hygiene routine in the morning contains different products, including soap, toothpaste, moisturizers, and hair products. While at the workplace, they can be exposed to some other HPCPs, including cosmetics, air fresheners, bleach, deodorants, paints, oven cleaners, insecticides, drain cleaners, disinfectants, and detergents [86]. HPCPs may contain various chemicals as additives to enhance their quality, performance, properties, *via*bility, and significance [16]. These additives can be toxic to human health, ecosystem, and aquatic organisms once they make their way into the water. Furthermore, most of these consumer products are usually persistent in water. They accumulate as emerging pollutants for many years, as stated by researchers who detected these HPCPs in water bodies at various concentration levels [87].

Daily use of these HPCPs may cause an "additive effect" due to a common ingredient in different products and a "cocktail effect" due to synergistic contact of various substances. The HPCPs having high water content can be easily contaminated by various microorganisms, another aspect of toxicity. In addition to some biocidal substances, various preservatives, including the mixture of 5-

chloro-2-methylisothiazol-3(2H)-one and 2-methylisothiazol-3 (2H)-one (CMIT/MIT), sodium benzoate, propylparaben, methylparaben, potassium sorbate, 1,4-dioxane and phenoxyethanol are commonly used in consumer products [88]. CMIT/MIT has been extensively used as a preservative in HPCPs since the 1980s. In 1985, the first case of skin allergy was reported for the products containing CMIT/MIT [89]. CMIT/MIT toxicity, European Union (EU) regulation has set the permissible limits up to 15 ppm for detergents, adhesives, paints, and cosmetics, whereas for industrial biocides, this limit is over 5,000 ppm [90]. Other preservatives in parabens have a lower risk to aquatic life due to lower half-life (~ 9.6 to 35.2 hrs). Propylparaben and methylparaben exist in sewage water in ample amounts having values up to 20,000 ng/L and 30,000 ng/L, respectively [91]. In addition to this, chloroxylenol (PCMX), Triclosan (TRC), and methylisothiazolinone (MIT) have been used frequently in handwashing and soaps owing to antimicrobial assets. At the same time, borax (BRX) is widely used in household cleaning items. These rinse-off products are resistant to degradation and wash away in drains, ultimately becoming a part of the aquatic ecosystem causing significant risk to marine organisms [92]. Capkin *et al*. [93], studied the chronic effects of PCMX, TRC, MIT and BRX on histopathology and genotoxicity of rainbow trout (*Oncorhynchus mykiss*) under a semi-static state for 40 days. DNA impairment to erythrocytes of the trout was detected from findings of the comet assay. Histopathological effects were observed in the liver, spleen, gills, and kidney. Necrotic hepatocytes, fat vacuoles and pyknotic nucleus in the liver, splenic necrosis and melanomacrophage centers in the spleen, epithelial necrosis, hyperplasia, and lamellar fusion in gills were observed. At the same time, glomerulus deterioration and capillaries dilation and the deterioration of epithelial cells and cloudy swelling were detected in the kidney.

As a crucial component, a massive concentration of surfactants is added to all high-volume rinse-off products. At the same time, the risk associated with surfactants depends on the final concentration entering the water body [94]. Surfactants are categorized into cationic, anionic, and non-ionic forms. Among them, the widely used surfactants are anionic, *i.e*., alcohol ethoxysulphates, sodium lauroyl sarcosinate, sodium, ammonium laureth sarcosinate, *etc*. These surfactants have lower bioaccumulation due to the high rate of biodegradability and consequently have a lower potential risk to aquatic organisms [95].

Agrochemicals

As emerging water pollutants, agrochemicals are crucial in today's agriculture system. These chemicals are represented as pesticides, fungicides, herbicides, and insecticides, depending on their main use in agriculture. For highly productive farming, 1–2.5 million tons of pesticides are used globally [96, 97]. In the 1980s, organochlorine pesticides (OCPs) were banned or restricted worldwide due to the properties of persistent organic pollutants [17]. On the other hand, synthetic pyrethroids (SPs) and organophosphates pesticides (OPPs) are deliberated as less persistent and widely used in pest control [98]. Hence the extensive use of these pesticides caused chronic and acute toxicity to various aquatic organisms, including mollusks, invertebrates, and fish [99]. Besides, OPPs easily distributed in the environment due to high water solubility through volatilization, abrasion, and dissolution [100]. Some widely used OPPs are chlorpyrifos-methyl, dimethoate, tolclofos-methyl, chlorpyrifos, acephate, isoxathion, fenitrothion, ebufos, profenphos, pirimiphos-methyl, and diazinon [101].

Derbalah *et al.* [101] conducted a study to evaluate the OPPs toxicity exposure ratio or RQ in river water throughout Japan. The results in terms of RQ values were calculated based on the mean concentration of OPPs in water. The highest and lowest RQ values were 258 and 0.006 for fenitrothion and tolclofos-methyl, respectively. Also, the diazinon showed a higher risk of having RQ > 1 for all the samples. United States Environmental Protection Agency (USEPA) categorized these OPPs in toxicity classes I to IV for having adverse impacts upon oral exposure or inhalation [100]. In a study, the toxicity of LC_{50} of dimethoate was investigated on *Labeo rohita* (Hamilton) and *C. catla* (Hamilton) for their behavioral response and blood hematological and biochemical parameters. Significant variations in behavior were observed, including uncoordinated movements, refusal of feeding, respiratory issues, muscle fasciculation, erratic swimming, body pigmentation, lethargy, enhanced mucus secretion, and loss of balance [102]. Similarly, the LC_{50} of chlorpyrifos at specific time intervals (24, 48, 72, and 96 hrs) was examined for frog *R. cyanophlyctis*, resulting in a high mortality rate that ultimately declined amphibian population [103].

The toxic effects of malathion in Nile tilapias (*Oreochromis niloticus*) resulted in considerable alterations in the liver, gills, brain, and kidney. Lamellar fusion, hyperplasia, and epithelial lifting were noticed in gills. Liver-associated alterations were blood sinusoids, narrow and dilated lumen, necrosis, hypertrophied epithelial cells, severe hemorrhage, and vacuolation of hepatocytes. Pyknosis and hyaline drops were observed in tubular epithelial cells in the kidneys. Erratic swimming, respiratory stress, and high mortality rate concerning malathion dose were also observed in fish [104].

RECOMMENDATIONS/ FUTURE PERSPECTIVES

We argue that upcoming research must strongly emphasize the characterization of emerging water pollutants to identify a clear connection between their intrinsic properties and associated concerned risks. A well-defined dose description in terms of mass, chemical composition, surface area, and morphology must be provided for all categories of emerging water pollutants. Before attaining this, it would be more significant to report the concerned dose concerning their concerning characteristics rather than selecting an irrelevant one. For example, in the case of NPs dose-response, the reference values are not available in the regulation guidelines of NPs. Therefore, in most NPs studies, the toxicity value of their respective metal ions has been used as a reference [105]. Future work should be done on setting guideline values for NPs dose-response information. Based on the previous literature, it is anticipated that various connecting factors must be considered for precise environmental risk assessment of these emerging pollutants. Moreover, standard reference data, testing procedures, and resulting tactics must be designed to interpret the tested results of these emerging pollutants.

CONCLUSION

The ever-increasing applications of advanced materials in consumer markets have escalated the production of emerging water pollutants, such as nanomaterials, pharmaceuticals compounds, personal care products, pesticides, *etc*. The toxicological effects of these emerging water pollutants on the human body and ecosystem have been a concern for researchers. Their toxicity depends on various intrinsic properties, including size, surface charge, chemistry, shape, coating, aggregation, dispersion, stability, concentration, mobility, and reactivity. The hazards linked with emerging water pollutants *in vivo* and *in vitro* levels have been identified, considering the entry route and putative targets. However, the effects of long-term exposure to these pollutants must be studied before they irreversibly contaminate the freshwater resources. Detailed research is required to understand the fate and behavior of these emerging water pollutants in different environmental matrices. Therefore, any conclusive statement based on the available literature would be deficient. However, standard methods need to be established in determining the hazardous impacts of all types of emerging pollutants.

CONSENT FOR PUBLICATION

Not applicable.

CONFLICT OF INTEREST

The author declares no conflict of interest, financial or otherwise.

ACKNOWLEDGEMENTS

Declared none.

REFERENCES

[1]   S. G. Gaurav, *Quantum Sensors: Quantum Entanglement for Communications and Beyond.*, 2019.

[2]   U. WWAP, *World Water Assessment Programme: The United Nations World Water Development Report 4: Managing Water under Uncertainty and Risk.* UNESCO: Paris, 2012.

[3]   R. Connor, A. Renata, C. Ortigara, E. Koncagül, S. Uhlenbrook, and B. M. Lamizana-Diallo, "The United Nations world water development report 2017. Wastewater: The untapped resource", *The United Nations World Water Development Report,* 2017.

[4]   A. Ahamad, S. Madhav, A.K. Singh, A. Kumar, and P. Singh, "Types of Water Pollutants: Conventional and Emerging", In: *Sensors in Water Pollutants Monitoring: Role of Material* Springer, 2020, pp. 21-41.
        [http://dx.doi.org/10.1007/978-981-15-0671-0_3]

[5]   G. Murnyak, J. Vandenberg, P.J. Yaroschak, L. Williams, K. Prabhakaran, and J. Hinz, "Emerging contaminants: presentations at the 2009 Toxicology and Risk Assessment Conference", *Toxicol. Appl. Pharmacol.,* vol. 254, no. 2, pp. 167-169, 2011.
        [http://dx.doi.org/10.1016/j.taap.2010.10.021] [PMID: 21034762]

[6]   Y. Tang, M. Yin, W. Yang, H. Li, Y. Zhong, L. Mo, Y. Liang, X. Ma, and X. Sun, "Emerging pollutants in water environment: Occurrence, monitoring, fate, and risk assessment", *Water Environ. Res.,* vol. 91, no. 10, pp. 984-991, 2019.
        [http://dx.doi.org/10.1002/wer.1163] [PMID: 31220374]

[7]   C. Peña-Guzmán, S. Ulloa-Sánchez, K. Mora, R. Helena-Bustos, E. Lopez-Barrera, J. Alvarez, and M. Rodriguez-Pinzón, "Emerging pollutants in the urban water cycle in Latin America: A review of the current literature", *J. Environ. Manage.,* vol. 237, pp. 408-423, 2019.
        [http://dx.doi.org/10.1016/j.jenvman.2019.02.100] [PMID: 30822645]

[8]   C.J. Houtman, "Emerging contaminants in surface waters and their relevance for the production of drinking water in Europe", *J. Integr. Environ. Sci.,* vol. 7, pp. 271-295, 2010.
        [http://dx.doi.org/10.1080/1943815X.2010.511648]

[9]   P. Holister, J-W. Weener, C. Roman, and T. Harper, "Nanoparticles", In: *Technology white papers* vol. 3. , 2003, pp. 1-11.

[10]   Q. Abbas, B. Yousaf, Amina, M.U. Ali, M.A.M. Munir, A. El-Naggar, J. Rinklebe, and M. Naushad, "Transformation pathways and fate of engineered nanoparticles (ENPs) in distinct interactive environmental compartments: A review", *Environ. Int.,* vol. 138, p. 105646, 2020.
        [http://dx.doi.org/10.1016/j.envint.2020.105646] [PMID: 32179325]

[11]   M.E. Vance, T. Kuiken, E.P. Vejerano, S.P. McGinnis, M.F. Hochella Jr, D. Rejeski, and M.S. Hull, "Nanotechnology in the real world: Redeveloping the nanomaterial consumer products inventory", *Beilstein J. Nanotechnol.,* vol. 6, pp. 1769-1780, 2015.
        [http://dx.doi.org/10.3762/bjnano.6.181] [PMID: 26425429]

[12]   X. He, H. Deng, W.G. Aker, and H-m. Hwang, *Regulation and safety of nanotechnology in the food and agriculture industry.* Food Applications of Nanotechnology, 2019, p. 12.
        [http://dx.doi.org/10.1201/9780429297038-23]

[13]   I. Bhatt, and B.N. Tripathi, "Interaction of engineered nanoparticles with various components of the environment and possible strategies for their risk assessment", *Chemosphere,* vol. 82, no. 3, pp. 308-317, 2011.
        [http://dx.doi.org/10.1016/j.chemosphere.2010.10.011] [PMID: 20980041]

[14]   L.J. Silva, C.M. Lino, L.M. Meisel, and A. Pena, "Selective serotonin re-uptake inhibitors (SSRIs) in the aquatic environment: an ecopharmacovigilance approach", *Sci. Total Environ.,* vol. 437, pp. 185-195, 2012.
[http://dx.doi.org/10.1016/j.scitotenv.2012.08.021] [PMID: 22940043]

[15]   A. Pereira, L. Silva, C. Laranjeiro, C. Lino, and A. Pena, "Selected pharmaceuticals in different aquatic compartments: Part I—Source, fate and occurrence", *Molecules,* vol. 25, no. 5, p. 1026, 2020.
[http://dx.doi.org/10.3390/molecules25051026] [PMID: 32106570]

[16]   M. Bilal, and H.M.N. Iqbal, "An insight into toxicity and human-health-related adverse consequences of cosmeceuticals - A review", *Sci. Total Environ.,* vol. 670, pp. 555-568, 2019.
[http://dx.doi.org/10.1016/j.scitotenv.2019.03.261] [PMID: 30909033]

[17]   Y. Chen, K. Yu, M. Hassan, C. Xu, B. Zhang, K.Y-H. Gin, and Y. He, "Occurrence, distribution and risk assessment of pesticides in a river-reservoir system", *Ecotoxicol. Environ. Saf.,* vol. 166, pp. 320-327, 2018.
[http://dx.doi.org/10.1016/j.ecoenv.2018.09.107] [PMID: 30278393]

[18]   S. Choi, M. Johnston, G-S. Wang, and C.P. Huang, "A seasonal observation on the distribution of engineered nanoparticles in municipal wastewater treatment systems exemplified by TiO_2 and ZnO", *Sci. Total Environ.,* vol. 625, pp. 1321-1329, 2018.
[http://dx.doi.org/10.1016/j.scitotenv.2017.12.326] [PMID: 29996429]

[19]   M. Zhang, J. Yang, Z. Cai, Y. Feng, Y. Wang, and D. Zhang, "Detection of engineered nanoparticles in aquatic environments: current status and challenges in enrichment, separation, and analysis", *Environ. Sci. Nano,* vol. 6, pp. 709-735, 2019.
[http://dx.doi.org/10.1039/C8EN01086B]

[20]   E. Emke, J. Sanchis, M. Farre, P. Bäuerlein, and P. De Voogt, "Determination of several fullerenes in sewage water by LC HR-MS using atmospheric pressure photoionisation", *Environ. Sci. Nano,* vol. 2, pp. 167-176, 2015.
[http://dx.doi.org/10.1039/C4EN00133H]

[21]   G. Cornelis, A. Forsberg-Grivogiannis, N. Sköld, S. Rauch, and J. Perez-Holmberg, "Sludge concentration, shear rate and nanoparticle size determine silver nanoparticle removal during wastewater treatment", *Environ. Sci. Nano,* vol. 4, pp. 2225-2234, 2017.
[http://dx.doi.org/10.1039/C7EN00734E]

[22]   F. Gómez-Rivera, J.A. Field, D. Brown, and R. Sierra-Alvarez, "Fate of cerium dioxide (CeO2) nanoparticles in municipal wastewater during activated sludge treatment", *Bioresour. Technol.,* vol. 108, pp. 300-304, 2012.
[http://dx.doi.org/10.1016/j.biortech.2011.12.113] [PMID: 22265985]

[23]   B. Petrie, R. Barden, and B. Kasprzyk-Hordern, "A review on emerging contaminants in wastewaters and the environment: current knowledge, understudied areas and recommendations for future monitoring", *Water Res.,* vol. 72, pp. 3-27, 2015.
[http://dx.doi.org/10.1016/j.watres.2014.08.053] [PMID: 25267363]

[24]   J.L. Wilkinson, P.S. Hooda, J. Swinden, J. Barker, and S. Barton, "Spatial distribution of organic contaminants in three rivers of Southern England bound to suspended particulate material and dissolved in water", *Sci. Total Environ.,* vol. 593-594, pp. 487-497, 2017.
[http://dx.doi.org/10.1016/j.scitotenv.2017.03.167] [PMID: 28360000]

[25]   D.R. Van Stempvoort, J.W. Roy, J. Grabuski, S.J. Brown, G. Bickerton, and E. Sverko, "An artificial sweetener and pharmaceutical compounds as co-tracers of urban wastewater in groundwater", *Sci. Total Environ.,* vol. 461-462, pp. 348-359, 2013.
[http://dx.doi.org/10.1016/j.scitotenv.2013.05.001] [PMID: 23738987]

[26]   R. Rodil, J.B. Quintana, E. Concha-Graña, P. López-Mahía, S. Muniategui-Lorenzo, and D. Prada-Rodríguez, "Emerging pollutants in sewage, surface and drinking water in Galicia (NW Spain)", *Chemosphere,* vol. 86, no. 10, pp. 1040-1049, 2012.

[http://dx.doi.org/10.1016/j.chemosphere.2011.11.053] [PMID: 22189380]

[27] D. Relić, A. Popović, D. Đorđević, and J. Čáslavský, "Occurrence of synthetic musk compounds in surface, underground, waste and processed water samples in Belgrade, Serbia", *Environ. Earth Sci.,* vol. 76, p. 122, 2017.
[http://dx.doi.org/10.1007/s12665-017-6441-z]

[28] H. Xie, X. Wang, J. Chen, X. Li, G. Jia, Y. Zou, Y. Zhang, and Y. Cui, "Occurrence, distribution and ecological risks of antibiotics and pesticides in coastal waters around Liaodong Peninsula, China", *Sci. Total Environ.,* vol. 656, pp. 946-951, 2019.
[http://dx.doi.org/10.1016/j.scitotenv.2018.11.449] [PMID: 30625680]

[29] A. Székács, M. Mörtl, and B. Darvas, "Monitoring pesticide residues in surface and ground water in Hungary: surveys in 1990–2015", *J. Chem.,* vol. 2015, 2015.
[http://dx.doi.org/10.1155/2015/717948]

[30] S.F. Hansen, "Regulation and risk assessment of nanomaterials: Too little, too late?", 2009.

[31] M.D. Hernando, A. Rodríguez, J.J. Vaquero, A.R. Fernández-Alba, and E. García, "Environmental Risk Assessment of Emerging Pollutants in Water: Approaches Under Horizontal and Vertical EU Legislation", In: *Critical Reviews in Environmental Science and Technology* vol. 41. , 2011, p. 699p. 731.

[32] B.D. Johnston, T.M. Scown, J. Moger, S.A. Cumberland, M. Baalousha, K. Linge, R. van Aerle, K. Jarvis, J.R. Lead, and C.R. Tyler, "Bioavailability of nanoscale metal oxides TiO(2), CeO(2), and ZnO to fish", *Environ. Sci. Technol.,* vol. 44, no. 3, pp. 1144-1151, 2010.
[http://dx.doi.org/10.1021/es901971a] [PMID: 20050652]

[33] Å. Helland, "Nanoparticles: a closer look at the risks to human health and the environment perceptions and precautionary measures of industry and regulatory bodies in Europe", 2004.

[34] M.N. Croteau, S.K. Misra, S.N. Luoma, and E. Valsami-Jones, "Bioaccumulation and toxicity of CuO nanoparticles by a freshwater invertebrate after waterborne and dietborne exposures", *Environ. Sci. Technol.,* vol. 48, no. 18, pp. 10929-10937, 2014.
[http://dx.doi.org/10.1021/es5018703] [PMID: 25110983]

[35] N. Ye, Z. Wang, S. Wang, and W.J.G.M. Peijnenburg, "Toxicity of mixtures of zinc oxide and graphene oxide nanoparticles to aquatic organisms of different trophic level: particles outperform dissolved ions", *Nanotoxicology,* vol. 12, no. 5, pp. 423-438, 2018.
[http://dx.doi.org/10.1080/17435390.2018.1458342] [PMID: 29658385]

[36] R.K. Shukla, A. Kumar, D. Gurbani, A.K. Pandey, S. Singh, and A. Dhawan, "TiO$_2$ nanoparticles induce oxidative DNA damage and apoptosis in human liver cells", *Nanotoxicology,* vol. 7, no. 1, pp. 48-60, 2013.
[http://dx.doi.org/10.3109/17435390.2011.629747] [PMID: 22047016]

[37] C.M. Sayes, R. Wahi, P.A. Kurian, Y. Liu, J.L. West, K.D. Ausman, D.B. Warheit, and V.L. Colvin, "Correlating nanoscale titania structure with toxicity: a cytotoxicity and inflammatory response study with human dermal fibroblasts and human lung epithelial cells", *Toxicol. Sci.,* vol. 92, no. 1, pp. 174-185, 2006.
[http://dx.doi.org/10.1093/toxsci/kfj197] [PMID: 16613837]

[38] J.J. Wang, B.J. Sanderson, and H. Wang, "Cyto- and genotoxicity of ultrafine TiO$_2$ particles in cultured human lymphoblastoid cells", *Mutat. Res.,* vol. 628, no. 2, pp. 99-106, 2007.
[http://dx.doi.org/10.1016/j.mrgentox.2006.12.003] [PMID: 17223607]

[39] E. Assadian, M.H. Zarei, A.G. Gilani, M. Farshin, H. Degampanah, and J. Pourahmad, "Toxicity of copper oxide (CuO) nanoparticles on human blood lymphocytes", *Biol. Trace Elem. Res.,* vol. 184, no. 2, pp. 350-357, 2018.
[http://dx.doi.org/10.1007/s12011-017-1170-4] [PMID: 29064010]

[40] L. Zhong, Y. Yu, H-z. Lian, X. Hu, H. Fu, and Y-j. Chen, "Solubility of nano-sized metal oxides

evaluated by using *in vitro* simulated lung and gastrointestinal fluids: implication for health risks"., *J. Nanopart. Res.,* vol. 19, p. 375, 2017.
[http://dx.doi.org/10.1007/s11051-017-4064-7]

[41]   S. Hackenberg, A. Scherzed, A. Technau, M. Kessler, K. Froelich, C. Ginzkey, C. Koehler, M. Burghartz, R. Hagen, and N. Kleinsasser, "Cytotoxic, genotoxic and pro-inflammatory effects of zinc oxide nanoparticles in human nasal mucosa cells *in vitro*", *Toxicol. in vitro,* vol. 25, no. 3, pp. 657-663, 2011.
[http://dx.doi.org/10.1016/j.tiv.2011.01.003] [PMID: 21232592]

[42]   R. Khaydarov, R. Khaydarov, O. Gapurova, and Y. Estrin, "A novel method of continuous fabrication of aqueous dispersions of silver nanoparticles", *Int. J. of Nanopar.,* vol. 3, pp. 77-91, 2010.
[http://dx.doi.org/10.1504/IJNP.2010.033223]

[43]   O. Water, *Edition of the drinking water standards and health advisories.*United States, 2012.

[44]   U. Epa, *"National recommended water quality criteria,"* United States Environmental Protection Agency. Office of Water, Office of Science and Technology, 2009.

[45]   N.R. Panyala, E.M. Peña-Méndez, and J. Havel, "Silver or silver nanoparticles: a hazardous threat to the environment and human health?", *J. of app. biomed.,* vol. 6, 2008.

[46]   S. Luoma, "Silver nanotechnologies and the environment: Old problems or new challenges? Woodrow Wilson International Center for Scholars, Project on Emerging Technologies, and PEW Charitable Trusts publication PEN15", In: *Silver nanotechnologies and the environment: Old problems or new challenges. Woodrow Wilson International Center for Scholars, Project on Emerging Nanotechnologies, The Pew Charitable Trusts* Washington, DC, 2008.

[47]   K.L. Garner, Y. Qin, S. Cucurachi, S. Suh, and A.A. Keller, "Linking exposure and kinetic bioaccumulation models for metallic engineered nanomaterials in freshwater ecosystems", *ACS Sustain. Chem.& Eng.,* vol. 6, pp. 12684-12694, 2018.
[http://dx.doi.org/10.1021/acssuschemeng.8b01691]

[48]   L.M. Skjolding, M. Winther-Nielsen, and A. Baun, "Trophic transfer of differently functionalized zinc oxide nanoparticles from crustaceans (Daphnia magna) to zebrafish (Danio rerio)", *Aquat. Toxicol.,* vol. 157, pp. 101-108, 2014.
[http://dx.doi.org/10.1016/j.aquatox.2014.10.005] [PMID: 25456224]

[49]   B. Collin, E. Oostveen, O.V. Tsyusko, and J.M. Unrine, "Influence of natural organic matter and surface charge on the toxicity and bioaccumulation of functionalized ceria nanoparticles in Caenorhabditis elegans", *Environ. Sci. Technol.,* vol. 48, no. 2, pp. 1280-1289, 2014.
[http://dx.doi.org/10.1021/es404503c] [PMID: 24372151]

[50]   T. Parsai, and A. Kumar, "Tradeoff between risks through ingestion of nanoparticle contaminated water or fish: Human health perspective", *Sci. Total Environ.,* vol. 740, p. 140140, 2020.
[http://dx.doi.org/10.1016/j.scitotenv.2020.140140] [PMID: 32927548]

[51]   S.B. Lovern, J.R. Strickler, and R. Klaper, "Behavioral and physiological changes in Daphnia magna when exposed to nanoparticle suspensions (titanium dioxide, nano-C60, and C60HxC70Hx)", *Environ. Sci. Technol.,* vol. 41, no. 12, pp. 4465-4470, 2007.
[http://dx.doi.org/10.1021/es062146p] [PMID: 17626453]

[52]   R.J. Griffitt, J. Luo, J. Gao, J.C. Bonzongo, and D.S. Barber, "Effects of particle composition and species on toxicity of metallic nanomaterials in aquatic organisms", *Environ. Toxicol. Chem.,* vol. 27, no. 9, pp. 1972-1978, 2008.
[http://dx.doi.org/10.1897/08-002.1] [PMID: 18690762]

[53]   E. Union, "Technical Guidance Document (TGD) on Risk Assessment of Chemical Substances following European Regulations and Directives, Parts I-IV", *Technical Report Number EUR 20418 EN/1-4,* 2003.

[54]   F. Gottschalk, T. Sonderer, R.W. Scholz, and B. Nowack, "Modeled environmental concentrations of

engineered nanomaterials ($TiO_{(2)}$, ZnO, Ag, CNT, Fullerenes) for different regions", *Environ. Sci. Technol.,* vol. 43, no. 24, pp. 9216-9222, 2009.
[http://dx.doi.org/10.1021/es9015553] [PMID: 20000512]

[55]   B. Nowack, and T.D. Bucheli, "Occurrence, behavior and effects of nanoparticles in the environment", *Environ. Pollut.,* vol. 150, no. 1, pp. 5-22, 2007.
[http://dx.doi.org/10.1016/j.envpol.2007.06.006] [PMID: 17658673]

[56]   C.O. Robichaud, D. Tanzil, and M.R. Wiesner, "Assessing life-cycle risks of nanomaterials", In: *Environmental Nanotechnology: Applications and Impacts of Nanomaterials,* 2007, pp. 481-522.

[57]   R. Prasad, V. Kumar, and K.S. Prasad, "Nanotechnology in sustainable agriculture: present concerns and future aspects", *Afr. J. Biotechnol.,* vol. 13, pp. 705-713, 2014.
[http://dx.doi.org/10.5897/AJBX2013.13554]

[58]   I.L. Bergin, and F.A. Witzmann, "Nanoparticle toxicity by the gastrointestinal route: evidence and knowledge gaps", *Int. J. Biomed. Nanosci. Nanotechnol.,* vol. 3, no. 1-2, pp. 163-210, 2013.
[http://dx.doi.org/10.1504/IJBNN.2013.054515] [PMID: 24228068]

[59]   B. Wang, W. Feng, M. Wang, T. Wang, Y. Gu, and M. Zhu, "Acute toxicological impact of nano-and submicro-scaled zinc oxide powder on healthy adult mice", *J. Nanopart. Res.,* vol. 10, pp. 263-276, 2008.
[http://dx.doi.org/10.1007/s11051-007-9245-3]

[60]   Q. Chaudhry, M. Scotter, J. Blackburn, B. Ross, A. Boxall, L. Castle, R. Aitken, and R. Watkins, "Applications and implications of nanotechnologies for the food sector", *Food Addit. Contam. Part A Chem. Anal. Control Expo. Risk Assess.,* vol. 25, no. 3, pp. 241-258, 2008.
[http://dx.doi.org/10.1080/02652030701744538] [PMID: 18311618]

[61]   M. Sajid, M. Ilyas, C. Basheer, M. Tariq, M. Daud, N. Baig, and F. Shehzad, "Impact of nanoparticles on human and environment: review of toxicity factors, exposures, control strategies, and future prospects", *Environ. Sci. Pollut. Res. Int.,* vol. 22, no. 6, pp. 4122-4143, 2015.
[http://dx.doi.org/10.1007/s11356-014-3994-1] [PMID: 25548015]

[62]   Y.K. Tak, S. Pal, P.K. Naoghare, S. Rangasamy, and J.M. Song, "Shape-dependent skin penetration of silver nanoparticles: does it really matter?", *Sci. Rep.,* vol. 5, p. 16908, 2015.
[http://dx.doi.org/10.1038/srep16908] [PMID: 26584777]

[63]   Y. Teow, P.V. Asharani, M.P. Hande, and S. Valiyaveettil, "Health impact and safety of engineered nanomaterials", *Chem. Commun. (Camb.),* vol. 47, no. 25, pp. 7025-7038, 2011.
[http://dx.doi.org/10.1039/c0cc05271j] [PMID: 21479319]

[64]   N.C. Mueller, and B. Nowack, "Exposure modeling of engineered nanoparticles in the environment", *Environ. Sci. Technol.,* vol. 42, no. 12, pp. 4447-4453, 2008.
[http://dx.doi.org/10.1021/es7029637] [PMID: 18605569]

[65]   B. Giese, F. Klaessig, B. Park, R. Kaegi, M. Steinfeldt, H. Wigger, A. von Gleich, and F. Gottschalk, "Risks, release and concentrations of engineered nanomaterial in the environment", *Sci. Rep.,* vol. 8, no. 1, p. 1565, 2018.
[http://dx.doi.org/10.1038/s41598-018-19275-4] [PMID: 29371617]

[66]   C. Coll, D. Notter, F. Gottschalk, T. Sun, C. Som, and B. Nowack, "Probabilistic environmental risk assessment of five nanomaterials (nano-TiO_2, nano-Ag, nano-ZnO, CNT, and fullerenes)", *Nanotoxicology,* vol. 10, no. 4, pp. 436-444, 2016.
[http://dx.doi.org/10.3109/17435390.2015.1073812] [PMID: 26554717]

[67]   C. Bouissou-Schurtz, P. Houeto, M. Guerbet, M. Bachelot, C. Casellas, A-C. Mauclaire, P. Panetier, C. Delval, and D. Masset, "Ecological risk assessment of the presence of pharmaceutical residues in a French national water survey", *Regul. Toxicol. Pharmacol.,* vol. 69, no. 3, pp. 296-303, 2014.
[http://dx.doi.org/10.1016/j.yrtph.2014.04.006] [PMID: 24768990]

[68]   S.M.L. de Souza, E.C. Vasconcelos, M. Dziedzic, and C.M.R. de Oliveira, "Environmental risk

assessment of antibiotics: an intensive care unit analysis", *Chemosphere,* vol. 77, no. 7, pp. 962-967, 2009.
[http://dx.doi.org/10.1016/j.chemosphere.2009.08.010] [PMID: 19744697]

[69]   J.L. Zhao, G.G. Ying, Y.S. Liu, F. Chen, J.F. Yang, L. Wang, X.B. Yang, J.L. Stauber, and M.S. Warne, "Occurrence and a screening-level risk assessment of human pharmaceuticals in the Pearl River system, South China", *Environ. Toxicol. Chem.,* vol. 29, no. 6, pp. 1377-1384, 2010.
[http://dx.doi.org/10.1002/*etc.*161] [PMID: 20821582]

[70]   P. Verlicchi, M. Al Aukidy, and E. Zambello, "Occurrence of pharmaceutical compounds in urban wastewater: removal, mass load and environmental risk after a secondary treatment--a review", *Sci. Total Environ.,* vol. 429, pp. 123-155, 2012.
[http://dx.doi.org/10.1016/j.scitotenv.2012.04.028] [PMID: 22583809]

[71]   P. Verlicchi, M. Al Aukidy, A. Galletti, M. Petrovic, and D. Barceló, "Hospital effluent: investigation of the concentrations and distribution of pharmaceuticals and environmental risk assessment", *Sci. Total Environ.,* vol. 430, pp. 109-118, 2012.
[http://dx.doi.org/10.1016/j.scitotenv.2012.04.055] [PMID: 22634557]

[72]   F. Sánchez-Bayo, S. Baskaran, and I.R. Kennedy, "Ecological relative risk (EcoRR): another approach for risk assessment of pesticides in agriculture", *Agric. Ecosyst. Environ.,* vol. 91, pp. 37-57, 2002.
[http://dx.doi.org/10.1016/S0167-8809(01)00258-4]

[73]   Z. Vryzas, C. Alexoudis, G. Vassiliou, K. Galanis, and E. Papadopoulou-Mourkidou, "Determination and aquatic risk assessment of pesticide residues in riparian drainage canals in northeastern Greece", *Ecotoxicol. Environ. Saf.,* vol. 74, no. 2, pp. 174-181, 2011.
[http://dx.doi.org/10.1016/j.ecoenv.2010.04.011] [PMID: 20553992]

[74]   I. Sebastine, and R. Wakeman, "Consumption and environmental hazards of pharmaceutical substances in the UK", *Process Saf. Environ. Prot.,* vol. 81, pp. 229-235, 2003.
[http://dx.doi.org/10.1205/095758203322299743]

[75]   P. Vazquez-Roig, V. Andreu, C. Blasco, and Y. Picó, "Risk assessment on the presence of pharmaceuticals in sediments, soils and waters of the Pego-Oliva Marshlands (Valencia, eastern Spain)", *Sci. Total Environ.,* vol. 440, pp. 24-32, 2012.
[http://dx.doi.org/10.1016/j.scitotenv.2012.08.036] [PMID: 23021792]

[76]   M. Galus, N. Kirischian, S. Higgins, J. Purdy, J. Chow, S. Rangaranjan, H. Li, C. Metcalfe, and J.Y. Wilson, "Chronic, low concentration exposure to pharmaceuticals impacts multiple organ systems in zebrafish", *Aquat. Toxicol.,* vol. 132-133, pp. 200-211, 2013.
[http://dx.doi.org/10.1016/j.aquatox.2012.12.021] [PMID: 23375851]

[77]   A.M.P.T. Pereira, L.J.G. Silva, C.S.M. Laranjeiro, L.M. Meisel, C.M. Lino, and A. Pena, "Human pharmaceuticals in Portuguese rivers: The impact of water scarcity in the environmental risk", *Sci. Total Environ.,* vol. 609, pp. 1182-1191, 2017.
[http://dx.doi.org/10.1016/j.scitotenv.2017.07.200] [PMID: 28787792]

[78]   S. Zhou, C. Di Paolo, X. Wu, Y. Shao, T-B. Seiler, and H. Hollert, "Optimization of screening-level risk assessment and priority selection of emerging pollutants - The case of pharmaceuticals in European surface waters", *Environ. Int.,* vol. 128, pp. 1-10, 2019.
[http://dx.doi.org/10.1016/j.envint.2019.04.034] [PMID: 31029973]

[79]   P. Kim, Y. Park, K. Ji, J. Seo, S. Lee, K. Choi, Y. Kho, J. Park, and K. Choi, "Effect of chronic exposure to acetaminophen and lincomycin on Japanese medaka (Oryzias latipes) and freshwater cladocerans Daphnia magna and Moina macrocopa, and potential mechanisms of endocrine disruption", *Chemosphere,* vol. 89, no. 1, pp. 10-18, 2012.
[http://dx.doi.org/10.1016/j.chemosphere.2012.04.006] [PMID: 22560975]

[80]   M. Galus, J. Jeyaranjaan, E. Smith, H. Li, C. Metcalfe, and J.Y. Wilson, "Chronic effects of exposure to a pharmaceutical mixture and municipal wastewater in zebrafish", *Aquat. Toxicol.,* vol. 132-133, pp. 212-222, 2013.

[http://dx.doi.org/10.1016/j.aquatox.2012.12.016] [PMID: 23351725]

[81]   F. Orias, and Y. Perrodin, "Characterisation of the ecotoxicity of hospital effluents: a review", *Sci. Total Environ.,* vol. 454-455, pp. 250-276, 2013.
        [http://dx.doi.org/10.1016/j.scitotenv.2013.02.064] [PMID: 23545489]

[82]   B.I. Escher, R. Baumgartner, M. Koller, K. Treyer, J. Lienert, and C.S. McArdell, "Environmental toxicology and risk assessment of pharmaceuticals from hospital wastewater", *Water Res.,* vol. 45, no. 1, pp. 75-92, 2011.
        [http://dx.doi.org/10.1016/j.watres.2010.08.019] [PMID: 20828784]

[83]   E.I. Iatrou, A.S. Stasinakis, and N.S. Thomaidis, "Consumption-based approach for predicting environmental risk in Greece due to the presence of antimicrobials in domestic wastewater", *Environ. Sci. Pollut. Res. Int.,* vol. 21, no. 22, pp. 12941-12950, 2014.
        [http://dx.doi.org/10.1007/s11356-014-3243-7] [PMID: 24981036]

[84]   Y. Jiang, M. Li, C. Guo, D. An, J. Xu, Y. Zhang, and B. Xi, "Distribution and ecological risk of antibiotics in a typical effluent-receiving river (Wangyang River) in north China", *Chemosphere,* vol. 112, pp. 267-274, 2014.
        [http://dx.doi.org/10.1016/j.chemosphere.2014.04.075] [PMID: 25048915]

[85]   M. Ågerstrand, and C. Rudén, "Evaluation of the accuracy and consistency of the Swedish environmental classification and information system for pharmaceuticals", *Sci. Total Environ.,* vol. 408, no. 11, pp. 2327-2339, 2010.
        [http://dx.doi.org/10.1016/j.scitotenv.2010.02.020] [PMID: 20206966]

[86]   E.M. Haugabrooks, "Chemicals: cosmetics and other consumer products", In: *Information Resources in Toxicology* Elsevier, 2020, pp. 159-164.
        [http://dx.doi.org/10.1016/B978-0-12-813724-6.00014-1]

[87]   S.D. Richardson, and S.Y. Kimura, "Emerging environmental contaminants: Challenges facing our next generation and potential engineering solutions", *Environmental Technology & Innovation,* vol. 8, pp. 40-56, 2017.
        [http://dx.doi.org/10.1016/j.eti.2017.04.002]

[88]   A. Panico, F. Serio, F. Bagordo, T. Grassi, A. Idolo, M. DE Giorgi, M. Guido, M. Congedo, and A. DE Donno, "Skin safety and health prevention: an overview of chemicals in cosmetic products", *J. Prev. Med. Hyg.,* vol. 60, no. 1, pp. E50-E57, 2019.
        [PMID: 31041411]

[89]   M.K. Kim, K-B. Kim, J.Y. Lee, S.J. Kwack, Y.C. Kwon, J.S. Kang, H.S. Kim, and B.M. Lee, "Risk assessment of 5-chloro-2-methylisothiazol-3 (2H)-one/2-methylisothiazol-3 (2H)-one (CMIT/MIT) used as a preservative in cosmetics", *Toxicol. Res.,* vol. 35, no. 2, pp. 103-117, 2019.
        [http://dx.doi.org/10.5487/TR.2019.35.2.103] [PMID: 31015893]

[90]   O. Aerts, A. Goossens, J. Lambert, and J-P. Lepoittevin, "Contact allergy caused by isothiazolinone derivatives: an overview of non-cosmetic and unusual cosmetic sources", *Eur. J. Dermatol.,* vol. 27, no. 2, pp. 115-122, 2017.
        [http://dx.doi.org/10.1684/ejd.2016.2951] [PMID: 28174143]

[91]   N.A. Vita, C.A. Brohem, A.D.P.M. Canavez, C.F.S. Oliveira, O. Kruger, M. Lorencini, and C.M. Carvalho, "Parameters for assessing the aquatic environmental impact of cosmetic products", *Toxicol. Lett.,* vol. 287, pp. 70-82, 2018.
        [http://dx.doi.org/10.1016/j.toxlet.2018.01.015] [PMID: 29408348]

[92]   D.E. Carey, and P.J. McNamara, "The impact of triclosan on the spread of antibiotic resistance in the environment", *Front. Microbiol.,* vol. 5, p. 780, 2015.
        [http://dx.doi.org/10.3389/fmicb.2014.00780] [PMID: 25642217]

[93]   E. Capkin, T. Ozcelep, S. Kayis, and I. Altinok, "Antimicrobial agents, triclosan, chloroxylenol, methylisothiazolinone and borax, used in cleaning had genotoxic and histopathologic effects on rainbow trout", *Chemosphere,* vol. 182, pp. 720-729, 2017.

[http://dx.doi.org/10.1016/j.chemosphere.2017.05.093] [PMID: 28531838]

[94]   M. Lechuga, M. Fernández-Serrano, E. Jurado, J. Núñez-Olea, and F. Ríos, "Acute toxicity of anionic and non-ionic surfactants to aquatic organisms", *Ecotoxicol. Environ. Saf.,* vol. 125, pp. 1-8, 2016.
[http://dx.doi.org/10.1016/j.ecoenv.2015.11.027] [PMID: 26650419]

[95]   C. Cowan-Ellsberry, S. Belanger, P. Dorn, S. Dyer, D. McAvoy, H. Sanderson, D. Versteeg, D. Ferrer, and K. Stanton, "Environmental safety of the use of major surfactant classes in North America", *Crit. Rev. Environ. Sci. Technol.,* vol. 44, no. 17, pp. 1893-1993, 2014.
[http://dx.doi.org/10.1080/10739149.2013.803777] [PMID: 25170243]

[96]   U.S. McKnight, J.J. Rasmussen, B. Kronvang, P.J. Binning, and P.L. Bjerg, "Sources, occurrence and predicted aquatic impact of legacy and contemporary pesticides in streams", *Environ. Pollut.,* vol. 200, pp. 64-76, 2015.
[http://dx.doi.org/10.1016/j.envpol.2015.02.015] [PMID: 25697475]

[97]   M.F.A. Jallow, D.G. Awadh, M.S. Albaho, V.Y. Devi, and B.M. Thomas, "Pesticide risk behaviors and factors influencing pesticide use among farmers in Kuwait", *Sci. Total Environ.,* vol. 574, pp. 490-498, 2017.
[http://dx.doi.org/10.1016/j.scitotenv.2016.09.085] [PMID: 27644027]

[98]   H. Li, F. Cheng, Y. Wei, M.J. Lydy, and J. You, "Global occurrence of pyrethroid insecticides in sediment and the associated toxicological effects on benthic invertebrates: An overview", *J. Hazard. Mater.,* vol. 324, no. Pt B, pp. 258-271, 2017.
[http://dx.doi.org/10.1016/j.jhazmat.2016.10.056] [PMID: 27825741]

[99]   V. Korkmaz, A. Güngördü, and M. Ozmen, "Comparative evaluation of toxicological effects and recovery patterns in zebrafish (Danio rerio) after exposure to phosalone-based and cypermethrin-based pesticides", *Ecotoxicol. Environ. Saf.,* vol. 160, pp. 265-272, 2018.
[http://dx.doi.org/10.1016/j.ecoenv.2018.05.055] [PMID: 29852429]

[100]   G.K. Sidhu, S. Singh, V. Kumar, D.S. Dhanjal, S. Datta, and J. Singh, "Toxicity, monitoring and biodegradation of organophosphate pesticides: a review", *Crit. Rev. Environ. Sci. Technol.,* vol. 49, pp. 1135-1187, 2019.
[http://dx.doi.org/10.1080/10643389.2019.1565554]

[101]   A. Derbalah, R. Chidya, W. Jadoon, and H. Sakugawa, "Temporal trends in organophosphorus pesticides use and concentrations in river water in Japan, and risk assessment", *J. Environ. Sci. (China),* vol. 79, pp. 135-152, 2019.
[http://dx.doi.org/10.1016/j.jes.2018.11.019] [PMID: 30784439]

[102]   M.I. Hussain, B. Kumar, and M. Ahmad, "Effect of organophosphate insecticide, Dimethoate on physiology of common carp, Catla catla (Hamilton) and Labeo rohita", *Int. J. Curr. Microbiol. Appl. Sci.,* vol. 5, pp. 322-341, 2016.
[http://dx.doi.org/10.20546/ijcmas.2016.505.034]

[103]   A.K. Srivastav, S. Srivastava, S.K. Srivastav, and N. Suzuki, "Acute toxicity of an organophosphate insecticide chlorpyrifos to an anuran, Rana cyanophlyctis", *Irani. J. of Toxico.,* vol. 11, pp. 45-49, 2017.
[http://dx.doi.org/10.29252/arakmu.11.2.45]

[104]   A. Subburaj, P. Jawahar, N. Jayakumar, A. Srinivasan, and B. Ahilan, "Histopathological investigations in liver and kidney of the fish, Oreochromis mossambicus (Tilapia) exposed to acute Malathion (EC 50%) toxicity", *J. Exp. Zool. India,* vol. 21, pp. 77-81, 2018.

[105]   M. Riediker, D. Zink, W. Kreyling, G. Oberdörster, A. Elder, U. Graham, I. Lynch, A. Duschl, G. Ichihara, S. Ichihara, T. Kobayashi, N. Hisanaga, M. Umezawa, T.J. Cheng, R. Handy, M. Gulumian, S. Tinkle, and F. Cassee, "Particle toxicology and health - where are we?", *Part. Fibre Toxicol.,* vol. 16, no. 1, p. 19, 2019.
[http://dx.doi.org/10.1186/s12989-019-0302-8] [PMID: 31014371]

CHAPTER 7

Fate of Emerging Water Pollutants

Muhammad Ashar Ayub[1], **Muhammad Zia ur Rehman**[1,*], **Wajid Umar**[2], **Mujahid Ali**[1] and **Zahoor Ahmad**[3]

[1] *Institute of Soil & Environmental Sciences, University of Agriculture, Faisalabad 38000, Pakistan*

[2] *Doctoral School of Environmental Science, Szent Istvan University, Gödöllő, Hungary*

[3] *Department of Botany, University of Central Punjab, Bhawalpur, Pakistan*

Abstract: Emerging pollutants reflect a major global water quality problem. When these compounds enter the environment, they cause significant environmental threats to aquatic and human health. Emerging water pollutants (EWPs) include new materials with no regulatory status butthey can adversely affect the environment and human health. Emerging water contaminants can be biological or synthetic that remain unregulated, and pose a potential threat. Major classes of such pollutants are pharmaceuticals, agro-chemicals, endocrine-disrupting chemicals (EDCs), industrial wastes, livestock wastes, synthetic nanomaterials, and petroleum products. These pollutants can enter the environment through numerous sources and pose severe threats to soil organisms, agriculture, aquatic life, and humans. Pharmaceutical waste, industrial effluents, cosmetic and cleansing products, household sanitation, discharge, and synthetic NPs enter water channels, agro-ecosystem, underground water (*via* seepage), *etc.*, posing a serious threat. These EWPs have different unknown and known effects on animals, plants, and human health, which must be viewed positively. This chapter summarizes the sources and classification of EWPs, their entry into the environment, and their fate. A major focus will be on the end sink of pollutants with potential threats and risk evaluation for plants and human health.

Keywords: Agro-ecosystem, Classification of EWPs, Effects of EWPs on humans, Emerging water pollutants, Endocrine-disrupting chemicals, Fate of EWPs, Food pollution, Of EWPs plants and animal health, Soil pollution, Sources of EWPs, Underground water, Water channels.

INTRODUCTION

Water pollution is an issue caused mainly by unregulated contamination of the water. This leads to the degradation of the global water quality and is correlated

* **Corresponding author Muhammad Zia ur Rehman:** Institute of Soil & Environmental Sciences, University of Agriculture, Faisalabad, 38000, Pakistan; E-mail:ziasindhu1399@gmail.com

Shaukat Ali Mazari, Nabisab Mujawar Mubarak & Nizamuddin Sabzoi (Eds.)

with the deaths of 3900 children every day and millions of people per year due to diseases caused by the consumption of contaminated water [1]. Today, one of the key issues is the careless release of new contaminants, such as pharmaceuticals and personal care products, to waterbodies where they remain unattended and move into the human food chain [2]. These emerging pollutants can be described as chemicals whose effects on the environment are unknown and spread worldwide in aquatic ecosystems because they have not been monitored [3, 4].

Personal care products (PCPs), pharmaceuticals, and many other synthetic organic substances have transformed daily life and are now a crucial part of a balanced society. Pharmaceutical-based disease management operations are an important part of developing and preserving healthy livestock and the human population. The PCPs and other pollutants of emerging concern (ECs) are recently noticed in all water bodies like wastewater, groundwater, precipitation, and even drinking water [5]. Also, these pollutants can be found in suspended solids and river sediments [6]. Almost all the ubiquitous effects of pharmaceutical products and other 'emerging pollutants' on the environment have become a major concern even at trace and ultra-trace levels (below ng/L) in recent years [7]. Many personal care products and pharmaceuticals often contain complex chemical structures. Processing plants are the main sources of these products (*e.g.*, paracetamol production in France in 2008 was 3303 tonnes). Prescription and non-prescription pharmaceuticals will finally be incorporated into our PCPs used in our everyday lives [7 - 9]. The efficient source of elimination of human PCPs is urine, which ultimately goes into water bodies (often untreated in developing countries) [10]. Pharmaceuticals are not completely consumed in the body, resulting in the introduction of a portion of the parental (or blended) derivative into the waste treatment facilities *via* biliary (feces) and renal (urine) routes after incomplete degradation [11]. In waste treatment plants or sewage treatment work (STW) and the aquatic setting, almost every human and veterinary pharmaceutical has been found at levels normally not exceeding upper ng/L, whether prescribed or not prescribed.

Personal care products are also responsible for releasing plasticizers used in product linings and packaging and the surfactant residues used in soaps and detergents, particular insecticides, disinfectants, and musk compounds used as fragrances. Recent issues regarding water-substantial contamination by the chemical products used in hydraulic shallow boiling can be seen as novel emerging pollutants (*e.g.*, surfactant naphthalene and clay stabilizer tetramethylammonium chloride) [12]. Most EWPC study reviews tend to be restricted to specific chemicals (*e.g.*, BPA) [13]. Materials are being used in PCPs engineered nanomaterials such as sun protection cosmetics and hip substitutes [14]. At the same time, the release of nanoparticles into the environment is associated with certain harm and concerns about their potential impact on the health of human beings. However, there remains limited awareness of their possible toxicity or damage [15]. Besides this, emerging pollutants include uncommonly controlled agricultural chemicals such as many pesticides, herbicides, and even recreational drug residues, pharmaceuticals, and their metabolites [16]. Several studies have been carried out to determine the toxicity of ECs to the environment. The studies report that veterinary and human drugs can be harmful to fish and a wide range of aquatic animals [17 - 19]. Pharmaceutical originated ECs like estrogens and steroids present in agricultural waste (*e.g.*, oestrone, estriol, and estradiol) were established to cause defects in fish and amphibian reproduction [20]. Besides these hormones, certain nanoparticles as ECs, can also disrupt aquatic life like algae growth [21].

Most of these pollutants (15-20 years ago) were unknown or unrecognized and have recently been identified to be pollutants that are important to possible environmental risks. Their fate is still somewhat unknown regarding water/wastewater treatment systems, designed structures, and the environment. Usually, these contaminants are bio-accumulative and bioactive and may be present and persistent. Most EWPs are not monitored, requiring regular monitoring and evaluation of major involvements in the disposal of water supplies. It is now obvious that a combined analysis and method for these classes of pollutants will require many disciplines, including chemical, biological, and technical fields [22]. This chapter focuses on the release of emerging contaminants, their routes of adulteration, and their fate.

CLASSIFICATION OF EMERGING WATER POLLUTANTS

The EWPs contain a broad range of synthetic and natural chemical components that are considered to pose a potential threat. However, there is still no adequate availability of information for each pollutant. These EWPs consist of compounds such as personal care products (PCPs) and endocrine-disrupting chemicals, hormones, pharmaceuticals, pesticides, surfactant metabolites, surfactants, flame retardants, steroids, nanomaterials, gasoline additives, and industrial additives [22]. The typical example of emerging pollutants in surface waters is now pharmaceuticals, nanomaterials, industrial chemicals, and PCPs, whose existence was established recently. The word "EWPs" is generalized and encompasses many chemical pollutants. Thus, chemical emissions may also be classified according to different properties in the following classes [22]:

- CMR: carcinogenic, mutagenic, toxic to reproduction
- PBT: persistent, bio accumulative, toxic
- EDCs: (toxic contaminants)

- POP: based on their environmental properties are persistent, organic, pollutant
- PCPs: personal care products and pharmaceuticals (on a daily usage base)
- Regulated Priority Pollutants
- vPvB: very persistent, very bio accumulative
- HPV: high production volume chemicals
- Toxins, Toxics, Toxicants (based on toxicity overall)
- Exotics, Xenobiotics (endogenous *vs.* foreign pollutants)

Fig. (**1**) shows various classes of emerging water pollutants.

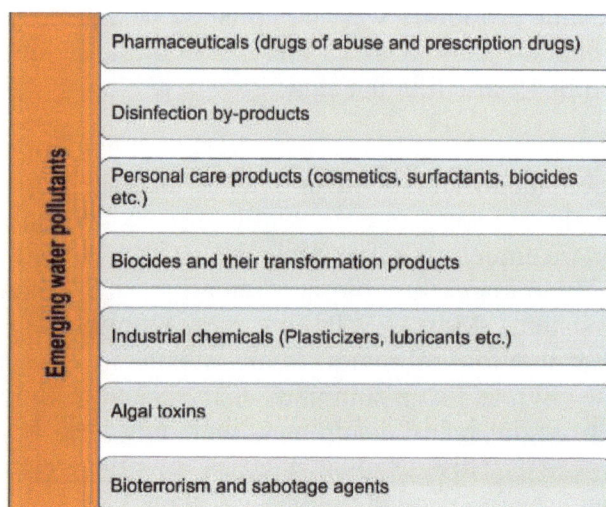

Fig. (1). Classification of EWPs.

EMERGING WATER POLLUTANTS

The two types of active ingredients in human care chemicals: pharmaceuticals and personal care products (collectively called PPCPs) have gained a little recognition. Still, collectively they constitute one of the main sources of chronic and recurrent incidents of ECs in the groundwater and surface water system worldwide [8]. These PPCPs contamination affects human life, which has become an important product for society, with global usage rates equal to the use of agricultural chemicals [17]. Many products are discarded or dumped *via* industrial and domestic sewage systems into aquatic systems. PPCPs have been spotted extensively in water treatment plants, wastewater treatment systems, groundwater, stormwater runoff, and even drinking water. The structure of all PPCPs is very complex, and the majority are combinations of many active chemicals as readily available commercial products. Therefore, the manufacturing facilities are the only traceable source for their production. After leaving the development centre and hitting the retail sector, they are too fragmented to track and evaluate pathways towards environmental exposure [23]. The average individual uses a variety of consumer chemicals, which form a PPCPs source, and each person makes an active contribution as such. All active sources are products like soap, detergents, home cleaners, disinfection agents, biocides, and cosmetics that directly enter water environments.

In consequence, unused and expired medicinal products have been disposed of in the drains, toilets and the release through excretion of unmetabolized medicine, medication overuse, agriculture's pest control, practices of veterinary and animal husbandry, all leading to active agents contaminating as PPCPs [6, 23]. These extremely persistent chemicals find their way back to humans through drinking water *via* the water cycle. They have been accused of causing reproductive defects, increased cancer rates, and resistance to antibiotic drugs.

The endocrine society identifies the EDCs as "an exogenous [non-natural] chemical or chemical mixture that interferes with any hormone-action element" [24]. EDCs disrupt the endocrine system either by mirroring or by obstructing the pathway of the hormonal structure. These are artificial chemicals that mimic or block hormones after entering the body, disrupting their normal activity. EDCs are widely recognized as estrogenic, either by imitating or blocking the path of immediate or oblique thyroid effects of natural oestrogen, androgens, or thyroids [7]. Engineered and natural EDCs, primarily by sewage treatment, are released into the atmosphere and ultimately introduced into groundwater, surface water, silt, and soil. Considering the above factors, many experiments on estrogenic compounds have been conducted. In untreated water, the amount of such compounds (ng L^{-1} or $\mu g\ L^{-1}$) is very low. The EDCs are of great concern because their long exposure duration and human impact have not been established. There are over 85,000 chemicals, thousands of which may only be EDCs. Triclosan has been widely used as an antimicrobial compound in a variety of PCPs for decades, such as soaps, skin care lotions, cosmetics, deodorants, creams, toothpaste, shampoos, and mouth washing machines [25]. Flammable polychlorinated biphenyl is a class of chemicals used for moulding textiles and plastics [26].

The external structure of the nanomaterial must be 1-100 nm. Nanomaterials are being used in various industries because of their high strength, thermal stability, high conductivity, and low permeability [27]. The use of nanotechnology in water and wastewater has increased significantly because of recent and continuous advances in nanotechnology. They are used for various social benefits in the manufacture of personal care products. However, nanomaterials are used to adulterate the aquatic environment and present threats to human beings and aquatic life in many fields, such as cosmetics, medicine, fashion, and materials [28]. Therefore, greater attention has also been paid to eliminating nanoparticles from water resources. Nanomaterials have a significant influence on the development of nanotechnology in water treatment for aquatic organisms and possibly humans. Nanomaterials can benefit plant growth, development, and reproduction at low concentrations. Aquatic species can also experience high toxic nanomaterial concentrations. In general, the effect on the aquatic environment of nanomaterials remains uncertain [28 - 30].

The group of EWPs includes phytoestrogens, endogenous natural steroidal, faecal, and sterols extractable from the human body, such as the androgenic hormones androgens, testosterone, and estrogens such as oestriol, progesterone, and oestrone. Synthetic androgens include natural, contraceptive substances and synthetic steroids and hormones used as free-acting in wastewater and effluent, including synthetic estrogenic compounds, nandrolone, oxandrolone, and xenoestrogens diethylstilboestrol. Most of these compounds are EDCs. The 4 most common compounds in groundwater are Bisphenol A, NP, E2, and Estrone [31].

Perfluorocarbons (PFCS) are anthropogenic compounds, such as oil and fat (water and heat resistant), used to produce resistant products/ coverings and packaging. The PFCs are surfactants, tensile, poisonous, bio-accumulative, and pollutants [32]. The main bond involves the basic one (C-F). Certain chemicals have been detected in different water resources, including groundwater, surface water, wastewater, and soil. Even though they are assumed to be EDCs, their impacts on the health of human beings and other living organisms are still unknown [25]. The effluents from wastewater treatment plants contain higher PFOA and PFOS, suggesting that PFOA and PFOS may be transferred to other fluorinated organic compounds during treatment processing [25].

SOURCES OF EMERGING WATER POLLUTANTS

In the last few decades, thousands of organic compounds have been classified as contaminants of emerging concerns (CECs), and many new ones are being added regularly into the environment [33]. They are responsible for various environmental problems, including soil, air, and water, such as groundwater or lakes, rivers, oceans, or even drinking water pollution. The wastewater and sludge residues of various EWPs can also be found [34]. The CECs can also be present in water, sand, soil, and biota at trace levels ranging from a few ng L^{-1} to a few μg L^{-1}. Several sources induce the presence of CECs in the environment.

Wastewater Treatment Plants

One of the most important sources of CECs is wastewater treatment plants (WWTPs). If the water treatment process is not administered accurately for emerging pollutants, these can make their way to waterbodies. Wastewater from diverse effluent sources (hospitals, industries, agricultural runoff) has a diverse mix of these pollutants, which are being dumped in waterbodies, especially in developing countries where WWTPs are not specialized in removing contaminants from water before their dumping. There are many other issues with WWTPs, limiting the removal of these pollutants like minute concentrations of pollutants and filtration methods resulting in various kinds of EWPs being released into waterbodies unattended, which may affect the human food chain [35]. The occurrence in wastewater and surface water bodies of CECs has been much better identified and characterized than in groundwater [36, 37]. However, in the aquatic environment, the major classes of micropollutants may be present and belong to the common CEC groups studied in other environmental areas with a pictorial view shown in Fig. (**2**). Major forms of EWPs present in wastewater are:

- Personal care products: insect repellents, disinfectants, products with ultraviolet (UV) block, and fragrances
- Pharmaceuticals: anticonvulsants, lipid regulators, NSAIDs, β-blockers, antibiotics, and stimulants
- Industrial chemicals: Pesticides, Fire retardants, plasticizers, herbicides, fungicides, and insecticides
- Hormones of the Steroids: oestrogens, surfactants that are not ionic.

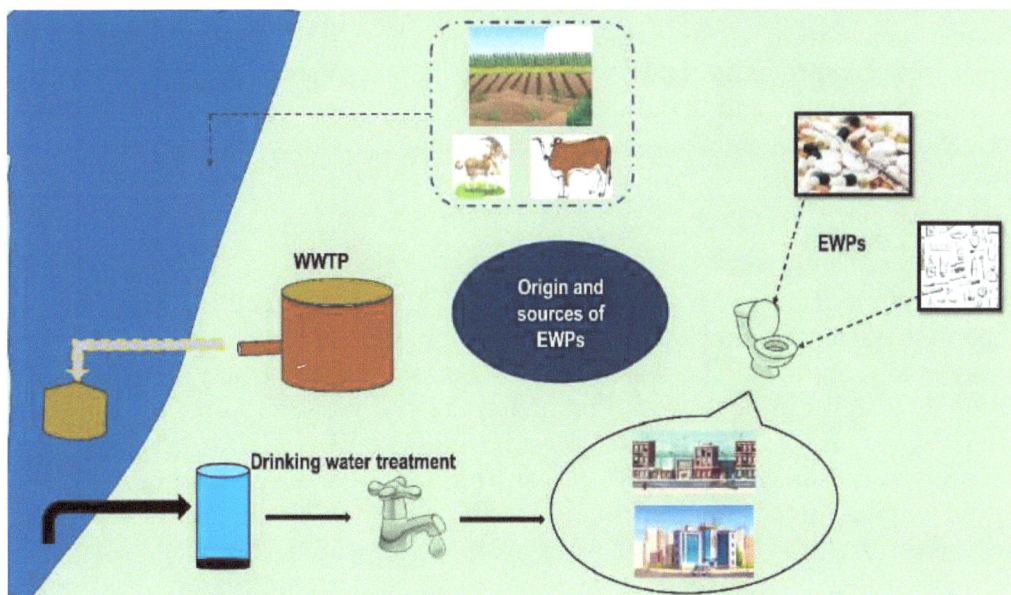

Fig. (2). Sources and fate of EWPs.

CECS IN WASTEWATER BIOSOLIDS

Once wastewater filtration is complete, most excluded pollutants accumulate in biosolid and sludge leftover. The sludge generated from these WWTPs contains nitrogen-stable biosolids, oxygen-rich in nutrients and carbon. Biosolids are also used in farm soils, including conditioners or fertilizers, which, if not treated, become a source of CECs. These biosolids are also being used in farm soils as soil conditioners or fertilizers, which, if not treated, become a source of CECs. Local research at the Tucson Roger Road Wastewater Recommendation Facility in (AZ, United States the US) studied the destiny of one class of compounds, PBDE. About 85-90% of PBDEs used as fire extinguishers were relocated to biosolids from wastewater during the treatment of wastewaters [38 - 40].

CECs from Agriculture and Livestock

Animal, plant, and microbial pollutants, such as degradation products, and natural toxins from man-made chemicals, are also generated in the natural world. The concern about EWPs risk is expressed in the rapid increase in scientific publications investigating the environmental effects of EWPs in the last decade [41 - 43]. When manure and slurry are applied in agricultural fields as a fertilizer for intensive livestock processing, veterinary medicines and their metabolites are released indirectly or directly through animals or pastures into the soil [44]. New EWPs may also intentionally be being applied to agricultural systems in the future, such as by increasing nanoparticles and nano pesticides [45]. In nanoparticulate form, existing pesticides can produce beneficial properties for the active ingredient to control pests. They have increased solubility, stability, plant absorption capacity, and higher pest toxicity [45]. Proper management strategies are required to manage agriculture and livestock waste-derived EWPs [46] effectively.

CECs in Landfill

Landfills are considered a major source of CECs. These facilities constitute an important waste repository for various sources, such as domestic, industrial, or commercial activities. The waste in sites consists of extremely heterogeneous materials which can pollute environmental factors, including soil, groundwater, surface water, air, and eventually human health (The EU 1999/31 / EC Directive). To effectively manage this EWPs threat from landfills, the following waste will not be admitted into a landfill: liquid waste, explosive waste, flammable waste, oxidizing waste, the clinical and other hospital, infectious waste, and used tires. Various kinds of pharmaceutical waste are also found in landfills which is another major contributor to EWPs, and effective management and pre-treatment of these are much required [47, 48]. Table **1** presents the types and sources of emerging water pollutants.

Table 1. Types and sources of emerging water pollutants.

Types of Emerging Pollutants	Sources of Emerging Pollutants	References
Pharmaceutical and personal care products (PPCPs)	Compost, Wastewater, and manure application are the sources of PPCPs.	[49]
	Hospital and household wastewater discharge, farming activities, application of manures, and effluent from the treatment plants.	[45, 46]
	PCPs include domestic sources, such as household waste disposal, landfills of urban solids, humans and animals, and excretions.	[9, 52]
	PCPs include commercial sources, such as pharmaceutical companies, the effluent of sewage and water treatment plants, and processes of manufacturing drugs.	[51]
	Waste and effluents of hospitals.	[50]
	Life enhancement products, detergents, cosmetics, sun-creams, perfumes, soap, toothpaste, *etc.*, are the sources of PCPs in the environment.	[9]
	Cosmetics and medicines products in landfills may contaminate the land, and water bodies are the sources PPCPs.	[53]
	Veterinary drugs and pesticides used in agriculture, resulting in groundwater contamination, are the sources PPCPs.	[54, 55]
	Urban wastewater inefficient purification and improper disposal play a role in contaminating water resources by PCPs.	[56, 57]
	Sewage effluent from wastewater purification locations (WWTPs) is the major attributable source of PCPs to water resources.	[58, 59]
Endocrine-disrupting chemicals	Water, sediments, biota, air, and soils are the sources of EDCs and possibly pose a major problem to the health of human beings.	[47, 60]
	Treated water is the major contributor to EDCs in the groundwater of the coastal region. The groundwater pollution in coastal areas also contaminates the seawater by discharging submarine groundwater.	[61]
	Extensive human-related activities in coastal areas, Coastal waters, and ecologically fragile regions are the main sources of EDC.	[62 - 64]

Types of Emerging Pollutants	Sources of Emerging Pollutants	References
Hormones and Steroids	Urban river, wastewater purification facilities (WWTP) release points, sewer overflows (CSOs). Dissolved and particulate steroid temporary and spatial distributions, testosterone),), medroxyprogesterone (MDRXYProg), progesterone Prog, norethindrone (Nore), estrone, levonorgestrel (Levo), estriol, oestradiol, and 17α ethinylestradiol are present in sewage, WWTP effluents, receiving river water and sediments, and in drinking water plant (DWP) intakes Steroids are present in raw sewage and treated wastewater. WWTP discharges, CSOs, river water and sediments and drinking water are sources of hormones and steroids.	[61]
	Some steroids sources are raw sewage, treated wastewater (WW), and river water.	[65]
	The most frequently detected steroids are estrogens, having higher concentrations of androgens and progesterone.	[66]
Perfluorochemicals (PFCs)	Industrial processes, consumer products, carpets protective coatings, cloth furniture, Polytetrafluoroethylene products, coatings and paper, and fire-fighting foams.	[67]
CECs from Agriculture and Livestock	Increased use of pesticides in the nanoscale	[44]
	large quantities of agrochemicals, nutrients	[68]
	Organic matter, including drug residues and sediments.	[69]
	Manure and slurry, intensive livestock processing, veterinary medicines, veterinary metabolites, and animal effluents are the main agricultural sources of EWP.	[45]

MONITORING OF EMERGING WATER POLLUTANTS

The EPs have also been detected and have raised serious questions about their severe environmental risks. However, because EPs occur in the environment in trace amounts, issues with enriching water samples, ecological contamination, and the complex analysis process remain hindrances for further studies of EPs. The major steps involved in evaluating and detecting emerging water pollutants include screening, sampling, preconcentration, improvement and optimization of solid-phase emission screening, and emerging methods like nanotechnology-based electrochemical sensors.

Screening

Suspect screening is a promising method of detecting unknown substances based on detailed data from multiple sources, including academic literature and online databases. For effective detection of emerging contaminants in water and soil, various techniques can be used, out of which high-resolution mass spectrometry (LC-HRMS) and solid phase emission (SPE) are the most prominent [70, 71].

Sampling and Preconcentration

Passive sampling and chromatography in the water at very low concentrations-tandem mass spectrometry in liquid water detect EPs. For a passive sampling of EWPs, Polar Organic Chemical Integrative Sampler (POCIS) can also be used [55]. Another way of detecting trace amounts of EWPs in water is fluid-liquid micro-extraction followed by electro-analysis, developed by Gabbana *et al.* [72]. Another method of EWPs determination was reported to use 1-butyl-3-methylimidazolium hexafluorophosphate [C4MIM] [PF6] ionic fluids (IL) and was found to be effective [73].

Improvement and Optimization of SPE

To analyze large numbers of EPs, SPE followed by LC and GC coupled with MS will consider a quick, accurate, and precise process. Current studies focus on optimizing and enhancing the adsorbent extraction process. A new trend of the miniaturization of the SPE system maintains extraction and preconcentration efficiency while reducing the use of organic solvents. New extraction systems mounted on on-site buoys are also essential to build. A first low-pressure micro-SPEs (m-SPE) model was developed to minimize the laboratory sample volumes and solvents compared to conventional SPE for *in-situ* EPs monitoring. In marine gliders and boats on robotic or automated systems, this low-pressure m-SPE method is anticipated [74]. In preparing the cartridge (a column of 6 ml vacuum) equipped with a bottom mixture of 350 mg Strata X-CW, Strata X-AW, and Isolute Env+ and a top mixture of 200 mg of Oasis HLB, with a ratio of 1/1.5 (X-AW / X-CW / Env+), a novel sample preparation method based on the process of SPE was developed, which proved better in recovering chemical sample width [75]. Testing and synthesizing hybrid SP cartridges of N-isopropyl acrylamide polygraph and their function is significant in developing SPE hybrid portable hybrid thermal responsive beds [76]. Compared to Oasis prime cartridges for SPE and Oasis HLB (up to 40 percent), 13 EPs in wastewater and eight EPs in surface water were recovered [77]. The benefit of a prime cartridge is to escape the conditions and decrease the preparation of the sample. The use of solvents of the same sorbent (N-vinylpyrrolidone and copolymer divinylbenzene) is found in both. This advantage is slight 'greener' by using the prime cartridge.

Emerging Methods

In matrix-rich aqueous samples, to increase the signal-to-noise ratio for target compounds, atmospheric pressure chemical ionization (APCI) and atmospheric

pressure photoionization (APPI) studies have been implemented to use more specific ionization to increase selectivity, decrease matrix interference [78, 79]. Liquid chromatography-tandem atmospheric pressure chemical ionization/atmospheric pressure photoionization with hybrid quadrupole/orbital trap mass spectrometry operated in high-resolution product scan mode (LC-APCI/APP-
-HRPS) for the reliable analytical measuring of medroxyprogesterone acetate in wastewater and surface water [80].

One of the key problems researchers are facing is how to increase policy makers' acceptability or take timely control steps. Gago-Ferrero *et al*. [81] reported that advanced LC-Suspect Screening Strategies (HRMS) were used in the 24-hour composite sampling of the selected compounds from the effluent of 3 key WWTP in Sweden. It showed that the classification of potentially dangerous substances could be used efficiently for priority purposes following recent developments in high-resolution mass spectrometry (HRMS). QuEChERS is a widely used technique in most pesticide testing facilities; it is fast, simple, inexpensive, reliable, rugged, and safe. Miossec *et al*. [82] found that the method can be modified and validated *via* a GC-MS (Gas Chromatography-Mass Spectrometry) to evaluate targets and EPs in sediments simultaneously. Changing the composition of the extraction solvent and enhancing the steps of cleaning and evaporation, the sample preparation was adapted. The process demonstrates strong linearity and repetitiveness.

Strong and lightweight material Bucky paper (BP) is a material constructed by carbon nanotube compression. Tomai *et al*. [83] announced that the Stir-disc SPE module oxidized BP was used for the first time as a sorbent membrane. Organic micropollutants were extracted in dynamic modus from water samples, and HPLC-MS/MS then analyzed the extracts. They showed a weak affinity to the compounds with log P <1, compounds with log P >1, showing recoveries of 50 to 100 percent depending on their log P and pKa, and compounds between 6 and 7.5 showed low yields independent of their log P.

Nanotechnology-based Electrochemical Sensors

Based on Metal and Metal Oxide NPs

Different NPs, such as AuNPs, with various morphological features [28], Ni NPs Metal-oxide [84], CdS NPs [85], TiO_2 nanotubes [86], and ZnO nano-tetrapod [87], used in sensing of the ECPs. Also, Au NPs with a tamer has been updated [88]. Xia *et al*. [89] stated that immobile estrogen receptors supported on Au NPs modified lipid membranes (bilayer) have been developed to monitor and detect estrogen-related chemicals in water samples. Alkasir *et al*. [84] used NPs as tyrosinase-immobilization matrices for BPA biosensing. The Ni NPs showed the greatest sensitivity compared to Fe_3O_4 NPs and AuNPs (BPA limits of detection (LOD) was 7.1 ~ 10−9 M).

Based on Other NMs

Additional NMs were also used in the ECPs electrochemical sensing. In the case of a selective water sample determination of BPA, for example, Co Te QDs and dendrimer poly (amidoamine) were used by Yin *et al*. [90]. The range of LODs from 1 to 5 was $10^{-9.}$ M. Wang *et al*. [91] documented a BPA electrochemical silica MCM-41-based sensor. The MCM-41 showed a greater signal enhancement ability with silica gel, activated carbon, graphite and CNT, supplying LOD 3.8 to 10-8 M. It is reported that the sensor array for water detection cancer-borne trihalomethane (THMs), a class of DBPs was developed. The sensor range has been developed to generate nanostructured films through the deposit on audio-interdigital electrodes of ducting and natural polymers. THMs were 0.02 mg/L in the LOD.

THE FATE OF EMERGING WATER POLLUTANTS

Transportation Through River Water *via* Adsorption on Suspended Solids and Sediments

Depending mainly on their sorption of suspended solid/sedimentary content, the PPCPs, and other ECs seem to have decreased at different rates in different rivers. For example, anti-epileptic carbamazepine tends to slow down in river water even more than galactoside, a musk compound typically used in soap and detergent fragrances [92]. This attenuation, however, does not necessarily signify the removal of the compound. PPCP/ECs adsorption on suspended substances may be the reason for attenuation in river water and sediments, as certain compounds are usually suspended rather than dissolved [93]. For example, tetracycline strongly suspends other antibiotic pollutants such as quinolones or sulfonamides. The transport of such compounds depends on solid adsorptive material in the water environment [94]. In deciding the environmental fate, the physiochemical properties of pollutants are more important than the adsorbing material properties [95].

Luo *et al*. [93] explained the CEC and the OM fraction of the solid material. The suspended sediment significantly impacts pollutant adsorption (increasing potential for exchange and increased organic matter) and attenuation. The reduction of contaminant adsorption into a solid material also depends on reducing the contact time with the pollutant. The increased river flow inhibits the adsorption of sediments and suspended solids more than the effect of tetracycline on sulphonamides, quinolones, and macrolide antibiotics adsorption [93].

Compounds with basic characteristics and hydrophobic tend to be ideally bound to solid materials which remain suspended and on sediments. There is often no detection of the dissolved aquatic zone of such preferentially bound contaminants with sediments and solids [11]. Water pollutants, such as monomethylates of nonylphenol, occur up to 9.5 and 4.7 times higher in suspended solid materials and sediments than in flowing water of the river [96]. Similarly, in the River Estuary of Columbia (Northwest USA), Gregg *et al.* [94] showed that PAH binding is up to 10 times higher than that seen on the water. Other common pollutants, such as gemfibrozil, erythromycin, diazepam, and metoprolol's, are present in aquatic settings rather than solid fractions, mainly if not exclusively dissolved [96].

However, while the chemicals are typically stored in a mass unit (*e.g.*, ng / g dry weight) with suspended material quantities, it must be pointed out that a more reliable distribution relation can be made, as the material suspended has been separated from its mass volume (*e.g.*, ng / L water separating the suspended material), studies have shown that ECs and other PPCP concentrations are less than suspended solids in river sediments and water [11]. These results, particularly concerning the degradation of the product's sediment and suspended solids adsorption, remain uncertain.

Transportation Through Atmosphere and Precipitation

For several volatile PPCPs and other ECs, the air has been recognized as a means of transport and a source [97, 98]. The emerging air and precipitate contaminants are often identified as alcoholic fluorotelomers (FTOHs). With world production Figs. (**1.1 - 1.4**) million kg/year, FTOHs have been widely produced mainly as acrylic polymers-based fluorotelomer with adhesive manufacturing properties, packaging of paper, and repellents. At 33 locations in Japan, the two most common atmospheric FTOH compounds, 8:2 FTOH and 6:2 FTOH, were recorded at < LOD-768 pg / m^3 and 32 to 247 pg / m^3, respectively [97]. In Japanese precipitation and surface waters, the FTOH compound was also estimated at 0.21 to 1.97 ng / L and 0.16 to 3.38 ng/L [98].

Through surface water or STW volatilization and atmospheric emissions, volatile emerging pollutants can enter the atmosphere during processing or incineration. Salgueiro-González *et al.* [99] proposed that many typical endocrine disruptive compounds may have less than 10 μm diameter associated with airborne particulate matter (PM). Indoor air PM contaminant levels are frequently significantly higher than outdoor air, resulting in poor ventilation and increased usage of plastics and spray cleaners for many other products [99]. Outdoor air levels of commonly found phthalates that have been reported as a level that trigger endocrine disruption are di-ethylhexyl phthalate (DEHP) (55 ng / m^3) [100], dibutyl phthalates (DBP) (30.5 ng/m^3) [101], and nonylphenol (NP) 45 ng/m^3 [102] respectively. DBP, DEHP, and NP were observed in indoor air at considerably higher levels of up to 1000 ng / m^3, 2300 ng / m^3, and 680 ng / m^3, respectively [101, 103]. To determine if these airborne endocrine-disrupting compounds can affect the ecosystem, further research on human health and environmental effects and potential sources of these pollutants are needed [99].

In precipitation, other pollutants such as atrazine herbicide have been identified separately from transition zones, which can enter the atmosphere by applying them as sprays to soils [104]. In addition to endocrine-disrupting compounds, 52 of the 89-households screened for 50-1500 ng / m^3, including plasticizers, emulsifiers, nonylphenols, and adhesives (4-tert-butylphenol) [103]. In the arable drift of sprays, some ECs, such as in the growth of fruit trees, these compounds may also be carried atmospherically. It is unknown to what degree these volatile compounds will migrate to the atmosphere until they reach the aquatic stage and are precipitously deposited on the surface.

Transportation to Groundwater Table

Most environmental research based on ECs focuses on waste and surface water, although it focuses less on groundwater [37]. ECs can leach from subsoil to groundwater and have been identified in aquifers since the 1990s [105]. ECs can come from many sources in groundwater, but wastewater is a primary source established. The principal sources are private wastewater treatment plants, animal waste lagoons, and wastewater leachates [31]. The controlled artificial refilling (*e.g.*, ingestion by banks) of partially treated wastewater may also be a significant source of ECs in groundwater [106]. Diffuse sources include waste, pesticides, biosolids, and atmospheric deposition of sewage sludge [106, 107]. Several studies in the US [108, 109], and Europe [37, 110] describe the occurrence of ECs in groundwater. According to Lapworth *et al.* [31], since 1993, a significant number of PCPs, and industrial and caffeine compounds (*e.g.*, carbamazepine and ibuprofen) (102 −104 ng/L) have been published. Previous studies showed that ECs are more abundant in surface water than groundwater [111, 112]. Moreover, the most common ECs listed in soil water vary from those in surface water [111, 112]. For example, Vulliet and Cren-Olive [112], reported that the chemical pollutants detected in surface water were not detected in groundwater in France. The main EC source in the aquatic world is wastewater effluent, released into surface water. However, in areas without or with a limited subsoil layer, groundwater systems are more susceptible to contamination by ECs [113]. ECs have demonstrated higher karst aquifers in the French, Italian, and United Kingdom groundwater [113, 114]. The exchange of surface water is another route by which ECs are transported to groundwater in addition to penetration through the surface layer [31]. Understanding ECs receptor pathways, sources, toxicity levels and mechanisms are still missing [39].

Transportation to Agricultural Soils *via* Effluent Irrigation

Climate changes and global warming have been widely recognized in the last few decades. Therefore, water supply and water management problems in semi-arid and arid areas are significant [115]. Agriculture, the major water user, will likely face the most threatened because of water shortages. Several comprehensive systems for the management of water resources are being introduced or taken into account, especially with a focus on the use of existing water supplies and new water resources, such as wastewater treatment (WWT) [116]. Consequently, TWW reuse is already a well-known method worldwide, primarily for irrigation [117]. While re-usage of TWW has gained interest as an alternative to crop nutrient requirements through irrigation, significant progress has been made in producing healthy treated re-use effluents (*e.g.*, chemical oxygen, nutrients, successful metal removal, demand down to low levels). TWW can also contain harmful chemicals that can significantly threaten the environment and health, often organic components and pathogens [118]. As a result, there are some serious concerns about the presence and eventual release of organic micro contaminants to the environment through TWW irrigation, given that many of these micro contaminants are relatively little awareness of their possible toxicity, including flame retardants, pharmaceuticals, personal care products, antibiotics, endocrine disrupting compounds, chemicals, plasticizers, transformation products, surfactants, pesticides, and pharmaceuticals, *etc.* Such contaminants are referred to as contaminants of emerging concern (CECs). In terms of the applications for recycling (*e.g.*, rinsing, floating water replenishment, surface storage for subsequent water), the available/applied treatment technology does not fully remove CECs because no combined knowledge is available on CECs' fate in biomass of activated sludge, the TWWs and agricultural environments *e.g.* ground/surface water, soil, and plants [118].

CECs are also released into the environment through various human activities, especially through pharmaceutical active substances (PhACs), including direct disposal of unused medicines or expired medicinal items, released from pharmaceutical production facilities and hospitals, and veterinary drugs [119]. Also, most PPCs in human and animal bodies are not fully absorbed and metabolized. At the same time, a substantial proportion (10% to 90%) of the dose administered is excreted within hours of application by urine and feces [120]. Thus, PhACs directly enter the atmosphere by adding animal manures and excesses through pastures [121]. The use of biosolids as soil modification and the re-use of TWW for irrigation contributes to the incorporation of CECs in agricultural settings [122]. In biosolid and manure modified soil, surface, and groundwater systems and sediments receiving TWW-irrigated fields, CECs are thus detected and rushed from these fields [113, 123, 124].

IMPACT OF EMERGING WATER POLLUTANTS

Water Quality

PPCPs are used for human and animal health. Human pharmaceuticals are of great importance, such as antimicrobials, antibiotics, antianxiety, analgesics, antidiabetics, antipsychotics, anticonvulsant/antiepileptics, and beta-blockers (β-blockers), cytostatic drugs, antihistamines, X-ray contrast media, hormonal and estrogens compounds, lipid-regulators antineoplastic, and stimulants. These compounds may be metabolized in the atmosphere directly or partially, leading to their eventual passage. There are several routes by which PPCPs can travel into our natural waters, for example, inappropriate disposal by individuals, hospital residues, and use of veterinary pharmaceuticals on animal farms such as animals, pigs, turkeys, poultry, *etc.* The issues related to human use of the PPCPs and their disposal in natural waters have been largely ignored in recent decades, particularly for antibiotics and steroids. A Geological Survey study conducted in 2002 demonstrated the concentration of PPCPs in water. PPCPs can be calculated in around 110 out of 139 susceptible streams in 30 states [125]. PPCPs were identified by untreated and refined water samples collected across Missouri, with caffeine detected at a concentration of 224 ng / L. Some water treatment plants in Germany have an average of 0.22 μg / L of aspirin, and an average concentration in some water treatment plants could be above 13 μg/L in Greece and Spain [126 - 128]. Benotti *et al.* [129] have screened various drug products, suspected endocrine-disrupting substances, and other nonregulated organic pollutants from 19 U.S. utilities. Due to the wide variety of PCPs transferred to our natural water resources, assessing and eliminating PCPs in water or drinking water sources is important because their environmental and human health risks are not obvious.

Natural organic matter or other chemicals (natural or man-made) by-products of water (DBPs) are potentially threatening to human health (DBPs). The N-nitrosamine group is one of the most widely researched DBPs since these compounds are extremely toxic. The most found nitrosamine in drinking water is N-nitroso dimethylamine (NDMA). It was initially found in Ontario, Canada, in chlorinated drinking water. According to research in Asia and the United States, in both water sources (surface water and groundwater) and drinking water, NDMA is present in low-ng / L [130, 131]. A detailed analysis of the development of NDMA and associated analogs on nitrosamines [132].

Chlorination of N-nitrosamine precursors in water with high concentrations may lead to higher NDMA formation. Recently, Wang *et al.* [133] developed nine N nitrosamines and four precursors of potable water, and the HPLC-MS / MS system for simultaneous quantitative analyses has been created. Their water MDLs varied from 0.05 µg / L to 5 µg / L without pre-concentration. For the water supply samples, the method to analyze nine MDLs ranging from 0.2-0.9 ng / L and 0.1-0.7 ng / L was developed for finished samples.

Plants and Human Health

Xenobiotics are substances not produced by living species, including personal care products, various toxins, pharmaceuticals, and pesticides in the ecosystem [134]. Xenobiotics can be extracted by processing, hospitals, and household waste [135]. Also, xenobiotic products can be disposed of in landfills and agricultural runoff and released into the environment, soil, and water bodies [136]. The potential effects on target and non-target organisms should be considered due to the variety of xenobiotics present in the environment. Aquatic species, plants, animals, and humans are not intended to trigger such xenobiotics. Plants can be exposed directly or indirectly to a wide variety of xenobiotics. Plant growth and physiology may be influenced after exposure to xenobiotics, and xenobiotics could be transformed, conjugated, and compartmentalized by some plants [137, 138]. Several techniques of assessing the effects of xenobiotics on plants have been applied, including evaluation of plant morphology and seed germination and direct xenobiotic concentration measurements in tissue [139, 140]. Cowpea (*Vigna unguiculata*) biotransformation products of ibuprofen have been discovered in research on the seeds, roots, and shoots. HRMS-LC is the parent identity and 46 metabolites, including hydroxylated derivatives and amino acid conjugates.

Another research investigated the morphological, physiological metabolite changes in lettuce (*Lactuca sativa*), exposed through the double-dimensional GC-HRMS eleven emerging concern contaminants. The concentrations of lettuce leaves were subsequently calculated for total benzophenone, butylated hydroxytoluene, phenazone, bisphenol A carbamazepine, 5-Methyl--H-benzotriazole, methylparaben, caffeine, 4-octylphenol, triclosan, and tris (2-Choroethyl) phosphate. Dosage-dependent decreases in stem width and the chlorophyll a and b concentration variance were observed in the morphological and physiological changes. The highest lettuce accumulation was found in pharmaceutical carbamazepine. After 34 days, the metabolite profiling induced major changes in the metabolites, which are affected by the carbohydrate metabolism, TCA cycle, the pentose phosphate pathways, and the glutathione pathway. The amino acid 5–oxyproline was most strongly upregulated and demonstrates increased activities in glutathione detoxification pathways as a metabolite intermediate for glutathione metabolism. Furthermore, lettuce's morphological and physiological modifications have also caused other metabolite modifications [141].

Some researchers [142] studied pesticides in rice leaves and seeds and transgene crossbreeding with the aim GC-LRMS approach to metabolism. The results showed that rice leaves were upregulated with antioxidants, phenols, and carbohydrates, while amino acids and other carbohydrates were downregulated with the rice crop. Organochlorine (Lindane and Chlordecone) pesticides were tested for their effect on the roots of maize. At least 26 metabolic pathways have been traced to 8 biochemical tracts in Kyoto Plant Metabolic Network and Gene Encyclopaedia.

When metabolomics in another nanoparticle and plant was investigated to understand spinach leaves metabolite changes after 7 days of a spray of foliar Cu $(OH)_2$ nano pesticides. Metabolomics has been shown to down-regulate metabolites such as tocopherol, threnodic acid, ferulic acid, ascorbic acid, and other phenolic compounds [143]. Cucumber is another food crop in which metabolite adjustments under exposure to nanoparticles have been studied. In one study, regulation of various fruit metal ions and molecules and regulation of citric acid, myo-inositol, ornithine, 1-kestosis, and cucumbers (up to 800 mg/kg) in one study were observed [144].

Pollutants from the PhAC, PCP, and EDC penetrate sources and can surpass reasonable limits. EDC pollutants and the widespread occurrence of ECs in water are highly likely to irrigate crops with contaminated water and pose a human health risk [145]. ECs may have detrimental impacts on wildlife, aquatic life, terrestrial animals, and humans. In animals and humans, endocrine chemistry destabilizes various reproductive and sexual defects. This subject matter will hinder the production and signalization of the endocrine system in pre-and post-natal life. The effects are permanent and often irreversible during growth. Water resource management for ECs is a crucial problem and needs to be addressed, particularly in delicate and increasingly growing ecosystems. But ECs' ecological results in the natural world vary from laboratory studies. Where several atmospheric factors are present, including pH, the form of soil or water, ionizable compounds, *etc.*, can affect the bioavailability of Emerging Contaminants [146, 147]. Using both structural and non-structural approaches, a comprehensive system-wide (source transfer date) reduction structure is required.

Aquatic Life

Human activities like industrial effluents, irrigation, urban waste management, and growing urbanization have posed a major threat to the freshwater environment [148, 149]. These pollution levels have impacted aquatic flora and fauna habitats [150]. It is, therefore, necessary to recognize key pollution activities and the origin and destiny of the marine environment, including the global distribution of water to freshwater organisms, and to ensure that ecosystem function [151, 152] To estimate the effect of emerging water contaminants on aquatic organisms, numerous studies have been performed.

Wang *et al.* [153] measured the endocrine-damaging effect on estrogenic receptors by tert-butylphenols, androgen receptors, thyroid hormone receptors, and freshwater aquatic toxicity. *The experiment used Photobacterium phosphorus, Vibrio fischeri, and Chlamydomonas reinhardtii as aquatic indicator species.* The exposure to the phenolic compounds demonstrated similar antagonistic responses to human oestrogen and androgen receptors. Morales *et al.* [154] studied the modulation of bisphenol expression and stress reactions in freshwater snail genes Physacuta following chemical exposure up to 96 hrs. The receptor and retinoid receptors were evaluated. 17β-Trenbolone endocrine disruptor that alters the ecological behavior in the freshwater environment of female Gambusia holbrooki mosquito fish. The results of the study showed that when exposed to endocrine disrupting chemicals, the behaviours of the fish were modified in several ways. Some of the changes include rapid feeding habits, less stimulus time, faster response to prey species and significant quantities of prey consumption [155]. Villa *et al.* [156] studied the behavioural impact of pharmaceuticals on larvae of the wild alpine species *Diamesa zernyi*; the study used a video monitoring device to analyze behavioural reactions due to prescription contamination (furosemide, trimethoprim, and ibuprofen). It was discovered from the study that due to the exposure to the chemicals, the larvae experienced changes in the normal distances traveled, the average speed of movement, and the pattern of body bending. However, an evaluation was performed on the toxicity of the chemicals.

The effects of oxytetracycline on biochemical pathways and neurotransmission levels of Oncorhynchus mykiss freshwater fish to produce energy have been investigated by Rodrigues *et al.* [157]. The study results revealed that AChE inhibition and changes in dehydrogenase activity of the eyes and gills were observed after exposure to oxytetracycline. Mechanisms for neurotransmission, including O-disruption, have also been found to interfere. In the case of freshwater fish, *Labeo rohita*, Renuka *et al.* [158] conducted a report on the reaction of erythromycin antioxidants up to 100 µg/L. The study found that behaviors in both SOD and CAT were decreasing and increasing in fingerlings. *L. Rohita* focused on concentration levels and duration of induction of erythromycin, the antioxidant activity on the livers was also more pronounced than on the fingerlings' gills. Parolini *et al.* [159] measured the Danio rerio embryo protein profiles tested at 0.3-1.0 µg/L concentration levels up to 9 hrs from cocaine and metabolites exposure (Benzoylecgonine and Ecgonine Methyl ester). Illegal drugs and their metabolites caused the fish protein to change its profile, modulating several functional proteins. Low cocaine and freshwater metabolites have been shown to cause changes in the action mechanisms of young zebrafish. According to Musee [160], triclosan and triclocarban, from personal care products to freshwater organisms in Pretoria, South Africa, have been carried out an environmental risk assessment. The study used a designed approach by evaluating the QR associated with Triclosan and Triclocarban toxicity for the concentration of such pollutants in freshwater ecosystems. The study found that the adverse effects of chemicals on the aquatic environment were observed because QR was calculated to be >1.

CONCLUSION

Several categories of EWPs reported worldwide but not limited to these, such as PPCPs, flame retardants, pesticides, surfactants, by-products from industries, and by-products from water treatment facilities, ionic liquids, and food additives. It is necessary to develop methodology and techniques to monitor these EWPs urgently. These techniques and methods must focus on the sources, fate, and impact of these pollutants of emerging concern. It is very difficult to manage EWPs. Regulations should be included in the national and global policies to manage the contaminants, which have been already studied, and the ones that need to be studied. For this purpose, it is necessary to include a science-based policy to monitor and deal with the affected sites. From a scientific perspective, it is essential to develop the latest and most effective techniques to overcome the hazard of these EWPs. There is not enough data available about the adequacy of currently available techniques. Few studies reported the effective removal of EWPs using current techniques, while some showed concern over the presently used methods to remove EWPs. More comprehensive techniques should be developed. To comprehensively analysed the impact of EWPs on the environment, chemical and biological techniques should be combined. Management-related practices should also be implemented, such as reducing the release of emerging pollutants, careful handling, proper disposal of contaminated waste, *etc*.

CONSENT FOR PUBLICATION

Not applicable.

CONFLICT OF INTEREST

The author declares no conflict of interest, financial or otherwise.

ACKNOWLEDGEMENTS

Declared none.

REFERENCES

[1]   S. Sharma, and A. Bhattacharya, "Drinking water contamination and treatment techniques", *Appl. Water Sci.,* vol. 7, no. 3, pp. 1043-1067, 2016.
        [http://dx.doi.org/10.1007/s13201-016-0455-7]

[2]   R.P. Schwarzenbach, "The Challenge of Micropollutants in Aquatic Systems", In: *Science (80-.)* vol. 313. , 2006, pp. 1072-1077.
        [http://dx.doi.org/10.1126/science.1127291]

[3]   T. Deblonde, C. Cossu-Leguille, and P. Hartemann, "Emerging pollutants in wastewater: a review of the literature", *Int. J. Hyg. Environ. Health,* vol. 214, no. 6, pp. 442-448, 2011.
        [http://dx.doi.org/10.1016/j.ijheh.2011.08.002] [PMID: 21885335]

[4]   S. Chowdhury, and R. Balasubramanian, "Recent advances in the use of graphene-family nanoadsorbents for removal of toxic pollutants from wastewater", *Adv. Colloid Interface Sci.,* vol. 204, pp. 35-56, 2014.
        [http://dx.doi.org/10.1016/j.cis.2013.12.005] [PMID: 24412086]

[5]   J.O. Tijani, O.O. Fatoba, O.O. Babajide, and L.F. Petrik, "Pharmaceuticals, endocrine disruptors, personal care products, nanomaterials and perfluorinated pollutants: a review", *Environ. Chem. Lett.,* vol. 14, no. 1, pp. 27-49, 2015.
        [http://dx.doi.org/10.1007/s10311-015-0537-z]

[6]   J. Wilkinson, P.S. Hooda, J. Barker, S. Barton, and J. Swinden, "Occurrence, fate and transformation of emerging contaminants in water: An overarching review of the field", *Environ. Pollut.,* vol. 231, no. Pt 1, pp. 954-970, 2017.
        [http://dx.doi.org/10.1016/j.envpol.2017.08.032] [PMID: 28888213]

[7]   S.A. Snyder, P. Westerhoff, Y. Yoon, and D.L. Sedlak, "Pharmaceuticals, Personal Care Products, and Endocrine Disruptors in Water: Implications for the Water Industry", *Environ. Eng. Sci.,* vol. 20, no. 5, pp. 449-469, 2003.
        [http://dx.doi.org/10.1089/109287503768335931]

[8]   K. Balakrishna, A. Rath, Y. Praveenkumarreddy, K.S. Guruge, and B. Subedi, "A review of the occurrence of pharmaceuticals and personal care products in Indian water bodies", *Ecotoxicol. Environ. Saf.,* vol. 137, pp. 113-120, 2017.
        [http://dx.doi.org/10.1016/j.ecoenv.2016.11.014] [PMID: 27915141]

[9]   Y. Yang, Y.S. Ok, K.H. Kim, E.E. Kwon, and Y.F. Tsang, "Occurrences and removal of pharmaceuticals and personal care products (PPCPs) in drinking water and water/sewage treatment plants: A review", *Sci. Total Environ.,* vol. 596-597, pp. 303-320, 2017.
        [http://dx.doi.org/10.1016/j.scitotenv.2017.04.102] [PMID: 28437649]

[10]   G.W. Olsen, S.C. Chang, P.E. Noker, G.S. Gorman, D.J. Ehresman, P.H. Lieder, and J.L. Butenhoff, "A comparison of the pharmacokinetics of perfluorobutanesulfonate (PFBS) in rats, monkeys, and humans", *Toxicology,* vol. 256, no. 1-2, pp. 65-74, 2009.
        [http://dx.doi.org/10.1016/j.tox.2008.11.008] [PMID: 19059455]

[11]   B.F. da Silva, A. Jelic, R. López-Serna, A.A. Mozeto, M. Petrovic, and D. Barceló, "Occurrence and distribution of pharmaceuticals in surface water, suspended solids and sediments of the Ebro river basin, Spain", *Chemosphere,* vol. 85, no. 8, pp. 1331-1339, 2011.
        [http://dx.doi.org/10.1016/j.chemosphere.2011.07.051] [PMID: 21880345]

[12]   B.C. Gordalla, U. Ewers, and F.H. Frimmel, "Hydraulic fracturing: a toxicological threat for groundwater and drinking-water?", *Environ. Earth Sci.,* vol. 70, no. 8, pp. 3875-3893, 2013.

[http://dx.doi.org/10.1007/s12665-013-2672-9]

[13] J. Corrales, L.A. Kristofco, W.B. Steele, B.S. Yates, C.S. Breed, E.S. Williams, and B.W. Brooks, "Global assessment of bisphenol a in the environment: Review and analysis of its occurrence and bioaccumulation", *Dose Response,* vol. 13, no. 3, p. 1559325815598308, 2015.
[http://dx.doi.org/10.1177/1559325815598308] [PMID: 26674671]

[14] V.L. Colvin, "The potential environmental impact of engineered nanomaterials", *Nat. Biotechnol.,* vol. 21, no. 10, pp. 1166-1170, 2003.
[http://dx.doi.org/10.1038/nbt875] [PMID: 14520401]

[15] M.R. Wiesner, G.V. Lowry, K.L. Jones, M.F. Hochella Jr, R.T. Di Giulio, E. Casman, and E.S. Bernhardt, "Decreasing uncertainties in assessing environmental exposure, risk, and ecological implications of nanomaterials", *Environ. Sci. Technol.,* vol. 43, no. 17, pp. 6458-6462, 2009.
[http://dx.doi.org/10.1021/es803621k] [PMID: 19764202]

[16] E.J. Rosi-Marshall, D. Snow, S.L. Bartelt-Hunt, A. Paspalof, and J.L. Tank, "A review of ecological effects and environmental fate of illicit drugs in aquatic ecosystems", *J. Hazard. Mater.,* vol. 282, pp. 18-25, 2015.
[http://dx.doi.org/10.1016/j.jhazmat.2014.06.062] [PMID: 25062553]

[17] C. G. Daughton, and T. A. Ternes, "Pharmaceuticals and personal care products in the environment: agents of subtle change?", *Environ. Health Perspect,* vol. 107, pp. 907-938, 1999.
[http://dx.doi.org/10.1289/ehp.99107s6907]

[18] K.D. Floate, K.G. Wardhaugh, A.B.A. Boxall, and T.N. Sherratt, "Fecal residues of veterinary parasiticides: nontarget effects in the pasture environment", *Annu. Rev. Entomol.,* vol. 50, no. 1, pp. 153-179, 2005.
[http://dx.doi.org/10.1146/annurev.ento.50.071803.130341] [PMID: 15471531]

[19] D. Pascoe, W. Karntanut, and C.T. Müller, "Do pharmaceuticals affect freshwater invertebrates? A study with the cnidarian Hydra vulgaris", *Chemosphere,* vol. 51, no. 6, pp. 521-528, 2003.
[http://dx.doi.org/10.1016/S0045-6535(02)00860-3] [PMID: 12615105]

[20] C.R. Tyler, S. Jobling, and J.P. Sumpter, "Endocrine disruption in wildlife: a critical review of the evidence", *Crit. Rev. Toxicol.,* vol. 28, no. 4, pp. 319-361, 1998.
[http://dx.doi.org/10.1080/10408449891344236] [PMID: 9711432]

[21] E. Kim, S-H. Kim, H-C. Kim, S.G. Lee, S.J. Lee, and S.W. Jeong, "Growth inhibition of aquatic plant caused by silver and titanium oxide nanoparticles", *Toxicol. Environ. Health Sci.,* vol. 3, no. 1, pp. 1-6, 2011.
[http://dx.doi.org/10.1007/s13530-011-0071-8]

[22] A. I. Stefanakis, and J. A. Becker, "A Review of Emerging Contaminants in Water", *Practice, Progress, and Proficiency in Sustainability.* pp. 55-80.
[http://dx.doi.org/10.4018/978-1-4666-9559-7.ch003]

[23] J.B. Ellis, "Pharmaceutical and personal care products (PPCPs) in urban receiving waters", *Environ. Pollut.,* vol. 144, no. 1, pp. 184-189, 2006.
[http://dx.doi.org/10.1016/j.envpol.2005.12.018]

[24] A. C. Gore, *Introduction to Endocrine-Disrupting Chemicals,* Humana press, pp. 3-8, .
[http://dx.doi.org/10.1007/1-59745-107-X_1]

[25] B.O. Clarke, and S.R. Smith, "Review of 'emerging' organic contaminants in biosolids and assessment of international research priorities for the agricultural use of biosolids", *Environ. Int.,* vol. 37, no. 1, pp. 226-247, 2011.
[http://dx.doi.org/10.1016/j.envint.2010.06.004] [PMID: 20797791]

[26] F. Rahman, K.H. Langford, M.D. Scrimshaw, and J.N. Lester, "Polybrominated diphenyl ether (PBDE) flame retardants", *Sci. Total Environ.,* vol. 275, no. 1-3, pp. 1-17, 2001.
[http://dx.doi.org/10.1016/S0048-9697(01)00852-X] [PMID: 11482396]

[27]   M.I. Sohail, "Environmental application of nanomaterials: A promise to sustainable future", *Eng. Nanomater. Phytonanotechnology Challenges Plant Sustain.,* vol. 87, p. 1, 2019.
[http://dx.doi.org/10.1016/bs.coac.2019.10.002]

[28]   Y. Zhu, X. Liu, Y. Hu, R. Wang, M. Chen, J. Wu, Y. Wang, S. Kang, Y. Sun, and M. Zhu, "Behavior, remediation effect and toxicity of nanomaterials in water environments", *Environ. Res.,* vol. 174, pp. 54-60, 2019.
[http://dx.doi.org/10.1016/j.envres.2019.04.014] [PMID: 31029942]

[29]   A-M.O. Mohamed, and E.K. Paleologos, "Emerging Pollutants: Fate, Pathways, and Bioavailability", In: *Fundamentals of Geoenvironmental Engineering,* 2018.
[http://dx.doi.org/10.1016/B978-0-12-804830-6.00010-7]

[30]   P.K. Seth, "Chemical contaminants in water and associated health hazards", In: *Water and Health,* 2014.
[http://dx.doi.org/10.1007/978-81-322-1029-0_22]

[31]   D.J. Lapworth, N. Baran, M.E. Stuart, and R.S. Ward, "Emerging organic contaminants in groundwater: A review of sources, fate and occurrence", *Environ. Pollut.,* vol. 163, pp. 287-303, 2012.
[http://dx.doi.org/10.1016/j.envpol.2011.12.034] [PMID: 22306910]

[32]   V.T. Nguyen, M. Reinhard, and G.Y-H. Karina, "Occurrence and source characterization of perfluorochemicals in an urban watershed", *Chemosphere,* vol. 82, no. 9, pp. 1277-1285, 2011.
[http://dx.doi.org/10.1016/j.chemosphere.2010.12.030] [PMID: 21208640]

[33]   J.M. Diamond, H.A. Latimer II, K.R. Munkittrick, K.W. Thornton, S.M. Bartell, and K.A. Kidd, "Prioritizing contaminants of emerging concern for ecological screening assessments", *Environ. Toxicol. Chem.,* vol. 30, no. 11, pp. 2385-2394, 2011.
[http://dx.doi.org/10.1002/etc.667] [PMID: 22002713]

[34]   "Contaminants of Emerging Environmental Concern", *American Society of Civil Engineers,* 2009.

[35]   M. Petrovic, "Fate and removal of pharmaceuticals and illicit drugs in conventional and membrane bioreactor wastewater treatment plants and by riverbank filtration", In: *Philos. Trans. R. Soc. A Math. Phys. Eng. Sci* vol. 367. , 2009, pp. 3979-4003.
[http://dx.doi.org/10.1098/rsta.2009.0105]

[36]   M.S. Kostich, A.L. Batt, S.T. Glassmeyer, and J.M. Lazorchak, "Predicting variability of aquatic concentrations of human pharmaceuticals", *Sci. Total Environ.,* vol. 408, no. 20, pp. 4504-4510, 2010.
[http://dx.doi.org/10.1016/j.scitotenv.2010.06.015] [PMID: 20619877]

[37]   A. Jurado, E. Vàzquez-Suñé, J. Carrera, M. López de Alda, E. Pujades, and D. Barceló, "Emerging organic contaminants in groundwater in Spain: a review of sources, recent occurrence and fate in a European context", *Sci. Total Environ.,* vol. 440, pp. 82-94, 2012.
[http://dx.doi.org/10.1016/j.scitotenv.2012.08.029] [PMID: 22985674]

[38]   V.K. Gupta, and I. Ali, "Wastewater Treatment by Biological Methods", In: *Environmental Water.* Elsevier, 2013, pp. 179-204.
[http://dx.doi.org/10.1016/B978-0-444-59399-3.00007-6]

[39]   L. Lamastra, M. Balderacchi, and M. Trevisan, "Inclusion of emerging organic contaminants in groundwater monitoring plans", *MethodsX,* vol. 3, pp. 459-476, 2016.
[http://dx.doi.org/10.1016/j.mex.2016.05.008] [PMID: 27366676]

[40]   A.S. Adeleye, J.R. Conway, K. Garner, Y. Huang, Y. Su, and A.A. Keller, "Engineered nanomaterials for water treatment and remediation: Costs, benefits, and applicability", *Chem. Eng. J.,* vol. 286, pp. 640-662, 2016.
[http://dx.doi.org/10.1016/j.cej.2015.10.105]

[41]   P.C.H. Li, E.J. Swanson, and F.A.P.C. Gobas, "Diazinon and its degradation products in agricultural water courses in British Columbia, Canada", *Bull. Environ. Contam. Toxicol.,* vol. 69, no. 1, pp. 59-65, 2002.

[http://dx.doi.org/10.1007/s00128-002-0010-0] [PMID: 12053258]

[42]   L.R. Zimmerman, R.J. Schneider, and E.M. Thurman, "Analysis and detection of the herbicides dimethenamid and flufenacet and their sulfonic and oxanilic acid degradates in natural water", *J. Agric. Food Chem.,* vol. 50, no. 5, pp. 1045-1052, 2002.
[http://dx.doi.org/10.1021/jf010779b] [PMID: 11853478]

[43]   A. Laganà, A. Bacaloni, I. De Leva, A. Faberi, G. Fago, and A. Marino, "Occurrence and determination of herbicides and their major transformation products in environmental waters", *Anal. Chim. Acta,* vol. 462, no. 2, pp. 187-198, 2002.
[http://dx.doi.org/10.1016/S0003-2670(02)00351-3]

[44]   A.B.A. Boxall, *"Veterinary Medicines in the Environment,"* Reviews of Environmental Contamination and Toxicology. Springer New York, 2004, pp. 1-91.
[http://dx.doi.org/10.1007/0-387-21729-0_1]

[45]   K. Lyons, and G. Scrinis, *"Under the regulatory radar? Nanotechnologies and their impacts for rural Australia,"* Tracking Rural Change: Community, Policy and Technology in Australia, New Zealand and Europe. ANU Press, 2009.
[http://dx.doi.org/10.22459/TRC.04.2009.08]

[46]   E. Topp, S.C. Monteiro, A. Beck, B.B. Coelho, A.B. Boxall, P.W. Duenk, S. Kleywegt, D.R. Lapen, M. Payne, L. Sabourin, H. Li, and C.D. Metcalfe, "Runoff of pharmaceuticals and personal care products following application of biosolids to an agricultural field", *Sci. Total Environ.,* vol. 396, no. 1, pp. 52-59, 2008.
[http://dx.doi.org/10.1016/j.scitotenv.2008.02.011] [PMID: 18377955]

[47]   J. Lu, J. Wu, P.J. Stoffella, and P. Chris Wilson, "Isotope dilution-gas chromatography/mass spectrometry method for the analysis of alkylphenols, bisphenol A, and estrogens in food crops", *J. Chromatogr. A,* vol. 1258, pp. 128-135, 2012.
[http://dx.doi.org/10.1016/j.chroma.2012.08.033] [PMID: 22947482]

[48]   J.R. Masoner, D.W. Kolpin, E.T. Furlong, I.M. Cozzarelli, J.L. Gray, and E.A. Schwab, "Contaminants of emerging concern in fresh leachate from landfills in the conterminous United States", *Environ. Sci. Process. Impacts,* vol. 16, no. 10, pp. 2335-2354, 2014.
[http://dx.doi.org/10.1039/C4EM00124A] [PMID: 25111596]

[49]   S. Keerthanan, C. Jayasinghe, J.K. Biswas, and M. Vithanage, "Pharmaceutical and Personal Care Products (PPCPs) in the environment: Plant uptake, translocation, bioaccumulation, and human health risks", *Crit. Rev. Environ. Sci. Technol.,* pp. 1-38, 2020.
[http://dx.doi.org/10.1080/10643389.2020.1753634]

[50]   E.N. Evgenidou, I.K. Konstantinou, and D.A. Lambropoulou, "Occurrence and removal of transformation products of PPCPs and illicit drugs in wastewaters: a review", *Sci. Total Environ.,* vol. 505, pp. 905-926, 2015.
[http://dx.doi.org/10.1016/j.scitotenv.2014.10.021] [PMID: 25461093]

[51]   R.P. Tasho, and J.Y. Cho, "Veterinary antibiotics in animal waste, its distribution in soil and uptake by plants: A review", *Sci. Total Environ.,* vol. 563-564, pp. 366-376, 2016.
[http://dx.doi.org/10.1016/j.scitotenv.2016.04.140] [PMID: 27139307]

[52]   L.M. Madikizela, S. Ncube, and L. Chimuka, "Uptake of pharmaceuticals by plants grown under hydroponic conditions and natural occurring plant species: A review", *Sci. Total Environ.,* vol. 636, pp. 477-486, 2018.
[http://dx.doi.org/10.1016/j.scitotenv.2018.04.297] [PMID: 29709865]

[53]   R.S. Al-Farsi, M. Ahmed, A. Al-Busaidi, and B.S. Choudri, "Translocation of pharmaceuticals and personal care products (PPCPs) into plant tissues: A review", *Emerg. Contam.,* vol. 3, no. 4, pp. 132-137, 2017.
[http://dx.doi.org/10.1016/j.emcon.2018.02.001]

[54]   B.P.S. Capece, G.L. Virkel, and C.E. Lanusse, "Enantiomeric behaviour of albendazole and

fenbendazole sulfoxides in domestic animals: pharmacological implications", *Vet. J.,* vol. 181, no. 3, pp. 241-250, 2009.
[http://dx.doi.org/10.1016/j.tvjl.2008.11.010] [PMID: 19124257]

[55]   A.K. Sarmah, M.T. Meyer, and A.B.A. Boxall, "A global perspective on the use, sales, exposure pathways, occurrence, fate and effects of veterinary antibiotics (VAs) in the environment", *Chemosphere,* vol. 65, no. 5, pp. 725-759, 2006.
[http://dx.doi.org/10.1016/j.chemosphere.2006.03.026] [PMID: 16677683]

[56]   D. Basu, and S.K. Gupta, "Biodegradation of 1,1,2,2-tetrachloroethane in Upflow anaerobic sludge blanket (UASB) reactor", *Bioresour. Technol.,* vol. 101, no. 1, pp. 21-25, 2010.
[http://dx.doi.org/10.1016/j.biortech.2009.06.074] [PMID: 19699084]

[57]   Q. Sun, M. Lv, M. Li, and C-P. Yu, "Personal Care Products in the Aquatic Environment in China", In: *The Handbook of Environmental Chemistry.* Springer International Publishing, 2014, pp. 73-94.
[http://dx.doi.org/10.1007/698_2014_284]

[58]   B.D. Blair, J.P. Crago, C.J. Hedman, and R.D. Klaper, "Pharmaceuticals and personal care products found in the Great Lakes above concentrations of environmental concern", *Chemosphere,* vol. 93, no. 9, pp. 2116-2123, 2013.
[http://dx.doi.org/10.1016/j.chemosphere.2013.07.057] [PMID: 23973285]

[59]   J-L. Liu, and M-H. Wong, "Pharmaceuticals and personal care products (PPCPs): a review on environmental contamination in China", *Environ. Int.,* vol. 59, pp. 208-224, 2013.
[http://dx.doi.org/10.1016/j.envint.2013.06.012] [PMID: 23838081]

[60]   J.M. Philip, U.K. Aravind, and C.T. Aravindakumar, "Emerging contaminants in Indian environmental matrices - A review", *Chemosphere,* vol. 190, pp. 307-326, 2018.
[http://dx.doi.org/10.1016/j.chemosphere.2017.09.120] [PMID: 28992484]

[61]   J. Lu, J. Wu, C. Zhang, and Y. Zhang, "Possible effect of submarine groundwater discharge on the pollution of coastal water: Occurrence, source, and risks of endocrine disrupting chemicals in coastal groundwater and adjacent seawater influenced by reclaimed water irrigation", *Chemosphere,* vol. 250, p. 126323, 2020.
[http://dx.doi.org/10.1016/j.chemosphere.2020.126323] [PMID: 32126332]

[62]   E. Gonzales-Gustavson, M. Rusiñol, G. Medema, M. Calvo, and R. Girones, "Quantitative risk assessment of norovirus and adenovirus for the use of reclaimed water to irrigate lettuce in Catalonia", *Water Res.,* vol. 153, pp. 91-99, 2019.
[http://dx.doi.org/10.1016/j.watres.2018.12.070] [PMID: 30703677]

[63]   C. Wu, A.L. Spongberg, J.D. Witter, M. Fang, and K.P. Czajkowski, "Uptake of pharmaceutical and personal care products by soybean plants from soils applied with biosolids and irrigated with contaminated water", *Environ. Sci. Technol.,* vol. 44, no. 16, pp. 6157-6161, 2010.
[http://dx.doi.org/10.1021/es1011115] [PMID: 20704212]

[64]   Z. Li, X. Xiang, M. Li, Y. Ma, J. Wang, and X. Liu, "Occurrence and risk assessment of pharmaceuticals and personal care products and endocrine disrupting chemicals in reclaimed water and receiving groundwater in China", *Ecotoxicol. Environ. Saf.,* vol. 119, pp. 74-80, 2015.
[http://dx.doi.org/10.1016/j.ecoenv.2015.04.031] [PMID: 25982733]

[65]   G-G. Ying, R.S. Kookana, and P. Dillon, "Sorption and degradation of selected five endocrine disrupting chemicals in aquifer material", *Water Res.,* vol. 37, no. 15, pp. 3785-3791, 2003.
[http://dx.doi.org/10.1016/S0043-1354(03)00261-6] [PMID: 12867347]

[66]   J. Gong, D. Duan, Y. Yang, Y. Ran, and D. Chen, "Seasonal variation and partitioning of endocrine disrupting chemicals in waters and sediments of the Pearl River system, South China", *Environ. Pollut.,* vol. 219, pp. 735-741, 2016.
[http://dx.doi.org/10.1016/j.envpol.2016.07.015] [PMID: 27431692]

[67]   J-M. Jian, Y. Guo, L. Zeng, L. Liang-Ying, X. Lu, F. Wang, and E.Y. Zeng, "Global distribution of perfluorochemicals (PFCs) in potential human exposure source-A review", *Environ. Int.,* vol. 108, pp.

51-62, 2017.
[http://dx.doi.org/10.1016/j.envint.2017.07.024] [PMID: 28800414]

[68]   A. N. Sharpley, "Future agriculture with minimized phosphorus losses to waters: Research needs and direction", *Ambio,* vol. 44, pp. S163-S179, 2015.
[http://dx.doi.org/10.1007/s13280-014-0612-x]

[69]   J. Wen, Z. Li, N. Luo, M. Huang, R. Yang, and G. Zeng, "Investigating organic matter properties affecting the binding behavior of heavy metals in the rhizosphere of wetlands", *Ecotoxicol. Environ. Saf.,* vol. 162, pp. 184-191, 2018.
[http://dx.doi.org/10.1016/j.ecoenv.2018.06.083] [PMID: 29990730]

[70]   M.A. Asghar, Q. Zhu, S. Sun, Y. Peng, and Q. Shuai, "Suspect screening and target quantification of human pharmaceutical residues in the surface water of Wuhan, China, using UHPLC-Q-Orbitrap HRMS", *Sci. Total Environ.,* vol. 635, pp. 828-837, 2018.
[http://dx.doi.org/10.1016/j.scitotenv.2018.04.179] [PMID: 29710606]

[71]   N. Hermes, K.S. Jewell, A. Wick, and T.A. Ternes, "Quantification of more than 150 micropollutants including transformation products in aqueous samples by liquid chromatography-tandem mass spectrometry using scheduled multiple reaction monitoring", *J. Chromatogr. A,* vol. 1531, pp. 64-73, 2018.
[http://dx.doi.org/10.1016/j.chroma.2017.11.020] [PMID: 29183669]

[72]   J.V. Gabbana, L.H. de Oliveira, G.C. Paveglio, and M.A.G. Trindade, "Narrowing the interface between sample preparation and electrochemistry: Trace-level determination of emerging pollutant in water samples after in situ microextraction and electroanalysis using a new cell configuration", *Electrochim. Acta,* vol. 275, pp. 67-75, 2018.
[http://dx.doi.org/10.1016/j.electacta.2018.04.134]

[73]   F. Abujaber, M. Zougagh, S. Jodeh, Á. Ríos, F.J. Guzmán Bernardo, and R.C. Rodríguez Martín-Doimeadios, "Magnetic cellulose nanoparticles coated with ionic liquid as a new material for the simple and fast monitoring of emerging pollutants in waters by magnetic solid phase extraction", *Microchem. J.,* vol. 137, pp. 490-495, 2018.
[http://dx.doi.org/10.1016/j.microc.2017.12.007]

[74]   B. Abaroa-Pérez, G. Sánchez-Almeida, J.J. Hernández-Brito, and D. Vega-Moreno, "In Situ Miniaturised Solid Phase Extraction (m-SPE) for Organic Pollutants in Seawater Samples", *J. Anal. Methods Chem.,* vol. 2018, p. 7437031, 2018.
[http://dx.doi.org/10.1155/2018/7437031] [PMID: 29805837]

[75]   V. Osorio, M. Schriks, D. Vughs, P. de Voogt, and A. Kolkman, "A novel sample preparation procedure for effect-directed analysis of micro-contaminants of emerging concern in surface waters", *Talanta,* vol. 186, pp. 527-537, 2018.
[http://dx.doi.org/10.1016/j.talanta.2018.04.058] [PMID: 29784397]

[76]   S.A. Jadhav, R. Nisticò, G. Magnacca, and D. Scalarone, "Packed hybrid silica nanoparticles as sorbents with thermo-switchable surface chemistry and pore size for fast extraction of environmental pollutants", *RSC Advances,* vol. 8, no. 3, pp. 1246-1254, 2018.
[http://dx.doi.org/10.1039/C7RA11869D] [PMID: 35540902]

[77]   M. Česen, D. Heath, M. Krivec, T. Košmrlj, T. Kosjek, and E. Heath, "Seasonal and spatial variations in the occurrence, mass loadings and removal of compounds of emerging concern in the Slovene aqueous environment and environmental risk assessment", *Environ. Pollut.,* vol. 242, no. Pt A, pp. 143-154, 2018.
[http://dx.doi.org/10.1016/j.envpol.2018.06.052] [PMID: 29966838]

[78]   M.F. Mirabelli, and R. Zenobi, "Solid-Phase Microextraction Coupled to Capillary Atmospheric Pressure Photoionization-Mass Spectrometry for Direct Analysis of Polar and Nonpolar Compounds", *Anal. Chem.,* vol. 90, no. 8, pp. 5015-5022, 2018.
[http://dx.doi.org/10.1021/acs.analchem.7b04514] [PMID: 29537821]

[79]　B. Yuan, J.P. Benskin, C-E.L. Chen, and Å. Bergman, "Determination of Chlorinated Paraffins by Bromide-Anion Attachment Atmospheric-Pressure Chemical Ionization Mass Spectrometry", *Environ. Sci. Technol. Lett.*, vol. 5, no. 6, pp. 348-353, 2018.
[http://dx.doi.org/10.1021/acs.estlett.8b00216]

[80]　O. Golovko, P. Šauer, G. Fedorova, H.K. Kroupová, and R. Grabic, "Determination of progestogens in surface and waste water using SPE extraction and LC-APCI/APPI-HRPS", *Sci. Total Environ.*, vol. 621, pp. 1066-1073, 2018.
[http://dx.doi.org/10.1016/j.scitotenv.2017.10.120] [PMID: 29102184]

[81]　P. Gago-Ferrero, A. Krettek, S. Fischer, K. Wiberg, and L. Ahrens, "Suspect Screening and Regulatory Databases: A Powerful Combination To Identify Emerging Micropollutants", *Environ. Sci. Technol.*, vol. 52, no. 12, pp. 6881-6894, 2018.
[http://dx.doi.org/10.1021/acs.est.7b06598] [PMID: 29782800]

[82]　C. Miossec, L. Lanceleur, and M. Monperrus, "Adaptation and validation of QuEChERS method for the simultaneous analysis of priority and emerging pollutants in sediments by gas chromatography—mass spectrometry", *Int. J. Environ. Anal. Chem.*, vol. 98, no. 8, pp. 695-708, 2018.
[http://dx.doi.org/10.1080/03067319.2018.1496245]

[83]　P. Tomai, A. Martinelli, S. Morosetti, R. Curini, S. Fanali, and A. Gentili, "Oxidized Buckypaper for Stir-Disc Solid Phase Extraction: Evaluation of Several Classes of Environmental Pollutants Recovered from Surface Water Samples", *Anal. Chem.*, vol. 90, no. 11, pp. 6827-6834, 2018.
[http://dx.doi.org/10.1021/acs.analchem.8b00927] [PMID: 29706074]

[84]　R.S.J. Alkasir, M. Ganesana, Y-H. Won, L. Stanciu, and S. Andreescu, "Enzyme functionalized nanoparticles for electrochemical biosensors: a comparative study with applications for the detection of bisphenol A", *Biosens. Bioelectron.*, vol. 26, no. 1, pp. 43-49, 2010.
[http://dx.doi.org/10.1016/j.bios.2010.05.001] [PMID: 20605712]

[85]　E. Valera, A. Muriano, I. Pividori, F. Sánchez-Baeza, and M-P. Marco, "Development of a Coulombimetric immunosensor based on specific antibodies labeled with CdS nanoparticles for sulfonamide antibiotic residues analysis and its application to honey samples", *Biosens. Bioelectron.*, vol. 43, pp. 211-217, 2013.
[http://dx.doi.org/10.1016/j.bios.2012.12.017] [PMID: 23313703]

[86]　B. Lu, M. Liu, H. Shi, X. Huang, and G. Zhao, "A Novel Photoelectrochemical Sensor for Bisphenol A with High Sensitivity and Selectivity Based on Surface Molecularly Imprinted Polypyrrole Modified TiO2Nanotubes", *Electroanalysis,* vol. 25, no. 3, pp. 771-779, 2013.
[http://dx.doi.org/10.1002/elan.201200585]

[87]　A. Qurashi, J.A. Rather, K. De Wael, B. Merzougui, N. Tabet, and M. Faiz, "Rapid microwave synthesis of high aspect-ratio ZnO nanotetrapods for swift bisphenol A detection", *Analyst (Lond.),* vol. 138, no. 17, pp. 4764-4768, 2013.
[http://dx.doi.org/10.1039/c3an00336a] [PMID: 23801275]

[88]　Y. Zhu, P. Chandra, K-M. Song, C. Ban, and Y-B. Shim, "Label-free detection of kanamycin based on the aptamer-functionalized conducting polymer/gold nanocomposite", *Biosens. Bioelectron.*, vol. 36, no. 1, pp. 29-34, 2012.
[http://dx.doi.org/10.1016/j.bios.2012.03.034] [PMID: 22542925]

[89]　W. Xia, Y. Li, Y. Wan, T. Chen, J. Wei, Y. Lin, and S. Xu, "Electrochemical biosensor for estrogenic substance using lipid bilayers modified by Au nanoparticles", *Biosens. Bioelectron.*, vol. 25, no. 10, pp. 2253-2258, 2010.
[http://dx.doi.org/10.1016/j.bios.2010.03.004] [PMID: 20353888]

[90]　H. Yin, Y. Zhou, S. Ai, Q. Chen, X. Zhu, X. Liu, and L. Zhu, "Sensitivity and selectivity determination of BPA in real water samples using PAMAM dendrimer and CoTe quantum dots modified glassy carbon electrode", *J. Hazard. Mater.*, vol. 174, no. 1-3, pp. 236-243, 2010.
[http://dx.doi.org/10.1016/j.jhazmat.2009.09.041] [PMID: 19782469]

[91]   F. Wang, J. Yang, and K. Wu, "Mesoporous silica-based electrochemical sensor for sensitive determination of environmental hormone bisphenol A", *Anal. Chim. Acta,* vol. 638, no. 1, pp. 23-28, 2009.
[http://dx.doi.org/10.1016/j.aca.2009.02.013] [PMID: 19298875]

[92]   K. Osenbrück, H.R. Gläser, K. Knöller, S.M. Weise, M. Möder, R. Wennrich, M. Schirmer, F. Reinstorf, W. Busch, and G. Strauch, "Sources and transport of selected organic micropollutants in urban groundwater underlying the city of Halle (Saale), Germany", *Water Res.,* vol. 41, no. 15, pp. 3259-3270, 2007.
[http://dx.doi.org/10.1016/j.watres.2007.05.014] [PMID: 17575997]

[93]   Y. Luo, L. Xu, M. Rysz, Y. Wang, H. Zhang, and P.J.J. Alvarez, "Occurrence and transport of tetracycline, sulfonamide, quinolone, and macrolide antibiotics in the Haihe River Basin, China", *Environ. Sci. Technol.,* vol. 45, no. 5, pp. 1827-1833, 2011.
[http://dx.doi.org/10.1021/es104009s] [PMID: 21309601]

[94]   T. Gregg, F.G. Prahl, and B.R.T. Simoneit, "Suspended particulate matter transport of polycyclic aromatic hydrocarbons in the lower columbia river and its estuary", *Limnol. Oceanogr.,* vol. 60, no. 6, pp. 1935-1949, 2015.
[http://dx.doi.org/10.1002/lno.10144]

[95]   J.L. Wilkinson, P.S. Hooda, J. Swinden, J. Barker, and S. Barton, "Spatial distribution of organic contaminants in three rivers of Southern England bound to suspended particulate material and dissolved in water", *Sci. Total Environ.,* vol. 593-594, pp. 487-497, 2017.
[http://dx.doi.org/10.1016/j.scitotenv.2017.03.167] [PMID: 28360000]

[96]   L. Patrolecco, S. Capri, S. De Angelis, R. Pagnotta, S. Polesello, and S. Valsecchi, "Partition of Nonylphenol and Related Compounds Among Different Aquatic Compartments in Tiber River (Central Italy)", *Water Air Soil Pollut.,* vol. 172, no. 1–4, pp. 151-166, 2006.
[http://dx.doi.org/10.1007/s11270-005-9067-9]

[97]   S. Oono, K.H. Harada, M.A.M. Mahmoud, K. Inoue, and A. Koizumi, "Current levels of airborne polyfluorinated telomers in Japan", *Chemosphere,* vol. 73, no. 6, pp. 932-937, 2008.
[http://dx.doi.org/10.1016/j.chemosphere.2008.06.069] [PMID: 18701130]

[98]   M.A.M. Mahmoud, A. Kärrman, S. Oono, K.H. Harada, and A. Koizumi, "Polyfluorinated telomers in precipitation and surface water in an urban area of Japan", *Chemosphere,* vol. 74, no. 3, pp. 467-472, 2009.
[http://dx.doi.org/10.1016/j.chemosphere.2008.08.029] [PMID: 19081600]

[99]   N. Salgueiro-González, M.J. López de Alda, S. Muniategui-Lorenzo, D. Prada-Rodríguez, and D. Barceló, ""Analysis and occurrence of endocrine-disrupting chemicals in airborne particles," TrAC", *Trends Analyt. Chem.,* vol. 66, pp. 45-52, 2015.
[http://dx.doi.org/10.1016/j.trac.2014.11.006]

[100] N. Salgueiro-González, M. López de Alda, S. Muniategui-Lorenzo, D. Prada-Rodríguez, and D. Barceló, "Determination of 13 estrogenic endocrine disrupting compounds in atmospheric particulate matter by pressurised liquid extraction and liquid chromatography-tandem mass spectrometry", *Anal. Bioanal. Chem.,* vol. 405, no. 27, pp. 8913-8923, 2013.
[http://dx.doi.org/10.1007/s00216-013-7298-y] [PMID: 24005601]

[101] I. Saito, A. Onuki, and H. Seto, "Indoor air pollution by alkylphenols in Tokyo", *Indoor Air,* vol. 14, no. 5, pp. 325-332, 2004.
[http://dx.doi.org/10.1111/j.1600-0668.2004.00250.x] [PMID: 15330792]

[102] M. Salapasidou, C. Samara, and D. Voutsa, "Endocrine disrupting compounds in the atmosphere of the urban area of Thessaloniki, Greece", *Atmos. Environ.,* vol. 45, no. 22, pp. 3720-3729, 2011.
[http://dx.doi.org/10.1016/j.atmosenv.2011.04.025]

[103] R.A. Rudel, D.E. Camann, J.D. Spengler, L.R. Korn, and J.G. Brody, "Phthalates, alkylphenols, pesticides, polybrominated diphenyl ethers, and other endocrine-disrupting compounds in indoor air

and dust", *Environ. Sci. Technol.,* vol. 37, no. 20, pp. 4543-4553, 2003.
[http://dx.doi.org/10.1021/es0264596] [PMID: 14594359]

[104] T. Hayes, K. Haston, M. Tsui, A. Hoang, C. Haeffele, and A. Vonk, "Atrazine-induced hermaphroditism at 0.1 ppb in American leopard frogs (Rana pipiens): laboratory and field evidence", *Environ. Health Perspect.,* vol. 111, no. 4, pp. 568-575, 2003.
[http://dx.doi.org/10.1289/ehp.5932] [PMID: 12676617]

[105] T. Heberer, "Occurrence, fate, and removal of pharmaceutical residues in the aquatic environment: a review of recent research data", *Toxicol. Lett.,* vol. 131, no. 1-2, pp. 5-17, 2002.
[http://dx.doi.org/10.1016/S0378-4274(02)00041-3] [PMID: 11988354]

[106] M. Stuart, D. Lapworth, E. Crane, and A. Hart, "Review of risk from potential emerging contaminants in UK groundwater", *Sci. Total Environ.,* vol. 416, pp. 1-21, 2012.
[http://dx.doi.org/10.1016/j.scitotenv.2011.11.072] [PMID: 22209399]

[107] L. Feng, E.D. van Hullebusch, M.A. Rodrigo, G. Esposito, and M.A. Oturan, "Removal of residual anti-inflammatory and analgesic pharmaceuticals from aqueous systems by electrochemical advanced oxidation processes. A review", *Chem. Eng. J.,* vol. 228, pp. 944-964, 2013.
[http://dx.doi.org/10.1016/j.cej.2013.05.061]

[108] K.K. Barnes, D.W. Kolpin, E.T. Furlong, S.D. Zaugg, M.T. Meyer, and L.B. Barber, "A national reconnaissance of pharmaceuticals and other organic wastewater contaminants in the United States--I) groundwater", *Sci. Total Environ.,* vol. 402, no. 2-3, pp. 192-200, 2008.
[http://dx.doi.org/10.1016/j.scitotenv.2008.04.028] [PMID: 18556047]

[109] M.J. Focazio, D.W. Kolpin, K.K. Barnes, E.T. Furlong, M.T. Meyer, S.D. Zaugg, L.B. Barber, and M.E. Thurman, "A national reconnaissance for pharmaceuticals and other organic wastewater contaminants in the United States--II) untreated drinking water sources", *Sci. Total Environ.,* vol. 402, no. 2-3, pp. 201-216, 2008.
[http://dx.doi.org/10.1016/j.scitotenv.2008.02.021] [PMID: 18433838]

[110] B. Lopez, P. Ollivier, A. Togola, N. Baran, and J-P. Ghestem, "Screening of French groundwater for regulated and emerging contaminants", *Sci. Total Environ.,* vol. 518-519, pp. 562-573, 2015.
[http://dx.doi.org/10.1016/j.scitotenv.2015.01.110] [PMID: 25782024]

[111] E.J. Tiedeken, A. Tahar, B. McHugh, and N.J. Rowan, "Monitoring, sources, receptors, and control measures for three European Union watch list substances of emerging concern in receiving waters - A 20year systematic review", *Sci. Total Environ.,* vol. 574, pp. 1140-1163, 2017.
[http://dx.doi.org/10.1016/j.scitotenv.2016.09.084] [PMID: 27741430]

[112] E. Vulliet, and C. Cren-Olivé, "Screening of pharmaceuticals and hormones at the regional scale, in surface and groundwaters intended to human consumption", *Environ. Pollut.,* vol. 159, no. 10, pp. 2929-2934, 2011.
[http://dx.doi.org/10.1016/j.envpol.2011.04.033] [PMID: 21570166]

[113] R. Meffe, and I. de Bustamante, "Emerging organic contaminants in surface water and groundwater: a first overview of the situation in Italy", *Sci. Total Environ.,* vol. 481, pp. 280-295, 2014.
[http://dx.doi.org/10.1016/j.scitotenv.2014.02.053] [PMID: 24602913]

[114] D.J. Lapworth, N. Baran, M.E. Stuart, K. Manamsa, and J. Talbot, "Persistent and emerging micro-organic contaminants in Chalk groundwater of England and France", *Environ. Pollut.,* vol. 203, pp. 214-225, 2015.
[http://dx.doi.org/10.1016/j.envpol.2015.02.030] [PMID: 25882715]

[115] H. March, O. Therond, and D. Leenhardt, "Water futures: Reviewing water-scenario analyses through an original interpretative framework", *Ecol. Econ.,* vol. 82, pp. 126-137, 2012.
[http://dx.doi.org/10.1016/j.ecolecon.2012.07.006]

[116] M. Milano, D. Ruelland, S. Fernandez, A. Dezetter, J. Fabre, and E. Servat, "Facing climatic and anthropogenic changes in the Mediterranean basin: What will be the medium-term impact on water stress?", *C. R. Geosci.,* vol. 344, no. 9, pp. 432-440, 2012.

[http://dx.doi.org/10.1016/j.crte.2012.07.006]

[117] Y. Zhang, and Y. Shen, "Wastewater irrigation: past, present, and future", *Wiley Interdiscip. Rev. Water,* vol. 6, no. 3, p. e1234, 2017.
[http://dx.doi.org/10.1002/wat2.1234]

[118] D. Fatta-Kassinos, I.K. Kalavrouziotis, P.H. Koukoulakis, and M.I. Vasquez, "The risks associated with wastewater reuse and xenobiotics in the agroecological environment", *Sci. Total Environ.,* vol. 409, no. 19, pp. 3555-3563, 2011.
[http://dx.doi.org/10.1016/j.scitotenv.2010.03.036] [PMID: 20435343]

[119] A. Grossberger, Y. Hadar, T. Borch, and B. Chefetz, "Biodegradability of pharmaceutical compounds in agricultural soils irrigated with treated wastewater", *Environ. Pollut.,* vol. 185, pp. 168-177, 2014.
[http://dx.doi.org/10.1016/j.envpol.2013.10.038] [PMID: 24286691]

[120] Y-L. Zhang, S-S. Lin, C-M. Dai, L. Shi, and X-F. Zhou, "Sorption-desorption and transport of trimethoprim and sulfonamide antibiotics in agricultural soil: effect of soil type, dissolved organic matter, and pH", *Environ. Sci. Pollut. Res. Int.,* vol. 21, no. 9, pp. 5827-5835, 2014.
[http://dx.doi.org/10.1007/s11356-014-2493-8] [PMID: 24443047]

[121] M. Pan, and L.M. Chu, "Adsorption and degradation of five selected antibiotics in agricultural soil", *Sci. Total Environ.,* vol. 545-546, pp. 48-56, 2016.
[http://dx.doi.org/10.1016/j.scitotenv.2015.12.040] [PMID: 26745292]

[122] O. Borgman, and B. Chefetz, "Combined effects of biosolids application and irrigation with reclaimed wastewater on transport of pharmaceutical compounds in arable soils", *Water Res.,* vol. 47, no. 10, pp. 3431-3443, 2013.
[http://dx.doi.org/10.1016/j.watres.2013.03.045] [PMID: 23591105]

[123] N. Gottschall, E. Topp, C. Metcalfe, M. Edwards, M. Payne, S. Kleywegt, P. Russell, and D.R. Lapen, "Pharmaceutical and personal care products in groundwater, subsurface drainage, soil, and wheat grain, following a high single application of municipal biosolids to a field", *Chemosphere,* vol. 87, no. 2, pp. 194-203, 2012.
[http://dx.doi.org/10.1016/j.chemosphere.2011.12.018] [PMID: 22300554]

[124] Y. Luo, W. Guo, H.H. Ngo, L.D. Nghiem, F.I. Hai, J. Zhang, S. Liang, and X.C. Wang, "A review on the occurrence of micropollutants in the aquatic environment and their fate and removal during wastewater treatment", *Sci. Total Environ.,* vol. 473-474, pp. 619-641, 2014.
[http://dx.doi.org/10.1016/j.scitotenv.2013.12.065] [PMID: 24394371]

[125] D.W. Kolpin, E.T. Furlong, M.T. Meyer, E.M. Thurman, S.D. Zaugg, L.B. Barber, and H.T. Buxton, "Pharmaceuticals, hormones, and other organic wastewater contaminants in U.S. streams, 1999-2000: a national reconnaissance", *Environ. Sci. Technol.,* vol. 36, no. 6, pp. 1202-1211, 2002.
[http://dx.doi.org/10.1021/es011055j] [PMID: 11944670]

[126] M.J. Martínez Bueno, "Pilot survey of chemical contaminants from industrial and human activities in river waters of Spain", *Int. J. Environ. Anal. Chem.,* vol. 90, no. 3–6, pp. 321-343, 2010.
[http://dx.doi.org/10.1080/03067310903045463]

[127] J.A. Dougherty, P.W. Swarzenski, R.S. Dinicola, and M. Reinhard, "Occurrence of herbicides and pharmaceutical and personal care products in surface water and groundwater around Liberty Bay, Puget Sound, Washington", *J. Environ. Qual.,* vol. 39, no. 4, pp. 1173-1180, 2010.
[http://dx.doi.org/10.2134/jeq2009.0189] [PMID: 20830904]

[128] H. Li, P.A. Helm, and C.D. Metcalfe, "Sampling in the Great Lakes for pharmaceuticals, personal care products, and endocrine-disrupting substances using the passive polar organic chemical integrative sampler", *Environ. Toxicol. Chem.,* vol. 29, no. 4, pp. 751-762, 2010.
[http://dx.doi.org/10.1002/etc.104] [PMID: 20821503]

[129] M.J. Benotti, R.A. Trenholm, B.J. Vanderford, J.C. Holady, B.D. Stanford, and S.A. Snyder, "Pharmaceuticals and endocrine disrupting compounds in U.S. drinking water", *Environ. Sci. Technol.,* vol. 43, no. 3, pp. 597-603, 2009.

[http://dx.doi.org/10.1021/es801845a] [PMID: 19244989]

[130] N.V. Huy, M. Murakami, H. Sakai, K. Oguma, K. Kosaka, M. Asami, and S. Takizawa, "Occurrence and formation potential of N-nitrosodimethylamine in ground water and river water in Tokyo", *Water Res.,* vol. 45, no. 11, pp. 3369-3377, 2011.
[http://dx.doi.org/10.1016/j.watres.2011.03.053] [PMID: 21514620]

[131] W. Wang, S. Ren, H. Zhang, J. Yu, W. An, J. Hu, and M. Yang, "Occurrence of nine nitrosamines and secondary amines in source water and drinking water: Potential of secondary amines as nitrosamine precursors", *Water Res.,* vol. 45, no. 16, pp. 4930-4938, 2011.
[http://dx.doi.org/10.1016/j.watres.2011.06.041] [PMID: 21843899]

[132] S.H. Mhlongo, B.B. Mamba, and R.W. Krause, "Review: Nitrosamines: A review on their prevalence as emerging pollutants and potential remediation options", *Water S.A.,* vol. 35, no. 5, 2009.
[http://dx.doi.org/10.4314/wsa.v35i5.49200]

[133] C. Wang, H. Shi, C.D. Adams, T. Timmons, and Y. Ma, "Investigation of removal of N-nitrosamines and their precursors in water treatments using activated carbon nanoparticles", *Int. J. Environ. Technol. Manag.,* vol. 16, no. 1/2, p. 34, 2013.
[http://dx.doi.org/10.1504/IJETM.2013.050682]

[134] E. Croom, "Metabolism of Xenobiotics of Human Environments", In: *Progress in Molecular Biology and Translational Science.* Elsevier, 2012, pp. 31-88.
[http://dx.doi.org/10.1016/B978-0-12-415813-9.00003-9]

[135] G. Byrns, "The fate of xenobiotic organic compounds in wastewater treatment plants", *Water Res.,* vol. 35, no. 10, pp. 2523-2533, 2001.
[http://dx.doi.org/10.1016/S0043-1354(00)00529-7] [PMID: 11394788]

[136] B. Nowack, and T.D. Bucheli, "Occurrence, behavior and effects of nanoparticles in the environment", *Environ. Pollut.,* vol. 150, no. 1, pp. 5-22, 2007.
[http://dx.doi.org/10.1016/j.envpol.2007.06.006] [PMID: 17658673]

[137] R.P. Singh, *A., Prasad, S. M., & Singh, "Plant Responses to Xenobiotics.* Springer Singapore, 2016.
[http://dx.doi.org/10.1007/978-981-10-2860-1]

[138] H. Sandermann Jr, "Plant metabolism of xenobiotics", *Trends Biochem. Sci.,* vol. 17, no. 2, pp. 82-84, 1992.
[http://dx.doi.org/10.1016/0968-0004(92)90507-6] [PMID: 1566333]

[139] I. Navarro, A. de la Torre, P. Sanz, M.Á. Porcel, J. Pro, G. Carbonell, and M.L. Martínez, "Uptake of perfluoroalkyl substances and halogenated flame retardants by crop plants grown in biosolids-amended soils", *Environ. Res.,* vol. 152, pp. 199-206, 2017.
[http://dx.doi.org/10.1016/j.envres.2016.10.018] [PMID: 27792944]

[140] P.A. Herklotz, P. Gurung, B. Vanden Heuvel, and C.A. Kinney, "Uptake of human pharmaceuticals by plants grown under hydroponic conditions", *Chemosphere,* vol. 78, no. 11, pp. 1416-1421, 2010.
[http://dx.doi.org/10.1016/j.chemosphere.2009.12.048] [PMID: 20096438]

[141] C. Hurtado, H. Parastar, V. Matamoros, B. Piña, R. Tauler, and J.M. Bayona, "Linking the morphological and metabolomic response of Lactuca sativa L exposed to emerging contaminants using GC × GC-MS and chemometric tools", *Sci. Rep.,* vol. 7, no. 1, p. 6546, 2017.
[http://dx.doi.org/10.1038/s41598-017-06773-0] [PMID: 28747703]

[142] K. Dreher, "Putting the Plant Metabolic Network Pathway Databases to Work: Going Offline to Gain New Capabilities", In: *Methods in Molecular Biology.* Humana Press, 2013, pp. 151-171.
[http://dx.doi.org/10.1007/978-1-62703-661-0_10]

[143] "Metabolomics Reveals Cu(OH)2 Nanopesticide-Activated Anti-oxidative Pathways and Decreased Beneficial Antioxidants in Spinach Leaves", *American Chemical Society (ACS),* 2016.

[144] "1H NMR and GC-MS Based Metabolomics Reveal Defense and Detoxification Mechanism of Cucumber Plant under Nano-Cu Stress", *American Chemical Society (ACS),* 2016.

[145] C. Hurtado, C. Domínguez, L. Pérez-Babace, N. Cañameras, J. Comas, and J.M. Bayona, "Estimate of uptake and translocation of emerging organic contaminants from irrigation water concentration in lettuce grown under controlled conditions", *J. Hazard. Mater.,* vol. 305, pp. 139-148, 2016.
[http://dx.doi.org/10.1016/j.jhazmat.2015.11.039] [PMID: 26651071]

[146] J.W. Nichols, B. Du, J.P. Berninger, K.A. Connors, C.K. Chambliss, R.J. Erickson, A.D. Hoffman, and B.W. Brooks, "Observed and modeled effects of pH on bioconcentration of diphenhydramine, a weakly basic pharmaceutical, in fathead minnows", *Environ. Toxicol. Chem.,* vol. 34, no. 6, pp. 1425-1435, 2015.
[http://dx.doi.org/10.1002/etc.2948] [PMID: 25920411]

[147] B. Du, S.P. Haddad, A. Luek, W.C. Scott, G.N. Saari, S.R. Burket, C.S. Breed, M. Kelly, L. Broach, J.B. Rasmussen, C.K. Chambliss, and B.W. Brooks, "Bioaccumulation of human pharmaceuticals in fish across habitats of a tidally influenced urban bayou", *Environ. Toxicol. Chem.,* vol. 35, no. 4, pp. 966-974, 2016.
[http://dx.doi.org/10.1002/etc.3221] [PMID: 26587912]

[148] F.J. Meijide, R.H. Da Cuña, J.P. Prieto, L.S. Dorelle, P.A. Babay, and F.L. Lo Nostro, "Effects of waterborne exposure to the antidepressant fluoxetine on swimming, shoaling and anxiety behaviours of the mosquitofish Gambusia holbrooki", *Ecotoxicol. Environ. Saf.,* vol. 163, pp. 646-655, 2018.
[http://dx.doi.org/10.1016/j.ecoenv.2018.07.085] [PMID: 30096666]

[149] S. Zhu, Z. Zhang, and D. Žagar, "Mercury transport and fate models in aquatic systems: A review and synthesis", *Sci. Total Environ.,* vol. 639, pp. 538-549, 2018.
[http://dx.doi.org/10.1016/j.scitotenv.2018.04.397] [PMID: 29800847]

[150] D.S. Schmeller, A. Loyau, K. Bao, W. Brack, A. Chatzinotas, F. De Vleeschouwer, J. Friesen, L. Gandois, S.V. Hansson, M. Haver, G. Le Roux, J. Shen, R. Teisserenc, and V.T. Vredenburg, "People, pollution and pathogens - Global change impacts in mountain freshwater ecosystems", *Sci. Total Environ.,* vol. 622-623, pp. 756-763, 2018.
[http://dx.doi.org/10.1016/j.scitotenv.2017.12.006] [PMID: 29223902]

[151] X. Liu, S. Lu, W. Guo, B. Xi, and W. Wang, "Antibiotics in the aquatic environments: A review of lakes, China", *Sci. Total Environ.,* vol. 627, pp. 1195-1208, 2018.
[http://dx.doi.org/10.1016/j.scitotenv.2018.01.271] [PMID: 30857084]

[152] X-M. Zhao, L.A. Yao, Q.L. Ma, G.J. Zhou, L. Wang, Q.L. Fang, and Z.C. Xu, "Distribution and ecological risk assessment of cadmium in water and sediment in Longjiang River, China: Implication on water quality management after pollution accident", *Chemosphere,* vol. 194, pp. 107-116, 2018.
[http://dx.doi.org/10.1016/j.chemosphere.2017.11.127] [PMID: 29197813]

[153] J. Wang, J. Wang, J. Liu, J. Li, L. Zhou, H. Zhang, J. Sun, and S. Zhuang, "The evaluation of endocrine disrupting effects of tert-butylphenols towards estrogenic receptor α, androgen receptor and thyroid hormone receptor β and aquatic toxicities towards freshwater organisms", *Environ. Pollut.,* vol. 240, pp. 396-402, 2018.
[http://dx.doi.org/10.1016/j.envpol.2018.04.117] [PMID: 29753247]

[154] M. Morales, P. Martínez-Paz, P. Sánchez-Argüello, G. Morcillo, and J.L. Martínez-Guitarte, "Bisphenol A (BPA) modulates the expression of endocrine and stress response genes in the freshwater snail Physa acuta", *Ecotoxicol. Environ. Saf.,* vol. 152, pp. 132-138, 2018.
[http://dx.doi.org/10.1016/j.ecoenv.2018.01.034] [PMID: 29407779]

[155] M.G. Bertram, M. Saaristo, J.M. Martin, T.E. Ecker, M. Michelangeli, C.P. Johnstone, and B.B.M. Wong, "Field-realistic exposure to the androgenic endocrine disruptor 17β-trenbolone alters ecologically important behaviours in female fish across multiple contexts", *Environ. Pollut.,* vol. 243, no. Pt B, pp. 900-911, 2018.
[http://dx.doi.org/10.1016/j.envpol.2018.09.044] [PMID: 30245452]

[156] S. Villa, V. Di Nica, T. Pescatore, F. Bellamoli, F. Miari, A. Finizio, and V. Lencioni, "Comparison of the behavioural effects of pharmaceuticals and pesticides on Diamesa zernyi larvae (Chironomidae)",

Environ. Pollut., vol. 238, pp. 130-139, 2018.
[http://dx.doi.org/10.1016/j.envpol.2018.03.029] [PMID: 29554561]

[157] S. Rodrigues, S.C. Antunes, A.T. Correia, and B. Nunes, "Oxytetracycline effects in specific biochemical pathways of detoxification, neurotransmission and energy production in Oncorhynchus mykiss", *Ecotoxicol. Environ. Saf.,* vol. 164, pp. 100-108, 2018.
[http://dx.doi.org/10.1016/j.ecoenv.2018.07.124] [PMID: 30098505]

[158] S. Renuka, S. Umamaheswari, C. Shobana, M. Ramesh, and R.K. Poopal, "Response of antioxidants to semisynthetic bacteriostatic antibiotic (erythromycin) concentrations: A study on freshwater fish", *Acta Ecol. Sin.,* vol. 39, no. 2, pp. 166-172, 2019.
[http://dx.doi.org/10.1016/j.chnaes.2018.08.002]

[159] M. Parolini, L. Bini, S. Magni, A. Rizzo, A. Ghilardi, C. Landi, A. Armini, L. Del Giacco, and A. Binelli, "Exposure to cocaine and its main metabolites altered the protein profile of zebrafish embryos", *Environ. Pollut.,* vol. 232, pp. 603-614, 2018.
[http://dx.doi.org/10.1016/j.envpol.2017.09.097] [PMID: 28993024]

[160] N. Musee, "Environmental risk assessment of triclosan and triclocarban from personal care products in South Africa", *Environ. Pollut.,* vol. 242, no. Pt A, pp. 827-838, 2018.
[http://dx.doi.org/10.1016/j.envpol.2018.06.106] [PMID: 30036836]

CHAPTER 8

Lifecycle Assessment of Emerging Water Pollutants

Nisa Waqar-Un[1,*]**, Luqman Riaz**[2]**, Aansa Rukya Saleem**[3]**, Samia Qadeer**[2]**, Tahir Hayat Malik**[4]**, Nazneen Bangash**[5]**, Talat Ara**[6] **and Audil Rashid**[7]

[1] *Center for Interdisciplinary Research in Basic Sciences (SA-CIRBS), International Islamic University Islamabad, Pakistan*

[2] *Department of Environmental Science, Pir Mehr Ali Shah-Arid Agriculture University Rawalpindi, Pakistan*

[3] *Department of Earth and Environmental Sciences, Bahria, University Islamabad, Pakistan*

[4] *Department of Environmental Sciences, COMSATS University Islamabad, Abbottabad Campus, Pakistan*

[5] *Department of Biosciences, COMSATS University Islamabad, Pakistan*

[6] *Department of Environmental Sciences, International Islamic University Islamabad, Pakistan*

[7] *Department of Botany, University of Gujrat, Pakistan*

Abstract: Emerging pollutants (EPs), also known as contaminants of emerging concern, include pharmaceuticals and personal care products (PPCPs), surfactants, plasticizers, pesticides, *etc.*, and the pharmaceuticals and personal care products are extensively used for therapeutic and non-therapeutic purposes in health care settings, livestock industry, and agriculture. Consumption and production of PPCPs have generated significant quantities of toxic waste in affluent entering the water streams, which poses a risk to aquatic life, public health, and the ecosystem. Given the potential toxicity impacts, continuous exposure to PPCPs is of critical concern. However, the concentrations of PPCPs in the environment are low. Efforts are being made to synergize efficient and cost-effective PPCPs removal technologies to remediate these pollutants from the environment. Still, the success rate is low because of their low concentration (ppb or ppt) and complex chemical structure. Common wastewater treatment technologies are not found efficient enough to attain their complete elimination from the aquatic matrix. Concurrently, ecological problems associated with water quality and aquatic life are aggravated in the prone areas, particularly in the developing world, owing to inadequate monitoring, data management, and treatment facilities. The lifecycle assessment (LCA) is an effective tool for efficient monitoring, quantification, and damage incurred by various stages from production to possible disposal. This chapter summarizes the LCA process of PPCPs, including the release and accumulation, to examine the impacts and associated risks to water quality, the aquatic environment, and ultimately human beings. Furthermore, the deep insight of

* Corresponding author Waqar-Un-Nisa: Center for Interdisciplinary Research in Basic Sciences (SA-CIRBS), International Islamic University Islamabad, Pakistan; E-mail:waqarunnisa@iiu.edu.pk

Shaukat Ali Mazari, Nabisab Mujawar Mubarak & Nizamuddin Sabzoi (Eds.)
All rights reserved-© 2022 Bentham Science Publishers

LCA will help to understand the kinetics of pollutants in environment exchange pools and help fill the existing knowledge gaps that would be a certain better step for management and remediation.

Keywords: Analgesics, Antibiotics, Bioaccumulation, Bioconcentration, Cradle to grave analysis, Emerging pollutants, Endocrine disruptors, Food additives, Hormones, ISO-14001, Lifecycle assessment, Lifecycle impact assessment, Lifecycle inventory, Monitoring, Personal care products, Pharmaceuticals, Steroids, Surfactants, Wastewater treatment, Water footprints.

INTRODUCTION

Emerging Pollutants (EP) are any synthetic/anthropogenic or naturally occurring chemicals or biological organisms (microorganisms) that are not usually monitored in the environment. Still, these compounds can potentially enter different environmental compartments and cause known or suspected adverse environmental and human health effects. More than 700 EPs are categorized into more than 20 classes based on origin. Emerging contaminants comprise a vast array of artificial chemicals in global use, such as disinfection byproducts, perfluorinated compounds, pharmaceuticals, gasoline additives, UV-filters, and manufactured nanomaterials which are important for the progress of modern society (Table **1**) [1]. Moreover, nanomaterials, 1,4-dioxane, and swimming pool disinfection by-products (DBPs) must be included in emerging pollutants [2]. At present, EPs are not monitored in routine examination by monitoring programs, and their interaction and fate in the environment are not yet understood completely. EPs in the environment may result from the point or diffuse pollution. The flow from source to sink depends on the nature of EP and the environmental compartment. In urban areas, EPs released from wastewater treatment plants are discharged into rivers, where they may be degraded, adsorbed, or transported through the medium. Whereas EPs are transported by air in rural areas, runoff, or leaching and contaminating the surface and groundwater [3]. The physicochemical properties of chemicals (*e.g.*, vapor pressure, water solubility, and polarity) decide their interaction in the environment. Thus, major sources of environmentally relevant emerging contaminants are primarily wastewater treatment plant effluents and secondarily terrestrial run-offs (from roads pavements, roofs, and agricultural land), including atmospheric deposition [2]. Due to their extensive use in industry, urbanization, transportation, and agriculture, the concentration of these non-biodegradable substances in the environment is increasing rapidly [1]. The EPs in the environment can be as high as a few hundred µg/L or as low as ng/L[4]. European commission-initiated the NORMAN project in 2005 to establish networking, research, and reference laboratories, integrating academia, industry, and government regulatory bodies with 70 participants from 20 countries and maintaining the largest records of emerging pollutants present in the environment [5]. The NORMAN list currently consists of over 1036 EP and their biotransformation products. Moreover, with advancements in the chemical industry and nanotechnology, various such pollutants will be introduced soon [6].

Table 1. Classification of emerging pollutants [2, 7, 8].

Categories	Compounds (Examples)
Drugs of abuse	Amphetamine, cocaine, tetrahydrocannabinol
Flame retardants	Chloroalkanes, polybrominated diphenyl ethers, (2-chloroethyl) phosphate, tetrabromo bisphenol A
Industrial additives and agents	Chelating agents, aromatic sulfonates
Gasoline additives	Dialkyl ethers, methyl-t-butyl ether
Food additive	Sucralose, triacetin
Wood preservative	2,4-Dinitrophenol
Plasticizers	Bisphenol
Personal-care products	Chelating agents (EDTA), aromatic sulfonates
Surfactants and their metabolites	Alkylphenol ethoxylates, alkylphenols, alkylphenol carboxylates
Steroids and hormones	Diethylstilbestrol, estriol, estrone
Fragrances, insect repellants, soaps, sunscreen agents	Nitro, polycyclic and macrocyclic musks; 2-benzyl-4-ch;orophenol
Pharmaceuticals Analgesics, antimicrobials (human and veterinary), antiepileptics, anti-tumoral drugs, blood-lipid regulators, cardiovascular drugs and b2-symphatomimetics, psychiatric drugs, X-ray contrast agents	Acetaminophen, acetylsalicylic acid, benzafibrate, carbamazepine, diclofenac, diazepam, iopromide, iopamidol, ibuprofen, aspirin, trimethoprim, erythromycin, lincomycin, ampicillin, doxycycline, amoxicillin, metoprolol, timolol, bisoprolol, clofibric acid, atorvastatin

The information about EPs' fate, behavior, and ecotoxicity are scarce. Currently, these EPs are not included in routine monitoring programs. However, more research studies are necessary to investigate the assessment, monitoring, and treatment techniques for rapid control action since drinking water is a scarce resource in the present world. This chapter focuses on the lifecycle assessment process of emerging water pollutants, including their release and accumulation, to examine the impacts and associated risks to water quality, the aquatic environment, and ultimately human beings.

PHARMACEUTICALS AND PERSONAL CARE PRODUCTS

Pharmaceutical and personal care products include any substance (product) used by individuals for personal health and cosmetic reasons. Furthermore, these products can be used by the agribusiness sector to enhance the growth and health of livestock. Pharmaceuticals and personal care products include over-the-counter drugs, veterinary drugs, nutritional supplements, and so on, while personal care products include fragrances, lotions, shampoos, detergents, sunscreens, cosmetics, *etc. Pharmaceuticals and personal care products* are together abbreviated as PPCPs.

Properties of PPCPs

Persistence

Conventional water treatment processes cannot easily remove pharmaceutical and personal care products. Due to their persistent nature, many PPCPs have been reported in drinking water [9]. This failure to completely remove PPCPs by wastewater treatment plants poses a potential threat to human health and aquatic organisms. Although not all PPCPs are persistent due to their continuous replenishment in the environment, they are considered "pseudo-persistent". Therefore, PPCPs that may have partially degraded would behave as persistent compounds in the background.

Bioaccumulation

Although the concentration of PPCPs detected in the freshwater environment is relatively low, most of them and their metabolites can adversely impact non-target aquatic organisms. Bioaccumulation of the carbamazepine (antiepileptic drug) by *Pseudokirchneriella subcapitata* (algae) and- *Thamnocephalus platyurus* (crustacean) with bioaccumulation factors of 2.2 and 12.6, respectively [10]. Furthermore, bioconcentration of pharmaceuticals by *Gambusia holbrooki* (mosquito fish) [11]. The bioconcentration factors measured for caffeine, diphenhydramine, diltiazem, carbamazepine, and ibuprofen were 2.0, 16, 16, 1.4, and 28, respectively. A high concentration of oxazepam in Eurasian perch fish was detected with a bioaccumulation factor of 12 [12]. Similarly, the accumulation of fluoxetine has been reported in snails with a bioaccumulation factor of 3000 [13]. Triclosan (TCS) and triclocarban (TCC) are added to soaps, cosmetics, toothpaste, detergents, and disinfectants due to their effective antimicrobial properties. Due to their structural stability and small quantities, their removal is difficult and detected in the aquatic environment, surface water, sediments, and humans—the presence of PPCP in algae collected around a WWTP in Texas [14]. The results revealed the presence of TCC, TCS, and its metabolite methyl-triclosan (M-TCS) in algae. The concentration was higher (50-400 ng/g) in fresh-weight algae and lower (50-200 ng/L) in water samples. The bioaccumulation factor for M-TCS, TCS, and TCC was (700-1500), (900-2100), and (1600-2700), respectively. The presence of Fluoroquinolones in fish species including rabbitfish, black porgy, and grey snapper found in marine aquaculture in a pearl river delta in China. The concentration of Fluoroquinolones in the liver was as high as 254.6 ng g^{-1} wet weight [15].

Toxicity

PPCPs in the aquatic environment are of paramount concern due to their ability to hamper the endocrine system's normal function, producing an unwanted effect. PPCPs also alter the parts of the endocrine system and subsequently cause adverse health effects in an organism and its sub-population and progeny, so they are also classified as endocrine disruptors. Furthermore, low concentrations of mixtures of PPCPs will also cause toxicity due to the synergistic effect. Similarly, the synergistic effect of 4-nonylphenol (NP) and estradiol results in vitellogenin production in the juvenile rainbow trout [16]. Moreover, 14 days of exposure of goldfish (*Carassius auratus*) to waterborne gemfibrozil resulted in a 50% reduction in plasma testosterone [17]. Another undesirable effect of extensive PPCPs usage involves the emergence of antibiotic resistance in pathogenic bacteria, due to which prevention and cure of many infectious diseases have become difficult [18].

According to recent studies, the toxicity of PPCPs changes depending on the organism exposed, length of exposure, PPCP concentration, and the development stage at which the organism is exposed. The presence of active compounds of PPCPs in the aquatic environment may change their toxicological properties under undesirable conditions. This was illustrated in a study when naproxen was subjected to photodegradation; the resulting products were more toxic to algae and rotifers, the parent compound [19].

ENVIRONMENTAL FATE AND BEHAVIOR OF PPCPS

The major sources of PPCPs to the environment are sewage treatment plants (STPs) [20], WWTPs, and landfill leaching [21]. Conventional wastewater treatment plants are not always able to completely remove PPCPs. Thus, they are commonly found in reclaimed surface water at concentrations ranging from ng L^{-1} to mg L^{-1} [22, 23].

The use of PPCPs in human and veterinary medicines has become widespread. After consuming these medicines, PPCPS are absorbed by the body and subsequently released into the sewage system or septic tank. Wastewater released from the production facility or the manufacturer goes directly into STPs and is the main source of PPCPs [21]. Sewerage treatment plants and wastewater treatment plants cannot completely remove PPCPs. Treated sludge from treatment plants carries PPCPs. This treated sludge is added to soil as fertilizer which, after the surface runoff, enters freshwater bodies or may leach into groundwater. Therefore, the major source of PPCPs includes sewerage treatment plants, wastewater treatment plants, and landfill leaching [2], which eventually loads the water bodies with harmful concentrations of PPCP. Veterinary drugs are released into the environment when animal wastes, either solid or liquid, are sprayed on the agricultural field as fertilizers. Other sources of PPCPs include shower waste, swimming, and washing sinks.

The general properties of PPCPs include low volatility, high polarity, and hydrophilicity. Due to these properties, PPCPs are transported into the environment through an aqueous medium and food [17].

Environmental Concerns

Pharmaceutical and personal care products, either human or veterinary, are present in the environment in very low concentrations. Therefore, studies have not been reported yet about the acute toxicity of environmental PPCPs to humans. However, long-term exposure to low concentrations of PPCPs has shown remarkable toxic effects in test organisms. Findings of long-term and low-dose exposure studies are more accurate and may reflect the toxicological effects of PPCPs on humans and the environment [1].

Many recent studies have reported pharmaceutical and personal care products' impact on microbes. Veterinary pharmaceuticals, such as antibiotics, anthelmintic, antiparasitic, nutritional supplements, and steroidal and nonsteroidal astringents, have been widely used for treatments worldwide. However, these pharmaceuticals are not completely absorbed in the guts of receiving animals. Thus, a large portion (30 -90%) of these pharmaceuticals are excreted into the environment [19], where these compounds may convert to prototypes [24]. These bioactive compounds will adversely affect the surrounding environment in the long run. Due to the increase in veterinary antibiotic (VA) use, a study was conducted to investigate the response of animal gut bacteria and antibiotic-resistant bacteria upon exposure to VA. The results showed that animal gut bacteria were five times more resistant to VA than common antibiotic-resistant bacteria [25]. In another study, the microbial community in streams was exposed to PPCPs, such as ciprofloxacin (an antibiotic) and cimetidine (an antacid and antihistamine), resulting in reduced respiration. Also, the microbial community composition was extremely different for ciprofloxacin [26].

Studies have shown that PPCPs can induce changes in the behavior and function of aquatic animals such as fish, frogs, and mussels. Particularly, hormones such as estradiol, estrone, and ethinyl estradiol at very low concentrations (ng L^{-1}) have been reported to cause the feminization of male fish [27]. Similarly, rare Chinese minnow (*Gobiocypris rarus*), Japanese medaka (*Oryzias latipes*), and male zebrafish (*Danio rerio*) exposed to 2–10 ng L^{-1} of 17α-ethinylestradiol (hormonal contraceptive) instigate vitellogenin production [28]. Similarly [29], conducted a study to investigate the response of zebrafish exposed to environmentally relevant concentration (3 ng/L) of artificial 17α-ethinylestradiol (EE2), and the results showed a stoppage of male gonad development in the test organism. Previously, low concentrations of PPCPs in the environment were not considered harmful. Still, after the perception has been changed, long-term exposure to low concentrations of PPCPs is considered potentially toxic to animals.

According to the National Institute of Environmental Health Sciences (NIEHS), endocrine disruptors may be a potential threat during organ and neural systems formation and the prenatal period [28].

Disinfectants commonly used in swimming pools and purifying drinking water protect humans from water-borne diseases. More than six hundred disinfectant byproducts DBP have been reported to date. Among these, trihalomethanes (THMs) have been associated with bladder and colorectal cancer [30, 31]. DBPs have also been considered potential risk factors for infertility, fetal loss, poor fetal development, and fetal anomalies. However, limited studies are available to justify the association between DBPs and the risks mentioned above; therefore, risk factors like smoking and genetic susceptibility should be ruled out before a validated relationship between DBPs and health risks.

Treatment Technologies for Pharmaceutical and Personal Care Products

Although PPCPs are found in very low concentrations in an aquatic environment due to their continuous replenishment into the climate and incomplete removal by conventional wastewater treatment plants, these chemicals are biologically active. Thus, long-term exposure to low concentration PPCPs may have adverse toxic effects on humans and aquatic organisms. Therefore, PPCPs removal has received much attention over the past few years. Different treatment technologies have been proposed to remove PPCPs from wastewater, including biological, chemical, and physical methods. Still, no such treatment method can remove the entire PPCPs family. A brief overview of these treatment methods is given below:

Physical Treatment Technologies

Adsorption is a frequently used physical process for removing PPCPs from an aqueous medium. Various adsorbents have been developed to enhance the removal of PPCPs from wastewater. These include activated carbon, carbon nanotubes, and graphene.

Activated Carbon

Activated carbon is conventionally used for the physical adsorption of pollutants. It is widely used in two forms that are powdered and granular activated carbon. Activated carbon is known to remove endocrine disrupting compounds [32]. After increasing the duration of operation, two main problems arise. The first is the reduction in adsorption capacity, and the second is the failure of activated carbon in complex wastewater systems [33]. The adsorption potential of activated carbon to target pollutants (PPCPs) mainly depends upon the charge and hydrophobicity of selected PPCPs [21].

Graphene and Graphene Oxide

Graphene is composed of carbon atoms covalently bonded through SP^2 hybridization and forms a honeycomb structure, while the oxidation of graphene comprises graphene oxide. Both can be used to remove PPCPs due to their remarkable properties and high surface area than activated carbon. The removal efficiency of graphene and its oxide depends on the physicochemical properties of target PPCPs. However, a high production cost is a limiting factor for the large-scale application of graphene and its oxides [33].

Nanomaterial

Carbon nanotubes are well known for efficient removal of PPCP such as triclosan, ibuprofen, caffeine, Ketoprofen and carbamazepine [32, 34]. The removal efficiency of carbon nanotubes depends upon the surface chemistry of nanotubes and the physicochemical properties of target PPCPs.

Membrane Filtration

Membrane filtration processes, including microfiltration (MF), nanofiltration (NF), ultrafiltration (UF) and reverse industrial and municipal treatment plants, have successfully used osmosis (RO) for the treatment of wastewater and purification of water. A membrane biological reactor is a membrane-based process tested to remove pharmaceutical compounds and is a potential candidate for future application in industrial and municipal wastewater treatments [35]. Two main units of MBR include a reactor tank and a filtration unit. Most often, modules in MBR include a hollow fiber (HF) and flat plate. MBR's main advantages include high hydraulic retention time, greater biomass concentration, reduced sludge retention time, improved disinfection capability, and a flexible operating system. Once the removal of pharmaceuticals is achieved *via* sorption membrane subsequently, the sludge generated requires immediate treatment to prevent leakage into an aquatic environment.

Biological Removal Methods

Microbial Degradation

Microbial degradation is a low oct method for removing/degrading organic pollutants. Microbes utilize targeted pollutants to gain metabolic processes, thereby detoxifying the polluted environment. Microbial degradation of PPCPs can be achieved using pure culture or mixed cultures. Pure cultures can be extracted from wastewater or sediments and sludge. Sometimes, extraction of pure culture becomes difficult; then, mixed cultures can be a substitute to achieve better results for the degradation of PPCPs [36]. reported removal of 17a-ethinylestradiol using a mixed culture approach. It has been reported that mixed cultures have a higher degradation capability for mixed PPCPs than pure cultures [37]. Activated sludge has also been reported for effective biodegradation of PPCPs [33].

Enzymatic Degradation

Oxidation of emerging pollutants by enzymes is a favorable alternative for the effective and sustainable degradation/removal of organic pollutants from water. Enzymes are biodegradable and ecofriendly catalysts applied for the degradation of recalcitrant pollutants. Peroxidases and laccases are two promising families for transforming PPCPs into non-toxic compounds. These enzymes work under a wide range of temperature and pH and utilize mild oxidants, such as oxygen laccase or peroxidases. The enzymatic treatment is considered tertiary because it removes persistent chemicals left after secondary treatment [38]. Complete removal of antibiotics sulfadimethoxine, sulfonamide, and sulfamonomethoxine was achieved by laccase enzyme using violuric acid as a mediator [39]. In another study, complete degradation of steroid hormones was achieved by manganese peroxidases [40]. The removal efficiency of PPCP by enzyme mainly depends on the nature of the enzyme, Ph, thermostability of enzyme and physicochemical properties of target PPCP.

Chemical Treatment Technologies

Advanced oxidation processes (AOPs) treatment techniques effectively remove PPCPs. In this method, organic pollutants under the influence of hydroxyl radical are converted to CO_2, H_2O, and mineral acid [8]. The hydroxyl group is located on the top of the oxidation power scale after fluorine and is highly reactive [38]. AOPs are generally categorized into (i) heterogeneous and (ii) homogeneous AOPs. Advanced oxidation processes have been extensively reported in the literature to degrade pharmaceuticals and endocrine-disrupting compounds (EDC). Inhomogeneous AOP, catalyst and reactant are in the same phase whereas, in heterogeneous AOP, both exist in different phases [8]. Some of the most used AOPs include photolysis, photocatalysis, sonolysis, radiolysis, ozonation, Fenton, photo Fenton, electrochemical degradation, ionizing irradiation, and UV treatment process [8, 33, 38].

Table 2. Classification of PPCP [7, 8, 21, 34, 41].

Group	Subgroup	Compounds	Endocrine-disrupting PPCPs
Pharmaceutical	**Human & veterinary Antibiotics** Tetracycline Beta-lactum, Aminoglycoside, Macrolide	Trimethoprim, Ampicillin, Amoxicillin Clarithromycin, Azithromycin Doxycycline Erythromycin dehydro-erythromycin, lincomycin, spiramycin, vancomycin, Gentamicin, Sulfamethoxazole, Sulfadimethoxine, Ciprofloxacin, Norfloxacin, Ofloxacin, Penicillin G-V, Chloramphenicol, Oxytetracycline	**Steroids** **Estrogen** 17 β-estradiol, estrone **Progestogens** Norenthindrone, progesterone **Estrogen antagonist** Tamoxifen **Androgen and Glucocorticoids** Testosetrone **Phytoestrogens** **Veterinary growth hormones** Zeranol, trenbolone acetate
	Hormones	Estrone (E1) Estradiol (E2) Ethinylestradiol (EE2)	-
	Analgesics and anti-inflammatory drugs	Diclofenac, Ibuprofen, Ketoprofen, Naproxen, Paracetamol, Acetaminophen, Acetylsalicylic acid	**Non-steroidal** **Antibiotics** Penicillin, amoxicillin, tetracyline **Analgesics** Paracetamol **Anti-inflammatories** Ibuprofen, naproxen, diclofanec

(Table 2) cont.....

Group	Subgroup	Compounds	Endocrine-disrupting PPCPs
	Opioids	Morphine, codeine, 6-acetylmorphine, morphine-6β-D-glucoronide, methadone, morphine-3β-D-glucoronide, oxycodone, hydrocodone, 2-ethyldene-1,5-dimethyl-3,3 diphenylpyrrolidine (EDDP), 6-acetylcodeine,	**Dermatological drugs** Hydrocostsone
	Illicit drugs	Cocaine and metabolites, Anhydroecgonine, Benzoylecgonine, Cocaethylene, Cocaine, ecgonine, Norbenzoylecgonine, Norcocaine, methylester,	-
	Antiepileptic drugs	Carbamazepine, Primidone	-
	Blood lipid regulators	Clofibrate, Gemfibrozil	-
	β-blockers	Metoprolol, Propanolol, Timolol, Bisoprolol	-
	Contrast media	Diatrizoate, Iothalamic acid, Iopromide	-
	Anti-cancer Cytostatic drugs	Ifosfamide Cyclophosphamide Methotrexate, Tamoxifen	-
	Cardiovascular	Atenolol, Enalapril	**Agents used on blood and blood-forming organs** Acetylsalicylic acid, pentoxifylline **Agents for treatment of heart and circulatory diseases** Clofibric acid
	CNS drugs (Anti-anxiety/hypnotic agents)	Diazepam Dimethyl-diazepam Carbamazepine	-
	Diuretics	Furosemide Hydrochlorothiazide	-
	Gastrointestinal	Omeprazole, Esomeprazole, Ranitidine	-
	Lipid regulators	Atorvastatin, Bezafibrate, Naproxen, Clofibrate, Paracetamol, gemfibrozil	-
	Erectile dysfunction drug	Sildenafil	-
	Bronchodilator	Sal butanol	-
	Antidiabetic	Metformin	-

(Table 2) cont.....

Group	Subgroup	Compounds	Endocrine-disrupting PPCPs
Personal care products	Disinfectants & Antiseptics	Triclosan Triclocarban Thymol	**Disinfectants Conservation agents Fragrances** Musk xylol, musk ketone, galaxolide, tonalide
	Synthetic musk/fragrances	Galaxolide (HHCB) Toxalide (AHTN)	-
	Insect repellents	N,N-diethyl-m-toluamide (DEET)	-
	Preservatives	Parabens	-
	Sunscreen UV filters	4-methyl-benzilidine-camphor (4MBC), 2-ethyl-hexy--4-trimethoxycinnamate (EHMC), 2-phenylbenzimidazole-5-sulfonic acid (PBSA), 4-methylbenzylidene camphor (4-MBC)	**UV screens** Benzophenone, homosalate, 4-methyl-benzilidine-camphor, octyl-methoxycinnamte

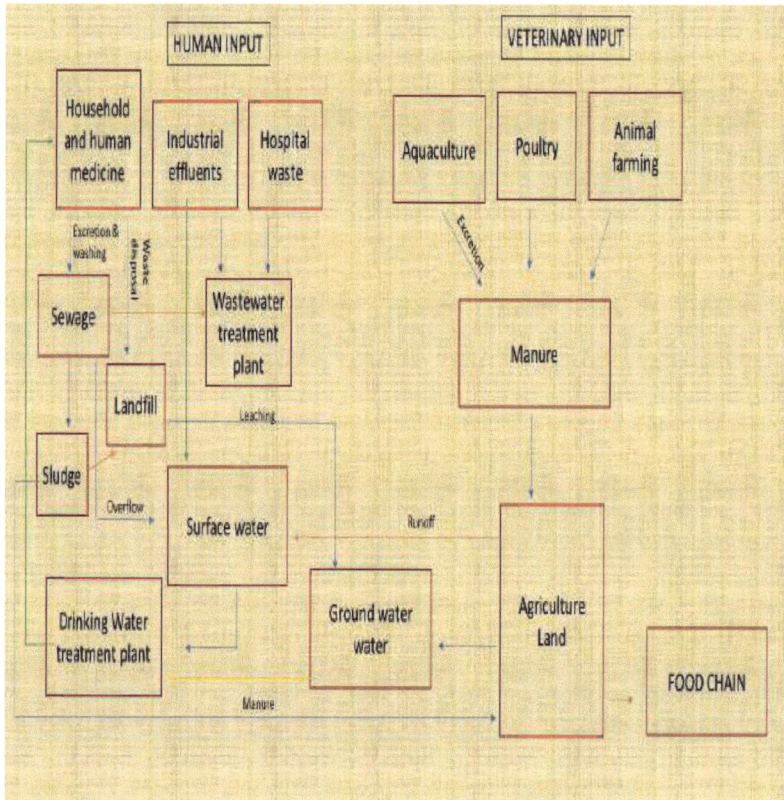

Fig. (1). Overview of environmental contamination with PPCP [2, 8, 21].

LIFECYCLE ASSESSMENT

The urban expansion has caused numerous environmental problems regarding natural resource consumption patterns, resource quantity and quality, process, and methods of resource distribution and supply chain network [42]. The environmental impact assessment process has been applied to identify and solve resource management issues in a new world. A lifecycle assessment process (LCA) was introduced in the era of the 1960s [43, 44]. In the beginning, LCA was performed in 1969 in a beverage factory [45]. The results were expressed on energy and material inflow accounts without considering the process effect on the environment and other socio-economic impacts of the activity. During the initial times, LCA was performed to reveal the priority or betterment claim about the product.

The grass root purpose of LCA is to develop a rational, multivariate, and comprehensive assessment of any process or a comparative analysis of different methods. The initial ISO 14040 and 14044 standards [46] explained a general methodology to perform LCA for any resource, product, or project, though it doesn't provide definite details for each discipline. After the year 2000 and particularly after the earth summit, the idea of sustainability gained the limelight, and LCA received more appreciation as an effective tool for environmental sustainability assessment. ISO explains that LCA is a comprehensive approach for collecting and assembling data about inputs and outputs at every step of a product system. It is also an assessment procedure for possible environmental impacts through potent contaminants or *via* new technology during each step of the product cycle [45].

Lifecycle assessment provides a dual opportunity to study a single product's life to review its impact on the environment and economic analysis. Concomitantly, it can be performed for a comparative study to examine the best process with the least impact on the economy and environment. The general structure of LCA in compliance with ISO 14044 standard comprises four salient stages, including a) goal and scope, highlighting the process's aim and limitations. b) Inventory analysis that collects and tabulatesn of input and output data of resources and energy. c) The third step is the lifecycle impact assessment (LCIA), which summarizes all possible and projected impacts on the environment through characterization and normalization of data. The final step is d) Data interpretation represents, discusses, and summarizes the collected information.

Goal & Scope

The purpose of the first stage is to collect information about the resource or product type, the procedure to carry out the process, the working limitations, limiting factors and the projected function of the product. The functional understanding unit, according to the ISO 14040, an operating unit is the 'a quantified unit output or efficiency of a methodology, system, and resource whose further use for calculation is defined as the reference unit [47]. During data collection, the inclusion and exclusion criteria should be determined before data collection [42]. Subsequently, it is necessary to know the possible environmental impacts of each step during the process. During the first stage of LCA, the involvement of selective variables is important based on cut-off criteria to maintain the range and quality of data that will influence the interpretation of the study. In the final report of any LCA, variable criteria, data number, and the range should be written with a possible justification of selection and repudiation of variables.

Lifecycle Inventory

According to ISO, inventory analysis, the second stage of LCA is defined as the 'phase of LCA which involves the collection, tabulation, arrange the assembling and quantification of all input and output steps related to a product or a system in its lifecycle [46]. The inventory analysis is a technical step to quantify all consumption of resources by various means and the release of waste or energy, as stated in the scope of LCA. Based on previous information, the four steps can be followed in inventory analysis [48]. First, it is feasible to develop a framework diagram. Second, data collection sources and methodology are based on scope and inclusion criteria [49]. Third, the background and relevant information in quantitative form. The fourth step is to translate the collected and studied information about data evaluation into a report.

Lifecycle Impact Analysis

The third step of the LCA method is to study the ecological and public health impact caused by the pollutant load in the environment over the entire lifecycle, which is previously identified in the inventory analysis. The ISO methodology is explicit that LCA's impact analysis comprises two kinds of data. It includes essential and optional elements to perform [46]. The essential element considers the classification and characterization of activities and their impacts. The optional elements in LCA include the normalization and weighting of collected data [42].

Lifecycle Analysis Interpretation

This is the final stage of LCA intended to provide a complete summary of data and results. Interpretation involves step by step appraisal of the LCA process to verify the accuracy of conclusions and validity of data regarding the goal and scope of the study. It also provides supportive and remarkable approaches to lessen environmental risks.

Lifecycle Assessment Evaluation for Wastewater

The wastewater treatment industry is responsible for ensuring the safe discharge of wastewater into the environment [50]. Wastewater treatment strategies reduce the hazardous environmental impacts. It requires the need to study a complete lifecycle of wastewater methodology. Lifecycle assessment (LCA) is gaining interest as an instrument for the environmental evaluation of wastewater, effluent chemicals, and treatment processes [51 - 53].

Along with the major organic and inorganic pollutants, personal care products are also released in our daily use, and wastewater is the main route of release for pharmaceuticals and personal care products [20]. Few studies have attempted to include micropollutants in the LCA of wastewater systems. Wenzel *et al.* [42] carried out a complete LCA of different advanced wastewater treatment options, considering the potential toxicity of heavy metals, endocrine disruptors, PAHs, phtalates, personal products, and detergents.

Use of LCA to Address the Change of Paradigm Shift in Wastewater Treatment

The progress and advancement in the LCA approach are limited to its applicability in developed countries, whereas the situation in developing countries is still despondent. According to different researchers, the developing countries should try to implement the LCA framework to successfully achieve the agenda 2030 for sustainable development [54]. A precisely premeditated LCA process can potentially provide systematic facts to manufacturers, end-users, and legislative experts to supplement the LCA application to ensure environmental protection [55]. LCA gives a *"cradle to grave"* picture of the treatment processes and methods. Therefore, a successful application of LCA to different wastewater treatment processes is expected to provide system improvements, modifications, or shifts towards the other processes.

Interestingly, WW is recognized as a resource, for instance, for energy, material, and recovery of nutrients. Thus, in the past decade, wastewater treatment has been fundamentally shifted from mere decontamination to resource recovery and wastewater reuse. This milestone is achieved by using different assessing tools that highlight the hidden potential of WW; one of the most widely applied tools is LCA. In some cases, the scope of LCA application is required to extend to the downstream receiving body and involvement of stakeholders, and the complication of the system can arise. Such indispensable factors fog the LCA application and question the desired benefits of environmental quality and standards. To address such issues [56], proposed an integrated approach of combining LCA with the Water Quality Model (WQM), Conjoint Analysis (CA), and Plackett–Burman (PB). Overall, LCA alone or in combination provides a complete and comprehensive way to ensure environmental sustainability and resource recovery.

Goal and Scope Definition

The term "Emerging contaminants" covers all the chemical entities in any form *viz*: solid, liquid or gas considered a threat to the environment or human health with considerably unforeseen impacts being published" [57]. This also includes any "novel approach, technology or substances" that predominantly provide a new exposure to human being or ecosystem from any point or non-point source. This term is usually used for but not limited to chemical substances solely; rather, this can be deployed for biological or electromagnetic contamination in any biological or physical system. The LCA of such contaminants includes their source distribution and fate.

Inventory Analysis

The inventory analysis is quantifiable inputs and outputs of the ECs. This will include how the ECs were produced, where they enter their baseline concentration, progressive concentration, impact on different physical and biological environments, and extent of their results; a brief inventory analysis with stepwise distribution is provided in Fig. (**2**) for PBDEs Polybrominated diphenyl ethers.

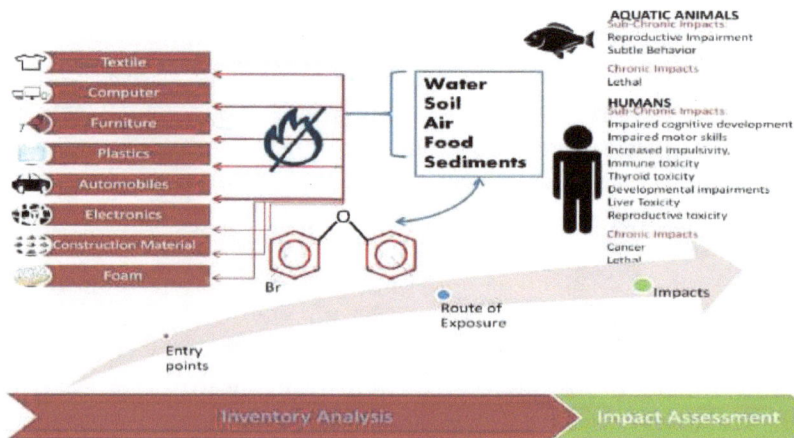

Fig. (2). Conceptual brief for the description of Inventory analysis and Impact assessment in LCA of PBDes. (*The steps are followed by the interpretation as discussed in the text*)

Impact Assessment

The impact assessment is critical to identify the impending hazards or risks posed by certain ECs on the environment in general. With the identification of routes of exposure and targeted organs/ organisms. For instance, the PBDEs are present in the air, water, and soil; however, the overall impacts on water bodies and associated life are under high pressure of deterioration. Fig. (**2**) Provides how the impacts can be classified for PBDEs. Impact analysis can be conducted through extensive surveys and data collection. However, ecosystem studies' background and baseline information are pivotal in impact assessment.

Similarly, the impact analysis of MTBE by Mennear [58] is provided by using a mouse as the subject that indicated its neurotoxicological sub-chronic effects. An extensive impact assessment of ECs on the biological system, particularly humans, is reviewed and presented [1]. Similarly, the distribution study based on toxicity also indicated high lethal to chronic sub lethal Impacts of PBDes and other ECs on human and aquatic life [59].

Interpretation

Overall, the material safety and data sheets so far have been developed for certain ECs indicate their inherent ruinous impacts on the environment and biological systems. This places most of the ECs are highly hazardous or highly toxic chemicals (https://www.osha.gov/Publications/OSHA3514.html). The LCA analysis of ECs also provides foresight to place novel chemicals as high-risk chemicals unless they are justified to be safe for the environment or personal use.

SCOPE OF LCA IN PROSPECTS OF EMERGING POLLUTANTS

LCA application to the targeted pollutants is simpler than applying LCA to the overall treatment process. The practical application of LCA on targeted pollutants can vary over a wide range of contaminants from nutrients to heavy metals, PCBs, PAHs, and many others, which will be beyond the scope of the current chapter. Therefore, we will discuss only the emerging pollutants (EPs) of prime importance in this chapter. The term emerging here refers to the new pollutants, particularly of persistent nature, found in the aquatic system and can impact or induce impacts of other contaminants. Few researchers identified EPs as entities detected in the environment but are currently not designated in any smaller group of pollutants like POPs that have bioaccumulative properties [60]. The list of these persistent pollutants is being actively updated with the addition and classification of new EPs. The efforts have now updated the list to 28 POPs [61] from initially classified 12 POPs [62].

In response to the EPs uncertainties, LCA offers a systematic process that considers all the stages without branched collaborative minor assessments [63]. Each EP is accurately monitored in LCA following the standard procedure as put forth by ISO 14040:2006 [46]. These pollutants pass through different phases of LCA, as discussed earlier in this chapter. In the past few years, the whole water cycle assessment has been considered with environmental assessment and treatment technologies [64]. LCA of the EPs focuses on the impacts of EP from effluent production, distribution, and treatment.

Furthermore, this is extended to the phases of treatment and production of by-products from those EPs [50]. The application of LCA for EPs is not only limited to the treatment of industrial WW; rather, in recent studies, it has also been applied to ensure drinking water safety. This includes the production and neutralizing stages of water production units [65] and other important factors like the quality and pertinent features of the water distribution system performance. Decide on the treatment facility for EP [66] highlighted the importance of a functional unit in the ion exchange process to monitor the initial and final concentration of the targeted pollutants. In the WW treatment units, the objective of the EPs treatment facility is designated to limit its toxicity to human health. At the same time, an opposite trend is observed in the literature, where most studies focus on comparing treatment facilities. These comparisons were made on economic feasibility and targeted pollutant reduction based on the initial and final concentration of the EPs [67]. The toxicity impacts and by-product production are usually overlooked in these studies. Setting the merits of LCA to tackle the EPs and their environmental implications is a critical process. However, few minimum criteria can be set to ensure the safe removal of EPs from the water bodies. These merits can be summarized as presented in Fig. (**3**).

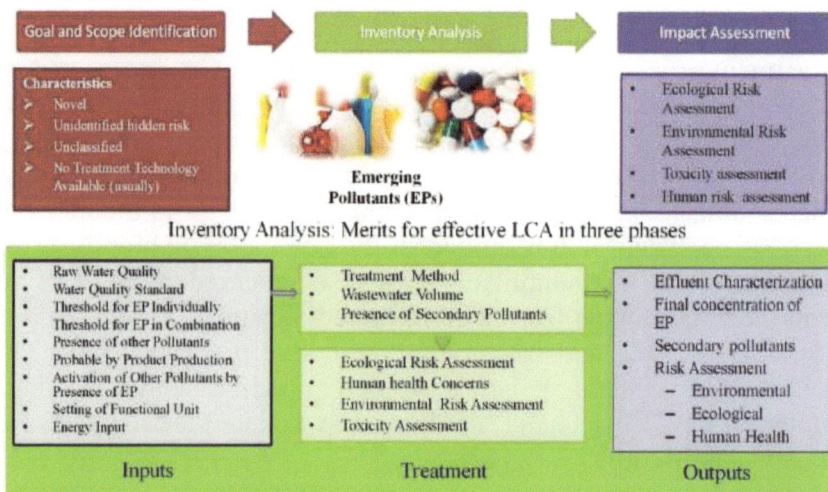

Goal and Scope Identification	Inventory Analysis	Impact Assessment
Characteristics ➤ Novel ➤ Unidentified hidden risk ➤ Unclassified ➤ No Treatment Technology Available (usually)	**Emerging Pollutants (EPs)**	• Ecological Risk Assessment • Environmental Risk Assessment • Toxicity assessment • Human risk assessment

Inventory Analysis: Merits for effective LCA in three phases

Inputs	Treatment	Outputs

• Raw Water Quality
• Water Quality Standard
• Threshold for EP Individually
• Threshold for EP in Combination
• Presence of other Pollutants
• Probable by Product Production
• Activation of Other Pollutants by Presence of EP
• Setting of Functional Unit
• Energy Input
	• Treatment Method
	• Wastewater Volume
	• Presence of Secondary Pollutants
	
	• Ecological Risk Assessment
	• Human health Concerns
	• Environmental Risk Assessment
	• Toxicity Assessment
		• Effluent Characterization
		• Final concentration of EP
		• Secondary pollutants
		• Risk Assessment
		  – Environmental
		  – Ecological
		  – Human Health

(Fig. 3) contd.....

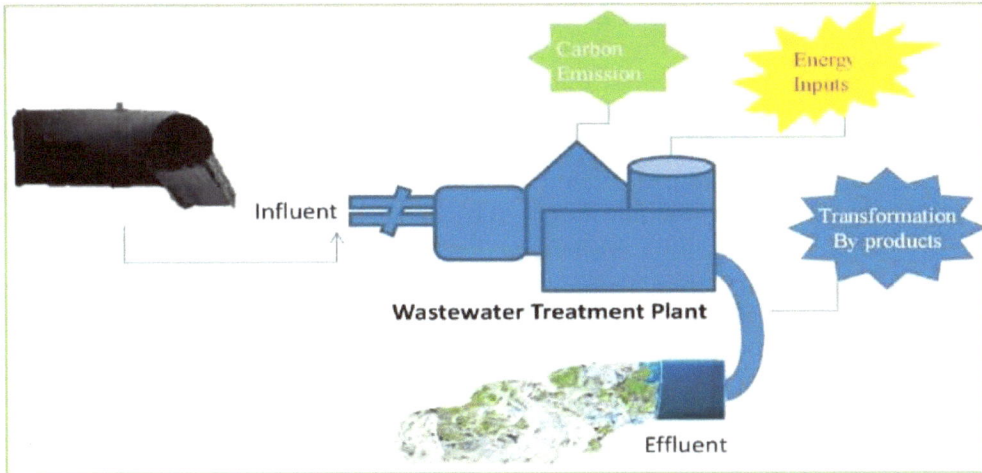

Fig. (3). Inventory Analysis: Merits for effective LCA in three phases.

IMPLEMENTATION OF LCA FOR IMPROVING DATA QUALITY AND IMPACT ASSESSMENT

In the past decade, it has been observed that LCA can be applied to enhance data quality. The application of LCA in the water footprint concept has broadened the application of LCA from mere treatment process compared to the impact assessment and water quality enhancement through improvement in data quality and quantity. The scope encompasses the water footprint and its environmental performance products. The application of LCA in water footprint assessment has limitations in terms of the regional possible environmental impact of water scarcity of water use, given regional differences in water scarcity [68]. The bright side of the LCA application to monitor water footprint is considering blue, green, and gray water footprints. Combining these three can provide a comprehensive set of data to monitor water quality, quantity, and impacts of water scarcity, both by contamination and environmental impacts. Such data can be employed to plan the water management facility and water decontamination process. However, the issue of sustainable, competent, and fair distribution of freshwater from the head source to the global level remains unaddressed. A regional success can be achieved to ensure public and ecological health. Applying LCA and water footprint in combination can provide a sound theoretical ground for allocating per unit each water type for human consumption. This can be subtracted from the available supply chain [69]. Thus, theoretically, the allocation unit can be modified based on the LCA of water supply chains. However, the practical approach to this concept is usually much more complicated as only considering the water supply chain or catchment unit capacity may overlook the environmental consequence of gray water. Thus, EPs and other important contaminants can be subsided, creating environmental complications and toxicity. This argument of increasing contamination was also put forth where they mentioned land use as an important parameter to monitor green water footprint [70].

Thus, in the current scenario where both water contamination and scarcity are worldwide known environmental implications, the application of LCA needs to be broadened from mere qualitative assessments of the water cycle to quantitative assessments. Such compressive data monitoring is only possible by integrating different fields that offer LCA application to the water resources management and provide a definite timeline of per unit water being consumed or converted. In this regard, the volume can be put forth as an initial consideration. With the progression of per unit water, the conversion of the water to green or gray requires qualitative assessments. These qualitative assessments must incorporate the risk and toxicity assessments to provide comprehensive data for water resource management. In the past few years, the quality of data from LCA has actively been improved by incorporating environmental impacts. Therefore, the term LCA under such circumstances has been modified to lifecycle impact assessment (LCIA). This modification has provided the origin and fate of contaminants and has also outwinged the impact of those contaminants on environmental systems and human health.

CONCLUSION

Excessive production and utilization of PPCPs in industries, agriculture, hospitals, and household levels release substantial quantities of PPCPs into the environment *via* landfill leaching or sewage treatment plants. Pharmaceutical and personal care products have been reported at concentrations ranging from ng/L to mg/L in reclaimed surface water. Although the concentrations are relatively low, due to their continuous input, they are transported through aqueous mediums and remain biologically active and induce hormonal changes in aquatic life, leading to multiple disorders such as the feminization of male fish, excessive production of vitellogenin, and preventing the growth of male gonads. Owing to their widespread occurrence in the environment and adverse effects on humans and other life forms, various physical, chemical, biological, and combinations of these methods have been employed to remove PPCPs from wastewater effectively. Activated carbon, graphene, graphene oxide, nanomaterials, and membrane filtration removed PPCPs effectively, but the high production cost is limiting for large-scale applications. In addition, advanced oxidation processes (AOPs) and microbial degradation also proved to be effective for the removal and degradation of PPCPs considering various factors.

The lifecycle assessment is an effective tool for monitoring, quantification, and damage incurred by various stages from production to possible disposal. LCA offers many benefits to wastewater treatment processes, including improved system efficiency in removing targeted pollutants, efficiency analysis at each step of the process, by-product synthesis, and policy development. Lifecycle assessment of the emerging contaminants focuses on their impacts from effluent production, distribution, and treatment and is extended to the phases of treatment and production of by-products. The objective of treatment facilities for emerging pollutants is to limit the toxicity of compounds within an acceptable range. However, much of the literature compares different treatment methods and their economic feasibilities. This trend has overlooked the toxicity impacts and byproduct production. Likewise, the application of LCA to water footprint assessment has limitations, depending on the effects of water scarcity in different regions.

Close collaboration between researchers and policymakers is needed to develop a holistic strategy to address the problem of emerging contaminants and prevent their obvious and unanticipated adverse effects in the years to come. Reducing the consumption of certain products and replacing them with less toxic compounds can be an effective strategy but only in the short term. For a more sustainable solution, new wastewater quality standards need to be developed, and the integration of industrial, agricultural, and municipal water consumers can yield beneficial outcomes. Research should develop hybrid systems to degrade and remove these contaminants from wastewater. In addition, more research should be devoted to the toxicology of emerging pollutants in various organisms and the development of reliable methods for toxicity testing at extremely low concentrations.

CONSENT FOR PUBLICATION

Not applicable.

CONFLICT OF INTEREST

The author declares no conflict of interest, financial or otherwise.

ACKNOWLEDGEMENTS

Declared none.

REFERENCES

[1]   M. Lei, L. Zhang, J. Lei, L. Zong, J. Li, Z. Wu, and Z. Wang, "Overview of Emerging Contaminants and Associated Human Health Effects", *BioMed Res. Int.,* vol. 2015, p. 404796, 2015.
[http://dx.doi.org/10.1155/2015/404796] [PMID: 26713315]

[2]   M.L. Farré, S. Pérez, L. Kantiani, and D. Barceló, "Fate and toxicity of emerging pollutants, their metabolites and transformation products in the aquatic environment", *Trends Analyt. Chem.,* vol. 27, no. 11, pp. 991-1007, 2008.
[http://dx.doi.org/10.1016/j.trac.2008.09.010]

[3]   V. Geissen, H. Mol, E. Klumpp, G. Umlauf, M. Nadal, and M. van der Ploeg, "Ritsema, C. J. Emerging pollutants in the environment: A challenge for water resource management", *Int. Soil Water Conserv. Res.,* p. 3, 2015.

[4]   M. Bilal, H.M.N. Iqbal, and D. Barceló, "Persistence of pesticides-based contaminants in the environment and their effective degradation using laccase-assisted biocatalytic systems", *Science of the Total Environment,* p. 133896, 2019.
[http://dx.doi.org/10.1016/j.scitotenv.2019.133896]

[5]   S. Dey, F. Bano, and A. Malik, "Pharmaceuticals and personal care product (PPCP) contamination—a global discharge inventory", *Pharmaceuticals and Personal Care Products: Waste Management and Treatment Technology.,* pp. 1-26, 2019.
[http://dx.doi.org/10.1016/B978-0-12-816189-0.00001-9]

[6]   S. Cassani, and P. Gramatica, "Identification of potential PBT behavior of personal care products by structural approaches", 2015.

[7]   G.F.C. Francisco, J. Isac-Garcia, and J.A. Dobado, *Emerging Pollutants: Origin, Structure, and Properties.* Jhon Wiley & Sons, 2018.

[8]   J.O. Tijani, O.O. Fatoba, and L.F. Petrik, "A Review of Pharmaceuticals and Endocrine-Disrupting Compounds: Sources, Effects, Removal, and Detections", *Water Air Soil Pollut.,* vol. 224, no. 11, p. 1770, 2013.
[http://dx.doi.org/10.1007/s11270-013-1770-3]

[9]   S.A. Snyder, "Occurrence, Treatment, and Toxicological Relevance of EDCs and Pharmaceuticals in Water", *Ozone Sci. Eng.,* vol. 30, no. 1, p. 65, 2008.
[http://dx.doi.org/10.1080/01919510701799278]

[10]   G. Vernouillet, P. Eullaffroy, A. Lajeunesse, C. Blaise, F. Gagné, and P. Juneau, "Toxic effects and bioaccumulation of carbamazepine evaluated by biomarkers measured in organisms of different trophic levels", *Chemosphere,* vol. 80, no. 9, pp. 1062-1068, 2010.
[http://dx.doi.org/10.1016/j.chemosphere.2010.05.010] [PMID: 20557923]

[11]   J. Wang, and P.R. Gardinali, "Uptake and depuration of pharmaceuticals in reclaimed water by mosquito fish (Gambusia holbrooki): a worst-case, multiple-exposure scenario", *Environ. Toxicol. Chem.,* vol. 32, no. 8, pp. 1752-1758, 2013.
[http://dx.doi.org/10.1002/etc.2238] [PMID: 23595768]

[12]   T. Brodin, S. Piovano, J. Fick, J. Klaminder, M. Heynen, and M. Jonsson, "Ecological effects of pharmaceuticals in aquatic systems--impacts through behavioural alterations. Philosophical Transactions of the Royal Society B: Biolog", *Sci,* vol. 369, no. 1656, p. 20130580, 2014.

[13]   B. Du, S.P. Haddad, W.C. Scott, C.K. Chambliss, and B.W. Brooks, "Pharmaceutical bioaccumulation by periphyton and snails in an effluent-dependent stream during an extreme drought", *Chemosphere,* vol. 119, pp. 927-934, 2015.
[http://dx.doi.org/10.1016/j.chemosphere.2014.08.044] [PMID: 25261960]

[14]   M.A. Coogan, R.E. Edziyie, T.W. La Point, and B.J. Venables, "Algal bioaccumulation of triclocarban, triclosan, and methyl-triclosan in a North Texas wastewater treatment plant receiving stream", *Chemosphere,* vol. 67, no. 10, pp. 1911-1918, 2007.
[http://dx.doi.org/10.1016/j.chemosphere.2006.12.027] [PMID: 17275881]

[15]   X. He, Z. Wang, X. Nie, Y. Yang, D. Pan, A.O.W. Leung, Z. Cheng, Y. Yang, K. Li, and K. Chen, "Residues of fluoroquinolones in marine aquaculture environment of the Pearl River Delta, South China", *Environ. Geochem. Health,* vol. 34, no. 3, pp. 323-335, 2012.

[http://dx.doi.org/10.1007/s10653-011-9420-4] [PMID: 21881861]

[16] K.L. Thorpe, T.H. Hutchinson, M.J. Hetheridge, M. Scholze, J.P. Sumpter, and C.R. Tyler, "Assessing the biological potency of binary mixtures of environmental estrogens using vitellogenin induction in juvenile rainbow trout (Oncorhynchus mykiss)", *Environ. Sci. Technol.,* vol. 35, no. 12, pp. 2476-2481, 2001.
[http://dx.doi.org/10.1021/es001767u] [PMID: 11432551]

[17] C. Mimeault, A.J. Woodhouse, X.S. Miao, C.D. Metcalfe, T.W. Moon, and V.L. Trudeau, "The human lipid regulator, gemfibrozil bioconcentrates and reduces testosterone in the goldfish, Carassius auratus", *Aquat. Toxicol.,* vol. 73, no. 1, pp. 44-54, 2005.
[http://dx.doi.org/10.1016/j.aquatox.2005.01.009] [PMID: 15892991]

[18] WHO, *Antibiotic Resistance: Multi – Country Public Awareness Survey.* World Health Organisation, 2015.

[19] M. Isidori, M. Lavorgna, A. Nardelli, A. Parrella, L. Previtera, and M. Rubino, "Ecotoxicity of naproxen and its phototransformation products", *Sci. Total Environ.,* vol. 348, no. 1-3, pp. 93-101, 2005.
[http://dx.doi.org/10.1016/j.scitotenv.2004.12.068] [PMID: 16162316]

[20] C.G. Daughton, and T.A. Ternes, "Pharmaceuticals and personal care products in the environment: agents of subtle change?", *Environ. Health Perspect.,* vol. 107, suppl. Suppl. 6, pp. 907-938, 1999.
[http://dx.doi.org/10.1289/ehp.99107s6907] [PMID: 10592150]

[21] R. Ebele, M. Campinas, J.L. Acero, and M.J. Rosa, "Investigating PPCP removal from wastewater by powdered activated carbon/ultrafiltration", *Water Air Soil Pollut.,* vol. 227, no. 6, pp. 1-4, 2016.

[22] W. Chen, J. Xu, S. Lu, W. Jiao, L. Wu, and A.C. Chang, "Fates and transport of PPCPs in soil receiving reclaimed water irrigation", *Chemosphere,* vol. 93, no. 10, pp. 2621-2630, 2013.
[http://dx.doi.org/10.1016/j.chemosphere.2013.09.088] [PMID: 24148973]

[23] S. Suárez, M. Carballa, F. Omil, and J.M. Lema, "How are pharmaceutical and personal care products (PPCPs) removed from urban wastewaters?", *Rev. Environ. Sci. Biotechnol.,* vol. 7, no. 2, pp. 125-138, 2008.
[http://dx.doi.org/10.1007/s11157-008-9130-2]

[24] R. Renner, "Do cattle growth hormones pose an environmental risk?", *Environ. Sci. Technol.,* vol. 36, no. 9, pp. 194A-197A, 2002.
[http://dx.doi.org/10.1021/es022301+] [PMID: 12026976]

[25] L.C. McDonald, M.J. Kuehnert, F.C. Tenover, and W.R. Jarvis, "Vancomycin-resistant enterococci outside the health-care setting: prevalence, sources, and public health implications", *Emerg. Infect. Dis.,* vol. 3, no. 3, pp. 311-317, 1997.
[http://dx.doi.org/10.3201/eid0303.970307] [PMID: 9284375]

[26] E.J. Rosi, H.A. Bechtold, D. Snow, M. Rojas, A.J. Reisinger, and J.J. Kelly, "Urban stream microbial communities show resistance to pharmaceutical exposure", *Ecosphere,* vol. 9, no. 1, 2018.
[http://dx.doi.org/10.1002/ecs2.2041]

[27] A.M. Vajda, L.B. Barber, J.L. Gray, E.M. Lopez, J.D. Woodling, and D.O. Norris, "Reproductive disruption in fish downstream from an estrogenic wastewater effluent", *Environ. Sci. Technol.,* vol. 42, no. 9, pp. 3407-3414, 2008.
[http://dx.doi.org/10.1021/es0720661] [PMID: 18522126]

[28] Q. Yang, X. Yang, J. Liu, Y. Chen, and S. Shen, "Effects of exposure to BPF on development and sexual differentiation during early life stages of zebrafish (Danio rerio)", *Comp. Biochem. Physiol. C Toxicol. Pharmacol.,* vol. 210, pp. 44-56, 2018.
[http://dx.doi.org/10.1016/j.cbpc.2018.05.004] [PMID: 29758382]

[29] M. Fenske, G. Maack, C. Schäfers, and H. Segner, "An environmentally relevant concentration of estrogen induces arrest of male gonad development in zebrafish, Danio rerio", *Environ. Toxicol.*

Chem., vol. 24, no. 5, pp. 1088-1098, 2005.
[http://dx.doi.org/10.1897/04-096R1.1] [PMID: 16110986]

[30] C.M. Villanueva, K.P. Cantor, J.O. Grimalt, N. Malats, D. Silverman, A. Tardon, R. Garcia-Closas, C. Serra, A. Carrato, G. Castaño-Vinyals, R. Marcos, N. Rothman, F.X. Real, M. Dosemeci, and M. Kogevinas, "Bladder cancer and exposure to water disinfection by-products through ingestion, bathing, showering, and swimming in pools", *Am. J. Epidemiol.,* vol. 165, no. 2, pp. 148-156, 2007.
[http://dx.doi.org/10.1093/aje/kwj364] [PMID: 17079692]

[31] C.C. Chang, S.C. Ho, L.Y. Wang, and C.Y. Yang, "Bladder cancer in Taiwan: relationship to trihalomethane concentrations present in drinking-water supplies", *J. Toxicol. Environ. Health A,* vol. 70, no. 20, pp. 1752-1757, 2007.
[http://dx.doi.org/10.1080/15287390701459031] [PMID: 17885932]

[32] F.F. Liu, J. Zhao, S. Wang, P. Du, and B. Xing, "Effects of solution chemistry on adsorption of selected pharmaceuticals and personal care products (PPCPs) by graphenes and carbon nanotubes", *Environ. Sci. Technol.,* vol. 48, no. 22, pp. 13197-13206, 2014.
[http://dx.doi.org/10.1021/es5034684] [PMID: 25353977]

[33] J. Wang, and S. Wang, "Removal of pharmaceuticals and personal care products (PPCPs) from wastewater: A review", *J. Environ. Manage.,* vol. 182, pp. 620-640, 2016.
[http://dx.doi.org/10.1016/j.jenvman.2016.07.049] [PMID: 27552641]

[34] Y.K. Liu, J. Hu, and J.L. Wang, "Fe2+ enhancing sulfamethazine degradation in aqueous solution by gamma irradiation", *Radiat. Phys. Chem.,* vol. 96, pp. 81-87, 2014.
[http://dx.doi.org/10.1016/j.radphyschem.2013.08.018]

[35] R. Lopez-Fernandez, L. Martınez, and S. Villaverde, "Membrane bioreactor for the treatment of pharmaceutical wastewater containing corticosteroids", *Desalination,* vol. 300, pp. 19-23, 2012.
[http://dx.doi.org/10.1016/j.desal.2012.05.032]

[36] W.O. Khunjar, S.A. Mackintosh, J. Skotnicka-Pitak, S. Baik, D.S. Aga, and N.G. Love, "Elucidating the relative roles of ammonia oxidizing and heterotrophic bacteria during the biotransformation of 17α-Ethinylestradiol and Trimethoprim", *Environ. Sci. Technol.,* vol. 45, no. 8, pp. 3605-3612, 2011.
[http://dx.doi.org/10.1021/es1037035] [PMID: 21428279]

[37] I.A. Vasiliadou, R. Molina, F. Martínez, and J.A. Melero, "Biological removal of pharmaceutical and personal care products by a mixed microbial culture: sorption, desorption and biodegradation", *Biochem. Eng. J.,* vol. 81, pp. 108-119, 2013.
[http://dx.doi.org/10.1016/j.bej.2013.10.010]

[38] E. Méndez, M.A. González-Fuentes, G. Rebollar-Perez, A. Méndez-Albores, and E. Torres, "Emerging pollutant treatments in wastewater: Cases of antibiotics and hormones", *J. Environ. Sci. Health Part A Tox. Hazard. Subst. Environ. Eng.,* vol. 52, no. 3, pp. 235-253, 2017.
[http://dx.doi.org/10.1080/10934529.2016.1253391] [PMID: 27901630]

[39] S.S. Weng, K.L. Ku, and H.T. Lai, "The implication of mediators for enhancement of laccase oxidation of sulfonamide antibiotics", *Bioresour. Technol.,* vol. 113, pp. 259-264, 2012.
[http://dx.doi.org/10.1016/j.biortech.2011.12.111] [PMID: 22257859]

[40] K. Suzuki, H. Hirai, H. Murata, and T. Nishida, "Removal of estrogenic activities of 17beta-estradiol and ethinylestradiol by ligninolytic enzymes from white rot fungi", *Water Res.,* vol. 37, no. 8, pp. 1972-1975, 2003.
[http://dx.doi.org/10.1016/S0043-1354(02)00533-X] [PMID: 12697240]

[41] F. Riva, S. Castiglioni, E. Fattore, A. Manenti, E. Davoli, and E. Zuccato, "Monitoring emerging contaminants in the drinking water of Milan and assessment of the human risk", *Int. J. Hyg. Environ. Health,* vol. 221, no. 3, pp. 451-457, 2018.
[http://dx.doi.org/10.1016/j.ijheh.2018.01.008] [PMID: 29366558]

[42] H. Wenzel, H.F. Larsen, J. Clauson-Kaas, L. Høibye, and B.N. Jacobsen, "Weighing environmental advantages and disadvantages of advanced wastewater treatment of micro-pollutants using

environmental life cycle assessment", *Water Sci. Technol.*, vol. 57, no. 1, pp. 27-32, 2008.
[http://dx.doi.org/10.2166/wst.2008.819] [PMID: 18192737]

[43]   H.A. Udo de Haes, "Applications of lifecycle assessment: expectations, drawbacks & perspectives", *J. Clean. Prod.*, vol. 1, no. 3–4, pp. 131-137, 1993.
[http://dx.doi.org/10.1016/0959-6526(93)90002-S]

[44]   P. Roeleveld, A. Klapwijk, P. Eggels, W. Rulkens, and W. Van Starkenburg, "Sustainability of municipal wastewater treatment", *Water Sci. Technol.*, vol. 35, no. 10, pp. 221-228, 1997.
[http://dx.doi.org/10.2166/wst.1997.0386]

[45]   M.T. Ahmed, "Lifecycle Analysis in wastewater: a sustainability perspective", In: *Waste Water Treatment and Reuse in the Mediterranean Region, Hdb Env Chem, 14.*, D. Barcelo´, M. Petrovic, Eds., Springer-Verlag: Berlin, Heidelberg, 2011, pp. 125-154.

[46]   International Standards Organization, *Environmental Management – Lifecycle Assessment – Principles and Framework* 2nd edition. , 2006.

[47]   S. Bai, X. Wang, X. Zhang, X. Zhaoa, and N. Ren, "Lifecycle assessment in wastewater treatment: influence of site-oriented normalization factors, lifecycle impact assessment methods, and weighting methods. Royal society of chemistry Advances", *RSC Advances,* vol. 7, pp. 26335-26341, 2017.
[http://dx.doi.org/10.1039/C7RA01016H]

[48]   S.S. Muthu, "Estimating the overall environmental impact of textile processing: lifecycle assessment (LCA) of textile products", In: *Assessing the Environmental Impact of Textiles and the Clothing Supply Chain* Woodhead Publishing, 2014, pp. 105-131.
[http://dx.doi.org/10.1533/9781782421122.105]

[49]   M.A. Curran, *Lifecycle Assessment: Principles and Practice.* US Environmental Protection Agency: Ohio, USA, 2006.

[50]   I. Muñoz, M. José Gómez, A. Molina-Díaz, M.A.J. Huijbregts, A.R. Fernández-Alba, and E. García-Calvo, "Ranking potential impacts of priority and emerging pollutants in urban wastewater through life cycle impact assessment", *Chemosphere,* vol. 74, no. 1, pp. 37-44, 2008.
[http://dx.doi.org/10.1016/j.chemosphere.2008.09.029] [PMID: 18951608]

[51]   X. Domènech, J.A. Ayllón, J. Peral, and J. Rieradevall, "How green is a chemical reaction? Application of LCA to green chemistry", *Environ. Sci. Technol.*, vol. 36, no. 24, pp. 5517-5520, 2002.
[http://dx.doi.org/10.1021/es020001m] [PMID: 12521184]

[52]   P. Saling, A. Kicherer, B. Dittrich-Krämer, R. Wittlinger, W. Zombik, I. Schmidt, W. Schrott, and S. Schmidt, "Eco-efficiency by BASF: the method", *Int. J. Lifecycle Ass.*, vol. 7, pp. 203-218, 2002.
[http://dx.doi.org/10.1007/BF02978875]

[53]   S. Hellweg, U. Fischer, M. Scheringer, and K. Hungerbühler, "Environmental assessment of chemicals: methods and application to a case study of organic solvents", *Green Chem.*, vol. 6, pp. 418-427, 2004.
[http://dx.doi.org/10.1039/B402807B]

[54]   U. GA., *Transforming our world: the 2030 Agenda for Sustainable Development.* Division for Sustainable Development Goals: New York, NY, USA, 2015.

[55]   R. Parra-Saldivar, M. Bilal, and H.M. Iqbal, "Lifecycle assessment in wastewater treatment technology", *Curr. Opin. Environ. Sci. Health,* vol. 13, pp. 80-84, 2020.
[http://dx.doi.org/10.1016/j.coesh.2019.12.003]

[56]   S. Bai, X. Zhang, Y. Xiang, X. Wang, X. Zhao, and N. Ren, "HIT.WATER scheme: An integrated LCA-based decision-support platform for evaluation of wastewater discharge limits", *Sci. Total Environ.*, vol. 655, pp. 1427-1438, 2019.
[http://dx.doi.org/10.1016/j.scitotenv.2018.11.209] [PMID: 30577134]

[57]   M. Gavrilescu, K. Demnerová, J. Aamand, S. Agathos, and F. Fava, "Emerging pollutants in the environment: present and future challenges in biomonitoring, ecological risks and bioremediation", *N.*

Biotechnol., vol. 32, no. 1, pp. 147-156, 2015.
[http://dx.doi.org/10.1016/j.nbt.2014.01.001] [PMID: 24462777]

[58]   J.H. Mennear, "Carcinogenicity studies on MTBE: critical review and interpretation", *Risk Anal.,* vol. 17, no. 6, pp. 673-681, 1997.
[http://dx.doi.org/10.1111/j.1539-6924.1997.tb01274.x] [PMID: 9463924]

[59]   E. Eljarrat, and D. Barcelo, "Priority lists for persistent organic pollutants and emerging contaminants based on their relative toxic potency in environmental samples", *Trends Analyt. Chem.,* vol. 22, no. 10, pp. 655-665, 2003.
[http://dx.doi.org/10.1016/S0165-9936(03)01001-X]

[60]   Norman Network, *Norman netw. Ref. Lab. Res. Centres Relat. Organ. Monit. Emerg. Environ. Subst,* 2016.

[61]   Stockholm Convention, "The 16 new POPs", In: *An Introduction to the Chemicals Added to the Stockholm Convention as Persistent Organic Pollutants by the Conference of the Parties,* 2017, p. 25.

[62]   "Stockholm Convention", *Stock. Conv. Persistent Org. Pollut. (POPs). Text Annex,* 2010.https://www.env.go.jp/chemi/pops/treaty/treaty_ en2009.pdf

[63]   F. Vince, E. Aoustin, P. Bréant, and F. Marechal, "LCA tool for the environmental evaluation of potable water production", *Desalination,* vol. 220, no. 1-3, pp. 37-56, 2008.
[http://dx.doi.org/10.1016/j.desal.2007.01.021]

[64]   P. Loubet, P. Roux, E. Loiseau, and V. Bellon-Maurel, "Life cycle assessments of urban water systems: a comparative analysis of selected peer-reviewed literature", *Water Res.,* vol. 67, pp. 187-202, 2014.
[http://dx.doi.org/10.1016/j.watres.2014.08.048] [PMID: 25282088]

[65]   E. Igos, A. Dalle, L. Tiruta-Barna, E. Benetto, I. Baudin, and Y. Mery, "Lifecycle Assessment of water treatment: what is the contribution of infrastructure and operation at unit process level?", *J. Clean. Prod.,* vol. 65, pp. 424-431, 2014.
[http://dx.doi.org/10.1016/j.jclepro.2013.07.061]

[66]   A. Amini, Y. Kim, J. Zhang, T. Boyer, and Q. Zhang, "Environmental and economic sustainability of ion exchange drinking water treatment for organics removal", *J. Clean. Prod.,* vol. 104, pp. 413-421, 2015.
[http://dx.doi.org/10.1016/j.jclepro.2015.05.056]

[67]   J.K. Choe, M.H. Mehnert, J.S. Guest, T.J. Strathmann, and C.J. Werth, "Comparative assessment of the environmental sustainability of existing and emerging perchlorate treatment technologies for drinking water", *Environ. Sci. Technol.,* vol. 47, no. 9, pp. 4644-4652, 2013.
[http://dx.doi.org/10.1021/es3042862] [PMID: 23484880]

[68]   A.M. Boulay, A.Y. Hoekstra, and S. Vionnet, "Complementarities of water-focused life cycle assessment and water footprint assessment", *Environ. Sci. Technol.,* vol. 47, no. 21, pp. 11926-11927, 2013.
[http://dx.doi.org/10.1021/es403928f] [PMID: 24147821]

[69]   A.Y. Hoekstra, "A critique on the water-scarcity weighted water footprint in LCA", *Ecol. Indic.,* vol. 66, pp. 564-573, 2016.
[http://dx.doi.org/10.1016/j.ecolind.2016.02.026]

[70]   B.G. Ridoutt, and S. Pfister, "A revised approach to water footprinting to make transparent the impacts of consumption and production on global freshwater scarcity", *Glob. Environ. Change,* vol. 20, no. 1, pp. 113-120, 2010.
[http://dx.doi.org/10.1016/j.gloenvcha.2009.08.003]

CHAPTER 9

Conventional Methods for Removal of Emerging Water Pollutants

Manoj Tripathi[1,*] and **Akanksha Verma**[1]

[1] Department of Physics and Materials Science and Engineering, Jaypee Institute of Information Technology, A-10, Sector 62 Noida 201309, India

Abstract: Water is essential for life and no creature can survive without clean and usable water. Most water is unusable as it contains salts and many other organic and inorganic impurities. Without taking these impurities out, the water available to us cannot be used. Different techniques can be adapted to purify the water and make it usable. The selection of the water purification technique can be made depending upon the water contamination, its loading, and other parameters. Based on the contamination and other parameters, chemical or physical techniques for water purification can be applied. Disinfection, desalination, coagulation, and chemical precipitation are common chemical methods used for water purification. For adsorption, membrane filtration is used to filter the pollutant out physically. Various char-based materials are synthesized and used for water purification using the adsorption route. Highly porous char materials can contain the contaminants into their pores and allow the clean water to pass through. The overflow of the adsorbent with the pollutant can be solved by using magnetic biochar as the contaminants can be taken out of the magnetic char-based adsorbent and reused. Thus, the process becomes more effective and efficient. The chapter talks about these processes and their limitations, and advantages over the others. It also describes different types of materials used for the water purification processes.

Keywords: Adsorption, Chemical precipitation, Coagulation, Emerging water pollutants, Filtration, Membrane filtration, Reverse osmosis, Water purification.

INTRODUCTION

From primary schools to science documentaries, Earth is referred to as 'the blue planet.' This analogy comes from water covering about 71% of the earth's surface. 326 million cubic miles of water on the planet, from which 97% of the earth's water is found in the oceans, but this source is not reliable due to the presence of excessive salt structure in water which makes it unsuitable for drinking and grow-

* **Corresponding author Manoj Tripathi:** Department of Physics and Materials Science and Engineering, Jaypee Institute of Information Technology, A-10, Sector 62 Noida 201309, India; E-mail:tripmanoj@gmail.com

Shaukat Ali Mazari, Nabisab Mujawar Mubarak & Nizamuddin Sabzoi (Eds.)

ing crops, and industrial uses except cooling. Only 3% of the earth's water can be considered fresh, but this is not the end of the story. From that 3%, only 2.5% of the earth's freshwater is unavailable for many reasons. The availability of fresh water is locked up in glaciers, polar ice caps, atmosphere, and soil, highly polluted in some places, or lies too far beneath the earth's surface to be extracted inexpensively. Thus, only a quarter of the earth's water is available as freshwater, which in real figures amounts to an average of 8.4 million liters (2.2 million gallons) for each person on earth.

We are facing a shortage of freshwater, and that fixed amount of water is being used by the whole world's population with the freedom of wasting it independently. What about that remaining percentage of water covering nearly the whole globe? By calling it "salty water of no use", we declare it to be polluted, so putting more pollutants in it does not make any difference. Things are not as simple and suitable as they seem. Our oceans, covering a larger amount of earth, are the world for aquatic animals, and aquatic life is equally important for our ecosystem. Still, micro-pollutants are inevitably dumped in our oceans, lakes, ponds, and other water bodies. An estimated 80% of wastewater is dumped into the water bodies. This situation does not seem alarming because water is a universal solvent that dissolves every toxic element from farms, industries, households, *etc*. Researchers are doing work from the recent past in technologies of extracting toxic elements from wastewater to minimize water pollution [18, from global water pollution]. In developing countries like India and others, deploying this technique is still challenging.

Any chemical, synthetic or natural, or any micro-organism that is not readily found in the environment but, if they do so, can cause adverse effects to human health and the ecosystem and be termed as emerging contaminants (ECs) defined by U.S geological survey. ECs are those chemicals that emerge from veterinary and human pharmaceuticals when released into the environment and are of great concern due to their toxic effect potentially. Treating wastewater effectively can create a huge bridge between water resource supply and a rise in the economy [1]. There are many parts of the world where freshwater is not available evenly. Here, the proportion of water is uneven with the ground proportion. As per the WHO (World Health Organization) report, in the year 2008, nearly 25,000 people died due to contamination in water [2]. The water to be used should be proportionate to the water to be supplied. The chain of water from aquifers to its uses should be in proportion. The ratio of using water to storage should be small if we want a sustainable environment. Moreover, some serious steps should be taken to suppress the scarcity of fresh drinking water. Behnam *et al*. reported that chlorinated phenoxyacid herbicides account for most pesticides used worldwide, and their presence in environmental waters is well documented [2].

As technology is rooting expeditiously and living standards and ways of using resources are changing, the emergence of new emerging contaminants is an obvious question to be answered presently [3]. Also, there is an urgent need to identify new emerging contaminates to process further in removing them from the environment. The global corporation is needed to be in the process of assembling this collection. The geological composition of aquifers is structured to lead to the main cause of the leaching of toxic elements into drinking water supplies. The main elements are arsenic, fluoride, selenium, and a few others, such as chromium and uranium, creating chaos. Arsenic has caused the greatest negative health effects among these geogenic contaminants and is a global concern [4].

WATER CONTAMINATION

Biological Contamination

There is a balance in nature weighted with natural habitats on one side and its consumption on the other, and this balance should be thriving rather than shifting needle to one side only. Biological organisms that can contaminate water are classified as Pathogenic microbes, viruses, parasites, and Protista. Microorganisms are the backbone of all infectious diseases, whatever the origin of the former. The source can be general or sporadic, but infection from them is inevitable. The microorganisms category is considered to be fungi, bacteria, viruses, and even parasites [5, 6]. Insects and animals are the main transmitters of these microorganisms, and sometimes contaminated food and water also play a contagious role in the human body and Environment. Typhoid, Chickenpox, and measles are some infectious diseases, and researchers have found that the origin of the emergence of these diseases is the pathogens present in the polluted water. These pathogens present in contaminated water are a global concern to draw a clear separation between groundwater to freshwater.

Water supplies include groundwater from aquifers, seas or oceans, lakes, ponds, or wastewater from sewage treatment and municipal wastewater treatment. There is a list of pathogenic microorganisms, including viruses, bacteria, micro-organisms, heavy metals, and protozoans. Researchers are not very much concerned about the number of toxins they produce. Still, the effect of these pathogenic factors is rancor, and it is alarming to manage freshwater management [7]. In highly developed countries, outbreaks caused by pathogenic *E. coli* and cryptosporidiosis are reported more often. Legionella pneumophila is increasing in a concerning manner in warm water supplies and air-conditioning systems of large buildings, such as hospitals. Outbreaks of typhoid fever occur only sporadically. The most petrified disease caused by pathogens is Alzheimer's disease, the leading cause of cognitive and behavioral impairment in industrialized societies [8]. The potential contribution of pathogenic microbes to aging and AD is becoming increasingly recognized [9].

Chemical Contamination

Chemical contaminants are organic or inorganic elements or compounds that include nitrogen, bleach, salts, pesticides, metals, toxins produced by bacteria, and human or animal drugs. Heavy metals have a worse effect on human health, including Fluoride, Arsenic, Lead, Cadmium, Mercury, petrochemicals, chlorinated solvents, pesticides, and nitrates [10 - 13]. Fluoride is an essential mineral in water. An appropriate ratio of the former is required in liquid for the beneficial effect as efficiency might be expensive on human health. It is added to the water from the external source to counterpart the deficiency and prevents tooth decay in children in their early growing age. Taken in larger amounts, it can cause several health risks. Contamination of arsenic to human health is worse because of its mobilization in groundwater and aquifers (*e.g.*, hydraulic fracturing). Hence, its contamination can affect many people [14].

Personal care products also fall in polluting the environment and water. Although these are the substances applied by the human body externally, what is put externally will remain external to the human body and released back into the environment gradually. These include perfumes, soaps, cosmetics, lotions & sunscreens [5]. The various chemicals found in PCPs have been described as potential hormone-disrupting substances [6, 15]. Due to their daily use by almost every human being *via* bathing, grooming, showering, and cleaning, many toxins are released into the environment and the water bodies [15]. Due to all this progression, these micro-pollutants are now present in the background at increasing concentrations [16]. Many PCPs are bioactive; the environment can cope with them and have the potential for bioaccumulation [17]. The path of pharmaceuticals and the main entry route of PCPs into the environment are similar, and that is municipal WWTP effluents.

Medical Contamination

Medical contaminants include active medications prescribed by doctors. When consumed by the human body, it gets transferred to wastewater to receiving waters and drinking waters successively; complete removal of these compounds by conventional treatment processes is tedious work. Department of Environmental Chemistry, IDAEA-CSIC, and partner institutions located in Spain collaborated to remove pharmaceuticals and illicit drugs through conventional and membrane bioreactor wastewater treatment. Besides these traditional methods, riverbank filtration, often described as a more 'natural' approach to treatment, was also tried to remove those unwanted compounds.

Pharmaceutical compounds are highly bioactive and, therefore, undesired effects in organisms cannot be excluded after their discharge into the aquatic environment where, owing to their polarity, they tend to be quite mobile [18]. Even though the presence of pharmaceuticals in wastewater and natural water could be expected from their large production and widespread use, only developments in analytical chemistry (LC-MS/MS) allowed the analysis of these compounds in the nanogram to microgram per liter range, which is typical for wastewater and aquatic systems [19]. The observed concentrations of human pharmaceuticals in raw sewage of up to several micrograms per liter confirm that municipal wastewater is the main pathway for discharge to the receiving water bodies [4].

Other Contamination

Radiological contaminants are radioactive chemical elements that emit ionizing radiation. Some naturally occurring radiological contaminants are isotopes of all radioactive elements found in the periodic table. The example can cover isotopes of cesium, plutonium, and uranium. Physical contaminants are the contaminants that degrade the physical appearance by altering its physical properties. Soil erosion from rivers, lakes, ponds, oceans, and other water bodies and suspension of organic materials in water bodies can alter physical appearance. It can be categorized as an example of physical contamination. Changes in water temperature adversely affect water quality and aquatic biota called thermal pollution. 90% of thermal pollution originates from human activities. Oxygen content gets reduced by increasing the temperature of the water. Some of the power plants release excessive direct heat to water bodies causing water to be polluted and deployed with less oxygen, for example, nuclear power plants, electric power plants, petroleum refineries, steel melting factories, coal fire power plants, and boilers in the industries. They release heat excessively and are directed towards the water bodies causing water body pollution by manipulating its physical, chemical, and biological characteristics from the precursor, disturbing reproductive cycles, respiratory and digestive rates, and other physiological changes causing difficulties for the aquatic life [20, 21].

Spilling of oil on the surface of the water also contributes to water pollution as water and oil are completely insoluble with each other and pouring oil accidentally like leakage of tanks carrying petrol, diesel, and other inorganic substitutes or directly into water bodies like seas and oceans pollute seawater to a great extent. Exploring oil from offshore also leads to water pollution [22, 23]. The spread of oil on the water's surface forms a thin layer of emulsion preventing the growth of aquatic animals by reducing the amount of oxygen and the amount of sunlight from entering the water bodies to reach marine creatures.

Acid rain contributes highly to water pollution by altering the pH level of the surrounding plants and other plant-like species. Excessive amounts of acids in these rains can harm or shorten the lifespan of that species by making it endangered in our environment. When acid falls to earth through precipitation and reaches the ground, it flows into clean waterways. As they carry their acidic compounds into water bodies, it harms the nature and physical properties and chemical properties of water by polluting them and degrading aquatic life by changing the pH level of water [24].

IMPACT OF POLLUTANTS ON WATER POLLUTION

Impact of domestic waste on water pollution

Domestic waste accounts for the destruction done by human life in a day-to-day pattern. More specifically, sewage waste that contributes more than 90% of household waste, including poisonous toxins and other insoluble waste, needs more management by the government and other agencies to avoid its involvement in groundwater and aquifers. If poorly managed, more than 70% can contribute to water pollution by domestic sewage waste [25]. Household waste from colonization includes plastic littering, micro inorganic grains of insoluble plastic wrapping, polyethylene, one-time used products mainly, disposable coffee cups, disposable cups, and plates, primarily used in *Dhabas* (undeveloped restaurants near highways with poor accommodation and less expensive food), and at some regional places like *gurudwaras* and Temples for *langar* (feeding needy for free). These create pollution in the environment and water [26].

Solid domestic waste is a heterogeneous mixture of solid waste, biodegradable food waste (garbage), paper, glass, metal items, rags, rubbish, *etc*. The garbage can be identified as organic waste, including leftovers from kitchens, food articles, vegetable peeling, fruit peeling, overcooked or undercooked undesirable food outlets from hotels and restaurants, *etc*. These all contribute to organic waste directly dumped in water bodies, but this organic but open and odorful waste is not organic for health and creates hazards quickly. The garbage density usually varies between 450 to 900 kg/m^3 [27].

Impact of Construction Industries on Water Pollution

Coastal areas represent beauty around water, and construction around them is necessary to fulfill basic needs. Construction and trolling water levels with concrete and dust and leftover from renewed buildings make it polluted and hazardous for the existence of marine life. Wood, concrete, steel, cement, aggregates, bricks, clay, and metal are some basic requirements for construction, and all these also contribute to deploying harmfulness to water bodies.

Sediment in daily practice creates every possible obstruction for many reasons. For example, these sediments sheet their layer in the water bodies by forming a cloud, restricting the penetration of sunlight, limiting the processes that depend on light and choking out aquatic life, but also entrained sediments fall out of suspension downstream resulting in accumulated particles, suffocating benthic organisms and potentially getting pollutants that may be attached to relocated residues. Because construction activities have been recognized as a major cause of serious detriment [28], few studies have been made to quantify the amount of accelerated erosion occurring from construction activities on a project site and the yields of sediment-laden runoff that directly enters surface water bodies or is discharged to storm drains. These studies have revealed that the water quality is highly deteriorating because of the practice of letting the polluted water go onto the surface water [28].

Impact of Agricultural waste on Water Pollution

The immense and urgent need for food, wood, shelter, and other essential resources can only be accomplished by good agricultural techniques for food, especially for good yield. Agriculture crop ground erodes with harmful inorganic substitutes, which is the key to the emergence of harmful contaminants. Antibiotics, steroids, hormones, and antimicrobial from animal feed and livestock, respectively [29 - 31]. Emerging contaminants released from agricultural practices contribute to the leaching of the aquatic environment, such as antibiotics and hormones as they were detected. *E.g.*, countries like the USA contribute more than 60% of their area as agricultural land. Wastewater treatment of field irrigation also contributes to water contamination by degrading water [32].

The use of fertilizers is inevitable when agricultural development is required. Polluting Fertilizers are loaded with chemicals with a proper ratio of raw material to enhance the crop yield and develop a profitable agriculture business. Still, this high proportion of chemicals present in fertilizers and other substitutes used in the fields is the main reason for degrading soil and polluting water. Nitrates, phosphates, pesticides, and polycyclic aromatic hydrocarbons are the main constituents of fertilizers and degrading and polluting water at an alarming level due to these intensified heavy metals from the list given above foreseen as present horrifying heavy metal due to their intensity of polluting underground water streams. Fertilizers are not solely responsible; animal waste, which includes animal dungs littered everywhere, threatens natural freshwater by leaching its contents and contributing to the flow [33, 34].

DIFFERENT POLLUTION REMOVAL TECHNIQUES

Many different techniques are used to purify the water and make it usable. Many of these are known from ancient times, and some methods have evolved with time. The newer techniques are more adaptable and offer better performance. Moreover, these more recent techniques, such as reverse osmosis and adsorption-based purification, are designed for the specific or targeted removal of impurities. This makes these techniques more effective and useful. The current section is aimed to discuss these water purification techniques.

Disinfection

It is one of the oldest known techniques for removing unrequited microbes from the water. In this technique, harmful bacteria and pathogens are removed from the water. Different types of micro-organisms are cleared in the process. Removing these harmful micro-organisms inhibits reproduction, and thus, water quality does not deteriorate further. Sterilization is also used for water purification. The problem with sterilization is that it kills both the harmful and harmless micro-organisms in water. Some of the micro-organisms present in water are good for health and cannot be considered nonclinical.

Disinfection is achieved using some chemicals and reagents called disinfectants. When mixed in water, these disinfectants selectively kill or deactivate certain micro-organisms. These disinfectants, in many cases, adversely affect the organic contaminants in water which are the nutrients for the infectious micro-organisms. One of the good qualities of these disinfectants is that they remain active for a long time and keep on neutralizing undesired microbes. Halogens are generally used as disinfectants for water [35]. Some of the common disinfectants used for the purification of water are chlorine (Cl), chlorine dioxide (ClO_2), hypo chloride (OCl) *etc*. Other halogens like bromine, iodine and their compounds, for example, bromine chloride (BrCl), and iodine chloride (ICl), may also be used as disinfectants. Recently, metal-based disinfectants have also gained popularity as these are more selective and have improved efficiency over halogen-based disinfectants (like copper and silver-based) [36]. However, these metal-based disinfectants are more expensive than halogen-based disinfectants. Since these metal-based disinfectants are more selective, therefore, are more suitable for a small plant and a specific purpose.

Moreover, a few aromatic compounds like phenol are also seen to be highly effective in killing or deactivating the micro-organisms of water. Many other compounds can kill harmful microbes like potassium permanganate ($KMnO_4$), hydrogen peroxide (H_2O_2), *etc*. As mentioned above, chlorine is the most widely used disinfectant among all these disinfectants.

Desalination

Desalination is when the excess salt present in water is removed. Desalination can be done using chemical and physical processes. Desalination appears to be a powerful method to convert highly saline seawater into usable water. Desalinated water is suitable for municipal, agricultural, or industrial applications. Effective desalination can solve water-related issues, especially in areas near the sea. There is one major problem associated with the desalination technique. When the water with high salt content is passed through the desalination plant, two different streams of water at the output are received. One of them contains water with an acceptable range of salts, while the other stream's water carries a heavier salt content. The desalination technique has recently enjoyed the attention of researchers. The technique has advanced, and a few newer methods have been developed for the post desalination process. These recent additional changes are intended to improve the quality and efficiency of the desalination process and can handle the water containing a high salt load. Thermal desalination technology and membrane desalination technology are freezing and ion exchange, but these technologies are not used on a large scale.

Thermal desalination is probably the most ancient technique for the desalination of water. In this technique, the salted water is boiled. The water vapors evaporated during the process are collected, and then these vapors are condensed to get the desalinated water. This process requires the burning of fuel to boil the water. Therefore, the energy consumption is high in the thermal desalination process. It is one of the major drawbacks of this technique. This thermal desalination process can be categorized into multistage flash (MSF) desalination, multi-effect desalination (MED), and vapor compression evaporation (VCE). All these techniques are used to reduce the energy consumption in the desalination process, which eventually improves the efficiency of the process.

Coagulation and Flocculation

If some impurity is mixed with the water, then the larger (heavier) particles settle down to the bottom of the water. If impurity particles are small and lighter, these particles do not settle at the bottom and are suspended in the water. Sometimes, these impurity particles are too small to pass through the membrane filters. Coagulation is a very effective technique to remove these contaminations in this situation. Coagulation is a completely chemical process for water purification. A chemical compound is used in the coagulation process, which binds the small and light particles into a floc. This floc is now larger and heavier. Thus, this floc can settle down to the bottom of the water and can easily be filtered out.

Quite often, the suspended particles in the water are negatively charged. These negatively charged ions repel each other and, thus, move away from each other. They do not combine to form heavier particles and remain suspended in the water. In coagulation, a suitable chemical compound is mixed with the water, which produces an effective positive charge. The effective positive charge produced by the newly added compound neutralizes the negative charge of the impurity ions. Thus, the impurity ions can combine to form a larger and heavier particle. This compound to attracts the negatively charged impurity ions. The negatively charged impurity ions come and form a floc [37].

The mixing of the coagulant in water can be rapid mixing or slow and steady. Generally, the rapid and vigorous mixing method is adopted to ensure a uniform and diverse mixing of the coagulant into the water. Slow mixing is also required once the rapid and vigorous mixing is done. This gentle mixing ensures collisions of impurity particles, leading to the formation of a floc. This process is called flocculation. One thing to be noted here is that the coagulation and the flocculation process go hand in hand. For a successful purification, both processes must occur. If either of the methods does not materialize, the purification will not occur. Once the coagulation and flocculation are complete, the impurity particles settle down to the bottom of the water, and then this water can be filtered to get pure water [38]. The flocculation process takes a longer time compared to coagulation. The time required depends mainly upon the particle size. Table **1** represents the flocculation time of different particles with varying particle sizes. The larger and heavier the particles, the lesser time is required to get these particles settled at the bottom. The suspended particles may contain different shapes, charges, types, or sources. The selection of appropriate chemical compounds is essential for the success of this process. The process also depends upon the Coagulant used, its dose, pH value of water, contamination type, *etc.* Therefore, selecting the appropriate coagulant is highly important [39].

Table 1. Particles' settling time after coagulation and flocculation [43].

Type of Particle	Particle Size (m)	Time required to settle the particles (at a depth of 1 m)
Virus	10^{-9}	200 years
	10^{-8}	20 years
	10^{-7}	2 years
Colloids	10^{-9}	200 years
	10^{-8}	20 years
	10^{-7}	2 years
Algae	10^{-6}	8 days
	10^{-5}	2 hours

(Table 1) cont.....

Type of Particle	Particle Size (m)	Time required to settle the particles (at a depth of 1 m)
Bacteria	10^{-6}	8 days
Clay	10^{-5}	2 hours
Fine sand	10^{-4}	2 minutes
Sand	10^{-3}	10 seconds
Gravel	10^{-2}	1 second

The coagulation-flocculation process has certain limitations. First, this process can be used only if the water contains immiscible impurities. This process cannot be used if the pollutant is chemically mixed with water. Even if the contaminant is omissible but is neutral, this process will not be of any help. Moreover, this process depends upon so many parameters mentioned above, so selecting the appropriate coagulant and the suitable dose is not easy. Afterward, by managing a combination of coagulant and dose, we can only remove about 60-70% of the impurity from water. Aluminum sulfate, ferric sulfate, ferric chloride, sodium aluminate, titanium, and zirconium-based salts are common coagulants used for water purification. Among these, aluminum and iron salts are the most widely used coagulants. However, newer techniques and materials have improved the efficacy of the process up to a certain extent. Permanganate oxidation-assisted coagulation has been reported to enhance the removal of natural organic material and algae [40]. Treatment of powdered activated carbon resulted in almost all the pollutants from the wastewater [41]. In another research work, synthetic hydroxylapatite and chitosan worked remarkably as coagulants and removed around 88-95% of Cu and Zn from the water [42]. Moreover, chitosan can improve the rate of coagulation as well.

Chemical Precipitation

Chemical precipitation is almost like the coagulation and flocculation process. This technology is widely used to remove heavy metals like arsenic or lead and radionuclides from the groundwater. This technique is generally used as a pretreatment technique for other water purification technique where the presence of the metals may cause trouble in the purification process. In this technique, the pH value of water is changed, and the heavy metals start to precipitate at the bottom of the water, from where they can be taken away. Carbonate, sulfate, sulfide, and hydroxide are generally used for chemical precipitation. These precipitants are mixed with water, and cause the precipitation of heavy metals. Even in an anaerobic environment, bacteria can chemically react with heavy metals and result in insoluble metal precipitation. Chemical precipitation occurs when the chemical reaction of the precipitant with soluble impurity (present in water) converts it into an insoluble salt or compound. If the soluble impurity content is high enough to make it supersaturated, that can also cause precipitation [44].

However, the precipitation technique may not be very useful if the water contains more than one metal impurity. Suppose there are two or more metal impurities. In that case, as we try to increase the chemical precipitation of one metal using a selected reagent, it may slow down the chemical precipitation of other metals. Continuous research in the field has derived many materials to remove specific chemical compounds very effectively. This precipitation method appears to be more suitable for extracting and eliminating selective metallic impurities from water. For example, the reduction of large polysaccharides was achieved by alcohol precipitation [45]. In another work reported by Shiba and Ntuli, phosphorus precipitation from the sewage water by leaching dry sewage sludge with 1 M sulfuric acid with a loading of 5% resulted in 82% phosphorus extraction [46].

Ozonation

Ozone (O_3) is a well-known gas and carries good disinfecting characteristics. The ozone is generated when a very high electric discharge of electric current occurs in the upper atmosphere. Ozone is highly unstable and gets converted to oxygen (O_2) quickly. However, there are few electrochemical methods available to produce ozone. Still, due to its instability, electric discharge is the only practical way to generate ozone at a laboratory or an industrial level [47]. Ozone is a strong oxidizing agent, and it is a very effective disinfectant. It can neutralize and deactivate micro-organisms like bacteria, fungi, and viruses in water. The main disinfectant mechanism is because of the cell lysis of the ozone molecules [48]. The neutralization or deactivation of the micro-organisms also causes damage to the generic gene structures and parasitic bacteria in the dead cell of these micro-organisms.

Ozone is more than 50% more effective than chlorine's most used disinfectant. Moreover, it has more than three times the micro-organism killing capacity than chlorine. Ozone has a high oxidizing capacity. It may oxidize the organic materials in bacterial membranes. It causes rupture of the cell, resulting in the death of a cell. This mechanism can damage almost all bacteria, viruses, and other harmful species with almost no immunity development [49]. One more advantageous factor with ozone is that the pH of ozone is 7. Therefore, the addition of ozone to the water does not change the pH value of water. Most of the other disinfectants do not have this advantage, and the use of those disinfectants may result in a slight change in the pH value of water. Mostly the disinfectants are alkali in nature, so the use of those disinfectants may make water more alkaline, while the ozonation does not have this issue.

Moreover, ozone for water purification is advantageous as this process does not involve using any harmful chemicals. Ozone is seen to be a very effective disinfectant to purify water. The cost involved in the generation of ozone is the major challenge. The high cost of ozone production limits its practical applicability [44, 50].

Reverse Osmosis

Reverse osmosis is a leading water purification technology that can remove ions, bacteria, colloids, micro-organisms, and other contaminants. French physicist Jean Antonie Nollet first observed reverse osmosis with the help of a semi-permeable membrane. Later in 1950, researchers of the University of California and the University of Florida, in a joint project, successfully purified seawater with the help of a semi-permeable membrane using the reverse osmosis process. Principally, reverse osmosis is a process in which the contaminated water is allowed to pass through a semi-permeable membrane by applying pressure. The semi-permeable membrane is porous. The membranes are designed so that the pore size allows the water molecules to pass through them but block the contaminants [51, 52]. The permeate in this process is purified and clean water.

The semi-permeable membrane is the most significant component in the reverse osmosis process. The quality and the pore size of the membrane will greatly affect the water purification ability of the process. The standard deviation of the pore size distribution of these membranes should be very low. Even a few larger pores can disrupt the whole process of purification. The membrane can block the contaminants based on their size and charge [53]. If the charge of the ions is more, it is more likely to be rejected from the membrane. Generally, the monovalent charge particles can pass through the RO membrane without blocking them, provided their size (or molecular weight) is smaller than that of water molecules. This is why sodium is generally not filtered by the RO membrane (a monovalent particle whose size is smaller than water).

Various kinds of membranes are being used in the reverse osmosis process. These include polymeric membranes, mixed matrix membranes (MMM), metal-oxid--based membranes, ceramic membranes, carbon nanotube (CNT) or graphene oxide (GO) based membranes, *etc*. The polymeric membranes are the most dominant membranes for the RO process as these are less expensive and are easier to fabricate. There are a variety of polymeric membranes which are used for water purification. Many commercially available RO purification units are equipped with these polymeric-based membranes. Thin films of these polymeric membranes are highly suitable and offer very good performance with a wide range of operating conditions. However, fouling is the major problem with polymeric RO membranes. The issue of fouling can be removed by introducing the materials having inorganic fillers in the organic matrix [54]. These types of membranes are called metal matrix membranes. These MMMs offer better mechanical strength and thermal stability [55].

Moreover, these MMMs are chemically more stable and have a longer life. Metal oxide membranes are also very useful for the RO process. These metal oxide membranes are generally made up of alumina, zirconia, titania, and their oxides. In search of new and inexpensive materials for the RO membranes, few ceramic membranes have been developed, which are very promising for water purification using the RO technique. Furthermore, carbon nanotube (CNT) and graphene oxide (GO) is also observed to be very suitable for the RO process [56, 57].

Adsorption

Adsorption is a phenomenon that can be used to remove both organic and inorganic types of pollutants from water. Adsorption is a surface phenomenon. Water purification using adsorption requires a highly porous organic material. When the polluted water meets the porous organic material, the solute (which generally is a pollutant), because of the intermolecular forces between the solid-liquid interface, gets deposited (more truly adsorbed) in the pores of the solid. The solid compound which is adsorbed on the solid surface is called adsorbate, and the solid on which the adsorbate is deposited is called adsorbent. The process of depositing adsorbate in the pore sites of the adsorbent is called adsorption. The molecules of adsorbent which lie in the inner layer are surrounded by their molecules in all directions, so their net force is zero. While the molecules which lie on the surface are not surrounded by their molecules in all directions, so these tend to attract other molecules. The precise characteristics and attraction force depend upon factors like electronegativity and other properties of involved molecules [58, 59].

For effective purification, the adsorbent should have a high adsorbate loading capacity. The high loading capacity can be achieved by using a highly porous material as an adsorbent. For this purpose, both natural and synthetic adsorbent materials are used. Charcoal, clay, clay-based minerals, and zeolite fall in natural adsorbents. Synthetic adsorbents are made from the treatment of agricultural waste, industrial waste, biomass waste, house waste, and other waste generated from different sources. The waste materials can be converted into highly porous materials using pyrolysis techniques. Many waste materials such as empty fruit bunch, rice straw, wheat straw, sugarcane, bagasse, oil palm shell, coconut shell, *etc.*, have been successfully converted to high porous materials [60 - 65]. These porous materials can be further used as adsorbents for water purification [66, 67]. Comparatively, natural adsorbents are less expensive than synthetic adsorbents. However, the pore size, distribution, BET surface area, and loading capacity are very high in synthetic adsorbents.

Activated carbon and other waste-derived char materials are promising materials for water purification through the adsorption route. Activated carbon was successfully used to remove Amoxicillin from the pharmaceutical effluent [68]. In another research work, activated carbon developed from sawdust was used to remove dye from wastewater [69]. Mohan *et al.* reported that methylene blue was effectively removed through the adsorption process using fly ash as an adsorbent [70].

Membranes

Membrane-based separation is one of the most used water purification techniques in places where the amount to be filtered is not too high. The membrane-based filtration process is not only efficient but also cost-effective. One of the major advantages of membrane-based filtration is that the membranes used in filters can be modified as per the requirement. Different materials are available which can serve as effective membranes for water filtration. Polymer-based membranes are very effective for water filtration and removing unwanted salts and impurities [71]. The membrane filtration is often a physical process, and the pore openings need to be adjusted according to the nature and size of the impurity. This membrane filter can also be used at the output of any processes mentioned above. These hybrid systems make the process more effective and efficient [72]. Newer materials have shown very good potential to be used as an effective membrane. The tiO$_2$-based film offered more than 99.9% salt rejection even with a high flux rate [73]. In another work, a direct contact membrane desalination film of PVDF coated TiO$_2$ removed 99.98% of impurity from the polluted water [74]. Other materials with polymer-based nanofiber membranes comprising PVDF-tetra butyl-ammonium chloride (TBAC) and fluorinated acrylate (FA) copolymer have a hierarchical tree-like structure nanofiber membrane (TINM) were successfully fabricated and tested [75]. Another polymer-based membrane was designed in which fluorinated polyurethane additive (FPA) was mixed with polysulfone (PSF). This membrane demonstrated a good response even at a flux of 53.8 kg/m^{2}h at a temperature as high as 60^0 C [76]. A brief description of membranes in terms of membrane pore size and detained species is given in Table **2**.

Table 2. Classification of water purification membranes regarding membrane pore size and detained species.

Membrane Type	Pore Size (nm)	Detained species
Micro-filtration membrane	10000 - 100	Large colloidal bacteria
Ultra-filtration membrane	100 - 2	Macro molecules, Proteins
Nano-filtration membrane	2 - 1	Multivalent salts
RO membrane	1 - 0.1	Monovalent salts

Membrane-based water filtration appears to be a good technique, but it suffers from two major issues. The first is the mechanical strength of the membranes, and the second is the de-carbonization of the membranes. Membranes used for the filtration process need to be thin yet mechanically strong to bear the load of a high flux of incoming water. The de-carbonization of the membranes is important as this will enhance the membrane's efficiency and life. Specially engineered bio-films are now available, which can tackle both of these problems and thus, make the process more efficient [77].

MATERIALS USED FOR POLLUTANT REMOVAL

Many researchers have tried to use different kinds of organic, inorganic, biological, and other materials for pollutant removal applications. These materials and compounds can be fed directly into the polluted water as disinfectants, coagulants, membranes, or adsorbents. Chlorine is a universally accepted and undoubtedly the most used disinfectant for water purification. Hypo chloride of chlorine and other halogen-based chlorides such as $BrCl$, ICl, $IOCl_2$, and OCl is also used to substitute the chlorine. Still, chlorine is highly effective and cost-effective and supersedes all other disinfectants. Ozone (O_3) can also be used to purify polluted water. Ozone has the potential to kill harmful bacteria, viruses, micro-organisms, and other biological pollutants and purify the water. Ozone is used for water purification in some European countries. Still, its synthesis cost is high, which prevents it from being a widely accepted and used chemical compound for water purification.

Activated carbon and biomass waste-derived char is used extensively for water purification. These materials are used as adsorbents and are used as water filtering membranes. The porous nature of the activated carbon and char materials makes them suitable for removing many heavy metals like As, Sb, Zn, Pb, *etc*. Char prepared from agricultural waste such as rice husk, wheat straw, corn stalk, coconut shell, *etc*., offers a large BET surface area and micro & macropores. The large BET surface area and porous nature of the char are highly recommended for the adsorbent materials for the water purification process. Table **3** lists a few chars derived from agricultural waste and their BET surface area. It is seen from Table 3 that the BET surface area is about 1000 m^2/g. The high BET surface area of char materials obviously can hold more pollutants and work for a longer time.

Table 3. Agriculture waste-derived magnetic char and their BET surface area.

Precursor	Treated with	BET Surface Area (m^2/g)	Reference
forestry wood waste	Fe_2O_3	227	[78]
Walnut shell	$FeSO_4$	138.09	[79]
sewage sludge	Fe_2O_3	65.47	[80]
Acon seed	$FeCl_2$	72.9	[81]
Wheat straw	Fe_3O_4	1416	[82]
pomelo peel	Fe_3O_4	1061	[83]
Sawdust	$FeCl_3$	1470	[84]
Rice straw	$FeCl_3$	1057	[85]
Empty fruit bunch	$FeCl_3$	890	[86]
Coconut shell	-	951.85	[87]

Many inorganic materials are also being used in the membrane filtration technique. These inorganic ceramic-based membranes are useful in water purification. Polymer-based membranes in which the metals are used as fillers are gaining acceptance for water filtration. TiO_2-based films are quite effective for water purification. Polymer-metal films are cost-effective, offer higher mechanical strength, and are less prone to corrosion. However, the loading capacity has always remained a problem with the films and membranes used for water purification. The magnetic biochar is a step forward in dealing with this problem. The magnetic biochar is the char treated with some Fe-based chemicals such as $FeCl_3$, $FeSO_4$, Fe_3O_4, Fe_3O_4, *etc* [67, 78]. These magnetic biochar provide extra strength to the film, and the laden can be recovered from it. Thus. The films made up of magnetic biochar can be reused, making them more efficient economically. Different agriculture waste-derived magnetic char have shown a good response and are used for water purification.

CONCLUSION

The scarcity of drinking water is a serious issue and should be a high priority. However, a huge portion of the earth carries water, but the water is not usable, especially not drinkable. Many conventional techniques are implanted to purify the polluted water. The contaminants present in water are bacteria, viruses, micro-organisms, different types of heavy metals and other impurities. Different water purification techniques have been developed based on the contaminants type and amount in water; by adopting a suitable approach, the contaminated water can be converted to usable water. The water purification techniques can broadly be divided into two categories. First, the chemical action on the pollutant purifies the water and the second one is in which the physical filtration is used to separate the contaminants from the polluted water. Both techniques have advantages and disadvantages depending upon the type of pollutant, amount of impurity, and potential usage of purified water; we can choose the purification techniques. Among the chemical reaction-based purification, disinfection, coagulation, precipitation, and ozonation are the most suitable techniques. Among these techniques, disinfection is a more widely accepted method.

Chlorine and compounds of other halogens are used as disinfectants to purify polluted water. These disinfectants can kill harmful bacteria and other micro-organisms effectively. However, many compounds can be used as disinfectants, but chlorine is the most popular and widely used disinfectant. Ozonation is also very effective in neutralizing living microorganisms, bacteria, and other pollutants but is expensive. It is not used generally. However, smaller units use this technique to achieve very high water purity. To remove water impurities, membranes and adsorption filtration are used. Different organic and inorganic materials are used in membranes. Other ceramics, metal-based, polymer-based, and Polymer metal composite-based membranes are used for the water purification process. TiO_2 is the most common material used for the filtration membrane.

Along with other advantages, it provides high mechanical strength to the membrane. Adsorption is one of the most effective techniques for purification. The porous adsorbent in the adsorption-based purification takes away all the solid contaminants and lets pure water pass through it. Other agricultural waste-derived char materials with high BET surface area and large porosity, such as coconut, rice husk, wheat husk, pinewood, coffee seeds, corn stalk, *etc.*, are being used as adsorbents for water purification. Magnetic biochar with a facility to recover the filtered material and reuse the adsorbent makes the process more effective and efficient. All these water purification techniques are suitable in their way. Moreover, the section of the purification technique should be done based on certain parameters like cost, nature of pollutant, its loading in water, pH of water, *etc*. An adequate selection of techniques will surely reduce the pollutant loading in water and make it more usable.

CONSENT FOR PUBLICATION

Not applicable.

CONFLICT OF INTEREST

The author declares no conflict of interest, financial or otherwise.

ACKNOWLEDGEMENTS

Declared none.

REFERENCES

[1]   M. Grassi, G. Kaykıoğlu, V. Belgiorno, and G. Lofrano, "Removal of emerging contaminants from water and wastewater by adsorption process", In: *Emerging compounds removal from wastewater: Natural and solar based treatments*, 2012, pp. 15-38.
[http://dx.doi.org/10.1007/978-94-007-3916-1_2]

[2]   H. Behnam, B. S, and Sabbagh. F, *Biological Contamination of the Water and Its Effects*, 2013.

[3]   A. Boxall, "New and Emerging Water Pollutants arising from Agriculture", 2012.

[4]   R. Schwarzenbach, T. Egli, T. Hofstetter, U. Gunten, and B. Wehrli, *Global Water Pollution and Human Health*. vol. 35. Ann Rev Environ Resour, 2010.

[5]   A. Pal, Y. He, M. Jekel, M. Reinhard, and K.Y-H. Gin, "Emerging contaminants of public health significance as water quality indicator compounds in the urban water cycle", *Environ. Int.,* vol. 71, pp. 46-62, 2014.
[http://dx.doi.org/10.1016/j.envint.2014.05.025] [PMID: 24972248]

[6]   E. Den Hond, M. Paulussen, T. Geens, L. Bruckers, W. Baeyens, F. David, E. Dumont, I. Loots, B. Morrens, B.N. de Bellevaux, V. Nelen, G. Schoeters, N. Van Larebeke, and A. Covaci, "Biomarkers of human exposure to personal care products: results from the Flemish Environment and Health Study (FLEHS 2007-2011)", *Sci. Total Environ.,* vol. 463-464, pp. 102-110, 2013.
[http://dx.doi.org/10.1016/j.scitotenv.2013.05.087] [PMID: 23792252]

[7]   T. Egli, W. Köster, and L. Meile, "Pathogenic microbes in water and food: changes and challenges", *FEMS Microbiol. Rev.,* vol. 26, no. 2, pp. 111-112, 2002.
[http://dx.doi.org/10.1111/j.1574-6976.2002.tb00603.x] [PMID: 12069876]

[8]   H.K. Brown, J. Hill, and R. Natale, "Caesarean section rates in Southwestern Ontario: changes over time after adjusting for important medical and social characteristics", *J. Obstet. Gynaecol. Can.,* vol. 36, no. 7, pp. 578-589, 2014.
[http://dx.doi.org/10.1016/S1701-2163(15)30537-5] [PMID: 25184976]

[9]   S. Poole, S.K. Singhrao, L. Kesavalu, M.A. Curtis, and S. Crean, "Determining the presence of periodontopathic virulence factors in short-term postmortem Alzheimer's disease brain tissue", *J. Alzheimers Dis.,* vol. 36, no. 4, pp. 665-677, 2013.
[http://dx.doi.org/10.3233/JAD-121918] [PMID: 23666172]

[10] F. Ferniza-García, A. Amaya-Chávez, G. Roa-Morales, and C.E. Barrera-Díaz, "Removal of Pb, Cu, Cd, and Zn Present in Aqueous Solution Using Coupled Electrocoagulation-Phytoremediation Treatment", *Int. J. Electrochem.,* vol. 2017, p. 7681451, 2017.
[http://dx.doi.org/10.1155/2017/7681451]

[11] H.M. Zwain, M. Vakili, and I. Dahlan, "Waste Material Adsorbents for Zinc Removal from Wastewater: A Comprehensive Review", *Int. J. Chem. Eng.,* vol. 2014, p. 347912, 2014.
[http://dx.doi.org/10.1155/2014/347912]

[12] W. Zhan, C. Xu, G. Qian, G. Huang, X. Tang, and B. Lin, "Adsorption of Cu(ii), Zn(ii), and Pb(ii) from aqueous single and binary metal solutions by regenerated cellulose and sodium alginate chemically modified with polyethyleneimine", *RSC Advances,* vol. 8, no. 33, pp. 18723-18733, 2018.
[http://dx.doi.org/10.1039/C8RA02055H] [PMID: 35541150]

[13] M. Doumer, A. Rigol, M. Vidal, and A. Mangrich, "Removal of Cd, Cu, Pb, and Zn from aqueous solutions by biochars", In: *Environmental science and pollution research international* vol. 23. , 2015.

[14] S. Murcott, *Arsenic Contamination in the World: An International Sourcebook 2012.* vol. 11. Water Intelligence Online, 2012.

[15] G.F. Birch, D.S. Drage, K. Thompson, G. Eaglesham, and J.F. Mueller, "Emerging contaminants (pharmaceuticals, personal care products, a food additive and pesticides) in waters of Sydney estuary, Australia", *Mar. Pollut. Bull.,* vol. 97, no. 1-2, pp. 56-66, 2015.
[http://dx.doi.org/10.1016/j.marpolbul.2015.06.038] [PMID: 26130525]

[16] J. Regueiro, M. Llompart, E. Psillakis, J.C. Garcia-Monteagudo, and C. Garcia-Jares, "Ultrasound-assisted emulsification-microextraction of phenolic preservatives in water", *Talanta,* vol. 79, no. 5, pp. 1387-1397, 2009.
[http://dx.doi.org/10.1016/j.talanta.2009.06.015] [PMID: 19635375]

[17] J.M. Brausch, and G.M. Rand, "A review of personal care products in the aquatic environment: environmental concentrations and toxicity", *Chemosphere,* vol. 82, no. 11, pp. 1518-1532, 2011.
[http://dx.doi.org/10.1016/j.chemosphere.2010.11.018] [PMID: 21185057]

[18] C. Houtman, "Emerging contaminants in surface waters and their relevance for the production of drinking water in Europe", *J. Integr. Environ. Sci.,* vol. 7, no. 4, pp. 271-295, 2010.
[http://dx.doi.org/10.1080/1943815X.2010.511648]

[19] R. Mancuso, F. Baglio, M. Cabinio, E. Calabrese, A. Hernis, R. Nemni, and M. Clerici, "Titers of herpes simplex virus type 1 antibodies positively correlate with grey matter volumes in Alzheimer's disease", *J. Alzheimers Dis.,* vol. 38, no. 4, pp. 741-745, 2014.
[http://dx.doi.org/10.3233/JAD-130977] [PMID: 24072067]

[20] F. D. Owa, "Water Pollution: Sources, Effects, Control and Management", *Mediterranean Journal of Social Sciences,* vol. 4, 2013.
[http://dx.doi.org/10.5901/mjss.2013.v4n8p65]

[21] H-J. Kim, G.J. Lee, A-J. Choi, T-H. Kim, T-i. Kim, and J-M. Oh, *Layered Double Hydroxide Nanomaterials Encapsulating Angelica gigas Nakai Extract for Potential Anticancer Nanomedicine.* vol. 9. , 2018.

[22] M.T. Rahman, T. Kameda, T. Miura, S. Kumagai, and T. Yoshioka, "Facile method for treating Zn, Cd, and Pb in mining wastewater by the formation of Mg–Al layered double hydroxide", *Int. J. Environ. Sci. Technol.,* vol. 17, no. 5, pp. 3023-3032, 2020.
[http://dx.doi.org/10.1007/s13762-020-02689-x]

[23] P.S. Kulkarni, S.D. Dhar, and S.D. Kulkarni, "A rapid assessment method for determination of iodate in table salt samples", *J. Anal. Sci. Technol.,* vol. 4, no. 1, p. 21, 2013.
[http://dx.doi.org/10.1186/2093-3371-4-21]

[24] A. Gupta, *Water Pollution-Sources, Effects And Control.* Pointer Publishers Jaipur, 2016.

[25] L.P. Cardoso, R. Celis, J. Cornejo, and J.B. Valim, "Layered double hydroxides as supports for the slow release of acid herbicides", *J. Agric. Food Chem.,* vol. 54, no. 16, pp. 5968-5975, 2006.
[http://dx.doi.org/10.1021/jf061026y] [PMID: 16881703]

[26] S.J.I.J.A.R.C.S. Kulkarni, "An Investigation on Organic Matter Removal from Wastewater by Local Groundnut Shell Derived Adsorbent from Salgar Budruk-Kinetic", *Isotherm and Parameter Studies,* vol. 4, pp. 41-47, 2017.

[27] P. R. C. Chhipa, "Impact of Solid Waste Disposal on Ground Water Quality in Different Disposal Site at Jaipur, India", *Int. J. of Eng. Sci. & Rese. Techno.,* vol. IJESRT, pp. 93-101, 2014.

[28] D.L. Houser, and H. Pruess, "The effects of construction on water quality: a case study of the culverting of Abram Creek", *Environ. Monit. Assess.,* vol. 155, no. 1-4, pp. 431-442, 2009.
[http://dx.doi.org/10.1007/s10661-008-0445-9] [PMID: 18629442]

[29]   W. Song, Y. Ding, C.T. Chiou, and H. Li, "Selected veterinary pharmaceuticals in agricultural water and soil from land application of animal manure", *J. Environ. Qual.,* vol. 39, no. 4, pp. 1211-1217, 2010.
[http://dx.doi.org/10.2134/jeq2009.0090] [PMID: 20830908]

[30]   S. Yang, and K. Carlson, "Evolution of antibiotic occurrence in a river through pristine, urban and agricultural landscapes", *Water Res.,* vol. 37, no. 19, pp. 4645-4656, 2003.
[http://dx.doi.org/10.1016/S0043-1354(03)00399-3] [PMID: 14568051]

[31]   T. Christian, R. Schneider, H. Färber, D. Skutlarek, M. Meyer, and H. Goldbach, "Determination of Antibiotic Residues in Manure, Soil, and Surface Waters", *Acta Hydrochim. Hydrobiol.,* vol. 31, no. 1, pp. 36-44, 2003.
[http://dx.doi.org/10.1002/aheh.200390014]

[32]   J.A. Pedersen, M.A. Yeager, and I.H. Suffet, "Xenobiotic organic compounds in runoff from fields irrigated with treated wastewater", *J. Agric. Food Chem.,* vol. 51, no. 5, pp. 1360-1372, 2003.
[http://dx.doi.org/10.1021/jf025953q] [PMID: 12590482]

[33]   O. Akinro, A. Oloruntade, and O. Imoukhuede, "Impacts of Agricultural Wastes on Groundwater Pollution in Lipakala Farms, Ondo Southwest Nigeria", *J. Environ. Earth Sci.,* vol. 2, pp. 7-12, 2012.

[34]   A. Akinro, A. Olufayo, and P. Oguntunde, "Crop Water Productivity of Plantain (Musa Sp) in a Humid Tropical Environment", *J. Eng. Sci. and Techno. Rev.,* vol. 5, 2012.
[http://dx.doi.org/10.25103/jestr.051.04]

[35]   P. Singh, P. Shandilya, P. Raizada, A. Sudhaik, A. Rahmani-Sani, and A. Hosseini-Bandegharaei, "Review on various strategies for enhancing photocatalytic activity of graphene based nanocomposites for water purification", *Arab. J. Chem.,* vol. 13, no. 1, pp. 3498-3520, 2020.
[http://dx.doi.org/10.1016/j.arabjc.2018.12.001]

[36]   T. Matsumoto, I. Tatsuno, and T. Hasegawa, "Instantaneous Water Purification by Deep Ultraviolet Light in Water Waveguide", *Escherichia Coli Bacteria Disinfection,* vol. 11, p. 968, 2019.

[37]   M. Ma, R. Liu, H. Liu, and J. Qu, "Effect of moderate pre-oxidation on the removal of Microcystis aeruginosa by KMnO4-Fe(II) process: significance of the in-situ formed Fe(III)", *Water Res.,* vol. 46, no. 1, pp. 73-81, 2012.
[http://dx.doi.org/10.1016/j.watres.2011.10.022] [PMID: 22078228]

[38]   M. Yan, D. Wang, S. You, J. Qu, and H. Tang, "Enhanced coagulation in a typical North-China water treatment plant", *Water Res.,* vol. 40, no. 19, pp. 3621-3627, 2006.
[http://dx.doi.org/10.1016/j.watres.2006.05.044] [PMID: 16904723]

[39]   M. Lapointe, C. Brosseau, Y. Comeau, and B. Barbeau, "Assessing Alternative Media for Ballasted Flocculation", *J. Environ. Eng.,* vol. 143, no. 11, p. 04017071, 2017.
[http://dx.doi.org/10.1061/(ASCE)EE.1943-7870.0001271]

[40]   K. Nielson, and D.W. Smith, *Ozone-enhanced electroflocculation in municipal wastewater treatment.* vol. 4. , 2005, pp. 65-76.

[41]   J.M. Younker, and M.E. Walsh, "Effect of adsorbent addition on floc formation and clarification", *Water Res.,* vol. 98, pp. 1-8, 2016.
[http://dx.doi.org/10.1016/j.watres.2016.03.044] [PMID: 27064206]

[42]   S. Kadouche, H. Lounici, K. Benaoumeur, N. Drouiche, M. Hadioui, and P. Sharrock, "Enhancement of Sedimentation Velocity of Heavy Metals Loaded Hydroxyapatite Using Chitosan Extracted from Shrimp Waste", *J. Polym. Environ.,* vol. 20, no. 3, pp. 848-857, 2012.
[http://dx.doi.org/10.1007/s10924-012-0440-7]

[43]   H. G. Peterson, "Rural Drinking Water And Waterborne Illness", 2002.

[44]   M. Chen, K. Shafer-Peltier, M. Veisi, S. Randtke, and E. Peltier, "Complexation and precipitation of scale-forming cations in oilfield produced water with polyelectrolytes", *Separ. Purif. Tech.,* vol. 222,

pp. 1-10, 2019.
[http://dx.doi.org/10.1016/j.seppur.2019.04.014]

[45]   G.Y. Koh, G. Chou, and Z. Liu, "Purification of a water extract of Chinese sweet tea plant (Rubus suavissimus S. Lee) by alcohol precipitation", *J. Agric. Food Chem.,* vol. 57, no. 11, pp. 5000-5006, 2009.
[http://dx.doi.org/10.1021/jf900269r] [PMID: 19419169]

[46]   N.C. Shiba, and F. Ntuli, "Extraction and precipitation of phosphorus from sewage sludge", *Waste Manag.,* vol. 60, pp. 191-200, 2017.
[http://dx.doi.org/10.1016/j.wasman.2016.07.031] [PMID: 27481032]

[47]   Y. Nomoto, T. Ohkubo, S. Kanazawa, and T. Adachi, "Improvement of ozone yield by a silent-surface hybrid discharge ozonizer", *IEEE Trans. Ind. Appl.,* vol. 31, no. 6, pp. 1458-1462, 1995.
[http://dx.doi.org/10.1109/28.475741]

[48]   N. K. Hunt, and B. J. Mariñas, "Kinetics of Escherichia coli inactivation with ozone", *Water Research,* vol. 31, pp. 1355-1362, .
[http://dx.doi.org/10.1016/S0043-1354(96)00394-6]

[49]   P. Li, C. Wu, Y. Yang, Y. Wang, S. Yu, S. Xia, and W. Chu, "Effects of microbubble ozonation on the formation of disinfection by-products in bromide-containing water from Tai Lake", *Separ. Purif. Tech.,* vol. 193, pp. 408-414, 2018.
[http://dx.doi.org/10.1016/j.seppur.2017.11.049]

[50]   F.J. Beltrán, and M. Checa, "Comparison of graphene oxide titania catalysts for their use in photocatalytic ozonation of water contaminants: Application to oxalic acid removal", *Chem. Eng. J.,* vol. 385, p. 123922, 2020.
[http://dx.doi.org/10.1016/j.cej.2019.123922]

[51]   P. Hadi, M. Yang, H. Ma, X. Huang, H. Walker, and B.S. Hsiao, "Biofouling-resistant nanocellulose layer in hierarchical polymeric membranes: Synthesis, characterization and performance", *J. Membr. Sci.,* vol. 579, pp. 162-171, 2019.
[http://dx.doi.org/10.1016/j.memsci.2019.02.059]

[52]   K. Ikehata, Y. Zhao, H.V. Kulkarni, Y. Li, S.A. Snyder, K.P. Ishida, and M.A. Anderson, "Water Recovery from Advanced Water Purification Facility Reverse Osmosis Concentrate by Photobiological Treatment Followed by Secondary Reverse Osmosis", *Environ. Sci. Technol.,* vol. 52, no. 15, pp. 8588-8595, 2018.
[http://dx.doi.org/10.1021/acs.est.8b00951] [PMID: 29916696]

[53]   M. Bassyouni, M.H. Abdel-Aziz, M.S. Zoromba, S.M.S. Abdel-Hamid, and E. Drioli, "A review of polymeric nanocomposite membranes for water purification", *J. Ind. Eng. Chem.,* vol. 73, pp. 19-46, 2019.
[http://dx.doi.org/10.1016/j.jiec.2019.01.045]

[54]   J.S. Beril Melbiah, P. Joseph, D. Rana, A. Nagendran, N. Nagendra Gandhi, and D.R. Mohan, "Customized antifouling polyacrylonitrile ultrafiltration membranes for effective removal of organic contaminants from aqueous stream", *J. Chem. Technol. Biotechnol.,* vol. 94, no. 3, pp. 859-868, 2019.
[http://dx.doi.org/10.1002/jctb.5833]

[55]   D. Vasanth, and A.D. Prasad, "Ceramic Membrane: Synthesis and Application for Wastewater Treatment-A Review", In: *Water Resources and Environmental Engineering II.* Singapore, 2019, pp. 101-106.
[http://dx.doi.org/10.1007/978-981-13-2038-5_10]

[56]   M. Thomas, and B. Corry, "A computational assessment of the permeability and salt rejection of carbon nanotube membranes and their application to water desalination", *Philos. Trans.- Royal Soc., Math. Phys. Eng. Sci.,* vol. 374, p. 20150020, 2016.

[57]   P. Zhang, J-L. Gong, G-M. Zeng, B. Song, W. Cao, H-Y. Liu, S-Y. Huan, and P. Peng, "Novel "loose" GO/MoS2 composites membranes with enhanced permeability for effective salts and dyes rejection at

low pressure", *J. Membr. Sci.,* vol. 574, pp. 112-123, 2019.
[http://dx.doi.org/10.1016/j.memsci.2018.12.046]

[58]   M. Tripathi, J.N. Sahu, and P. Ganesan, "Effect of process parameters on production of biochar from biomass waste through pyrolysis: A review", *Renew. Sustain. Energy Rev.,* vol. 55, pp. 467-481, 2016.
[http://dx.doi.org/10.1016/j.rser.2015.10.122]

[59]   Q.U. Jiuhui, "Research progress of novel adsorption processes in water purification: a review", *J. Environ. Sci. (China),* vol. 20, no. 1, pp. 1-13, 2008.
[http://dx.doi.org/10.1016/S1001-0742(08)60001-7] [PMID: 18572516]

[60]   S. Wang, Y. Boyjoo, and A. Choueib, "A comparative study of dye removal using fly ash treated by different methods", *Chemosphere,* vol. 60, no. 10, pp. 1401-1407, 2005.
[http://dx.doi.org/10.1016/j.chemosphere.2005.01.091] [PMID: 16054909]

[61]   M. Tripathi, J.N. Sahu, P. Ganesan, and T.K. Dey, "Effect of temperature on dielectric properties and penetration depth of oil palm shell (OPS) and OPS char synthesized by microwave pyrolysis of OPS", *Fuel,* vol. 153, pp. 257-266, 2015.
[http://dx.doi.org/10.1016/j.fuel.2015.02.118]

[62]   H. Nam, W. Choi, D.A. Genuino, and S.C. Capareda, "Development of rice straw activated carbon and its utilizations", *J. Environ. Chem. Eng.,* vol. 6, no. 4, pp. 5221-5229, 2018.
[http://dx.doi.org/10.1016/j.jece.2018.07.045]

[63]   M. Farahani, S.R.S. Abdullah, S. Hosseini, S. Shojaeipour, and M. Kashisaz, "Adsorption-based Cationic Dyes using the Carbon Active Sugarcane Bagasse", *Procedia Environ. Sci.,* vol. 10, pp. 203-208, 2011.
[http://dx.doi.org/10.1016/j.proenv.2011.09.035]

[64]   M. Tripathi, P.B. Ganesan, and J.N. Sahu, "Microwave pyrolysis of OPS to synthesize micro porous OPS char: Effect of process parameters", *AIP Conf. Proc.,* vol. 2136, p. 040019, 2019.
[http://dx.doi.org/10.1063/1.5120933]

[65]   M. Tripathi, A. Bhatnagar, N.M. Mubarak, J.N. Sahu, and P. Ganesan, "RSM optimization of microwave pyrolysis parameters to produce OPS char with high yield and large BET surface area", *Fuel,* vol. 277, p. 118184, 2020.
[http://dx.doi.org/10.1016/j.fuel.2020.118184]

[66]   H. Li, L. Li, T. Nguyen, G.T. Rochelle, and J. Chen, "Characterization of Piperazine/2-Aminomethylpropanol for Carbon Dioxide Capture", *Energy Procedia,* vol. 37, pp. 340-352, 2013.
[http://dx.doi.org/10.1016/j.egypro.2013.05.120]

[67]   Y. Zhang, M. Z. Joel, Y. He, D. Weathersby, F. Han, and G. Rimal, "Synthesis of Fe_2O_3/biochar nanocomposites by microwave method for magnetic energy-storage concentration cells", *Materials Letters: X,* vol. 3, p. 100020, 2019.
[http://dx.doi.org/10.1016/j.mlblux.2019.100020]

[68]   E.K. Putra, R. Pranowo, J. Sunarso, N. Indraswati, and S. Ismadji, "Performance of activated carbon and bentonite for adsorption of amoxicillin from wastewater: mechanisms, isotherms and kinetics", *Water Res.,* vol. 43, no. 9, pp. 2419-2430, 2009.
[http://dx.doi.org/10.1016/j.watres.2009.02.039] [PMID: 19327813]

[69]   P.K. Malik, "Dye removal from wastewater using activated carbon developed from sawdust: adsorption equilibrium and kinetics", *J. Hazard. Mater.,* vol. 113, no. 1-3, pp. 81-88, 2004.
[http://dx.doi.org/10.1016/j.jhazmat.2004.05.022] [PMID: 15363517]

[70]   D. Mohan, K.P. Singh, G. Singh, and K. Kumar, "Removal of Dyes from Wastewater Using Flyash, a Low-Cost Adsorbent", *Ind. Eng. Chem. Res.,* vol. 41, no. 15, pp. 3688-3695, 2002.
[http://dx.doi.org/10.1021/ie010667+]

[71]   J.R. Werber, C.O. Osuji, and M. Elimelech, "Materials for next-generation desalination and water purification membranes", *Nat. Rev. Mater.,* vol. 1, no. 5, p. 16018, 2016.

[http://dx.doi.org/10.1038/natrevmats.2016.18]

[72]   N.C. Darre, and G.S. Toor, "Desalination of Water: a Review", *Curr. Pollut. Rep.,* vol. 4, no. 2, pp. 104-111, 2018.
[http://dx.doi.org/10.1007/s40726-018-0085-9]

[73]   J. Guo, B.J. Deka, K-J. Kim, and A.K. An, "Regeneration of superhydrophobic TiO2 electrospun membranes in seawater desalination by water flushing in membrane distillation", *Desalination,* vol. 468, p. 114054, 2019.
[http://dx.doi.org/10.1016/j.desal.2019.06.020]

[74]   L-F. Ren, F. Xia, V. Chen, J. Shao, R. Chen, and Y. He, "TiO2-FTCS modified superhydrophobic PVDF electrospun nanofibrous membrane for desalination by direct contact membrane distillation", *Desalination,* vol. 423, pp. 1-11, 2017.
[http://dx.doi.org/10.1016/j.desal.2017.09.004]

[75]   Z. Li, Y. Liu, J. Yan, K. Wang, B. Xie, Y. Hu, W. Kang, and B. Cheng, "Electrospun polyvinylidene fluoride/fluorinated acrylate copolymer tree-like nanofiber membrane with high flux and salt rejection ratio for direct contact membrane distillation", *Desalination,* vol. 466, pp. 68-76, 2019.
[http://dx.doi.org/10.1016/j.desal.2019.05.005]

[76]   M. Khayet, C. García-Payo, and T. Matsuura, "Superhydrophobic nanofibers electrospun by surface segregating fluorinated amphiphilic additive for membrane distillation", *J. Membr. Sci.,* vol. 588, p. 117215, 2019.
[http://dx.doi.org/10.1016/j.memsci.2019.117215]

[77]   A.G. Fane, "A grand challenge for membrane desalination: More water, less carbon", *Desalination,* vol. 426, pp. 155-163, 2018.
[http://dx.doi.org/10.1016/j.desal.2017.11.002]

[78]   Y. Sun, I.K.M. Yu, D.C.W. Tsang, X. Cao, D. Lin, L. Wang, N.J.D. Graham, D.S. Alessi, M. Komárek, Y.S. Ok, Y. Feng, and X.D. Li, "Multifunctional iron-biochar composites for the removal of potentially toxic elements, inherent cations, and hetero-chloride from hydraulic fracturing wastewater", *Environ. Int.,* vol. 124, pp. 521-532, 2019.
[http://dx.doi.org/10.1016/j.envint.2019.01.047] [PMID: 30685454]

[79]   X. Tao, T. Huang, and B. Lv, "Synthesis of Fe/Mg-Biochar Nanocomposites for Phosphate Removal", *Materials (Basel),* vol. 13, no. 4, p. 816, 2020.
[http://dx.doi.org/10.3390/ma13040816] [PMID: 32054049]

[80]   J. Wang, Z. Liao, J. Ifthikar, L. Shi, Z. Chen, and Z. Chen, "One-step preparation and application of magnetic sludge-derived biochar on acid orange 7 removal via both adsorption and persulfate based oxidation", *RSC Advances,* vol. 7, no. 30, pp. 18696-18706, 2017.
[http://dx.doi.org/10.1039/C7RA01425B]

[81]   R.P. Mohubedu, P.N.E. Diagboya, C.Y. Abasi, E.D. Dikio, and F. Mtunzi, "Magnetic valorization of biomass and biochar of a typical plant nuisance for toxic metals contaminated water treatment", *J. Clean. Prod.,* vol. 209, pp. 1016-1024, 2019.
[http://dx.doi.org/10.1016/j.jclepro.2018.10.215]

[82]   T. Ai, X. Jiang, Q. Liu, L. Lv, and S. Dai, "Single-component and competitive adsorption of tetracycline and Zn(ii) on an NH4Cl-induced magnetic ultra-fine buckwheat peel powder biochar from water: studies on the kinetics, isotherms, and mechanism", *RSC Advances,* vol. 10, no. 35, pp. 20427-20437, 2020.
[http://dx.doi.org/10.1039/D0RA02346A] [PMID: 35517772]

[83]   V.H. Nguyen, H.T. Van, V.Q. Nguyen, X.V. Dam, L.P. Hoang, and L.T. Ha, "Magnetic FeO4 Nanoparticle Biochar Derived from Pomelo Peel for Reactive Red 21 Adsorption from Aqueous Solution", *J. Chem.,* vol. 2020, p. 3080612, 2020.
[http://dx.doi.org/10.1155/2020/3080612]

[84]   X. Zhu, F. Qian, Y. Liu, S. Zhang, and J.M. Chen, *Environmental performances of hydrochar-derived*

magnetic carbon composite affected by its carbonaceous precursor. vol. 5. RSC Adv, 2015.

[85] Y. Qiu, Z. Zheng, Z. Zhou, and G.D. Sheng, "Effectiveness and mechanisms of dye adsorption on a straw-based biochar", *Bioresour. Technol.,* vol. 100, no. 21, pp. 5348-5351, 2009.
[http://dx.doi.org/10.1016/j.biortech.2009.05.054] [PMID: 19540756]

[86] A. Kundu, J. Sahu, E. Abdullah, and J. Natesan, *Synthesis of palm oil empty fruit bunch magnetic pyrolytic char impregnating with FeCl3 by microwave heating technique.* vol. 61. Biomass and Bioenergy, 2014.

[87] Z. Hao, C. Wang, Z. Yan, H. Jiang, and H. Xu, "Magnetic particles modification of coconut shell-derived activated carbon and biochar for effective removal of phenol from water", *Chemosphere,* vol. 211, pp. 962-969, 2018.
[http://dx.doi.org/10.1016/j.chemosphere.2018.08.038] [PMID: 30119027]

CHAPTER 10

Advanced Membrane Processes for the Removal of Emerging Water Pollutants

Arbab Tufail[1,*]

[1] *Strategic Water Infrastructure Laboratory, School of Civil, Mining and Environmental Engineering, University of Wollongong, Wollongong, NSW 2522, Australia*

Abstract: This chapter demonstrates the source and pathway of emerging contaminants (ECs) and their removal by advanced membrane technologies. These ECs are naturally occurring or synthetic organic pollutants, including pharmaceuticals and personal care products, estrogens, industrial chemicals, UV filters, pesticides, and endocrine-disrupting chemicals ubiquitously detected in wastewater and wastewater-impacted surface waterbodies. Emerging contaminants have detrimental effects on aquatic flora and fauna and may affect human health. Due to the persistent nature of ECs, they are resistant to conventional wastewater treatments. Moreover, different physicochemical processes have shown ineffectiveness for the removal of ECs. Therefore, there is a need for robust wastewater treatment processes such as advanced membrane technologies that can effectively remove these ECs. Advanced membrane technologies use membranes that separate ECs from the solution and include forward osmosis, reverse osmosis, nanofiltration, ultrafiltration, microfiltration, catalytic membranes, and membrane bioreactors. Briefly, the focus of this chapter is to provide an overview of different membrane separation technologies and illustrate various examples of ECs removal.

Keywords: Catalytic membranes, Ceramic membrane, Emerging contaminants, Membrane bioreactor, Forward osmosis, Hollow fiber, Membrane fouling, Nanofiltration, Polymer membranes, Rejection of contaminants, Reverse osmosis, Ultrafiltration, Water reuse.

INTRODUCTION

Current demographic, technological, and economic advancement throughout the world has knowingly and unknowingly altered the environment we live in. Our actions affect the global environment, impacting the quality and quantity of freshwater resources [1]. Among all the available water on the earth (97% salt

* **Corresponding author Arbab Tufail**: Strategic Water Infrastructure Laboratory, School of Civil, Mining and Environmental Engineering, University of Wollongong, Wollongong, NSW 2522, Australia; E-mail: at742@uowmail.edu.au

Shaukat Ali Mazari, Nabisab Mujawar Mubarak & Nizamuddin Sabzoi (Eds.)

water and 2% ice), only 1% is present as freshwater, which mainly occurs in groundwater, rivers, and lakes and is used to meet water demands [2]. Water is required for drinking, irrigation, fisheries, and food production. It is also needed on an industrial scale for energy generation-hydropower, fossil fuel extraction, and a coolant in nuclear power plants. The food industry accounts for 70% of water use globally, and the energy sector consumes 10% of available water resources [3, 4]. With an ever-growing population and higher living standards, water demand increases. Under the water-energy-food nexus approach, three tiers are well connected, and impact in one sector can significantly affect the performance of other sectors [5, 6]. Today, the world is worried about water scarcity due to increasing water demand and its impacts on the food and energy sector. In recent years, water demand on a global scale has exceeded twice that of the population. Moreover, urban development, the spread of infectious disease, and the detection of emerging contaminants also stressed the current water supply [1]. Logically, to sustain life, we need to practice "3R," which is reducing (preserving existing water resources), reusing (wastewater use) and recycling (treatment of wastewater).

During the last few decades, many emerging contaminants, including pharmaceuticals and personal care products (PPCPs), industrial chemicals, endocrine-disrupting chemicals, and pesticides, have been detected in soil and water, including drinking water, surface water, and groundwater [7]. Emerging contaminants are found in a few nanograms to micrograms per liter in wastewater and wastewater impacted surface water bodies [8]. Antibiotics, a particular class of PPCPs, are extensively used by humans to cure animal diseases and are ubiquitously detected in wastewater. They are highly persistent and biologically active compounds that cannot be completely metabolized and thus continuously enter the environment through urine and feces [9]. Emerging contaminants have detrimental effects on humans and aquatic lives due to continuous exposure. Therefore, their complete removal from wastewater is essential for safe disposal in the environment. Current wastewater treatment plants are not designed to remove emerging pollutants from wastewater. Thus, complete elimination of emerging contaminants cannot be achieved through activated sludge treatment [10]. Different physicochemical and advanced oxidation processes are being used to remove these pollutants. Membrane separation processes have been considered an efficient treatment for removing emerging contaminants [11].

Membrane technology is a promising candidate for the removal of emerging contaminants. The membrane process generally involves feed that passes through the membrane and retains solute particles in the feed reactor by different phenomena, including adsorption, size excursion, differential interaction, and charge repulsion. Based on particle size and membrane types, membrane technologies can be classified as microfiltration (MF), ultrafiltration (UF), nanofiltration (NF), and reverse osmosis (RO). MF with a nominal pore size of 100-1000 nm and UF with less than 100 nm can reject some macro and micro molecules. They are applied to separate suspended solids, emulsions, large collides, proteins, bacteria, and viruses [12].

NF and RO are used to remove dyes, ionic species, and emerging contaminants. Membrane processes are used in biotechnology and wastewater treatment from different resources such as textile plant effluent, hospital wastewater, and refinery wastewater. The application of membrane technologies for bioprocesses involves replacing conventional separation processes such as centrifugation, evaporation, and demineralization in the dairy industry. Other applications include separating protein/peptides, recycling cleaning solutions, impurities removal, filtration of fermentation broths, and extraction and purification of micro chemicals. For instance, centrifugation is used in drug manufacturing industries to clarify vaccine and cell culture and involves high capital and maintenance costs. Membrane processes can be used as an alternative to centrifugation for vaccine production [13]. Recently, Muniandi *et al.* [14] used tangential flow filtration for clarification during the manufacturing processes of tetanus toxin. Various examples of tangential flow and normal membrane filtration in vaccine manufacturing industries are available in the literature. During product recovery and purification, membrane filtration processes such as MF, UF, and NF are also used. González *et al.* [13] studied the lactic acid recovery from whey fermentation broths and used UF for cell recycling and RO for pre-concentration. Also, membrane processes have been used to extract and purify pharmaceuticals due to their simple operation. In a study, the ultrafiltration membrane was used in the solvent extraction process to remove bio-emulsifier from benzylpenicillin-containing fermentation broth [15].

Membrane processes also remove emerging contaminants from wastewater received from different resources. MF is only effective for the removal of bacteria and viruses. At the same time, UF can only remove some ECs, and removal depends on the physicochemical properties of contaminants and membrane type [16, 17]. Nanofiltration can significantly remove the ECs compared to MF and UF membranes [18, 19]. Radjenović *et al.* [20] reported greater than 85% removal of selected pharmaceuticals, namely, ketoprofen, diclofenac, acetaminophen, propyphenazone, sotalol, metoprolol, carbamazepine, sulfamethoxazole, gemfibrozil, and hydrochlorothiazide in the NF system. Another study reported 99% removal of diclofenac through the nanofiltration membrane process [21]. Recently, nanotechnology-based membrane processes have gained much attraction for the removal of ECs. In this technology, nanomaterials such as carbon nanotubes, polymers, silver, and silica have been used to fabricate membrane that is very effective for the enhanced elimination of ECs. Nanotechnology-based membranes have higher thermal stability, better mechanical and antifouling properties, and higher fluxes and permeability [22, 23]. Besides, they have demonstrated efficient removal of recalcitrant ECs in several studies.

This chapter aims to provide a systematic overview of the occurrence and sources of emerging contaminants present in wastewater and their removal through various membrane technologies. This chapter consists of three parts. The authorities, types, and routes of emerging pollutants present in wastewater have been explained in depth in the first part. The second part deals with working principles and recent advancements in membrane technologies, namely forward osmosis, reverse osmosis, nanofiltration, ultrafiltration, microfiltration, membrane bioreactor, and catalytic membranes. Moreover, their performances in the removal of ECs have also been elucidated. In the third part, challenges associated with different membrane technologies have been discussed. In the end, a brief overview of the chapter and future direction in membrane technologies is presented.

SOURCES AND PATHWAYS OF EMERGING CONTAMINANTS TO WATER

Emerging contaminants in the environmental system have been considered a worldwide public health issue and environmental concern. Emerging contaminants are an anthropogenic substance that is natural or synthetic. They are classified as pharmaceuticals and personal care products, pesticides, surfactants, UV filters, industrial chemicals, endocrine-disrupting chemicals, and industrial chemicals [7, 8]. Pharmaceutical production has been increasing daily due to the high demand for human use and cure of epidemic breakout in aquaculture activities and livestock. Overall, the consumption of pharmaceuticals is less than 15 g/capita. Yr. The main sources of ECs include sewage wastewater, agriculture, hospital wastewater, industrial waste effluent, livestock activities, and solid waste dumping. Sources and pathways of emerging contaminants in groundwater, seawater, and surface water are shown in Fig. (**1**).

Emerging contaminants in wastewater have been extensively studied and found in varying environmental concentrations. Their concentration varies between 1-10 ng/L, but some pharmaceuticals and personal care products are present in high concentration due to their high demand and supply [7]. ECs such as triclosan, naproxen, atenolol, and diclofenac are present in higher concentrations in wastewater. Moreover, few medicines do not require a prescription, such as caffeine and ibuprofen, and their high concentration can be attributed to ease of availability. In addition to this, some local diseases can also be a reason for the increased consumption of emerging contaminants in a region. The high concentration of pesticides in wastewater can be attributed to seasonal use, which, after rainfall, leach into wastewater [9]. The average concentration range of most ECs lies between 0.001-1 µg/L in wastewater. Pesticides and steroid hormones show less concentration (less than 1 µg/L) in WWTP effluent. The concentration of ECs in WWTPs depends on several factors such as water consumption per person daily, metabolism, production rate, environmental persistence, and size of treatment plants. Moreover, conventional wastewater treatment processes are not effective for the higher removal of persistent emerging contaminants [24, 25].

Therefore, wastewater treatment plants (WWTPs) are the main source of emerging contaminants in surface water. The average concentration of personal care products (410 ng/L) and industrial chemicals (1150 ng/L) is very high in surface water. The concentration of pharmaceuticals varies from 70-140 ng/L, which is moderate. The concentration level of steroid hormones varies from 1-10 ng/L and can easily cause estrogenic effects.

Groundwater is also considered the main source of water supply for domestic purposes, industrial use, and agricultural land in many countries. Overall, groundwater represents 97% of freshwater resources utilized for different purposes [25]. Groundwater is being used on a large scale, and over-exploitation of these resources has caused degradation of groundwater quality. Many studies have reported the presence of ECs in groundwater, and these ECs can enter the groundwater from different resources, including surface water, industrial wastewater (food processing plants and manufacturing plants), stormwater run-off, and percolation of pesticide from agricultural land, landfill leachate, and leakage from septic tanks (Fig. **1**). The concentration of ECs due to landfill leachate and leakage from septic tanks lies between 10-103 ng/L and 10-104 ng/L, respectively. Pesticides in agricultural areas may pollute groundwater as they can percolate with irrigation water [9]. For example, Clara *et al.* [26] investigated the presence of carbamazepine in groundwater. ECs can also enter into groundwater through artificial recharge by surface water and reclaimed water, and later is of poor quality and has a short residence time in soil [25]. According to an estimate, more than 80% of untreated wastewater is directly disposed of into the ocean without any prior treatment and is considered a major source of ECs in seawater [27]. According to some statistics, 150 tons of ECs are discharged into the ocean *via* the Yangtze River in China [28].

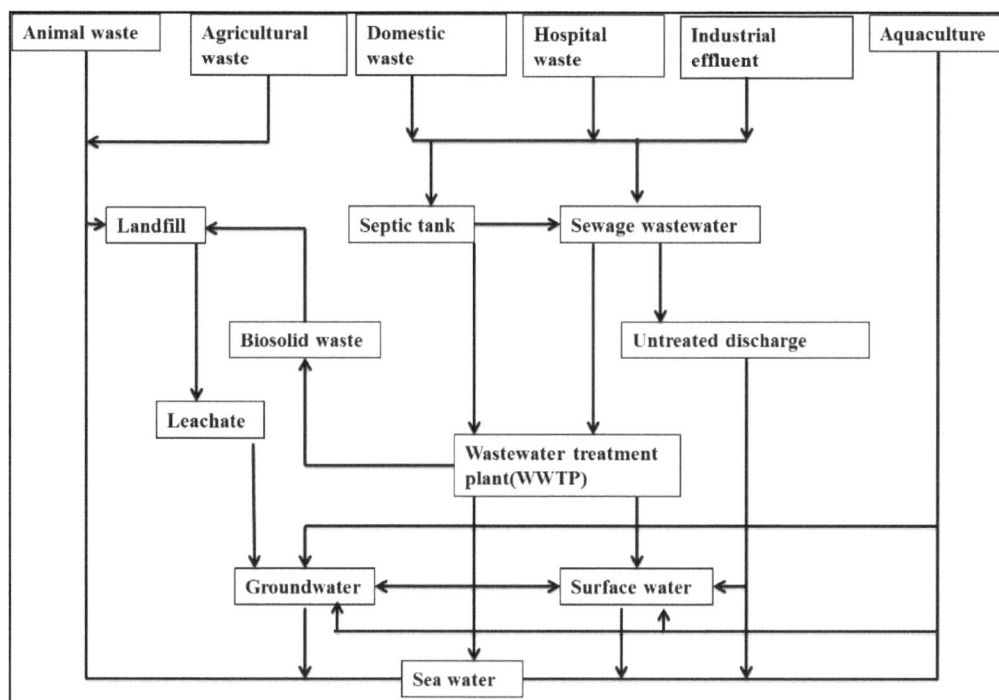

Fig. (1). Different sources of emerging contaminants.

In addition to this, many Asian countries are using septic tanks for waste collection, which is then directly released into the ocean without treatment [29]. Moreover, aquaculture is another big industry that involves more than 500 million people, and it consumes different pharmaceuticals to overcome outbreaks of diseases. Grigorakis & Rigos [30] have reported that 75% of pharmaceuticals that have been used during aquaculture are directly discharged into the ocean *via* excretion and food pellets. In the same way, groundwater and surface water are also sources of seawater contamination by surface runoff, leaching of solid waste, and agriculture runoff. Moreover, solid waste is directly dumped in the ocean in many countries. The concentration of ECs in seawater varies between 0.1-1000ng/L and is greater than the predicted no-effect concentration (PNEC), 10 ng/L [25].

ADVANCED MEMBRANE TECHNOLOGIES

Membrane separation processes are used for water or wastewater treatment and include nanofiltration, microfiltration, ultrafiltration, forward osmosis (FO), reverse osmosis, and direct contact membranes (DCMD). Membranes can be either pressure-driven membranes (NF, UF, MF, and RO) or diffusive membranes

(DCMD) [31]. They can be categorized based on the pore size of the membrane material used in the system. Fig. (**2**) illustrates the different types of membranes and their ability to remove various physical or chemical components from the water.

Type of contaminants	Emerging contaminants Dyes Soluble salts	Small organic		Bacteria Algae Giardia
Molecular weight (Da)	200	2,000	20,000	200,000
Pore size (nm)	1	10	100	1000
Separation process	Reverse osmosis DCMD	Nanofiltration	Ultrafiltration	Microfiltration

Fig. (2). Overview of membrane separation processes.

Membrane materials are usually of two types, namely polymeric and ceramic [32]. Both have been shown to have high efficiency in rejecting emerging contaminants [15, 23]. Generally, membranes remove contaminants through different mechanisms, namely adsorption, size exclusion, or electrostatic repulsion. Neutral solute transport through the membrane *via* a size-exclusion mechanism [23]. Electrostatic repulsion of attraction is based on the charge on the membrane surface and ions present in the feed solution. The adsorption phenomenon describes the interaction between charged species and charges on the membrane surface that originates from the ionization of ionic groups at the surface of the membrane and within its pores. Ionic groups can be acidic or basic, and their dissociation is strongly dependent on the system's pH [31, 33]. Rejection of emerging contaminants depends on the physicochemical properties of pollutants and the properties of the membrane [15]. Physicochemical properties include the molecular weight of impurities, hydrophobicity, solubility, and charge. Hydrophobicity mainly controls contaminants' transport through a membrane and their interaction with the membrane. Also, membrane hydrophobicity, permeability, water chemistry, and operating parameters are important in contaminant rejection [20, 31].

Forward Osmosis Membrane Technology

Forward osmosis is a membrane technology based on natural osmotic pressure created when feed solution and draw solution at different concentrations are separated by membrane [34]. There are two operation modes of FO: active membrane layer facing feed solution and active membrane layer facing draw solution. FO technology is considered cost-effective and sustainable compared to classical membrane technologies such as membrane distillation and reverse osmosis [35]. This membrane technology has been utilized for wastewater treatment, food technology, and desalination. It is also applied as an alternative to NF or RO processes to remove emerging contaminants. Removal of pollutants depends on various factors, including molecular weight, hydrophobicity, solubility, and charge of contaminants and water chemistry (*i.e.,* temperature, pH, and presence of organic or inorganic species) [36, 37].

Table 1 illustrates the rejection of different emerging contaminants in FO technology. For example, Kim *et al.* [38] investigated the rejection of four emerging contaminants (atenolol, atrazine, primidone, and caffeine) having different structural properties and molecular weight in the FO process using other draw solutions, namely monoammonium phosphate, diammonium phosphate fertilizers, and potassium chloride. The authors observed that in the presence of monoammonium phosphate draw solution, the flux is governed by physicochemical properties of emerging contaminants and positively charged hydrophobic emerging contaminants showed higher flux. The effect of molecular weight became dominant when potassium chloride was used as a draw solution, and the transfer of atenolol having higher molecular weight was hindered [38]. Madsen *et al.* [39] studied the effect of two different FO membrane (HTI membrane and aquaporin membrane) types on the removal of pesticides. Results indicated that the aquaporin membrane could reject 97% of atrazine, 2,6-dichlorobenzamide, and desethyl- desisopropyl-atrazine. The authors mentioned that rejection in HTI was dominated by steric hindrance while rejection in aquaporin membrane was governed by contaminants diffusion through the membrane. The former membrane resulted in better flux compared to the later membrane. Another study reported the rejection of bisphenol A and methyl paraben using Aquaporin Inside™ FO hollow fiber module. This membrane resulted in 95% rejection of methyl paraben and 99% rejection of bisphenol A [39]. The authors reported that rejection is based on the adsorption capacity of the membrane surface and the molecular size of contaminants. Lower molecular weight methylparaben easily diffused through the membrane, thus showing the lowest adsorption [39].

Table 1. Rejection of emerging contaminants in forwarding osmosis membrane.

Membrane	Target Contaminant	Operating Conditions	Solution		Rejection (%)	Reference
			Feed	Draw		
Commercial cellulose-based FO	Bisphenol A	$[C_0] =$ 500µg/L; HRT = 1h; T = 25°C	Organic solute in Milli-Q water	NaCl	78	[37]
Commercial cellulose-based FO	Triclosan	$[C_0] =$ 500µg/L; HRT = 1h; T = 25°C	Organic solute in Milli-Q water	NaCl	80	[37]
Commercial cellulose-based FO	Diclofenac	$[C_0] =$ 500µg/L; HRT = 1h; T = 25°C	Organic solute in Milli-Q water	NaCl	100	[37]
Cellulose triacetate	Atrazine	$[C_0] =$ 1mg/L; HRT = 3h; T = 25°C	Organic solute in Milli-Q water	NaCl	97	[39]
Cellulose triacetate	2,6-dichlorobenzamide	$[C_0] =$ 1mg/L; HRT = 3h; T = 25°C	Organic solute in Milli-Q water	NaCl	97	[39]
Cellulose triacetate	Desethyl-desisopropyl-atrazine	$[C_0] =$ 1mg/L; HRT = 3h; T = 25°C	Organic solute in Milli-Q water	NaCl	97	[39]
Biomimetic aquaporin (AqP)	Atrazine	$[C_0] =$ 1mg/L; HRT = 3h; T = 25°C	Organic solute in Milli-Q water	NaCl	97	[39]
Biomimetic aquaporin (AqP)	2,6-dichlorobenzamide	$[C_0] =$ 1mg/L; HRT = 3h; T = 25°C	Organic solute in Milli-Q water	NaCl	97	[39]
Biomimetic aquaporin (AqP)	Desethyl-desisopropyl-atrazine	$[C_0] =$ 1mg/L; HRT = 3h; T = 25°C	Organic solute in Milli-Q water	NaCl	97	[39]

(Table 1) cont.....

Membrane	Target Contaminant	Operating Conditions	Solution		Rejection (%)	Reference
Cellulose from HTI	Pyridine	[C0] = 50mg/L; HRT = 12h; Flux = 80Lh⁻1; T = 25°C	Coking wastewater	Na₂SO₄	49.41	[41]
Poten	Pyridine	[C0] = 50mg/L; HRT = 12h; Flux = 80Lh⁻1; T = 25°C	Coking wastewater	Na₂SO₄	73.77	[41]
Cellulose from HTI	Indole	[C0] = 50mg/L; HRT = 12h; Flux = 80Lh⁻1; T = 25°C	Coking wastewater	Na₂SO₄	48.69	[41]
Poten	Indole	[C0] = 50mg/L; HRT = 12h; Flux = 80Lh⁻1; T = 25°C	Coking wastewater	Na₂SO₄	71.69	[41]
Cellulose triacetate with embedded polyester screen	Tetracycline	$[C_0]$ = 500µg/L; Time = 100 days; T = 25°C	Secondary effluent of an anoxic–oxic (AO) system	NaCl	83	[42]
Cellulose triacetate with a cast nonwoven	Tetracycline	$[C_0]$ = 500µg/L; Time = 100 days; T = 25°C	Secondary effluent of an anoxic–oxic (AO) system	NaCl	69	[42]
Thin-film composite with embedded polyester screen	Tetracycline	$[C_0]$ = 500µg/L; Time = 100 days; T = 25°C	Secondary effluent of an anoxic–oxic (AO) system	NaCl	44	[42]
Cellulose triacetate with embedded polyester screen support	Acetaminophen	$[C_0]$ = 1mg/L; HRT = 12h; T = 25°C	Organic solute in Milli-Q water	NaCl	29.1	[43]

(Table 1) cont.....

Membrane	Target Contaminant	Operating Conditions	Solution		Rejection (%)	Reference
Cellulose triacetate with embedded polyester screen support	Carbamazepine	$[C_0] =$ 1mg/L; HRT = 12h; T = 25°C	Organic solute in Milli-Q water	NaCl	83	[43]
Cellulose triacetate with embedded polyester screen support	Ciprofloxacin	$[C_0] =$ 1mg/L; HRT = 12h; T = 25°C	Organic solute in Milli-Q water	NaCl	94.8	[43]
Cellulose triacetate with embedded polyester screen support	Sulfamethoxazole	$[C_0] =$ 1mg/L; HRT = 12h; T = 25°C	Organic solute in Milli-Q water	NaCl	84.4	[43]

Jin *et al.* [40] studied the rejection of four pharmaceuticals (carbamazepine, diclofenac, ibuprofen, and naproxen) by forwarding osmosis using two different commercially available membranes, namely cellulose triacetate (CTA) based membrane and thin-film composite polyamide (TFC) based membrane. They reported that rejection behaviour was governed by contaminants' membrane interfacial properties, feed properties, and physicochemical properties. TFC polyamide membrane showed good performance overall with higher flux, stability at different pH, and rejection greater than 94% [40]. For the CTA-based membrane, the rejection was strongly affected by the interaction between pharmaceuticals and the membrane surface under an acidic medium. At alkaline pH, rejection of contaminants was in the following order: diclofenac (99%) > carbamazepine (95%) > ibuprofen (93%) ≈ naproxen (93%) [40]. Removal of indole and pyridine has been investigated using poten FO membrane and CTA membrane by [41]. Results illustrated that the poten FO membrane exhibited better performance and higher rejection of contaminants than the CTA membrane.

Reverse Osmosis Technology

Reverse osmosis is a pressure-driven membrane technology widely applied in water desalination and tertiary wastewater treatment. This technology is important because its advantages include better performance for selective contaminant rejection, high water permeability, and small footprints that allow more chances of combining with other treatment processes [20, 44].

Reverse osmosis uses a semi-permeable membrane that keeps the contaminants in the feed tank and only allows liquid to pass through the membrane to permeate the tank. This technology rejects contaminants *via* size exclusion, charge exclusion, and interactions between feed and membrane [20, 45]. The performance of RO depends on feed characteristics, reaction conditions, and the type of membrane used in the system. Generally, two membrane modules are used, namely hollow fiber and spiral wound. Hollow fiber modules can produce permeate at a higher rate than other modules but have higher chances of membrane fouling. The membrane can be either a polymer membrane or a composite membrane consisting of two or more different layers [46]. The function group of the polymer membrane determines the charge on the membrane surface, and the adsorption of contaminants onto the membrane surface is determined by its hydrophobicity [45].

RO technology has wide applications, including food processing, pharmaceuticals, desalination, textile, dairy wastewater, and emerging contaminants [20, 47 - 49]. Table **2** represents the performance of the RO system in the treatment of various emerging contaminants. Al-Rifai *et al.* [50] investigated the fate of 11 emerging contaminants present in wastewater treatment plants by RO system and reported almost 97% of it. Another study assessed the performance of the RO system and NF system for the removal of five ECs (sulfamethoxazole, trimethoprim, ciprofloxacin, dexamethasone, and febantel) in real pharmaceuticals wastewater. They reported that the RO system showed higher ECs removal than the NF membrane system. Another study reported the removal of five contaminants by RO membrane and NF membrane [51]. Results showed that the RO system demonstrates more rejection (98%) than NF (90%). The authors mentioned that electrostatic and hydrophobic effects less influenced rejection in RO membrane than in NF membrane. Radjenović *et al.* [20] studied the performance of RO membrane for the removal of a wide range of emerging contaminants present in full-scale drinking water treatment plants using groundwater. They reported that ECs with larger molecular weight were effectively retained (>85%) in the system.

Table 2. Emerging contaminants removal by reverse osmosis technology.

Membrane Type	Target Contaminant	Water Matrix	Rejection (%)	Reference
XLE	Ibuprofen	A synthetic feed with the addition of Humic Acid	100	[52]
Spiral-wound LCF1-4040		Secondary effluent	96.9	[53]
Flat sheet UTC 60		Milli.Q water	95	[29]
Flat sheet UTC 70U		Milli.Q water	97	[29]
Koch		Groundwater	100	[54]
X20	Gemfibrozil	Raw Lake Ontario water	97	[55]
X20		MBR effluent	98	[55]
Spiral-wound LCF1-4040		Secondary effluent	98	[53]
Flat sheet UTC 60		Milli.Q water	95	[29]
Flat sheet UTC 70U		Milli.Q water	97	[29]
Spiral-wound LCF1-4040	Bezafibrate	Secondary effluent	100	[53]
Full scale BW30LE-440		Groundwater	99	[20]
TFC-HR		Milli.Q water	90	[56]
TFC-HR		Milli.Q water	100	[56]
X20	17-a-estradiol	Raw Lake Ontario water	98	[55]
XLE	Erythromycin	Milli.Q water	83	[57]
SC-3100		Milli.Q water	29	[57]
Flat sheet UTC 60		Milli.Q water	55	[29]
XLE	Primidone	Milli.Q water	87	[57]
SC-3100		Milli.Q water	85	[57]
RO-XLE		Milli.Q water	84	[57]
X20	Carbamazepine	Raw Lake Ontario water	91	[55]
X20		Secondary effluent	97	[55]
XLE		Milli.Q water	99	[52]
XLE		Milli.Q water	98	[52]
XLE		A synthetic feed with the addition of Humic Acid	100	[52]
XLE		Synthetic feed with addition of Humic Acid	89	[52]
Full scale BW30LE-440	Metoprolol	Groundwater	76	[20]

(Table 2) cont.....

Membrane Type	Target Contaminant	Water Matrix	Rejection (%)	Reference
Full scale BW30LE-440	Sotolol	Groundwater	99	[20]
Spiral-wound LCF1-4040	Atenolol	Secondary effluent	99.5	[53]
Spiral-wound LCF1-4040	Oflaxacin	Secondary effluent	95	[53]
XLE	Sulfamethazine	Milli.Q water	90	-
X20		Raw Lake Ontario water	87.9	[55]
X20		MBR effluent	97.9	[55]
XLE	Sulfadiazine	A synthetic feed with the addition of Humic Acid and silica	100	[52]
XLE		A synthetic feed with the addition of Humic Acid and silica	99	[52]
XLE		A synthetic feed with the addition of Humic Acid	98	[52]
XLE		A synthetic feed with the addition of Humic Acid	99	[52]
XLE		Synthetic feed with addition of Humic Acid	100	[52]
Full scale BW30LE-440	Trimethoprim	Groundwater	99	[20]
Saehan		Tertiary effluent	98.7	-
XLE		A synthetic feed with the addition of Humic Acid	100	[52]
XLE		A synthetic feed with the addition of silica	100	[52]
XLE		Synthetic feed with addition of Humic Acid and silica	100	[52]
Flat sheet UTC 70U	Ketoprofen	Milli.Q water	97	[29]
TFC-HR		Milli.Q water	98	[56]
TFC-HR		Milli.Q water	99	[56]
XLE	Diclofenac	Milli.Q water	95	[20]
Spiral-wound RE8040-FL		MBR effluent	99	[58]
Full scale BW30LE-440		Groundwater	99	[20]
Flat sheet UTC 60		Milli.Q water	95	[29]

Membrane Type	Target Contaminant	Water Matrix	Rejection (%)	Reference
TFC-HR	Naproxen	Milli.Q water	98	[56]
TFC-HR		Milli.Q water	99	[56]
Full scale BW30LE-440		Groundwater	100	[20]
Flat sheet UTC 60		Milli.Q water	95	[29]

Nano Filtration Membranes

Nanofiltration membranes have emerged as an efficient and cost-effective way to remove sediments, toxic metals, and pathogens. They are pressure (5-40 bar) driven membranes with pore sizes ranging between 0.5 nm-10 nm and having properties between reverse osmosis (RO) and ultrafiltration membranes [59]. The key characteristics of these membranes are an effective rejection of divalent ions and a higher flux compared to RO membranes. Based on material, nanofiltration membranes are classified as organic, inorganic, or hybrid organic-inorganic membranes [59]. Organic membranes include polymers (polyvinylidene fluoride, polysulfone, and polydimethylsiloxane), graphene, and nanotubes, while inorganic membranes are ceramic. Hybrid NF membranes are made up of a top layer with a polymer such as polyvinylidene fluoride, polytetrafluoroethylene, and inorganic material at the bottom [60, 61]. In the last few decades, NF has been elucidating promising results for the rejection of pharmaceuticals and other emerging contaminants. Rejection in NF occurs through various effects, namely, steric, transport, and dielectric, which have been explained elsewhere [59]. Several studies have reported the rejection of different emerging contaminants in NF membranes, as shown in Table **3**. However, other studies have shown that NF membranes are not effective for removing all types of emerging contaminants. For example, Yoon *et al.* [62] studied the removal of 27 pharmaceuticals and endocrine-disrupting chemicals by NF membranes and reported 30-90% removal for all the contaminants except naproxen, which showed only < 10% removal. Radjenović *et al.* [20] also studied removing a wide range of pharmaceuticals in full-scale drinking water treatment plants using NF membranes. Authors stated that nine among 12 pharmaceuticals achieved excellent removal (> 85%) in NF membrane while moderate reduction (around 50%) was observed for mefenamic acid, acetaminophen, and gemfibrozil. They mentioned that contaminant removal depends on the pore size and surface properties of the NF membrane. When a membrane with a molecular weight cut off (MWCO) greater than 200 Da is used, then lower molecular weight (< 200 Da) pharmaceuticals pass through the membrane. Adsorption mechanism of contaminants on the surface of membrane also affects their removal efficiency.

When membranes with less polar properties are applied for the treatment, hydrophobic contaminants show a high adsorption rate. Hydrophobicity of contaminants also plays an important role in adsorption, and the effect of this property is more noticeable with larger pore size membranes as large pores facilitate the way for the contaminant to the adsorption sites on the membrane surface and inside the membrane pores. For example, Verliefde *et al.* [63] reported that when a negatively charged NF membrane is used, the removal of positively charged or neutral pharmaceuticals reduces with the increase in the hydrophobic character of the contaminant. Nonetheless, the authors did not find any correlation between the hydrophobicity of pollutants and their removal through a charge on the membrane surface [63].

Table 3. Rejection of emerging contaminants in nanofiltration membrane.

Membrane	Target Contaminant	Operating Conditions	Water Matrix	Rejection (%)	Reference
NF270	Acetaminophen	$[C_0]$ = 1.12µg/L; HRT = 24h; T = 24°C; pH = 8; P = 100psi	Raw Lake Ontario water	31.2	[68]
NF90	Acetaminophen	$[C_0]$ = 0.05µg/L; Time = 15h; T = 17°C; pH = 5.6	Full-scale drinking water treatment plant	44.6	[20]
NF	Acetaminophen	$[C_0]$ = 50µg/L; T = 24°C; pH = 7.5; P = 445kPa	-	44.4	[69]
NF90	Diclofenac	$[C_0]$ = 0.05µg/L; Time = 15h; T = 17°C; pH = 5.6	Full-scale drinking water treatment plant	99.9	[20]
NF270	Carbamazepine	$[C_0]$ = 1.13µg/L; HRT = 24h; T = 24°C; pH = 8; P = 100psi	Raw Lake Ontario water	51.7	[68]
NF	Carbamazepine	$[C_0]$ = 50µg/L; T = 24°C; pH = 7.5; P = 445kPa	-	63.4	[69]
NF270	Gemfibrozil	$[C_0]$ = 1.13µg/L; HRT = 24h; T = 24°C; pH = 8; P = 100psi	Raw Lake Ontario water	85.7	[68]
NF90	Gemfibrozil	$[C_0]$ = 750µg/L; HRT = 20h; T = 24°C; pH = 9.5; P = 1800kPa	Milli-Q water	97.8	[70]
NF270	Gemfibrozil	$[C_0]$ = 750µg/L; HRT = 20h; T = 24°C; pH = 9.5; P = 1800kPa	Milli-Q water	31.4	[70]

Membrane	Target Contaminant	Operating Conditions	Water Matrix	Rejection (%)	Reference
NF90	Sulfamethoxazole	$[C_0]$ = 750μg/L; HRT = 20h; T = 20˚C; pH = 8; P = 1800kPa	Milli-Q water	99	[71]
NF90	Ibuprofen	$[C_0]$ = 750μg/L; HRT = 20h; T = 20˚C; pH = 8; P = 1800kPa	Milli-Q water	99	[71]
NF90	Carbamazepine	$[C_0]$ = 750μg/L; HRT = 20h; T = 20˚C; pH = 8; P = 1800kPa	Milli-Q water	96	[71]
NF90	Triclosan	$[C_0]$ = 750μg/L; HRT = 20h; T = 20˚C; pH = 8; P = 1800kPa	Milli-Q water	100	[71]
NF270	Sulfamethoxazole	$[C_0]$ = 750μg/L; HRT = 20h; T = 20˚C; pH = 8; P = 1800kPa	Milli-Q water	80	[71]
NF270	Ibuprofen	$[C_0]$ = 750μg/L; HRT = 20h; T = 20˚C; pH = 8; P = 1800kPa	Milli-Q water	80	[71]
NF270	Carbamazepine	$[C_0]$ = 750μg/L; HRT = 20h; T = 20˚C; pH = 8; P = 1800kPa	Milli-Q water	70	[71]
NF270	Triclosan	$[C_0]$ = 750μg/L; HRT = 20h; T = 20˚C; pH = 8; P = 1800kPa	Milli-Q water	90	[71]
NF90	Norfloxacin	$[C_0]$ = 50mg/L; HRT = 20h; T = 25˚C; pH = 6.5; P = 12 bar	Deionized water	99.5	[72]
NF270	Norfloxacin	$[C_0]$ = 50mg/L; HRT = 30min; T = 25˚C; pH = 6.5; P = 17.3 bar	Deionized water	93.6	[72]

The performance of NF also depends on the pH of the solution as an ionic character of emerging contaminant changes with it. Thus, a change in pH can significantly affect the removal of contaminants by changing their ionic character. Another parameter that affects the ionic character of the impurity is the acid dissociation constant (pKa). Vergili [64] studied the removal of three contaminants, namely carbamazepine, diclofenac, and ibuprofen, from a drinking water treatment plant through the NF membrane. The author reported that 30-39% removal was achieved for ionic diclofenac and ibuprofen, while neutral carbamazepine showed 55-61% removal [64]. Some studies have also investigated the interaction between the contaminant and NF membrane in removing pollutants [65, 66]. Further, natural organic matter (NOM) in the treating water also affects the contaminant removal either due to the fouling mechanism or interaction between contaminant-NOM [67].

Ultrafiltration Membrane Technology

Benchold introduced the term ultrafiltration (UF) in 1907, and it has been in practice for more than a century to mechanically separate material from the mixture. It has applications in cell harvesting, chemical recovery, juice concentration, dairy production, medical use, and wastewater reclamation [73 - 75]. Particularly, it is a well-known separation and disinfection process and is often used for wastewater treatment, fouling, and bacteria removal. The ultrafiltration process is preferred over conventional separation technologies due to low energy requirement, mild temperature control, and no/less chemical need [76, 77]. In the separation process, hydrostatic pressure induces the movement of liquid through a semipermeable membrane [78].

The separation of emerging contaminants in the ultrafiltration process is limited due to a specific molecular weight cut-off (MWCO) by a particular membrane used in the system. In ultrafiltration technology, molecular weight cut-off ranges between 1 to 200 KDa. Other factors that govern the removal of the emerging contaminants include physicochemical properties of ECs, such as the molecular weight of pollutants and their surface charge at the neutral pH and Log D value and hydrodynamic condition of the system. The key mechanism involved in the ECs separations is size exclusion, but the interaction between solution particles and membrane may affect the system efficiency [78 - 80]. For instance, Yoon *et al.* [62] investigated the removal of ECs having a molecular weight of 0.4 kDa by using a 100 KDa membrane in the UF system. The authors reported that ECs were not affectively blocked by the UF membrane and passed through it. Snyder *et al.* [54] reported similar results during the investigation of removing pharmaceuticals and endocrine-disrupting chemicals by using a UF membrane. Membrane surface adoption also plays an important role in the removal of ECs. Contaminants with $K_{ow} < 3$ are high and cannot adsorb on the membrane surface, thus pass-through membrane pores. However, pollutants with $K_{ow} > 3$ can easily adsorb on the surface of the membrane. Fernández *et al.* [81] investigated the interaction between the properties of contaminants and the UF membrane. Authors reported that contaminants with low K_{ow} value, including diclofenac, ibuprofen, and naproxen, were poorly removed while carbamazepine with a high K_{ow} value was 70% removed.

Table **4** demonstrates the removal of emerging contaminants by the UF technology. Studies also reported that membrane fouling changes the surface properties of the membrane and reduces pore size, which in turn may help improve the removal of the emerging contaminants. For instance, Garcia-Ivars *et al.* [82] assessed the contaminant removal efficiency during the treatment of WWTP secondary effluent by using a UF membrane. They reported that organic matter and inorganic compounds formed a fouling layer on the surface of the ceramic membrane and acted as a barrier for contaminants, and improved their removal. Another study reported that the removal of pollutants is higher in the presence of a fouling layer compared to those obtained for clean membrane [58]. The presence of organic matter such as humus also improved the contaminant removal through complex formation in the system. Organic matter can form complexes due to hydrogen bonding or electrostatic attraction between organic matter and polar moieties of emerging contaminants [82]. These complexes are rejected by charge repulsion between them and the membrane surface [82].

Table 4. Rejection of emerging contaminants in ultrafiltration membrane.

Membrane Type	Target Contaminant	Operating Conditions	Water Matrix	Rejection (%)	Reference
Zenon ZeeWeedTM 1000 (ZW1000) UF pilot system, nominal membrane pore size 0.02μm	Ibuprofen	$[C_0] = 39$ ng/L	Secondary wastewater effluent	7.7	[54]
Three seven-channel ceramic UF membranes	-	$[C_0] = 1000$ ng/L ; Pore size = 1, 5, 8 kDa; pH = 7	Spiked WWTP secondary effluent	60 35 25	[82]
Hollow-fiber polyvinylidene fluoride (PVDF) UF membrane	Diclofenac	$[C_0] = 126$ ng/L ; Pore size = kDa; pH =	Secondary wastewater effluent	44.9	[58]
Zenon ZeeWeedTM 1000 (ZW1000) UF pilot system, nominal membrane pore size 0.02μm	-	$[C_0] = 38$ ng/L	Secondary wastewater effluent	2.6	[54]
Three seven-channel ceramic UF membranes	-	$[C_0] = 300$ ng/L ; Pore size = 1, 5, 8 kDa; pH = 7	Spiked WWTP secondary effluent	60 37 27	[82]
Zenon ZeeWeedTM 1000 (ZW1000) UF pilot system, nominal membrane pore size 0.02μm	Naproxen	$[C_0] = 24$ ng/L	Secondary wastewater effluent	12.5	[54]

Membrane Type	Target Contaminant	Operating Conditions	Water Matrix	Rejection (%)	Reference
Three seven-channel ceramic UF membranes	-	$[C_0]$ = 300 ng/L ; Pore size = 1, 5, 8 kDa; pH = 7	Spiked WWTP secondary effluent	55 38 13	[82]
Three seven-channel ceramic UF membranes	Acetaminophen	$[C_0]$ = 1000 ng/L ; Pore size = 1, 5, 8 kDa; pH = 7	Spiked WWTP secondary effluent	44 30 23	[82]
Zenon ZeeWeedTM 1000 (ZW1000) UF pilot system, nominal membrane pore size 0.02μm	-	$[C_0]$ = 18 ng/L	Spiked WWTP secondary effluent	5.6	[54]
Three seven-channel ceramic UF membranes	Sulfamethoxazole	$[C_0]$ = 1000 ng/L ; Pore size = 1, 5, 8 kDa; pH = 7	Spiked WWTP secondary effluent	45 38 13	[82]
Hollow-fiber polyvinylidene fluoride (PVDF) UF membrane	-	$[C_0]$ = 155.6 ng/L	Secondary wastewater effluent	29.9	[58]
Zenon ZeeWeedTM 1000 (ZW1000) UF pilot system, nominal membrane pore size 0.02μm	-	$[C_0]$ = 66 ng/L	Secondary wastewater effluent	4.5	[54]
Zenon ZeeWeedTM 1000 (ZW1000) UF pilot system, nominal membrane pore size 0.02μm	Erythromycin	$[C_0]$ = 289 ng/L	Secondary wastewater effluent	15.2	[54]
Zenon ZeeWeedTM 1000 (ZW1000) UF pilot system, nominal membrane pore size 0.02μm	Trimethoprim	$[C_0]$ = 138 ng/L	Secondary wastewater effluent	18.1	[54]
Three seven-channel ceramic UF membranes	-	$[C_0]$ = 300 ng/L ; Pore size = 1, 5, 8 kDa; pH = 7	Spiked WWTP secondary effluent	33 29 15	[82]
Hollow-fiber polyvinylidene fluoride (PVDF) UF membrane	Atenolol	$[C_0]$ = 206.6 ng/L	Secondary wastewater effluent	5.9	[58]

(Table 4) cont.....

Membrane Type	Target Contaminant	Operating Conditions	Water Matrix	Rejection (%)	Reference
Hollow-fiber polyvinylidene fluoride (PVDF) UF membrane	Carbamazepine	$[C_0] = 105.5$ ng/L	Secondary wastewater effluent	8.1	[58]
Zenon ZeeWeedTM 1000 (ZW1000) UF pilot system, nominal membrane pore size 0.02μm	-	$[C_0] = 191$ ng/L	Secondary wastewater effluent	15.7	[54]
Zenon ZeeWeedTM 1000 (ZW1000) UF pilot system, nominal membrane pore size 0.02μm	Estrone	$[C_0] = 98$ ng/L	Secondary wastewater effluent	90.8	[54]
Zenon ZeeWeedTM 1000 (ZW1000) UF pilot system, nominal membrane pore size 0.02μm	Estriol	$[C_0] = 87$ ng/L	Secondary wastewater effluent	100	[54]
Zenon ZeeWeedTM 1000 (ZW1000) UF pilot system, nominal membrane pore size 0.02μm	Ethinyl estradiol	$[C_0] = 87$ ng/L	Secondary wastewater effluent	100	[54]
Zenon ZeeWeedTM 1000 (ZW1000) UF pilot system, nominal membrane pore size 0.02μm	Progesterone	$[C_0] = 87$ ng/L	Secondary wastewater effluent	100	[54]
Zenon ZeeWeedTM 1000 (ZW1000) UF pilot system, nominal membrane pore size 0.02μm	Testosterone	$[C_0] = 81$ ng/L	Secondary wastewater effluent	71.6	[54]
Zenon ZeeWeedTM 1000 (ZW1000) UF pilot system, nominal membrane pore size 0.02μm	Gemfibrozil	$[C_0] = 82$ ng/L	Secondary wastewater effluent	-8.5	[54]

Microfiltration Technology

Microfiltration is a low-pressure-driven membrane technology, and operating pressure varies between 1-5 bars in most operations. Usually, a microfiltration membrane has a pore size ranging between 0.1-5 μm and separates suspended particles, bacteria, protein, pigments, and paints [83, 84]. This membrane has applications in different fields, including pharmaceuticals, biotechnology, food, desalination, and wastewater treatment [85 - 88].

Separation by MF technology is based on the size-exclusion mechanism, and removal depends on the MWCO of the membrane and the size of particles present in the feed. MF membrane is manufactured by forming a thin polymer layer onto the surface of the hollow fiber and mostly uses hydrophilic polymers, namely polyvinylidene fluoride (PVDF), polysulfone (PSF), polytetrafluoroethylene (PTFE), polypropylene (PP), and nylon. These membranes can operate either in crossflow mode or dead-end flow mode. In the crossflow mode, a part of treated water is continuously returned to the system inlet. In contrast, in dead-end flow mode, all the source water passes through the membrane and periodically by batches is discharged on solid concentration around membranes. Ba *et al.* [89] assessed the performance of three emerging contaminants by using a polysulfone hollow fiber MF membrane system. Authors reported up to 85% for acetaminophen, mefenamic acid, and carbamazepine.

Table **5** demonstrates the removal of the emerging contaminants by microfiltration membrane technology. MF membranes can remove a large range of suspended particles and, therefore, are often applied as a pre-treatment along with other membrane technologies, including NF, UF, and RO, to improve the process efficiency [86, 90]. For instance, Rodriguez-Mozaz *et al.* [91] investigated the removal of 28 pharmaceuticals and 20 pesticides in the WWTP effluents using the MF-RO system. The authors reported that the combined system could remove the pharmaceuticals and pesticides from permeating completely. Another study reported the combination of laccase enzymes with microfiltration technology to remove emerging contaminants [89]. The authors reported that the combination of laccase with MF membrane almost completely (99.99%) eliminates the ECs. Campinas *et al.* [92] assessed the performance of MF membrane and powdered activated carbon to remove 36 pharmaceuticals present in the surface water. The combined powdered activated carbon/MF membrane was more than 75% removed.

Table 5. Emerging contaminants removal by microfiltration technology.

Target Contaminant	Membrane Material	Membrane Arrangement	Rejection (%)	Reference
Triclosan	Carbon nanotubes	Single-walled	99	[88]
	Carbon nanotubes	Multi-walled	100	[88]

(Table 5) cont.....

Target Contaminant	Membrane Material	Membrane Arrangement	Rejection (%)	Reference
Acetaminophen	Ceramic (ZrO_2/TiO_2)	Tubular	30	[92]
	Carbon nanotubes	Single-walled	60	[88]
	Carbon nanotubes	Multi-walled	78	[88]
	Polysulfone	Hollow fiber	100	[91]
Ibuprofen	Carbon nanotubes	Single-walled	45	[88]
	Carbon nanotubes	Multi-walled	70	[88]
	Polyelectrolyte complex with pore size 0.378 μm	Up-flow	99.8	[93]
	Polysulfone	Hollow fiber	100	[91]
Ephedrine	Polyelectrolyte complex with pore size 0.378 μm	Up-flow	99	[93]
Propranolol	Polyelectrolyte complex with pore size 0.378 μm	Up-flow	99.8	[93]
	Polysulfone	Hollow fiber	100	[91]
	Ceramic ($ZrO2/TiO_2$)	Tubular	30	[92]
Diclofenac	Polysulfone	Hollow fiber	100	[91]
	Ceramic ($ZrO2/TiO_2$)	Tubular	36	[92]
Sulfamethoxazole	Polysulfone	Hollow fiber	100	[91]
Naproxen	Polysulfone	Hollow fiber	100	[91]
Carbamazepine	Polysulfone	Hollow fiber	100	[91]
	Ceramic ($ZrO2/TiO_2$)	Tubular	45	[92]
Gemfibrozil	Polysulfone	Hollow fiber	100	[91]
Mefenamic acid	Polysulfone	Hollow fiber	100	[91]
Ketoprofen	Polysulfone	Hollow fiber	100	[91]
Atrazine	Polysulfone	Hollow fiber	100	[91]
	Ceramic ($ZrO2/TiO_2$)	Tubular	38	[92]
Diazinon	Polysulfone	Hollow fiber	100	[91]
Malathion	Polysulfone	Hollow fiber	100	[91]
Diuron	Polysulfone	Hollow fiber	100	[91]
Linuron	Polysulfone	Hollow fiber	100	[91]
	Ceramic ($ZrO2/TiO_2$)	Tubular	40	[92]
Alachlor	Polysulfone	Hollow fiber	100	[91]
	Ceramic ($ZrO2/TiO_2$)	Tubular	40	[92]

Membrane Bioreactors

Membrane bioreactor is a hybrid treatment combining the conventional activated sludge process with low-pressure membrane technology, thus operating as a signal unit and eliminating the need for a secondary clarifier in wastewater treatment [94]. MBRs are classified into two types based on their configuration: (i) side stream MBR and (ii) submerged MBR. Compared to the conventional activated sludge process, MBR is considered a more effective treatment for sludge reduction, almost suspended solids-free permeate, enhanced removal of pathogens, and production of high-quality effluent [95]. In MBR, longer sludge retention time and autotrophic bacteria result in high nitrogen removal [94, 96]. Therefore, MBRs can be a robust treatment for the removal of organic and inorganic pollutants by three different phenomena: (i) biodegradation with sludge, (ii) adsorption through a high concentration of biomass and larger surface area, and (iii) separation through the membrane. The growth of nitrifying bacteria increases with high sludge retention time and helps remove emerging biodegradable contaminants [17]. The application of MBR to remove emerging contaminants has gained interest in the last few decades. It releases various pollutants, including pharmaceuticals and personal care products, endocrine-disrupting chemicals, industrial chemicals, pesticides, and fire retardants.

Table **6** demonstrates the removal efficiency of different emerging contaminants in the MBR system. Literature shows that physicochemical properties such as chemical structure and hydrophobicity appeared to influence the biodegradation and adsorption mechanisms in MBR, thus effect the removal of emerging contaminants. Tadkaew *et al.* [95] investigated the removal of 40 pollutants in MBR and reported a correlation between the chemical structure of contaminants and their physicochemical properties on contaminants removals. Authors stated that hydrophobic (Log D > 3.2) emerging contaminants with electron-donating groups ($-NH_2$, -OH, -R, *etc.*) functional groups were well removed (> 85%) while hydrophilic and moderately hydrophobic (Log D < 3.2) contaminants with electron-withdrawing groups ($-SO_2$, -X, -COOH, *etc.*) are comparatively less removed (< 20%) in MBR system. Another study investigated the solid (sludge) and aqueous phase removal of 29 contaminants, including pharmaceuticals, steroids hormones, phytoestrogens, UV filters, and pesticides in the MBR system [97]. The authors also stated the connection between biodegradation/adsorption of emerging contaminants and their physicochemical properties. They reported that the removal of hydrophobic contaminants mainly occurred through adsorption, and biodegradation was the main mechanism for eliminating hydrophilic contaminants. Pollutants that are either moderately hydrophilic or hydrophobic were significantly accumulated in the solids (sludge) phase [97].

Table 6. Removal of emerging contaminants in membrane bioreactor.

Target Contaminant	Operating Conditions	Water Matrix	Rejection (%)	Comments	Reference
Acetaminophen	$[C_0] = 5\mu g/L$; HRT = 24h; T = 22°C; pH = 7.2-7.5	Lab scale	87	MBR could effectively remove hydrophobic contaminants	[96]
Acetaminophen	$[C_0] = 9.9\mu g/L$; HRT = 15h; T = 20°C; pH = 7.2	WWTP Terrassa	99.8	-	[24]
Diclofenac	$[C_0] = 5\mu g/L$; HRT = 24h; T = 22°C; pH = 7.2-7.5	Lab scale	15	-	[96]
Diclofenac	$[C_0] = 1.32\mu g/L$; HRT = 15h; T = 20°C; pH = 7.2	WWTP Terrassa	65.8	-	[24]
Diclofenac	$[C_0] = 5\mu g/L$; HRT = 26h; T = 26°C; pH = 7.3	Lab scale	26	Both biodegradation and adsorption govern the removal of TrOC by MBR	[97]
Ketoprofen	$[C_0] = 5\mu g/L$; HRT = 24h; T = 22°C; pH = 7.2-7.5	Lab scale	66	-	[96]
Ketoprofen	$[C_0] = 1.08\mu g/L$; HRT = 15h; T = 20°C; pH = 7.2	WWTP Terrassa	43.9	-	[24]
Ketoprofen	$[C_0] = 5\mu g/L$; HRT = 26h; T = 26°C; pH = 7.3	Lab scale	94	Biodegradation is the most important removal mechanism of hydrophilic	[97]
Naproxen	$[C_0] = 5\mu g/L$; HRT = 24h; T = 22°C; pH = 7.2-7.5	Lab scale	45	-	[96]
Naproxen	$[C_0] = 0.46\mu g/L$; HRT = 15h; T = 20°C; pH = 7.2	WWTP Terrassa	90.7	-	[24]
Naproxen	$[C_0] = 5\mu g/L$; HRT = 26h; T = 26°C; pH = 7.3	Lab scale	82	compounds Adsorption aids the degradation of hydrophobic compounds	[97]
Ibuprofen	$[C_0] = 5\mu g/L$; HRT = 24h; T = 22°C; pH = 7.2-7.5	Lab scale	96	-	[96]
Ibuprofen	$[C_0] = 21.7\mu g/L$; HRT = 15h; T = 20°C; pH = 7.2	WWTP Terrassa	99.2	-	[24]
Ibuprofen	$[C_0] = 5\mu g/L$; HRT = 26h; T = 26°C; pH = 7.3	Lab scale	99	-	[97]

(Table 6) cont.....

Target Contaminant	Operating Conditions	Water Matrix	Rejection (%)	Comments	Reference
Carbamazepine	$[C_0] = 5\mu g/L$; HRT = 24h; T = 22°C; pH = 7.2-7.5	Lab scale	32	-	[96]
Carbamazepine	$[C_0] = 0.156\mu g/L$; HRT = 15h; T = 20°C; pH = 7.2	WWTP Terrassa	10	-	[24]
Carbamazepine	$[C_0] = 5\mu g/L$; HRT = 26h; T = 26°C; pH = 7.3	Lab scale	58	-	[97]
Carbamazepine	$[C_0] = 2\mu g/L$; HRT = 24h; T = 20°C; pH = 8	Lab scale	13.2	-	[102]
Metronidazole	$[C_0] = 5\mu g/L$; HRT = 24h; T = 22°C; pH = 7.2-7.5	Lab scale	40	-	[96]
Atenolol	$[C_0] = 2\mu g/L$; HRT = 24h; T = 20°C; pH = 8	Lab scale	96.9	-	[102]
Trimethoprim	$[C_0] = 2\mu g/L$; HRT = 24h; T = 20°C; pH = 8	Lab scale	36.4	-	[102]
Gemfibrozil	$[C_0] = 5\mu g/L$; HRT = 24h; T = 22°C; pH = 7.2-7.5	Lab scale	98	-	[96]

Since removal in the MBR system depends on biodegradation and adsorption phenomena and is highly affected by operational parameters such as sludge retention time, hydraulic retention time, biomass concentration, pH, and temperature, therefore, many studies have been reported on parameter optimization for the removal of emerging contaminants in the MBR system. For example, Shariati *et al.* [98] showed that initial contaminant concentration, chemical oxygen demand, and mixed liquor suspended solids (MLSS) are the important parameters for removing acetaminophen in MBR treatment. In their study, contaminant concentration dropped from 1000ppm to below the detection limit in the permeate samples within 24h. Acetaminophen structure offers less steric hindrance and provides easy access to bacterial decay [98]. Maeng *et al.* [99] studied the effect of sludge retention time (SRT) on removing gemfibrozil and ketoprofen. They reported that removal of gemfibrozil and ketoprofen improved from 41 to 88% and 64 to 90%, respectively, when SRT increased from 20 to 80 days [99]. Tadkaew *et al.* [100] reported that ketoprofen demonstrated higher removal at pH 5 and 9 than pH 7 [100]. Hai *et al.* [101] investigated the effect of temperature on the removal of different contaminants. They reported that removals of moderately hydrophobic contaminants were significantly affected by temperature above 20 °C, while removals of the most hydrophobic contaminants were stable even at 35 °C temperature [101].

Catalytic Membranes

In recent years, catalytic membranes have received much attention due to their characteristics and involve both separations through the membrane and chemical reactions in one unit. These membranes are considered potential treatment technology in biotechnology, pharmaceuticals, energy, and wastewater treatment [103, 104].

These membranes can be classified as organic (polymeric) or inorganic (ceramic and metal), depending on process requirements. The polymeric membrane can be used at a temperature ranging from room temperature to 150 °C, while the inorganic membrane operates at very high temperatures (300-1000 °C). Generally, an inorganic membrane has an asymmetric structure with a dense top and porous base. The most used material in the preparation of inorganic membranes is alumina, silica, zirconia, and metal oxides. Ceramic membranes are formed through the sol-gel process, and porosity is improved by post-treatment with a strong alkali solution. Inorganic membranes are well suited for wastewater treatment due to their high mechanical, chemical, and thermal strength. These membranes can withstand harsh chemical cleaning, and flux can be easily recovered after fouling. Organic membranes also have an asymmetric structure with porous surfaces and less porous bases of similar material. Separation in organic membrane occurs through the surface layer while the base provides a smoother path for water. Cellulose acetate is a commonly employed polymer. Other membrane types include sulfonated, polyether sulfone, polyacrylonitrile, polysulfone, polyvinylidene fluoride, and polytetrafluoroethylene. Another organic membrane is a thin-film composite membrane consisting of a porous polymeric base and ultra-thin top layer. Thin-film composite membranes have been formed from polymers such as polyamide, polyether, polyuria, and amide. The polymeric membrane can be manufactured in different shapes, including hollow, spiral, and flat sheets). The hybrid membrane is a blend of organic and inorganic components and has been considered a potential candidate for wastewater treatment [105 - 108].

Guo *et al.* [109] investigated the removal of emerging contaminants (benzophenone-3) by using a ceramic membrane fabricated with $CuMn_2O_4$ particles. They reported that a modified membrane improved the removal of contaminants. Scarati *et al.* [110] assessed the performance of CuO-coated ceramic membrane to remove 1, 4-dioxane and reported around 50% removal. Table **7** illustrates the removal of emerging contaminants by catalytic membranes.

Table 7. Emerging contaminants removal by catalytic membranes.

Membrane and material for modification	Contaminant	Rejection (%)	Reference
TiO_2-GO-modified ceramic	Ibuprofen	71	[111]
TiO_2-GO-modified ceramic	Diclofenac	79	[111]
TiO_2-GO-modified ceramic	Carbamazepine	26	[111]
TiO_2-GO-modified ceramic	Naproxen	74	[111]
$CuMn_2O_4$-modified ceramic	Benzophenone 3	76.6	[109]
CeO_2/Al_2O_3	bisphenol A	80	[112]
Mn_2O_3/Al_2O_3		98	[112]
CeO_2/Al_2O_3	Benzotriazole	57	[112]
Mn_2O_3/Al_2O_3		55	[112]
CeO_2/Al_2O_3	Clofibric acid	40	[112]
Mn_2O_3/Al_2O_3		55	[112]
Ni foam/PVDF	Nitrobenzene	85	[113]
Nano-TiO_2/PVDF		59.5	[114]
PAC/PVDF	DEET	39	[115]
Fh [(2)]-AC/PVDF		60	[115]
Fe_2O_3/CéRAM	Salicylic acid	95	[116]

CHALLENGES RELATED TO MEMBRANE TECHNOLOGIES

The main challenges related to advanced membrane technologies during the treatment of wastewater are: 1) membrane fouling that increases the system operational cost and 2) incomplete removal of emerging contaminants which may need further treatment before discharge into the environment. Both the challenges are explained in the following sub-section.

Membrane fouling decreases membrane permeability, reduces membrane life, increases energy demand, and changes the permeate quality. In short, fouling increases the production cost of water. At the laboratory scale, membrane fouling at constant pressure results in flux decline. In contrast, system pressure usually increases on an industrial scale for a water recycling plant as the membrane fouling reduces the membrane permeability to maintain water production. The increase in pressure causes a rise in energy costs. Membrane fouling in wastewater treatment plants differs from desalination due to low salt contents and high organic contaminants. Therefore, the types and effects of membrane fouling in other membrane separation technologies have been explained in this section.

Membrane fouling can be of two types: 1) scaling due to crystallization of inorganic salts and 2) cake formation due to biological substances or organic matter. Organic fouling is the main contributor to membrane fouling during wastewater treatment because it has a high concentration of organic matter and microorganisms. Scaling can cause detrimental effects on the membrane separation process, including membrane damage and reduction in membrane permeability. Primary salts involved in scaling are calcium sulphate and calcium carbonate [117, 118]. Different fouling types also enhance the effect on each other. For instance, cake formation due to biofilm has increased the scaling.

Moreover, chemical membrane and membrane compaction damage is also considered membrane fouling. Although fouling declines the permeate flux, the fouling impact on the salt rejection may be beneficial or detrimental. For instance, Fujioka *et al.* [117] studied the effect of fouling on the rejection of salt and N-nitrosamine in the presence of model organic and colloidal foulants by NF and RO membranes. They reported the rejection of salt and N-nitrosamine increased in the presence of fouling. Another study reported similar results during an assessment of RO and NF membranes to remove organic contaminants from secondary effluent [119]. However, the removal of hydrophobic neutral emerging contaminant (primidone) decreased. Membrane fouling can be reduced through membrane cleaning, membrane modification, or pre-treatment of wastewater [120, 121].

The removal of emerging contaminants is the primary goal of designing wastewater treatment systems. Membrane separation is one of the steps involved in the removal of pollutants. Different studies have shown that RO systems can effectively (>99%) remove some of the emerging impurities depending on the nature of contaminants. However, it cannot completely remove all types of pollutants. Thus, relying on a single membrane system to eliminate ECs is not recommended, especially when water contains boron, urea, formaldehyde, and methanol. Therefore, combining membrane systems with biological or physicochemical processes is recommended [122, 123].

CONCLUSION

The demand for membrane separation technologies is increasing for water and wastewater treatment. Moreover, emerging contaminants in the wastewater and their detrimental effect on aquatic lives have increased the need for suitable wastewater treatment technologies for their complete elimination. Different membrane technologies, namely NF, RO, FO, and MBR, have emerged as efficient water treatment processes. However, several challenges are involved in membrane separation technologies that require attention in the future. Some challenges include enhancing emerging contaminants removal, preventing membrane fouling for wastewater treatment, increasing membrane resistance to cleaning agents, improving membrane permeability, and reducing energy costs. To achieve better water quality, polymer membrane technology is very crucial. Improvement in the membrane material, membrane modification, or multilayer membrane may increase the membrane performance. Moreover, the membrane fouling phenomenon and interaction of emerging contaminants will help in membrane development for drinking water treatment.

CONSENT FOR PUBLICATION

Not applicable.

CONFLICT OF INTEREST

The author declares no conflict of interest, financial or otherwise.

ACKNOWLEDGEMENTS

Declared none.

REFERENCES

[1]   W.J. Cosgrove, and D.P. Loucks, "Water management: Current and future challenges and research directions", *Water Resour. Res.,* vol. 51, no. 6, pp. 4823-4839, 2015.
[http://dx.doi.org/10.1002/2014WR016869]

[2]   J. Mahlknecht, R. González-Bravo, and F.J. Loge, "Water-energy-food security: A Nexus perspective of the current situation in Latin America and the Caribbean", *Energy,* vol. 194, p. 116824, 2020.
[http://dx.doi.org/10.1016/j.energy.2019.116824]

[3]   F. FAO, "The future of food and agriculture–Trends and challenges", *Annual Report,* 2017.

[4]   I. IEA, *"World energy outlook 2011," Int.* vol. Vol. 666. Energy Agency, 2011.

[5]   W. Majewski, "World Water Day 2014–Water & Energy", *Acta Energetica,* 2014.
[http://dx.doi.org/10.12736/issn.2300-3022.2014209]

[6]   J. Dai, S. Wu, G. Han, J. Weinberg, X. Xie, X. Wu, X. Song, B. Jia, W. Xue, and Q. Yang, "Water-energy nexus: A review of methods and tools for macro-assessment", *Appl. Energy,* vol. 210, pp. 393-408, 2018.
[http://dx.doi.org/10.1016/j.apenergy.2017.08.243]

[7]   P. Gao, Y. Ding, H. Li, and I. Xagoraraki, "Occurrence of pharmaceuticals in a municipal wastewater treatment plant: mass balance and removal processes", *Chemosphere,* vol. 88, no. 1, pp. 17-24, 2012.
[http://dx.doi.org/10.1016/j.chemosphere.2012.02.017] [PMID: 22494528]

[8]   B. Kasprzyk-Hordern, R.M. Dinsdale, and A.J. Guwy, "The occurrence of pharmaceuticals, personal care products, endocrine disruptors and illicit drugs in surface water in South Wales, UK", *Water Res.,* vol. 42, no. 13, pp. 3498-3518, 2008.
[http://dx.doi.org/10.1016/j.watres.2008.04.026] [PMID: 18514758]

[9]   A.K. Sarmah, M.T. Meyer, and A.B.A. Boxall, "A global perspective on the use, sales, exposure pathways, occurrence, fate and effects of veterinary antibiotics (VAs) in the environment", *Chemosphere,* vol. 65, no. 5, pp. 725-759, 2006.

[http://dx.doi.org/10.1016/j.chemosphere.2006.03.026] [PMID: 16677683]

[10]   R. Rosal, A. Rodríguez, J.A. Perdigón-Melón, A. Petre, E. García-Calvo, M.J. Gómez, A. Agüera, and A.R. Fernández-Alba, "Occurrence of emerging pollutants in urban wastewater and their removal through biological treatment followed by ozonation", *Water Res.,* vol. 44, no. 2, pp. 578-588, 2010.
        [http://dx.doi.org/10.1016/j.watres.2009.07.004] [PMID: 19628245]

[11]   J.L. Sotelo, A.R. Rodríguez, M.M. Mateos, S.D. Hernández, S.A. Torrellas, and J.G. Rodríguez, "Adsorption of pharmaceutical compounds and an endocrine disruptor from aqueous solutions by carbon materials", *J. Environ. Sci. Health B,* vol. 47, no. 7, pp. 640-652, 2012.
        [http://dx.doi.org/10.1080/03601234.2012.668462] [PMID: 22560026]

[12]   S. Alzahrani, and A.W. Mohammad, "Challenges and trends in membrane technology implementation for produced water treatment: A review", *J. Water Process Eng.,* vol. 4, pp. 107-133, 2014.
        [http://dx.doi.org/10.1016/j.jwpe.2014.09.007]

[13]   M.I. González, S. Alvarez, F.A. Riera, and R. Álvarez, "Lactic acid recovery from whey ultrafiltrate fermentation broths and artificial solutions by nanofiltration", *Desalination,* vol. 228, no. 1-3, pp. 84-96, 2008.
        [http://dx.doi.org/10.1016/j.desal.2007.08.009]

[14]   C. Muniandi, K.R. Mani, and S. Rathinasamy, "Large scale recovery of tetanus toxin and toxoid from fermentation broth by microporous tangential flow filtration", *Int. J. Biotechnol. Mol. Biol. Res.,* vol. 4, no. 2, pp. 28-37, 2013.
        [http://dx.doi.org/10.5897/IJBMBR12.014]

[15]   S.Z. Li, X.Y. Li, Z.F. Cui, and D.Z. Wang, "Application of ultrafiltration to improve the extraction of antibiotics", *Separ. Purif. Tech.,* vol. 34, no. 1-3, pp. 115-123, 2004.
        [http://dx.doi.org/10.1016/S1383-5866(03)00185-0]

[16]   B.D. McCloskey, H.B. Park, H. Ju, B.W. Rowe, D.J. Miller, B.J. Chun, K. Kin, and B.D. Freeman, "Influence of polydopamine deposition conditions on pure water flux and foulant adhesion resistance of reverse osmosis, ultrafiltration, and microfiltration membranes", *Polymer (Guildf.),* vol. 51, no. 15, pp. 3472-3485, 2010.
        [http://dx.doi.org/10.1016/j.polymer.2010.05.008]

[17]   A. Melo-Guimarães, F.J. Torner-Morales, J.C. Durán-Álvarez, and B.E. Jiménez-Cisneros, "Removal and fate of emerging contaminants combining biological, flocculation and membrane treatments", *Water Sci. Technol.,* vol. 67, no. 4, pp. 877-885, 2013.
        [http://dx.doi.org/10.2166/wst.2012.640] [PMID: 23306268]

[18]   J.L. Acero, F.J. Benitez, F.J. Real, and E. Rodriguez, "Elimination of selected emerging contaminants by the combination of membrane filtration and chemical oxidation processes", *Water Air Soil Pollut.,* vol. 226, no. 5, p. 139, 2015.
        [http://dx.doi.org/10.1007/s11270-015-2404-8]

[19]   X. Wei, Z. Wang, F. Fan, J. Wang, and S. Wang, "Advanced treatment of a complex pharmaceutical wastewater by nanofiltration: Membrane foulant identification and cleaning", *Desalination,* vol. 251, no. 1-3, pp. 167-175, 2010.
        [http://dx.doi.org/10.1016/j.desal.2009.08.005]

[20]   J. Radjenović, M. Petrović, F. Ventura, and D. Barceló, "Rejection of pharmaceuticals in nanofiltration and reverse osmosis membrane drinking water treatment", *Water Res.,* vol. 42, no. 14, pp. 3601-3610, 2008.
        [http://dx.doi.org/10.1016/j.watres.2008.05.020] [PMID: 18656225]

[21]   A.R.D. Verliefde, S.G.J. Heijman, E.R. Cornelissen, G. Amy, B. Van der Bruggen, and J.C. van Dijk, "Influence of electrostatic interactions on the rejection with NF and assessment of the removal efficiency during NF/GAC treatment of pharmaceutically active compounds in surface water", *Water Res.,* vol. 41, no. 15, pp. 3227-3240, 2007.
        [http://dx.doi.org/10.1016/j.watres.2007.05.022] [PMID: 17583761]

[22] S. Sarkar, A. Sarkar, and C. Bhattacharjee, "10 - Nanotechnology-based membrane-separation process for drinking water purification", In: *Water Purification,* A.M. Grumezescu, Ed., Academic Press, 2017, pp. 355-389.

[23] M.M. Pendergast, and E.M.V. Hoek, "A review of water treatment membrane nanotechnologies", *Energy Environ. Sci.,* vol. 4, no. 6, pp. 1946-1971, 2011.
[http://dx.doi.org/10.1039/c0ee00541j]

[24] J. Radjenović, M. Petrović, and D. Barceló, "Fate and distribution of pharmaceuticals in wastewater and sewage sludge of the conventional activated sludge (CAS) and advanced membrane bioreactor (MBR) treatment", *Water Res.,* vol. 43, no. 3, pp. 831-841, 2009.
[http://dx.doi.org/10.1016/j.watres.2008.11.043] [PMID: 19091371]

[25] D.J. Lapworth, N. Baran, M.E. Stuart, and R.S. Ward, "Emerging organic contaminants in groundwater: A review of sources, fate and occurrence", *Environ. Pollut.,* vol. 163, pp. 287-303, 2012.
[http://dx.doi.org/10.1016/j.envpol.2011.12.034] [PMID: 22306910]

[26] M. Clara, B. Strenn, and N. Kreuzinger, "Carbamazepine as a possible anthropogenic marker in the aquatic environment: investigations on the behaviour of Carbamazepine in wastewater treatment and during groundwater infiltration", *Water Res.,* vol. 38, no. 4, pp. 947-954, 2004.
[http://dx.doi.org/10.1016/j.watres.2003.10.058] [PMID: 14769414]

[27] A. Pal, K.Y-H. Gin, A.Y-C. Lin, and M. Reinhard, "Impacts of emerging organic contaminants on freshwater resources: review of recent occurrences, sources, fate and effects", *Sci. Total Environ.,* vol. 408, no. 24, pp. 6062-6069, 2010.
[http://dx.doi.org/10.1016/j.scitotenv.2010.09.026] [PMID: 20934204]

[28] C. Wu, X. Huang, J.D. Witter, A.L. Spongberg, K. Wang, D. Wang, and J. Liu, "Occurrence of pharmaceuticals and personal care products and associated environmental risks in the central and lower Yangtze river, China", *Ecotoxicol. Environ. Saf.,* vol. 106, pp. 19-26, 2014.
[http://dx.doi.org/10.1016/j.ecoenv.2014.04.029] [PMID: 24836873]

[29] H. Ozaki, N. Ikejima, Y. Shimizu, K. Fukami, S. Taniguchi, R. Takanami, R.R. Giri, and S. Matsui, "Rejection of pharmaceuticals and personal care products (PPCPs) and endocrine disrupting chemicals (EDCs) by low pressure reverse osmosis membranes", *Water Sci. Technol.,* vol. 58, no. 1, pp. 73-81, 2008.
[http://dx.doi.org/10.2166/wst.2008.607] [PMID: 18653939]

[30] K. Grigorakis, and G. Rigos, "Aquaculture effects on environmental and public welfare - the case of Mediterranean mariculture", *Chemosphere,* vol. 85, no. 6, pp. 899-919, 2011.
[http://dx.doi.org/10.1016/j.chemosphere.2011.07.015] [PMID: 21821276]

[31] M. Taheran, S.K. Brar, M. Verma, R.Y. Surampalli, T.C. Zhang, and J.R. Valero, "Membrane processes for removal of pharmaceutically active compounds (PhACs) from water and wastewaters", *Sci. Total Environ.,* vol. 547, pp. 60-77, 2016.
[http://dx.doi.org/10.1016/j.scitotenv.2015.12.139] [PMID: 26789358]

[32] A.I. Schäfer, I. Akanyeti, and A.J. Semião, "Micropollutant sorption to membrane polymers: a review of mechanisms for estrogens", *Adv. Colloid Interface Sci.,* vol. 164, no. 1-2, pp. 100-117, 2011.
[http://dx.doi.org/10.1016/j.cis.2010.09.006] [PMID: 21106187]

[33] A.I. Schäfer, I. Akanyeti, and A.J.C. Semião, "Micropollutant sorption to membrane polymers: a review of mechanisms for estrogens", *Adv. Colloid Interface Sci.,* vol. 164, no. 1-2, pp. 100-117, 2011.
[http://dx.doi.org/10.1016/j.cis.2010.09.006] [PMID: 21106187]

[34] S. Zhao, L. Zou, C.Y. Tang, and D. Mulcahy, "Recent developments in forward osmosis: Opportunities and challenges", *J. Membr. Sci.,* vol. 396, pp. 1-21, 2012.
[http://dx.doi.org/10.1016/j.memsci.2011.12.023]

[35] S. Qi, C.Q. Qiu, Y. Zhao, and C.Y. Tang, "Double-skinned forward osmosis membranes based on layer-by-layer assembly—FO performance and fouling behavior", *J. Membr. Sci.,* vol. 405-406, pp.

20-29, 2012.
[http://dx.doi.org/10.1016/j.memsci.2012.02.032]

[36] P.S. Goh, A.F. Ismail, B.C. Ng, and M.S. Abdullah, "Recent progresses of forward osmosis membranes formulation and design for wastewater treatment", *Water,* vol. 11, no. 10, p. 2043, 2019.
[http://dx.doi.org/10.3390/w11102043]

[37] M. Xie, L.D. Nghiem, W.E. Price, and M. Elimelech, "Comparison of the removal of hydrophobic trace organic contaminants by forward osmosis and reverse osmosis", *Water Res.,* vol. 46, no. 8, pp. 2683-2692, 2012.
[http://dx.doi.org/10.1016/j.watres.2012.02.023] [PMID: 22402269]

[38] Y. Kim, S. Li, S. Phuntsho, M. Xie, H.K. Shon, and N. Ghaffour, "Understanding the organic micropollutants transport mechanisms in the fertilizer-drawn forward osmosis process", *J. Environ. Manage.,* vol. 248, p. 109240, 2019.
[http://dx.doi.org/10.1016/j.jenvman.2019.07.011] [PMID: 31310933]

[39] H.T. Madsen, N. Bajraktari, C. Hélix-Nielsen, B. Van der Bruggen, and E.G. Søgaard, "Use of biomimetic forward osmosis membrane for trace organics removal", *J. Membr. Sci.,* vol. 476, pp. 469-474, 2015.
[http://dx.doi.org/10.1016/j.memsci.2014.11.055]

[40] X. Jin, J. Shan, C. Wang, J. Wei, and C.Y. Tang, "Rejection of pharmaceuticals by forward osmosis membranes", *J. Hazard. Mater.,* vol. 227-228, pp. 55-61, 2012.
[http://dx.doi.org/10.1016/j.jhazmat.2012.04.077] [PMID: 22640821]

[41] Z. Li, L. Jiang, and C. Tang, "Investigation on removing recalcitrant toxic organic polluters in coking wastewater by forward osmosis", *Chin. J. Chem. Eng.,* vol. 28, no. 1, pp. 122-135, 2020.
[http://dx.doi.org/10.1016/j.cjche.2019.07.011]

[42] Y. Zheng, M. Huang, L. Chen, W. Zheng, P. Xie, and Q. Xu, "Comparison of tetracycline rejection in reclaimed water by three kinds of forward osmosis membranes", *Desalination,* vol. 359, pp. 113-122, 2015.
[http://dx.doi.org/10.1016/j.desal.2014.12.009]

[43] D-Q. Cao, X-X. Yang, W-Y. Yang, Q-H. Wang, and X-D. Hao, "Separation of trace pharmaceuticals individually and in combination via forward osmosis", *Sci. Total Environ.,* vol. 718, p. 137366, 2020.
[http://dx.doi.org/10.1016/j.scitotenv.2020.137366] [PMID: 32092521]

[44] P. Chelme-Ayala, D.W. Smith, and M.G. El-Din, "Membrane concentrate management options: a comprehensive critical review", *J. Environ. Eng. Sci.,* vol. 8, no. 3, pp. 326-339, 2013.
[http://dx.doi.org/10.1680/jees.2013.0033]

[45] C. Bellona, and J.E. Drewes, "The role of membrane surface charge and solute physico-chemical properties in the rejection of organic acids by NF membranes", *J. Membr. Sci.,* vol. 249, no. 1-2, pp. 227-234, 2005.
[http://dx.doi.org/10.1016/j.memsci.2004.09.041]

[46] C.J. Gabelich, K.P. Ishida, and R.M. Bold, "Testing of water treatment copolymers for compatibility with polyamide reverse osmosis membranes", *Environ. Prog.,* vol. 24, no. 4, pp. 410-416, 2005.
[http://dx.doi.org/10.1002/ep.10111]

[47] K. Häyrynen, J. Langwaldt, E. Pongrácz, V. Väisänen, M. Mänttäri, and R.L. Keiski, "Separation of nutrients from mine water by reverse osmosis for subsequent biological treatment", *Miner. Eng.,* vol. 21, no. 1, pp. 2-9, 2008.
[http://dx.doi.org/10.1016/j.mineng.2007.06.003]

[48] G. Maragliano, and P. Moss, "The development of a high flow seawater membrane. A case history of one of the first applications using high flow seawater elements in a plant producing process and boiler feed water for ENEL (now EDIPOWER) at San Filippo del Mela power plant in Italy", *Desalination,* vol. 184, no. 1-3, pp. 247-252, 2005.
[http://dx.doi.org/10.1016/j.desal.2005.01.016]

[49] H. Zhang, S. Fang, C. Ye, M. Wang, H. Cheng, H. Wen, and X. Meng, "Treatment of waste filature oil/water emulsion by combined demulsification and reverse osmosis", *Separ. Purif. Tech.,* vol. 63, no. 2, pp. 264-268, 2008.
[http://dx.doi.org/10.1016/j.seppur.2008.05.012]

[50] J.H. Al-Rifai, H. Khabbaz, and A.I. Schäfer, "Removal of pharmaceuticals and endocrine disrupting compounds in a water recycling process using reverse osmosis systems", *Separ. Purif. Tech.,* vol. 77, no. 1, pp. 60-67, 2011.
[http://dx.doi.org/10.1016/j.seppur.2010.11.020]

[51] K.P.M. Licona, L.R.O. Geaquinto, J.V. Nicolini, N.G. Figueiredo, S.C. Chiapetta, A.C. Habert, and L. Yokoyama, "Assessing potential of nanofiltration and reverse osmosis for removal of toxic pharmaceuticals from water", *J. Water Process Eng.,* vol. 25, pp. 195-204, 2018.
[http://dx.doi.org/10.1016/j.jwpe.2018.08.002]

[52] Y-L. Lin, "Effects of organic, biological and colloidal fouling on the removal of pharmaceuticals and personal care products by nanofiltration and reverse osmosis membranes", *J. Membr. Sci.,* vol. 542, pp. 342-351, 2017.
[http://dx.doi.org/10.1016/j.memsci.2017.08.023]

[53] A.M. Urtiaga, G. Pérez, R. Ibáñez, and I. Ortiz, "Removal of pharmaceuticals from a WWTP secondary effluent by ultrafiltration/reverse osmosis followed by electrochemical oxidation of the RO concentrate", *Desalination,* vol. 331, pp. 26-34, 2013.
[http://dx.doi.org/10.1016/j.desal.2013.10.010]

[54] S.A. Snyder, S. Adham, A.M. Redding, F.S. Cannon, J. DeCarolis, J. Oppenheimer, E.C. Wert, and Y. Yoon, "Role of membranes and activated carbon in the removal of endocrine disruptors and pharmaceuticals", *Desalination,* vol. 202, no. 1-3, pp. 156-181, 2007.
[http://dx.doi.org/10.1016/j.desal.2005.12.052]

[55] A.M. Comerton, R.C. Andrews, D.M. Bagley, and P. Yang, "Membrane adsorption of endocrine disrupting compounds and pharmaceutically active compounds", *J. Membr. Sci.,* vol. 303, no. 1-2, pp. 267-277, 2007.
[http://dx.doi.org/10.1016/j.memsci.2007.07.025]

[56] P. Xu, J.E. Drewes, C. Bellona, G. Amy, T-U. Kim, M. Adam, and T. Heberer, "Rejection of emerging organic micropollutants in nanofiltration-reverse osmosis membrane applications", *Water Environ. Res.,* vol. 77, no. 1, pp. 40-48, 2005.
[http://dx.doi.org/10.2175/106143005X41609] [PMID: 15765934]

[57] K. Kimura, S. Toshima, G. Amy, and Y. Watanabe, "Rejection of neutral endocrine disrupting compounds (EDCs) and pharmaceutical active compounds (PhACs) by RO membranes", *J. Membr. Sci.,* vol. 245, no. 1-2, pp. 71-78, 2004.
[http://dx.doi.org/10.1016/j.memsci.2004.07.018]

[58] K. Chon, J. Cho, and H.K. Shon, "A pilot-scale hybrid municipal wastewater reclamation system using combined coagulation and disk filtration, ultrafiltration, and reverse osmosis: removal of nutrients and micropollutants, and characterization of membrane foulants", *Bioresour. Technol.,* vol. 141, pp. 109-116, 2013.
[http://dx.doi.org/10.1016/j.biortech.2013.03.198] [PMID: 23611699]

[59] A.W. Mohammad, Y.H. Teow, W.L. Ang, Y.T. Chung, D.L. Oatley-Radcliffe, and N. Hilal, "Nanofiltration membranes review: Recent advances and future prospects", *Desalination,* vol. 356, pp. 226-254, 2015.
[http://dx.doi.org/10.1016/j.desal.2014.10.043]

[60] N. Hilal, H. Al-Zoubi, N.A. Darwish, A.W. Mohamma, and M. Abu Arabi, "A comprehensive review of nanofiltration membranes:Treatment, pretreatment, modelling, and atomic force microscopy", *Desalination,* vol. 170, no. 3, pp. 281-308, 2004.
[http://dx.doi.org/10.1016/j.desal.2004.01.007]

[61] B. Van der Bruggen, "Chemical modification of polyethersulfone nanofiltration membranes: A review", *J. Appl. Polym. Sci.,* vol. 114, no. 1, pp. 630-642, 2009.
[http://dx.doi.org/10.1002/app.30578]

[62] Y. Yoon, P. Westerhoff, S.A. Snyder, E.C. Wert, and J. Yoon, "Removal of endocrine disrupting compounds and pharmaceuticals by nanofiltration and ultrafiltration membranes", *Desalination,* vol. 202, no. 1-3, pp. 16-23, 2007.
[http://dx.doi.org/10.1016/j.desal.2005.12.033]

[63] A.R.D. Verliefde, E.R. Cornelissen, S.G.J. Heijman, I. Petrinic, T. Luxbacher, G.L. Amy, B. Van der Bruggen, and J.C. van Dijk, "Influence of membrane fouling by (pretreated) surface water on rejection of pharmaceutically active compounds (PhACs) by nanofiltration membranes", *J. Membr. Sci.,* vol. 330, no. 1-2, pp. 90-103, 2009.
[http://dx.doi.org/10.1016/j.memsci.2008.12.039]

[64] I. Vergili, "Application of nanofiltration for the removal of carbamazepine, diclofenac and ibuprofen from drinking water sources", *J. Environ. Manage.,* vol. 127, pp. 177-187, 2013.
[http://dx.doi.org/10.1016/j.jenvman.2013.04.036] [PMID: 23708199]

[65] L.D. Nghiem, A.I. Schäfer, and M. Elimelech, "Pharmaceutical retention mechanisms by nanofiltration membranes", *Environ. Sci. Technol.,* vol. 39, no. 19, pp. 7698-7705, 2005.
[http://dx.doi.org/10.1021/es0507665] [PMID: 16245847]

[66] T. Heberer, D. Feldmann, K. Reddersen, H-J. Altmann, and T. Zimmermann, "Production of Drinking Water from Highly Contaminated Surface Waters: Removal of Organic, Inorganic, and Microbial Contaminants Applying Mobile Membrane Filtration Units", *Acta Hydrochim. Hydrobiol.,* vol. 30, no. 1, pp. 24-33, 2002.
[http://dx.doi.org/10.1002/1521-401X(200207)30:1<24::AID-AHEH24>3.0.CO;2-O]

[67] A.H.M.A. Sadmani, R.C. Andrews, and D.M. Bagley, "Impact of natural water colloids and cations on the rejection of pharmaceutically active and endocrine disrupting compounds by nanofiltration", *J. Membr. Sci.,* vol. 450, pp. 272-281, 2014.
[http://dx.doi.org/10.1016/j.memsci.2013.09.017]

[68] A.M. Comerton, R.C. Andrews, D.M. Bagley, and C. Hao, "The rejection of endocrine disrupting and pharmaceutically active compounds by NF and RO membranes as a function of compound and water matrix properties", *J. Membr. Sci.,* vol. 313, no. 1-2, pp. 323-335, 2008.
[http://dx.doi.org/10.1016/j.memsci.2008.01.021]

[69] Y. Yoon, P. Westerhoff, S.A. Snyder, and E.C. Wert, "Nanofiltration and ultrafiltration of endocrine disrupting compounds, pharmaceuticals and personal care products", *J. Membr. Sci.,* vol. 270, no. 1-2, pp. 88-100, 2006.
[http://dx.doi.org/10.1016/j.memsci.2005.06.045]

[70] L.D. Nghiem, and S. Hawkes, "Effects of membrane fouling on the nanofiltration of pharmaceutically active compounds (PhACs): Mechanisms and role of membrane pore size", *Separ. Purif. Tech.,* vol. 57, no. 1, pp. 176-184, 2007.
[http://dx.doi.org/10.1016/j.seppur.2007.04.002]

[71] L.D. Nghiem, P.J. Coleman, and C. Espendiller, "Mechanisms underlying the effects of membrane fouling on the nanofiltration of trace organic contaminants", *Desalination,* vol. 250, no. 2, pp. 682-687, 2010.
[http://dx.doi.org/10.1016/j.desal.2009.03.025]

[72] D.I. de Souza, E.M. Dottein, A. Giacobbo, M.A.S. Rodrigues, M.N. de Pinho, and A.M. Bernardes, "Nanofiltration for the removal of norfloxacin from pharmaceutical effluent", *J. Environ. Chem. Eng.,* vol. 6, no. 5, pp. 6147-6153, 2018.
[http://dx.doi.org/10.1016/j.jece.2018.09.034]

[73] X. Shi, G. Tal, N.P. Hankins, and V. Gitis, "Fouling and cleaning of ultrafiltration membranes: A review", *J. Water Process Eng.,* vol. 1, pp. 121-138, 2014.

[http://dx.doi.org/10.1016/j.jwpe.2014.04.003]

[74] K.M. Barry, T.G. Dinan, and P.M. Kelly, "Pilot scale production of a phospholipid-enriched dairy ingredient by means of an optimised integrated process employing enzymatic hydrolysis, ultrafiltration and super-critical fluid extraction", *Innov. Food Sci. Emerg. Technol.,* vol. 41, pp. 301-306, 2017.
[http://dx.doi.org/10.1016/j.ifset.2017.04.004]

[75] X. Zhou, J. Liang, Y. Zhang, H. Zhao, Y. Guo, and S. Shi, "Separation and purification of α-glucosidase inhibitors from Polygonatum odoratum by stepwise high-speed counter-current chromatography combined with Sephadex LH-20 chromatography target-guided by ultrafiltration-HPLC screening", *J. Chromatogr. B Analyt. Technol. Biomed. Life Sci.,* vol. 985, pp. 149-154, 2015.
[http://dx.doi.org/10.1016/j.jchromb.2015.01.030] [PMID: 25682336]

[76] R. Lu, C. Zhang, M. Piatkovsky, M. Ulbricht, M. Herzberg, and T.H. Nguyen, "Improvement of virus removal using ultrafiltration membranes modified with grafted zwitterionic polymer hydrogels", *Water Res.,* vol. 116, pp. 86-94, 2017.
[http://dx.doi.org/10.1016/j.watres.2017.03.023] [PMID: 28324709]

[77] G.J. Gentile, M.C. Cruz, V.B. Rajal, and M.M. Fidalgo de Cortalezzi, "Electrostatic interactions in virus removal by ultrafiltration membranes", *J. Environ. Chem. Eng.,* vol. 6, no. 1, pp. 1314-1321, 2018.
[http://dx.doi.org/10.1016/j.jece.2017.11.041]

[78] S. Al Aani, T.N. Mustafa, and N. Hilal, "Ultrafiltration membranes for wastewater and water process engineering: A comprehensive statistical review over the past decade", *J. Water Process Eng.,* vol. 35, p. 101241, 2020.
[http://dx.doi.org/10.1016/j.jwpe.2020.101241]

[79] I.N. Widiasa, G.R. Harvianto, H. Susanto, T. Istirokhatun, and T.W. Agustini, "Searching for ultrafiltration membrane molecular weight cut-off for water treatment in recirculating aquaculture system", *J. Water Process Eng.,* vol. 21, pp. 133-142, 2018.
[http://dx.doi.org/10.1016/j.jwpe.2017.12.006]

[80] C. Fonseca Couto, L.C. Lange, and M.C. Santos Amaral, "A critical review on membrane separation processes applied to remove pharmaceutically active compounds from water and wastewater", *J. Water Process Eng.,* vol. 26, pp. 156-175, 2018.
[http://dx.doi.org/10.1016/j.jwpe.2018.10.010]

[81] R.L. Fernández, J.A. McDonald, S.J. Khan, and P. Le-Clech, "Removal of pharmaceuticals and endocrine disrupting chemicals by a submerged membrane photocatalysis reactor (MPR)", *Separ. Purif. Tech.,* vol. 127, pp. 131-139, 2014.
[http://dx.doi.org/10.1016/j.seppur.2014.02.031]

[82] J. Garcia-Ivars, J. Durá-María, C. Moscardó-Carreño, C. Carbonell-Alcaina, M-I. Alcaina-Miranda, and M-I. Iborra-Clar, "Rejection of trace pharmaceutically active compounds present in municipal wastewaters using ceramic fine ultrafiltration membranes: Effect of feed solution pH and fouling phenomena", *Separ. Purif. Tech.,* vol. 175, pp. 58-71, 2017.
[http://dx.doi.org/10.1016/j.seppur.2016.11.027]

[83] W. Eykamp, "Chapter 1 Microfiltration and ultrafiltration", In: *Membrane Science and Technology.,* R.D. Noble, S.A. Stern, Eds., Elsevier, 1995, pp. 1-43.
[http://dx.doi.org/10.1016/S0927-5193(06)80003-3]

[84] N. Uzal, L. Yilmaz, and U. Yetis, "Microfiltration/ultrafiltration as pretreatment for reclamation of rinsing waters of indigo dyeing", *Desalination,* vol. 240, no. 1-3, pp. 198-208, 2009.
[http://dx.doi.org/10.1016/j.desal.2007.10.092]

[85] C. Rouquié, L. Dahdouh, J. Ricci, C. Wisniewski, and M. Delalonde, "Immersed membranes configuration for the microfiltration of fruit-based suspensions", *Separ. Purif. Tech.,* vol. 216, pp. 25-33, 2019.
[http://dx.doi.org/10.1016/j.seppur.2019.01.062]

[86]   S.F. Anis, R. Hashaikeh, and N. Hilal, "Reverse osmosis pretreatment technologies and future trends: A comprehensive review", *Desalination,* vol. 452, pp. 159-195, 2019.
[http://dx.doi.org/10.1016/j.desal.2018.11.006]

[87]   P. Saini, V.K. Bulasara, and A.S. Reddy, "Performance of a new ceramic microfiltration membrane based on kaolin in textile industry wastewater treatment", *Chem. Eng. Commun.,* vol. 206, no. 2, pp. 227-236, 2019.
[http://dx.doi.org/10.1080/00986445.2018.1482281]

[88]   Y. Wang, J. Zhu, H. Huang, and H-H. Cho, "Carbon nanotube composite membranes for microfiltration of pharmaceuticals and personal care products: Capabilities and potential mechanisms", *J. Membr. Sci.,* vol. 479, pp. 165-174, 2015.
[http://dx.doi.org/10.1016/j.memsci.2015.01.034]

[89]   S. Ba, J.P. Jones, and H. Cabana, "Hybrid bioreactor (HBR) of hollow fiber microfilter membrane and cross-linked laccase aggregates eliminate aromatic pharmaceuticals in wastewaters", *J. Hazard. Mater.,* vol. 280, pp. 662-670, 2014.
[http://dx.doi.org/10.1016/j.jhazmat.2014.08.062] [PMID: 25218263]

[90]   S.F. Anis, R. Hashaikeh, and N. Hilal, "Flux and salt rejection enhancement of polyvinyl(alcohol) reverse osmosis membranes using nano-zeolite", *Desalination,* vol. 470, p. 114104, 2019.
[http://dx.doi.org/10.1016/j.desal.2019.114104]

[91]   S. Rodriguez-Mozaz, M. Ricart, M. Köck-Schulmeyer, H. Guasch, C. Bonnineau, L. Proia, M.L. de Alda, S. Sabater, and D. Barceló, "Pharmaceuticals and pesticides in reclaimed water: Efficiency assessment of a microfiltration-reverse osmosis (MF-RO) pilot plant", *J. Hazard. Mater.,* vol. 282, pp. 165-173, 2015.
[http://dx.doi.org/10.1016/j.jhazmat.2014.09.015] [PMID: 25269743]

[92]   M. Campinas, R.M.C. Viegas, R. Coelho, H. Lucas, and M.J. Rosa, "Adsorption/Coagulation/Ceramic Microfiltration for Treating Challenging Waters for Drinking Water Production", *Membranes (Basel),* vol. 11, no. 2, p. 91, 2021.
[http://dx.doi.org/10.3390/membranes11020091] [PMID: 33514022]

[93]   S.S.M. Hassan, H.I. Abdel-Shafy, and M.S.M. Mansour, "Removal of pharmaceutical compounds from urine via chemical coagulation by green synthesized ZnO-nanoparticles followed by microfiltration for safe reuse", *Arab. J. Chem.,* vol. 12, no. 8, pp. 4074-4083, 2019.
[http://dx.doi.org/10.1016/j.arabjc.2016.04.009]

[94]   C. Li, C. Cabassud, and C. Guigui, "Evaluation of membrane bioreactor on removal of pharmaceutical micropollutants: a review", *Desalination and Water Treatment,* vol. 55, 2014.

[95]   N. Tadkaew, F.I. Hai, J.A. McDonald, S.J. Khan, and L.D. Nghiem, "Removal of trace organics by MBR treatment: the role of molecular properties", *Water Res.,* vol. 45, no. 8, pp. 2439-2451, 2011.
[http://dx.doi.org/10.1016/j.watres.2011.01.023] [PMID: 21388651]

[96]   L.N. Nguyen, F.I. Hai, J. Kang, W.E. Price, and L.D. Nghiem, "Removal of emerging trace organic contaminants by MBR-based hybrid treatment processes", *Int. Biodeterior. Biodegradation,* vol. 85, pp. 474-482, 2013.
[http://dx.doi.org/10.1016/j.ibiod.2013.03.014]

[97]   K.C. Wijekoon, F.I. Hai, J. Kang, W.E. Price, W. Guo, H.H. Ngo, and L.D. Nghiem, "The fate of pharmaceuticals, steroid hormones, phytoestrogens, UV-filters and pesticides during MBR treatment", *Bioresour. Technol.,* vol. 144, pp. 247-254, 2013.
[http://dx.doi.org/10.1016/j.biortech.2013.06.097] [PMID: 23871927]

[98]   F.P. Shariati, M.R. Mehrnia, B.M. Salmasi, M. Heran, C. Wisniewski, and M.H. Sarrafzadeh, "Membrane bioreactor for treatment of pharmaceutical wastewater containing acetaminophen", *Desalination,* vol. 250, no. 2, pp. 798-800, 2010.
[http://dx.doi.org/10.1016/j.desal.2008.11.044]

[99] S.K. Maeng, B.G. Choi, K.T. Lee, and K.G. Song, "Influences of solid retention time, nitrification and microbial activity on the attenuation of pharmaceuticals and estrogens in membrane bioreactors", *Water Res.,* vol. 47, no. 9, pp. 3151-3162, 2013.
[http://dx.doi.org/10.1016/j.watres.2013.03.014] [PMID: 23582351]

[100] N. Tadkaew, M. Sivakumar, S.J. Khan, J.A. McDonald, and L.D. Nghiem, "Effect of mixed liquor pH on the removal of trace organic contaminants in a membrane bioreactor", *Bioresour. Technol.,* vol. 101, no. 5, pp. 1494-1500, 2010.
[http://dx.doi.org/10.1016/j.biortech.2009.09.082] [PMID: 19864128]

[101] F.I. Hai, K. Tessmer, L.N. Nguyen, J. Kang, W.E. Price, and L.D. Nghiem, "Removal of micropollutants by membrane bioreactor under temperature variation", *J. Membr. Sci.,* vol. 383, no. 1-2, pp. 144-151, 2011.
[http://dx.doi.org/10.1016/j.memsci.2011.08.047]

[102] A.A. Alturki, N. Tadkaew, J.A. McDonald, S.J. Khan, W.E. Price, and L.D. Nghiem, "Combining MBR and NF/RO membrane filtration for the removal of trace organics in indirect potable water reuse applications", *J. Membr. Sci.,* vol. 365, no. 1-2, pp. 206-215, 2010.
[http://dx.doi.org/10.1016/j.memsci.2010.09.008]

[103] L. Giorno, and E. Drioli, "Biocatalytic membrane reactors: applications and perspectives", *Trends Biotechnol.,* vol. 18, no. 8, pp. 339-349, 2000.
[http://dx.doi.org/10.1016/S0167-7799(00)01472-4] [PMID: 10899815]

[104] B. Hallstrom, G. Tragardh, and J. Nilsson, "Membrane technology in the food industry", *Engineering and Food,* vol. 3, pp. 194-208, 1990.

[105] H. Zhu, R. Jiang, L. Xiao, Y. Chang, Y. Guan, X. Li, and G. Zeng, "Photocatalytic decolorization and degradation of Congo Red on innovative crosslinked chitosan/nano-CdS composite catalyst under visible light irradiation", *J. Hazard. Mater.,* vol. 169, no. 1-3, pp. 933-940, 2009.
[http://dx.doi.org/10.1016/j.jhazmat.2009.04.037] [PMID: 19477069]

[106] R.V. Kumar, L. Goswami, K. Pakshirajan, and G. Pugazhenthi, "Dairy wastewater treatment using a novel low cost tubular ceramic membrane and membrane fouling mechanism using pore blocking models", *J. Water Process Eng.,* vol. 13, pp. 168-175, 2016.
[http://dx.doi.org/10.1016/j.jwpe.2016.08.012]

[107] C.F. Bennani, B. Ousji, and D.J. Ennigrou, "Reclamation of dairy wastewater using ultrafiltration process", *Desalination Water Treat.,* vol. 55, no. 2, pp. 297-303, 2015.
[http://dx.doi.org/10.1080/19443994.2014.913996]

[108] S. Psaltou, and A. Zouboulis, "Catalytic Ozonation and Membrane Contactors—A Review Concerning Fouling Occurrence and Pollutant Removal", *Water,* vol. 12, no. 11, p. 2964, 2020.
[http://dx.doi.org/10.3390/w12112964]

[109] Y. Guo, Z. Song, B. Xu, Y. Li, F. Qi, J-P. Croue, and D. Yuan, "A novel catalytic ceramic membrane fabricated with CuMn2O4 particles for emerging UV absorbers degradation from aqueous and membrane fouling elimination", *J. Hazard. Mater.,* vol. 344, pp. 1229-1239, 2018.
[http://dx.doi.org/10.1016/j.jhazmat.2017.11.044] [PMID: 29198887]

[110] G. Scaratti, A. De Noni Júnior, H.J. José, and R. de Fatima Peralta Muniz Moreira, "1,4-Dioxane removal from water and membrane fouling elimination using CuO-coated ceramic membrane coupled with ozone", *Environ. Sci. Pollut. Res. Int.,* vol. 27, no. 18, pp. 22144-22154, 2020.
[http://dx.doi.org/10.1007/s11356-019-07497-6] [PMID: 31916160]

[111] C. Li, W. Sun, Z. Lu, X. Ao, C. Yang, and S. Li, "Systematic evaluation of TiO_2-GO-modified ceramic membranes for water treatment: Retention properties and fouling mechanisms", *Chem. Eng. J.,* vol. 378, p. 122138, 2019.
[http://dx.doi.org/10.1016/j.cej.2019.122138]

[112] W.J. Lee, Y. Bao, X. Hu, and T-T. Lim, "Hybrid catalytic ozonation-membrane filtration process with

CeOx and MnOx impregnated catalytic ceramic membranes for micropollutants degradation", *Chem. Eng. J.,* vol. 378, p. 121670, 2019.
[http://dx.doi.org/10.1016/j.cej.2019.05.031]

[113] K. Li, L. Xu, Y. Zhang, A. Cao, Y. Wang, H. Huang, and J. Wang, "A novel electro-catalytic membrane contactor for improving the efficiency of ozone on wastewater treatment", *Appl. Catal. B,* vol. 249, pp. 316-321, 2019.
[http://dx.doi.org/10.1016/j.apcatb.2019.03.015]

[114] Y. Zhu, H. Zhang, and X. Zhang, "Study on Catalytic Ozone Oxidation with Nano-TiO_2 Modified Membrane for Treatment of Municipal Wastewater", *Asian J. Chem.,* vol. 26, no. 13, pp. 3871-3874, 2014.
[http://dx.doi.org/10.14233/ajchem.2014.15980]

[115] Y. Li, and K.L. Yeung, "Polymeric catalytic membrane for ozone treatment of DEET in water", *Catal. Today,* vol. 331, pp. 53-59, 2019.
[http://dx.doi.org/10.1016/j.cattod.2018.06.005]

[116] B.S. Karnik, S.H. Davies, M.J. Baumann, and S.J. Masten, S. H. D. Bhavana S. Karnik, "Melissa J. Baumann, and Susan J. Masten, "Use of Salicylic Acid as a Model Compound to Investigate Hydroxyl Radical Reaction in an Ozonation–Membrane Filtration Hybrid Process", *Environ. Eng. Sci.,* vol. 24, no. 6, pp. 852-860, 2007.
[http://dx.doi.org/10.1089/ees.2006.0156]

[117] T. Fujioka, S.J. Khan, J.A. McDonald, R.K. Henderson, Y. Poussade, J.E. Drewes, and L.D. Nghiem, "Effects of membrane fouling on N-nitrosamine rejection by nanofiltration and reverse osmosis membranes", *J. Membr. Sci.,* vol. 427, pp. 311-319, 2013.
[http://dx.doi.org/10.1016/j.memsci.2012.09.055]

[118] D.E. Potts, R.C. Ahlert, and S.S. Wang, "A critical review of fouling of reverse osmosis membranes", *Desalination,* vol. 36, no. 3, pp. 235-264, 1981.
[http://dx.doi.org/10.1016/S0011-9164(00)88642-7]

[119] P. Xu, J.E. Drewes, T-U. Kim, C. Bellona, and G. Amy, "Effect of membrane fouling on transport of organic contaminants in NF/RO membrane applications", *J. Membr. Sci.,* vol. 279, no. 1-2, pp. 165-175, 2006.
[http://dx.doi.org/10.1016/j.memsci.2005.12.001]

[120] H. Ivnitsky, I. Katz, D. Minz, G. Volvovic, E. Shimoni, E. Kesselman, R. Semiat, and C.G. Dosoretz, "Bacterial community composition and structure of biofilms developing on nanofiltration membranes applied to wastewater treatment", *Water Res.,* vol. 41, no. 17, pp. 3924-3935, 2007.
[http://dx.doi.org/10.1016/j.watres.2007.05.021] [PMID: 17585989]

[121] H. Huang, K. Schwab, and J.G. Jacangelo, "Pretreatment for low pressure membranes in water treatment: a review", *Environ. Sci. Technol.,* vol. 43, no. 9, pp. 3011-3019, 2009.
[http://dx.doi.org/10.1021/es802473r] [PMID: 19534107]

[122] M.P. del Pino, and B. Durham, "Wastewater reuse through dual-membrane processes: opportunities for sustainable water resources", *Desalination,* vol. 124, no. 1-3, pp. 271-277, 1999.
[http://dx.doi.org/10.1016/S0011-9164(99)00112-5]

[123] T. Asano, F. Burton, and H. Leverenz, *Water reuse: issues, technologies, and applications.* McGraw-Hill Education, 2007.

CHAPTER 11

Osmotic and Filtration Processes for the Removal of Emerging Water Pollutants

Kamran Manzoor[1] and **Sher Jamal Khan**[1,*]

[1] *Institute of Environmental Sciences and Engineering (IESE), School of Civil and Environmental Engineering (SCEE), National University of Sciences and Technology (NUST), Sector H-12, Islamabad, Pakistan*

Abstract: Emerging pollutants (EPs) in water and wastewater are one of the global water quality challenges and have substantially adverse and serious effects on ecosystems and human health. However, the presence of these EP's is generally in minute quantities ranging from microgram per liter to nanogram per liter in the environment. These emerging water pollutants may contain endocrine-disrupting compounds (EDCs), personal care and pharmaceutical products, surfactants, hormones, steroids, *etc*. EPs can also be generated from the synthesis of new chemicals and their by-products in industries. Considering the potential impact of these EPs, an appropriate and effective wastewater treatment approach is needed, which can remove the wide variety of these EPs. Membrane technologies have gained more attention in water filtration processes as membrane technology can remove the emerging water and wastewater pollutants with different membranes. The presence of the membrane barrier is one of the main advantages of the membrane filtration process, which offers a wide variety of supplementary adsorption mechanisms for EPs. The pressure-driven membrane filtration processes include micro-filtration (MF), nano-filtration (NF), ultra-filtration (UF), and reverse osmosis (RO). In contrast, the osmotically driven membrane filtration processes (ODMFP) include pressure retarded osmosis (PRO) and forward osmosis (FO) only. This chapter will review the major characteristics, advancements, and principles of NF, RO, ODMFP, and other emerging membrane filtration technology for treating EPs in water and wastewater.

Keywords: Emerging pollutants, Filtration processes, Forward osmosis, Osmotically driven membranes, Pressure-driven membranes.

INTRODUCTION

Due to the innovative analytical techniques and advanced analytical instrumentations in present days, many chemicals with very few concentrations can be

* **Corresponding author Sher Jamal Khan:** Institute of Environmental Sciences and Engineering (IESE), School of Civil and Environmental Engineering (SCEE), National University of Sciences and Technology (NUST), Sector H-12, Islamabad, Pakistan; Tel: +92-51-90854353; E-mails: s.jamal@iese.nust.edu.pk and sherjamal77@gmail.com

Shaukat Ali Mazari, Nabisab Mujawar Mubarak & Nizamuddin Sabzoi (Eds.)

detected within nature and drinking waters. These chemicals are called emerging pollutants (EPs) which may not have instantaneous toxic effects on humans, although these are present in natural and drinking water bodies at low concentrations. Hence, it is very important to find out the attention of these EPs in water reservoirs and natural sources to avoid greater risks to human health, the economy, and the environment. At very low concentrations of some emerging pollutants, their direct adverse effects on human health and ecology have not been observed. Still, long-term accumulation can be a public concern that may affect human health and ecology [1]. The major pathways of many EPs from different sources in the aquatic environment are industrial, agricultural, hospital waste, and municipal wastewater, as shown in Fig. (**1**).

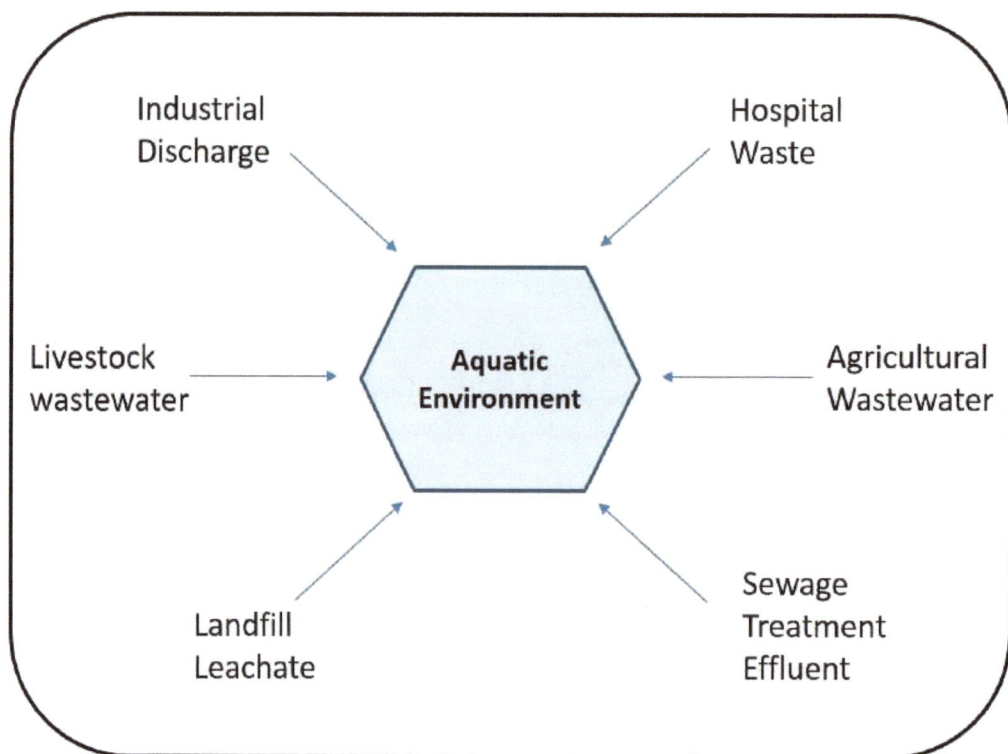

Fig. (1). Pathways of emerging pollutants from different sources in the aquatic environment.

Depending upon the use, properties, and origin, EPs may be classified into different groups, normally comprising endocrine-disrupting compounds (EDCs), personal care and pharmaceutical products, nanoparticles, surfactants, hormones, steroids, *etc*. The examples of some major EPs present in an aquatic environment are mentioned in Table **1**. EPs can also be generated in industries during the synthesis of new chemicals and their by-products [2]. Only some EPs have acidic and basic functional groups, while most are polar compared to the traditional contaminants. The analytical finding and removal methods can create unique challenges coupled with the occurrence at trace levels (*i.e.*, in very small quantities from microgram per liter to nanogram per liter in the environment). The source of these EPs can be found in the manufacturing plants. But their paths of exposure to the environment after discharging from the manufacturing plants become less understood, more complex, and vary between rural and urban environments. Most EPs are presently not regulated. Hence, regular monitoring of their possible presence in the effluent discharge and water supplies is required [2].

Table 1. Some major emerging pollutants present in an aquatic environment (Adopted from [2]).

Emerging Pollutant Class	Examples
Personal Care Items	Poly-siloxane Glaxolides
Hormones	Estrone Endocrine
Pharmaceutical	Caffeine Ibuprofen
Livestock and human antibiotics	Trimethoprim Lincomycin

Due to the rapid population growth, change in climate, and high consumption of water, experts are proposing and implementing water recycling methods to meet the increasing demand. The fate of EPs in water reservoirs and water recycling intensity is an important concern according to the use of EPs. It is also remarkable that the guidelines for water recycling are more stringent than drinking water. Environmental/chemical engineers and scientists must understand the removal mechanisms of EPs, and the design of more effective and explicit water and wastewater treatment processes must be understood by environmental/chemical engineers scientists [3, 4].

Membrane Filtration Technologies

A membrane can be defined as a thin layer/selective barrier of semi-permeable matter which can separate solute in a solution. Only those materials can pass through the pores of the membranes that have smaller pore sizes than the pores of the membrane while stopping others. Membrane filtration is generally used to remove microorganisms, natural organic materials, and emerging pollutants (Fig. **2**). Conventional treatment systems, including activated sludge process, sand filtration, trickling filtration, and coagulation/flocculation, are mostly ineffective and not considered to remove EPs. Thus, in osmotically driven membrane filtration processes (ODMFP), including forwarding osmosis (FO), while in pressure caused membrane filtration processes, including reverse osmosis (RO), nano-filtration (NF), ultra-filtration (UF), and micro-filtration (MF) are needed for the removal of emerging pollutants and water recycling applications [5 - 8].

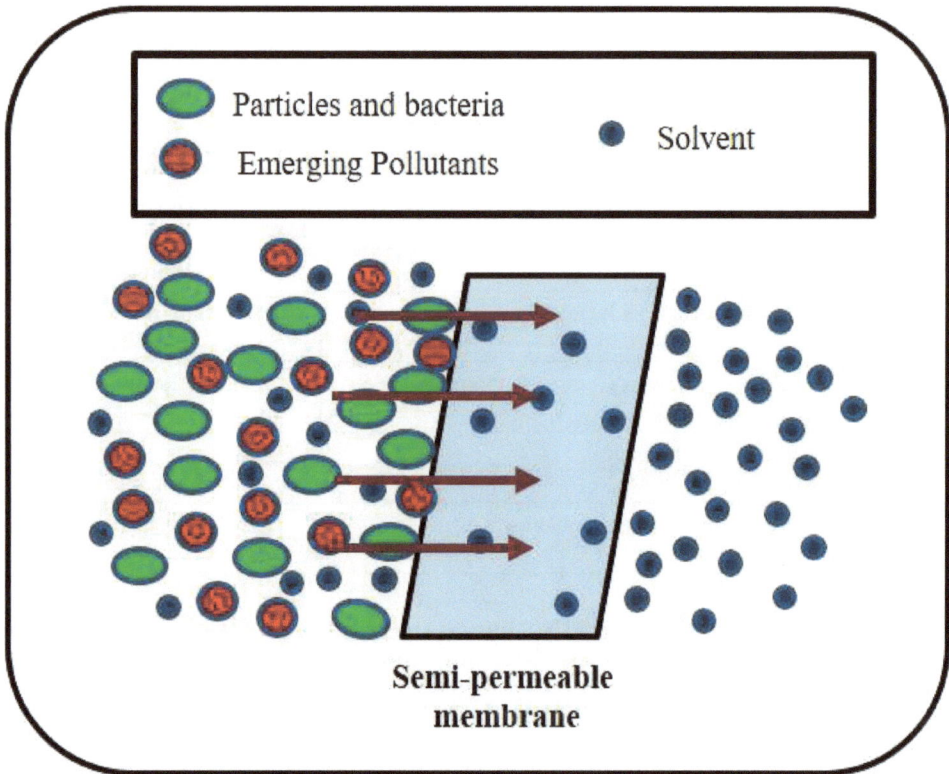

Fig. (2). Schematic diagram of a semi-permeable membrane.

Classification and Retention Mechanism of Membrane Filtration

The classification of the membrane depends upon its nature, possibilities of use, structure, and transfer mechanism. Membranes can be found in porous structures with different pore sizes and non-porous structures, liquid or solid, asymmetric or symmetric, and heterogeneous or homogenous. Size is the main parameter in a porous membrane that determines the efficiency and effectiveness of the membrane. Membranes are manufactured in hollow fiber, flat sheet, and tubular form and can further be classified according to the membrane porosity, separation, geometry, driving force, and fabrication material. Depending upon the pressure, distinct diffusivity and solubility of EPs constituents in solution allow size separation of ions and molecules [9]. Transport through a membrane can be analyzed in three different ways, *i.e.*, dense, ionic charge, and porous, and is conditioned both by the chemical and physical properties of substances that pass through the membrane and the structure and nature of membranes. The transport of some compounds is much quicker than others by selecting a membrane for the separation process and presenting a unique profile of weaknesses and strengths.

Characteristics of the Membrane Filtration Processes

Membrane filtration processes permit the elimination of pollutants at lower concentrations like EPs. They also separate the contaminants, either dispersed or dissolved in colloidal form. They can also be combined with other treatment processes for improved water quality and the removal of emerging contaminants. In the trans-membrane passage of compounds, the efficiency of the membrane filtration processes and morphology and surface chemistry of the membranes play a significant role. Surface modification of membranes is a promising approach as they can provide different chemical functionalities and energies. Commonly, the main purpose of modifying the surface is to increase the selectivity and enhance the chemical resistance (fouling resistance or swelling and solvent resistance) to control the pore size.

Osmotically Driven Membrane Filtration Processes

The difference in osmotic pressure in membrane filtration processes is important between the bulk solution and the permeate side. Thus, this section will briefly discuss the concept of osmosis, osmotic pressure difference, and osmotically driven membrane filtration processes (ODMFP).

Osmosis and Osmotic Pressure Difference

The terms osmotic pressure and osmosis are shown in Fig. (**3**). At the start, two different concentration solutions are separated using a semi-permeable membrane that only allows the solvent's transportation and not the solute. The concentration of solute C1 is greater than the concentration of solute C2. But after the time 't', the solvent starts to flow from the lower concentrated solute side to the higher focused solute through the membrane, as shown by 't_1'. In that way, the more concentrated solute solution tends to dilute by the solvent flow. This flow of solvent is called osmosis. Because of this osmosis, the liquid level on the concentrated solute side increases, resulting in the buildup of hydrostatic pressure difference. Ultimately, the hydrostatic pressure difference (concentrated solute side liquid level) reaches a certain level and attains the equilibrium after t_2. At equilibrium, the hydrostatic pressure difference between the two solutions is called osmotic pressure. If hydrostatic pressure becomes identical to the osmotic pressure initially applied to the solution side C_1 at time t = 0, the flow of osmosis is prevented [12].

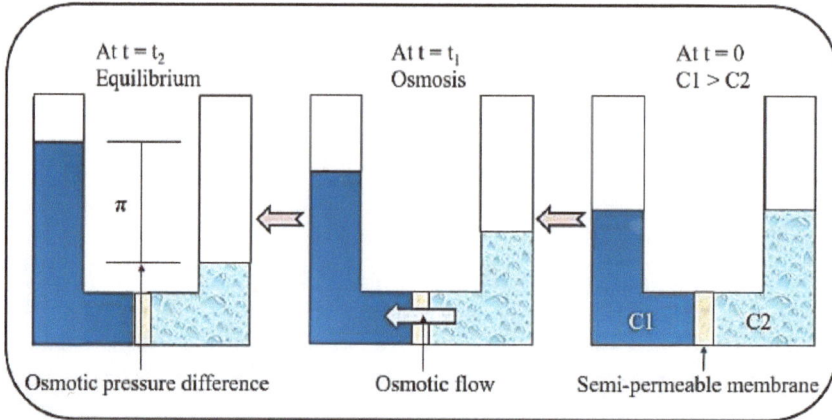

Fig. (3). Osmosis and Osmotic Pressure Description (Adopted from [12]).

Forward osmosis (FO) is an osmotically driven membrane process that requires no hydrostatic pressure. In this process, the two solutions of different concentrations are separated by a semi-permeable membrane that only allows the solvent's transportation. An osmotic pressure gradient is a driving force for the effective separation of the contaminants by artificially creating a high concentrated draw solution across the membrane that poses (induces) the flow of water molecules from the feed solution into the focused draw solution side.

Various factors affect the osmotically driven membrane filtration processes. One is the difference in osmotic pressure in FO membrane filtration processes between the bulk solution and the permeate side [13]. Other important factors affecting the osmotically driven membrane filtration processes are shown in Fig. (4).

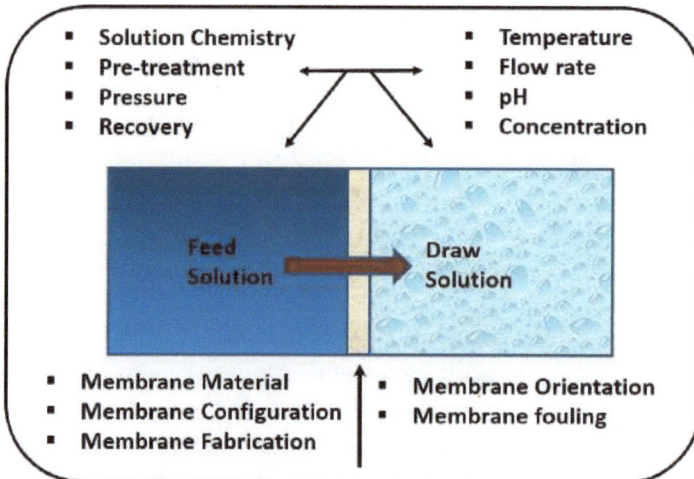

Fig. (4). Factors affecting osmotically driven membrane filtration processes (Adopted from [16]).

Pressure retarded osmosis (PRO) is when the solvent is separated from an additional concentrated solution, but the osmotic pressure difference becomes higher under the applied hydraulic pressure. By using a semi-permeable membrane, the solvent allows the passing of the concentrated solution side by osmosis. PRO process possesses intermediate characteristics lying between RO and FO processes. Water diffuses from a feed solution (FS) of low osmotic pressure into a pressurized draw solution (DS) of high osmotic pressure through the membrane. The difference in osmotic pressure between the DS and FS should be greater than the hydrostatic pressure on the DS side to transport the water. The PRO technique can apply electric power generation from the salinity gradient energy by depressurizing the diluted seawater through a generator or hydro turbine.

Pressure-assisted forward osmosis (AFO) is a process in which pressure is applied at the feed side to increase the water flux and hence further improve the FO performance. In an AFO process, the medium pressure pump is installed in a conventional FO process. The advantage of this method is the additional hydraulic pressure applied to a system that results in water transport in both mechanisms, *i.e.*, by the osmotic pressure (FO technique) and flux driven by hydraulic pressure (RO technique).

The direction of permeate water (flux) in the AFO, FO, PRO, and RO processes is described in Fig. (**5**).

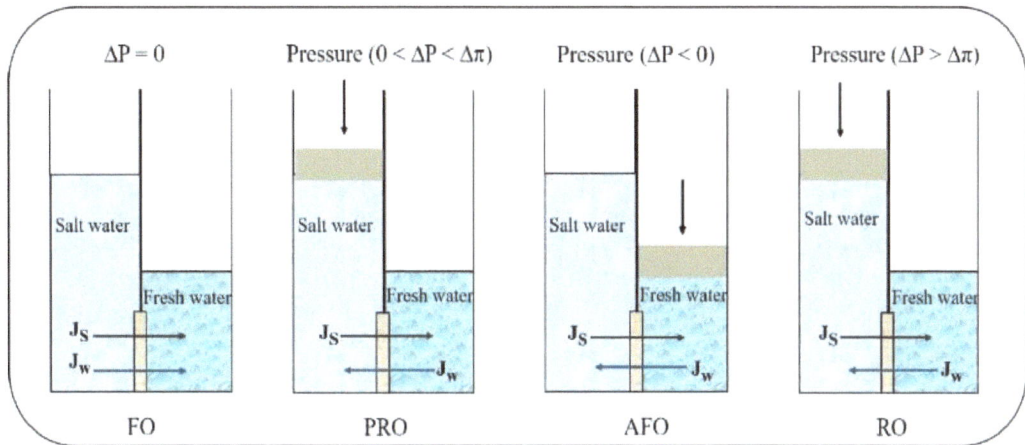

Fig. (5). Osmotic-driven membrane filtration processes. $\Delta\pi$ is the osmotic pressure difference between the solutions; ΔP is hydraulic pressure applied; J_s is reverse salt flux; J_w is water flux (Adopted from [17]).

Removal of EPs by Forwarding Osmosis and Pressure Retarded Osmosis

A specific emphasis was made on the forward osmosis process for EPs which can potentially be a major platform for the next generation of water and wastewater treatment technologies. For wastewater treatment, FO has been extensively investigated in recent years. Compared with the other membrane filtration technologies, the main advantages of FO are the removal of various EPs from wastewater streams to produce high-quality permeate in the presence of osmotic driving force without any hydraulic pressure difference [18, 19]. FO has a less fouling tendency compared to other membrane filtration processes. For downstream processes like thermal distillation and RO, FO can also be used as an advanced pretreatment [20].

FO provides a sustainable water treatment by operating at zero hydrostatic pressure. Still, the process (except for seawater desalination) has not yet achieved commercial success, perhaps due to a lack of suitable FO filtration membranes. For the efficient FO process, the essential components include (a) an effective draw solution (DS), (b) an efficient DS recovery process, and (c) the membranes which are less prone to vulnerability to the internal concentration of membrane [5]. On a bench and pilot scale, the rejection of EPs using the FO-RO hybrid process was investigated by Hancock *et al.* [21] and compared its efficiency with the FO system. They studied the various EPs such as positively and negatively charged, hydrophobic and non-ionic species on both systems. For bisphenol-A, RO showed a better rate of rejection than the FO. In contrast, for triclosan pesticides, oxybenzone, methylparaben, and amitriptyline, the FO showed an improved rejection rate greater than 99% than RO. Thus, the result revealed that by increasing the molecular weight of EPs, non-ionic rejection compounds were also significantly increased.

Xie *et al.* [22] first demonstrated the retarded forward diffusion (*i.e.*, the bi-directional solute transport in the FO process). They investigated the impact of DS diffusion on EPs rejection. From the DS to the feed, forward diffusion of EPs could be hindered by the reverse diffusion of solutes by examining various DSs. Hence, EPs rejection can be enhanced by a high reverse salt flux. In another study, Alturki *et al.* [23] performed the rejection of various types of EPs by cellulose triacetate (CTA) membranes. Due to their asymmetric nature, these membranes were used either in the pressure retarded osmosis (PRO) mode or FO mode. They also examined that a higher water flux can be produced using the PRO mode, while the EPs rejection rate was notably lower than in the FO mode.

A novel method was investigated by combining the FO process with activated sludge (hybrid system) landfill leachate treatment containing various EPs for pilot-scale and bench-scale applications. The major advantage of this approach compared to other methods is that a higher rejection was attained. High rejection is due to the enhanced retention time of small EPs by the FO process in a biological reactor [5]. Furthermore, bench-scale short-term studies specified that the osmotic membrane bioreactor (OMBR) could be a better alternative to produce high-quality water [24].

Removal of EPs by Osmotic Membrane Bioreactor

The membrane bioreactor process is an integrated membrane separation and activated sludge process, widely used for over 50 years. Hollow fiber or flat sheet UF or MF membranes are commonly used for MBR processes [25]. The advantages of MBR were reviewed by Besha *et al.* [26] and compared with the activated sludge process. They demonstrated that the MBR could produce better quality effluent, a high solid retention time (SRT), decreased reactor volume, elevated mixed liquid suspended solids (MLSS), easier operation, and low sludge yield. However, some limitations are still present in this process, including high energy consumption, membrane fouling, and low removal efficacy of poorly biodegradable EPs like carbamazepine, atrazine, and diclofenac.

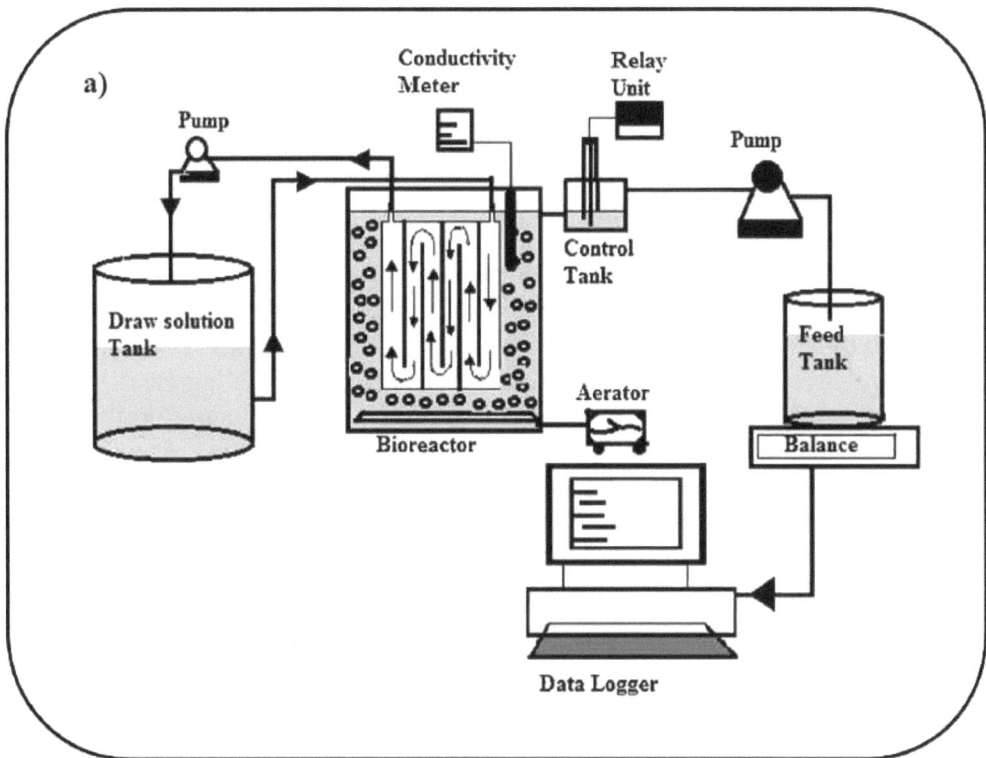

(Fig. 6) contd.....

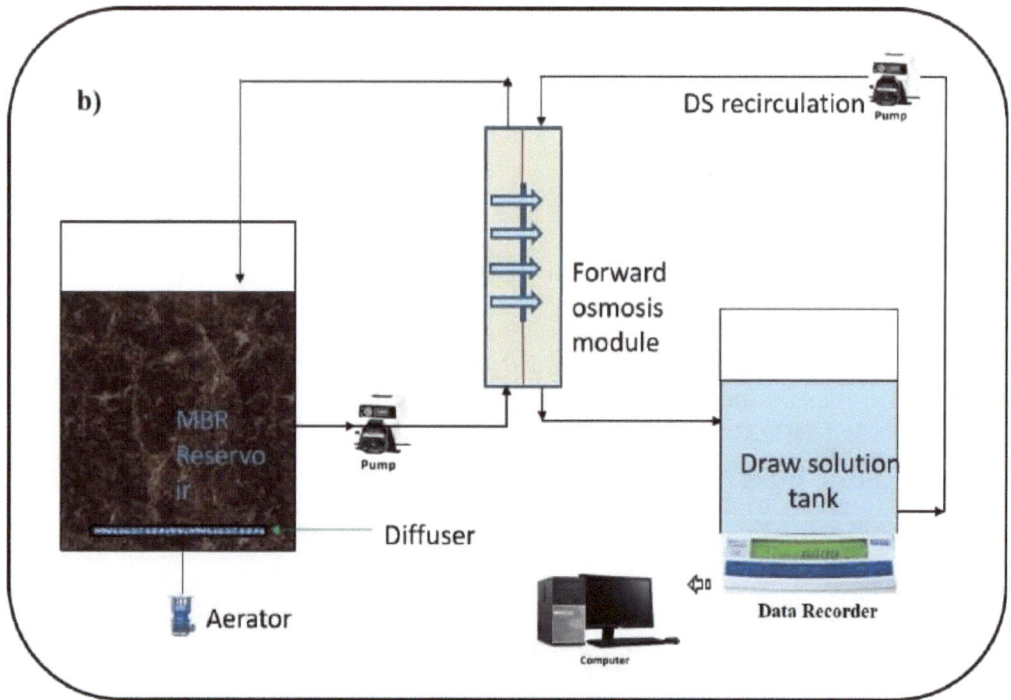

Fig. (6). Schematic of OMBR **a**) Submerged OMBR (adopted from [30]) **b**) Sidestream OMBR.

In a novel osmotic membrane bioreactor (OMBR), MF or UF membranes are replaced by FO membranes with either cellulose triacetate (CTA) membrane or a new generation thin-film composite (TFC) membrane in MBR to encourage wastewater treatment and reuse (Fig. **6**). Water from the feed side in a biological reactor is forced to transfer to the DS side under osmotic pressure *via* a semi-permeable membrane. Hence, activated sludge, emerging pollutants, and solids are rejected by the membrane. Challenges in the OMBR system are due to salinity buildup, which may impact bacterial activity and flux. Fouling in FO and MF modules was the same, but the advantage of osmotic backwashing in the OMBR proved to be an effective chemical-free cleaning solution [33]. Lay *et al.* [34] investigated the removal of pharmaceutical compounds over an experimental duration of 73 days in an OMBR system and removed greater than 96% of pharmaceutical compounds. They demonstrated that the OMBR system mostly attained excellent water quality permeate due to the retention property of the FO membrane. The pharmaceutical components permitted these compounds to be concentrated in the bioreactor due to a high rejection which may enhance biodegradation. In the overall OMBR performance, the biological reactor played a key role in achieving a high product water quality and other process parameters.

Alturki *et al*. [35] reported the potential and challenges in the OMBR process for removing EPs. They investigated the removal of 50 EPs and found that the removal of 80% of 25 out of the 27 EPs was achieved with a range of molecular weight greater than 266 g/mol. At the same time, the removal of the remaining 23 EP compounds was scattered. They also found that stable flux was obtained on a continuous operation after four days. Still, deterioration in the biological reactor may be due to the salinity buildup in the bioreactor. In a recent study, Raghavan *et al*. [36] studied the removal of 12 antibiotics of class 5 in OMBR and their effect on the system's performance. They found that a high antibiotic removal achieved 77.7 to 99.8%, with a high rejection rate of greater than 90% by the FO membrane. Each antibiotic with a concentration of 500 ng/L had negligible effects on the overall OMBR performance. All antibiotics except for roxithromycin and ciprofloxacin showed significant biodegradation removal in a bioreactor. To remove different classes of antibiotics, OMBR was found to be an effective process. Another study investigated the OMBR performance using a novel biomimetic aquaporin forward osmosis membrane. In this study, 30 emerging pollutants were removed, and membrane performance was examined. In OMBR operation, aquaporin forward osmosis membrane showed excellent contaminant removal and much lower salt permeability and thus lower reverse salt flux, resulting in a lower severe salinity buildup in the bioreactor. OMBR removed all 30 EPs over 85% regardless of their various properties. This stable and high removal of contaminants of emerging pollutants in the OMBR process indicated the compatibility and stability of the aquaporin forward osmosis membrane with an activated sludge treatment combination [37].

Pressure Driven Membrane Filtration Processes

Removal of EPs by Reverse Osmosis

Reverse osmosis is a pressure-driven process that only allows water molecules to permeate through semi-permeable membranes and has the potential to remove smaller particles and dissolved solids (*e.g*., pesticides, viruses, bacteria, organic and inorganic substances, and cysts) that are less than 0.1 nm. The permeate flow ratio and applied pressure range from 10 to 100 LMH and 2000 to 7000 kPa [38]. Some limitations of the RO process are faced, which include high operating and capital costs due to the high pressure, prone to fouling, and the requirement of a high level of pretreatment [39]. By using the low-pressure RO membranes, the retention effectiveness of various personal care products, pharmaceuticals, and disruptive endocrine chemicals was examined by several methods such as diffusion, simultaneous adsorption, and size exclusion, and it was found that the size exclusion method was relatively dominant as compared to other applied methods [40]. Abdelmelek *et al*. [41] evaluated the removal of pharmaceutical compounds by RO, and then retentates of RO were further treated by the advanced oxidation process (AOP). Treatment results of the AOP revealed that the combination process (*i.e.,* hybrid) could remove the pharmaceutical components efficiently even in both inorganic and organic compounds present in the influent.

Al-Rifai *et al.* [3] have made research endeavors to evaluate the evacuation efficiencies of different micro- EPs in wastewater recycling processes under various processing conditions. Evacuation efficiencies of the eleven other pharmaceuticals and two disruptive endocrine chemicals were inspected by RO and MF systems. The BPA components in wastewater effluent (6 - 23 ug/L) were in less quantity, while salicylic acid (11 - 38 ug/L) was found in abundance. Additionally, the decrease in concentration for all the EPs from primary to secondary treatment was observed drastically. Greater than 97% retention was acquired in recycled water (except BPA), resulting in less than 0.1 ug/L of concentration in the product water. Moreover, greater than 0.5 ug/L concentration was observed for BPA in product water, and its presence shows serious concern.

The combination of RO and MBR processes was employed for the secondary and raw sewage effluent treatments [42]. The overall retention rate of 20 pharmaceutical components was more than 90% in an RO-MBR process [43], while alone RO showed a very efficient rate of retention for various micro-pollutants like metoprolol, atenolol, erythromycin, and clarithromycin under the detection limit of ≤ 10 ng/L [44], while for anti-inflammatories, psychiatric control, and antibiotics greater than 90% [45], metoprolol, sulfamethoxazole, and sotalol greater than 98% [46], and carbamazepine greater than 99% [47]. The key retention mechanisms of EPs in RO membranes contain hydrophobic interaction (adsorption), electrostatic interactions (repulsion), and steric hindrance between the membrane and EPs [48]. Also, relatively hydrophilic pharmaceutical components have a high retention rate of> 99%. At the same time, no adsorption can occur on hydrophilic compounds on the polymeric membrane matrix [49] as MWCO of RO membrane (TR70-4021-HF) is about 100 Da, a potential removal mechanism involved in size exclusion (steric hindrance). Further retention of some pharmaceutical compounds may be affected by the electrostatic interaction (repulsion or attraction) in an RO membrane due to their charge, which includes a negative charge of SMX and the positive charge of macrolide antibiotics [43].

Removal of EPs by Nano-filtration

NF has the properties between reverse osmosis and ultrafiltration, which can remove those ions contributing at a lower pressure than RO and having a pore size of 0.001 μm and molecular weight cut-off (MWCO) from 200 to 1000 Daltons

(Da). It has been used in various interesting applications such as water and wastewater treatment processes. An effective pretreatment is needed for NF to remove some heavily polluted water [50, 51]. A combination of three mechanisms, such as sieving, electrostatic repulsion, and adsorption, are responsible for separating EPs compounds through NF. The separation mechanism of the NF membrane can vary from compound to compound and is strictly correlated to EPs, as shown in Table 2. For the separation of concerned emerging water pollutants from waste streams, NF has been proven to be a promising alternative and can achieve a>90% of removal rate.

Using the two different commercial polyamide NF membranes (NF-270 and NF-90), López-Muñoz *et al.* [54] studied the removal of various pharmaceutics EPs such as ranitidine hydrochloride, nicotine, diclofenac sodium, sulfamethoxazole, and 4-acetamidoantipyrine. Solute retention through the NF-270 membrane appeared to have a low range value from 75% (nicotine) to 95% (ranitidine hydrochloride), while NF-90 showed greater than 95%. This study further found that the EPs rejection rate and permeate flux were dependent on the EPs, transmembrane pressure, and physicochemical properties of the membrane.

Ruíz *et al.* [55] performed a case study by using four different nanofiltration membranes (NF-90 MWCO of 200Da, NF- 270 MWCO of 200Da, 302986 MWCO of 200Da, and TFC-SR3 MWCO of Da) and compared them to remove various EPs including bisphenol-A, 4-nonylphenol, triclosan, butyl-benzy--phthalate, and carbamazepine from urban and agricultural wastewater. Only one membrane NF-270 was possibly selected after applying the critical flux concept for agricultural wastewater with the operating condition of 800 kPa pressure and 87.3 LMH flux. Water solubility decreased the retention of bisphenol-A while removal of hydrophobic compounds of EPs was increased due to adsorption on the membrane surface. To reduce the permeability loss, a pretreatment step with 150% of lime was also employed in this study to control the hardness scaling.

Table 2. Separation Mechanism of Nanofiltration membrane for the removal of EPs.

Properties	Mechanism
Membrane properties	Pore size Permeability Surface charge Hydrophobicity
Physico-chemical properties	Solubility Molecular size Hydrophobicity Polarity

(Table 2) cont.....

Properties	Mechanism
Membrane operating conditions	Trans-membrane pressure Flux Recovery/rejection Feed quality

Using the NF membranes system, an investigation has been done for the removal patterns of four EPs (natural hormonal steroids), including progesterone, estrone, estradiol, and testosterone. Two NF membranes were used with different characteristics, *i.e.*, salt rejection and permeability. Results demonstrated that adsorption of EPs on the membrane surface is the key removal mechanism at the initial phase of the filtration process. Due to the limited adsorption capacity of the membrane, the final retention was stabilized when the adsorption of hormones achieved equilibrium on the membrane surface. In the later filtration phase, the overall retention of EPs was lower than expected based on size-exclusion phenomena. This behavior was attributed to adsorption (partitioning) and the diffusion of hormonal molecules in the polymeric membrane phase, which eventually resulted in lower retention [56].

Removal of EPs by Ultra-filtration

UF is also a pressure-driven filtration process in which compounds having a molecular size between 0.005 to 0.1 μm are separated [57]. With low energy consumption, UF membranes are highly remarkable water filters in removing suspended matters, macromolecules, and pathogenic microorganisms compared to others [58]. But still, UF offers some disadvantages, including its inefficacy in eliminating dissolved substances (inorganic) from water and sustaining a high-pressure water flow with regular cleaning [59]. Contrary to the MF process, UF has a better solute rejection system for separating macro-EPs. Considering the advantage mentioned above, the success of the UF process has been demonstrated in treating EPs from any wastewater, as many EPs fall under the macro-molecules range category [60]. Nevertheless, UF may not generate ultrapure water with other organics in wastewater to remove macro-EPs. In this situation, UF can be combined with the NF commercial membranes to achieve the anticipated levels of water purity by eliminating EPs secondary effluents [61].

Benitez *et al.* [62] used the UF-NF hybrid system to separate four pharmaceutical contaminants from the secondary effluents such as naproxen, phenacetin, amoxicillin, and metoprolol. This study revealed that nano-filtration accomplished the maximum retention of 80% except for phenacetin. While treating the secondary effluent, it was observed that the membrane fouling and decline in flux were higher for the UF. The removal of several EPs containing various pesticides and pharmaceutical compounds present in the municipal secondary effluents was studied by Acero *et al.* [61]. Separation was attained on the membrane surface mostly through the adsorption of pollutants. For the UF membranes, adsorption was the basic mechanism for the retention of micropollutants. The electrostatic repulsion and the size exclusion at elevated pH were considerably prominent for the NF membranes. Thus, the mutual NF and UF hybrid system is essential for treating secondary effluent.

To separate the EPs from domestic and industrial wastewater, micellar compounds with UF have been cost-effective. Various micelles comprising Triton X 100, cetyltrimethylammonium bromide, sodium dodecyl sulfate, cetylpyridinium chloride, and tween-20 enhanced the performance of UF for the removal of pharmaceutical contaminants, including isoproturon, caffeine, metoprolol, antipyrine, flumequine, sulfamethoxazole, ketorolac, and diclofenac. Cationic micelles like cetylpyridinium chloride and cetyltrimethylammonium bromide removed the hydrophilic and negatively charged pharmaceutical contaminants. Furthermore, cetylpyridinium chloride showed up to 95% separation among all the micellar [63].

Removal of EPs by Micro-filtration

MF is a pressure-driven filtration process in which compounds having a molecular size of 0.1 to 0.4 μm (like microparticles) are separated [64]. MF filters can also be used as pretreatment for reverse osmosis and nano-filtration to minimize their fouling potential [65]. The main disadvantage of MF is that the emerging pollutants cannot be eliminated directly. MF with other processes has been commonly employed to remove complex water, wastewater, and emerging industrial pollutants. TiO_2-based composite membranes removed pharmaceutical emerging contaminants such as diclofenac, methylene blue, and ibuprofen with polymers of polyvinylidene difluoride using titanium tetra-isopropoxide [66]. The MF-RO hybrid system removed about 98% of highly concentrated pharmaceutical emerging pollutants such as erythromycin, atenolol, carbamazepine, and pesticides from the wastewater treatment plant [67]. In another study, Ragab *et al.* [68] investigated the hybrid MF systems to separate the different EPs types. A spiral-wound arrangement was employed, having the zeolite imidazolate metal-organic frame nanoparticles combined with a polytetrafluorethylene double-layer polymer membrane. They showed that the rejection was 95% by giving high water flux for the EPs even at the low operating pressures. Such an approach is good for separating hormonal EPs from wastewater.

Endocrine disrupters such as 17-beta-estradiol, ethynylestradiol, bisphenol-A, and estrone are present in domestic wastewater, and their presence, even in very small quantities, can seriously damage the human endocrine [69, 70]. Han *et al.* [71] tried and employed the laboratory scale crossflow micro-filtration system to solve this above-stated problem. They used a series of membranes prepared from cellulose acetate, polyamide-66, regenerated cellulose, polyethersulfone, and nitrocellulose. They observed that for polyamide-66 membrane having a pore size of 0.2 μm resulted in sorption capacity of 0.44 μg/cm² (81 L/m²) for estrone, 0.82 μg/cm² (150 L/m²) for 17-beta-estradiol, 1.23 μg/cm² (208 L/m²) for ethynylestradiol, and 0.32 μg/cm² (69 L/cm²) for bisphenol-A. The experiment showed that the membrane performance was severely affected by adsorption on the surface of some EPs at elevated concentrations and can produce membrane fouling due to the presence of organic matter in the feed. With the polyamide membrane filter through hydrogen bonding, the overall EPs exhibited a steady interaction indicating the efficient removal from the feed.

CONCLUSION

This book chapter discusses removal processes using osmotically and pressure-based membrane filtration for emerging pollutants. It was verified that effective removal was observed for the emerging contaminants from water and wastewater streams using different membrane filtration technologies. However, the main challenges for membrane filtration technology are salt diffusion and membrane fouling in FO and OMBR, toxicity, membrane sensitivity, and membrane fouling in pressure-driven membranes. Therefore, researchers and scientists are developing new membrane types and membrane modification methods that can be more effective for high-quality products and process limitations. Overall, it is well developed that membrane filtration processes can be a promising solution for emerging pollutants. Performances of various technologies used in water and wastewater treatment plants are summarized in Table **3** based on the examined specific EPs literature review.

Table 3. Unit processes and operations used for EPs removal (Adopted from [72]).

Group	Classification	FO	RO	NF	UF	MF
Endocrine disruptive compounds	Industrial chemicals	F-E	E	E	P-F	P
	Pesticides	F-E	E	G	P-F	P
	Steroids	F-E	E	G	P-F	P
	Metals	F-E	E	G	P-F	P
	Inorganics	F-E	E	G	P-F	P
Pharmaceutical compounds	Antibiotics	F-E	E	E	P-F	P
	Anti-depressant	F-E	E	G-E	P-F	P
	Anti-inflammatories	F-E	E	G-E	P-F	P
	Xray-contrast media	F-E	E	G-E	P-F	P
	Psychiatric control	F-E	E	G-E	P-F	P

(Table 3) cont.....

Group	Classification	FO	RO	NF	UF	MF
Personal care products	Surfactants/detergents	F-E	E	E	P-F	P
	Antimicrobial	F-E	E	G-E	P-F	P
	Sunscreens	F-E	E	G-E	P-F	P
	Synthetic scents	F-E	E	G-E	P-F	P

E = Excellent (> 90%), G = Good (70 - 90%), F = Fair (40 – 70%), L = Low (20 – 40%), P = Poor (< 20%)

CONSENT FOR PUBLICATION

Not applicable.

CONFLICT OF INTEREST

The author declares no conflict of interest, financial or otherwise.

ACKNOWLEDGEMENTS

Declared none.

REFERENCES

[1]   Y. Ma, "Assessment and Removal of Emerging Water Contaminants", *J. Environ. Anal. Toxicol.,* vol. 02, no. 07, 2012.
[http://dx.doi.org/10.4172/2161-0525.S2-003]

[2]   J.Q. Jiang, Z. Zhou, and V.K. Sharma, "Occurrence, transportation, monitoring and treatment of emerging micro-pollutants in waste water - A review from global views", *Microchem. J.,* vol. 110, pp. 292-300, 2013.
[http://dx.doi.org/10.1016/j.microc.2013.04.014]

[3]   J.H. Al-Rifai, H. Khabbaz, and A.I. Schäfer, "Removal of pharmaceuticals and endocrine disrupting compounds in a water recycling process using reverse osmosis systems", *Separ. Purif. Tech.,* vol. 77, no. 1, pp. 60-67, 2011.
[http://dx.doi.org/10.1016/j.seppur.2010.11.020]

[4]   C.M. Park, "Occurrence and Removal of Engineered Nanoparticles in Drinking Water Treatment and Wastewater Treatment Processes", *Separ. Purif. Rev.,* vol. 46, no. 3, pp. 255-272, 2017.
[http://dx.doi.org/10.1080/15422119.2016.1260588]

[5]   S.P. Dharupaneedi, S.K. Nataraj, M. Nadagouda, K.R. Reddy, S.S. Shukla, and T.M. Aminabhavi, "Membrane-based separation of potential emerging pollutants", *Sep. Purif. Technol,* vol. 210, pp. 850-866, 2019.
[http://dx.doi.org/10.1016/j.seppur.2018.09.003]

[6]   B.D. Blair, J.P. Crago, C.J. Hedman, R.J. Treguer, C. Magruder, L.S. Royer, and R.D. Klaper, "Evaluation of a model for the removal of pharmaceuticals, personal care products, and hormones from wastewater", *Sci. Total Environ.,* vol. 444, pp. 515-521, 2013.
[http://dx.doi.org/10.1016/j.scitotenv.2012.11.103] [PMID: 23295178]

[7]   T. Kistemann, E. Rind, A. Rechenburg, C. Koch, T. Classen, S. Herbst, I. Wienand, and M. Exner, "A comparison of efficiencies of microbiological pollution removal in six sewage treatment plants with different treatment systems", *Int. J. Hyg. Environ. Health,* vol. 211, no. 5-6, pp. 534-545, 2008.
[http://dx.doi.org/10.1016/j.ijheh.2008.04.003] [PMID: 18565791]

[8]   F. Beyer, "Long-term performance and fouling analysis of full-scale direct nanofiltration (NF)

installations treating anoxic groundwater", *J. Membr. Sci.,* vol. 468, pp. 339-348, 2014.
[http://dx.doi.org/10.1016/j.memsci.2014.06.004]

[9]   M.A.A. El-Ghaffar, and H.A. Tieama, *A Review of Membranes Classifications, Configurations, Surface Modifications, Characteristics and its Applications in Water Purification,* 2017. http://www.Sciencepublishinggroup.com

[10]   M.N.A. Seman, M. Khayet, and N. Hilal, "Nanofiltration thin-film composite polyester polyethersulfone-based membranes prepared by interfacial polymerization", *J. Membr. Sci.,* vol. 348, no. 1–2, pp. 109-116, 2010.
[http://dx.doi.org/10.1016/j.memsci.2009.10.047]

[11]   C.S. Ong, B. Al-Anzi, W.J. Lau, P.S. Goh, G.S. Lai, A.F. Ismail, and Y.S. Ong, "Anti-fouling double-skinned forward osmosis membrane with zwitterionic brush for oily wastewater treatment", *Sci. Rep.,* vol. 7, no. 1, p. 6904, 2017.
[http://dx.doi.org/10.1038/s41598-017-07369-4] [PMID: 28761159]

[12]   T. Lind, and G. Mackay, "Downstream processing", *Nor. Oil Policies,* pp. 62-77, 2017.
[http://dx.doi.org/10.2307/j.ctt1w6tc79.8]

[13]   G. Blandin, P. Le-Clech, E. Cornelissen, A.R.D. Verliefde, J. Comas, and I. Rodriguez-Roda, "Can osmotic membrane bioreactor be a realistic solution for water reuse?", In: *npj Clean Water* vol. 1. , 2018.
[http://dx.doi.org/10.1038/s41545-018-0006-x]

[14]   C.F. Wan, and T.S. Chung, "Osmotic power generation by pressure retarded osmosis using seawater brine as the draw solution and wastewater retentate as the feed", *J. Membr. Sci.,* vol. 479, pp. 148-158, 2015.
[http://dx.doi.org/10.1016/j.memsci.2014.12.036]

[15]   F. Helfer, C. Lemckert, and Y.G. Anissimov, "Osmotic power with Pressure Retarded Osmosis: Theory, performance and trends - A review", *J. Membr. Sci.,* vol. 453, pp. 337-358, 2014.
[http://dx.doi.org/10.1016/j.memsci.2013.10.053]

[16]   W.Y. Chia, "Factors affecting the performance of membrane osmotic processes for bioenergy development", *Energies,* vol. 13, no. 2, 2020.
[http://dx.doi.org/10.3390/en13020481]

[17]   J. Korenak, S. Basu, M. Balakrishnan, C. Hélix-Nielsen, and I. Petrinic, "Forward osmosis in wastewater treatment processes", *Acta Chim. Slov.,* vol. 64, no. 1, pp. 83-94, 2017.
[http://dx.doi.org/10.17344/acsi.2016.2852] [PMID: 28380231]

[18]   G. Blandin, A.R.D. Verliefde, J. Comas, I. Rodriguez-Roda, and P. Le-Clech, "Efficiently combining water reuse and desalination through forward osmosis-reverse osmosis (FO-RO) hybrids: A critical review", *Membranes (Basel),* vol. 6, no. 3, p. E37, 2016.
[http://dx.doi.org/10.3390/membranes6030037] [PMID: 27376337]

[19]   B.D. Coday, B.G.M. Yaffe, P. Xu, and T.Y. Cath, "Rejection of trace organic compounds by forward osmosis membranes: a literature review", *Environ. Sci. Technol.,* vol. 48, no. 7, pp. 3612-3624, 2014.
[http://dx.doi.org/10.1021/es4038676] [PMID: 24552278]

[20]   T. Cath, A. Childress, and M. Elimelech, "Forward osmosis: Principles, applications, and recent developments", *J. Membr. Sci.,* vol. 281, no. 1–2, pp. 70-87, 2006.
[http://dx.doi.org/10.1016/j.memsci.2006.05.048]

[21]   N.T. Hancock, P. Xu, D.M. Heil, C. Bellona, and T.Y. Cath, "Comprehensive bench- and pilot-scale investigation of trace organic compounds rejection by forward osmosis", *Environ. Sci. Technol.,* vol. 45, no. 19, pp. 8483-8490, 2011.
[http://dx.doi.org/10.1021/es201654k] [PMID: 21838294]

[22]   M. Xie, L.D. Nghiem, W.E. Price, and M. Elimelech, "Comparison of the removal of hydrophobic trace organic contaminants by forward osmosis and reverse osmosis", *Water Res.,* vol. 46, no. 8, pp.

2683-2692, 2012.
[http://dx.doi.org/10.1016/j.watres.2012.02.023] [PMID: 22402269]

[23] A.A. Alturki, J.A. McDonald, S.J. Khan, W.E. Price, L.D. Nghiem, and M. Elimelech, "Removal of trace organic contaminants by the forward osmosis process", *Separ. Purif. Tech.*, vol. 103, pp. 258-266, 2013.
[http://dx.doi.org/10.1016/j.seppur.2012.10.036]

[24] J-J. Qin, "Optimization of Operating Conditions in Forward Osmosis for Osmotic Membrane Bioreactor", *Open Chem. Eng. J.*, vol. 3, no. 1, pp. 27-32, 2009.
[http://dx.doi.org/10.2174/1874123100903010027]

[25] L. Li, W. Shi, and S. Yu, "Research on forward osmosis membrane technology still needs improvement in water recovery and wastewater treatment", *Water (Switzerland)*, vol. 12, no. 1, pp. 1-27, 2020.
[http://dx.doi.org/10.3390/w12010107]

[26] A.T. Besha, A.Y. Gebreyohannes, R.A. Tufa, D.N. Bekele, E. Curcio, and L. Giorno, "Removal of emerging micropollutants by activated sludge process and membrane bioreactors and the effects of micropollutants on membrane fouling: A review", *J. Environ. Chem. Eng.*, vol. 5, no. 3, pp. 2395-2414, 2017.
[http://dx.doi.org/10.1016/j.jece.2017.04.027]

[27] L.G. Shen, Q. Lei, J.R. Chen, H.C. Hong, Y.M. He, and H.J. Lin, "Membrane fouling in a submerged membrane bioreactor: Impacts of floc size", *Chem. Eng. J.*, vol. 269, pp. 328-334, 2015.
[http://dx.doi.org/10.1016/j.cej.2015.02.002]

[28] N. Tadkaew, F.I. Hai, J.A. McDonald, S.J. Khan, and L.D. Nghiem, "Removal of trace organics by MBR treatment: the role of molecular properties", *Water Res.*, vol. 45, no. 8, pp. 2439-2451, 2011.
[http://dx.doi.org/10.1016/j.watres.2011.01.023] [PMID: 21388651]

[29] L. Zhou, W. Zhuang, X. Wang, K. Yu, S. Yang, and S. Xia, "Potential effects of loading nano zero valent iron discharged on membrane fouling in an anoxic/oxic membrane bioreactor", *Water Res.*, vol. 111, pp. 140-146, 2017.
[http://dx.doi.org/10.1016/j.watres.2017.01.007] [PMID: 28068534]

[30] M. Adnan, S.J. Khan, and K. Manzoor, "N. P. %J P. S. Hankins, and E. Protection, "Performance evaluation of fertilizer draw solutions for forward osmosis membrane bioreactor treating domestic wastewater,""",

[31] Y. Wu, "Metagenomic insights into the influence of salinity and cytostatic drugs on the composition and functional genes of microbial community in forward osmosis anaerobic membrane bioreactors", *Chem. Eng. J.*, vol. 326, pp. 462-469, 2017.
[http://dx.doi.org/10.1016/j.cej.2017.05.172]

[32] X. Wang, V.W.C. Chang, and C.Y. Tang, "Osmotic membrane bioreactor (OMBR) technology for wastewater treatment and reclamation: Advances, challenges, and prospects for the future", *J. Membr. Sci.*, vol. 504, pp. 113-132, 2016.
[http://dx.doi.org/10.1016/j.memsci.2016.01.010]

[33] G. Blandin, C. Gautier, M. Sauchelli Toran, H. Monclús, I. Rodriguez-Roda, and J. Comas, "Retrofitting membrane bioreactor (MBR) into osmotic membrane bioreactor (OMBR): A pilot scale study", In: *Chem. Eng. J* vol. 339. , 2018, pp. 268-277.
[http://dx.doi.org/10.1016/j.cej.2018.01.103]

[34] W.C.L. Lay, "Effect of Pharmaceuticals on the Performance of a Novel Osmotic Membrane Bioreactor (OMBR)", *Sep. Sci. Technol.*, vol. 47, no. 4, pp. 543-554, 2012.
[http://dx.doi.org/10.1080/01496395.2011.630249]

[35] A. Alturki, J. McDonald, S.J. Khan, F.I. Hai, W.E. Price, and L.D. Nghiem, "Performance of a novel osmotic membrane bioreactor (OMBR) system: flux stability and removal of trace organics", *Bioresour. Technol.*, vol. 113, pp. 201-206, 2012.

[http://dx.doi.org/10.1016/j.biortech.2012.01.082] [PMID: 22342586]

[36]   D.S. Srinivasa Raghavan, G. Qiu, and Y.P. Ting, "Fate and removal of selected antibiotics in an osmotic membrane bioreactor", In: *Chem. Eng. J* vol. 334. , 2018, pp. 198-204.
[http://dx.doi.org/10.1016/j.cej.2017.10.026]

[37]   W. Luo, "Biomimetic aquaporin membranes for osmotic membrane bioreactors: Membrane performance and contaminant removal", In: *Bioresour. Technol* vol. 249. , 2018, pp. 62-68.
[http://dx.doi.org/10.1016/j.biortech.2017.09.170]

[38]   P. Safety, M. Racar, D. Dolar, and K. Kosutic, *Application of UF / NF / RO membranes for treatment and reuse of rendering plant wastewater Application of UF / NF / RO membranes for treatment and reuse of rendering plant wastewater,* 2016.
[http://dx.doi.org/10.1016/j.psep.2016.11.015]

[39]   G. Liu, "Graphene oxide/triethanolamine modified titanate nanowires as photocatalytic membrane for water treatment", *Chem. Eng. J.,* vol. 320, pp. 74-80, 2017.
[http://dx.doi.org/10.1016/j.cej.2017.03.024]

[40]   H. Ozaki, N. Ikejima, Y. Shimizu, K. Fukami, S. Taniguchi, R. Takanami, R.R. Giri, and S. Matsui, "Rejection of pharmaceuticals and personal care products (PPCPs) and endocrine disrupting chemicals (EDCs) by low pressure reverse osmosis membranes", *Water Sci. Technol.,* vol. 58, no. 1, pp. 73-81, 2008.
[http://dx.doi.org/10.2166/wst.2008.607] [PMID: 18653939]

[41]   S.B. Abdelmelek, J. Greaves, K.P. Ishida, W.J. Cooper, and W. Song, "Removal of pharmaceutical and personal care products from reverse osmosis retentate using advanced oxidation processes", *Environ. Sci. Technol.,* vol. 45, no. 8, pp. 3665-3671, 2011.
[http://dx.doi.org/10.1021/es104287n] [PMID: 21384915]

[42]   E. Dialynas, and E. Diamadopoulos, "Integration of a membrane bioreactor coupled with reverse osmosis for advanced treatment of municipal wastewater", *Desalination,* vol. 238, no. 1–3, pp. 302-311, 2009.
[http://dx.doi.org/10.1016/j.desal.2008.01.046]

[43]   D. Dolar, M. Gros, S. Rodriguez-Mozaz, J. Moreno, J. Comas, I. Rodriguez-Roda, and D. Barceló, "Removal of emerging contaminants from municipal wastewater with an integrated membrane system, MBR-RO", *J. Hazard. Mater.,* vol. 239-240, pp. 64-69, 2012.
[http://dx.doi.org/10.1016/j.jhazmat.2012.03.029] [PMID: 22476093]

[44]   A. Joss, C. Baenninger, P. Foa, S. Koepke, M. Krauss, C.S. McArdell, K. Rottermann, Y. Wei, A. Zapata, and H. Siegrist, "Water reuse: >90% water yield in MBR/RO through concentrate recycling and CO2 addition as scaling control", *Water Res.,* vol. 45, no. 18, pp. 6141-6151, 2011.
[http://dx.doi.org/10.1016/j.watres.2011.09.011] [PMID: 21959090]

[45]   S. Snyder, P. Westerhoff, Y. Yoon, and D. Sedlak, "Disruptors in Water : Implications for the Water Industry", *Environ. Eng. Sci.,* vol. 20, no. 5, pp. 449-469, 2003.
[http://dx.doi.org/10.1089/109287503768335931]

[46]   J. Radjenović, M. Petrović, F. Ventura, and D. Barceló, "Rejection of pharmaceuticals in nanofiltration and reverse osmosis membrane drinking water treatment", *Water Res.,* vol. 42, no. 14, pp. 3601-3610, 2008.
[http://dx.doi.org/10.1016/j.watres.2008.05.020] [PMID: 18656225]

[47]   S. Gur-Reznik, I. Koren-Menashe, L. Heller-Grossman, O. Rufel, and C.G. Dosoretz, "Influence of seasonal and operating conditions on the rejection of pharmaceutical active compounds by RO and NF membranes", *Desalination,* vol. 277, no. 1–3, pp. 250-256, 2011.
[http://dx.doi.org/10.1016/j.desal.2011.04.029]

[48]   C. Bellona, J.E. Drewes, P. Xu, and G. Amy, "Factors affecting the rejection of organic solutes during NF/RO treatment--a literature review", *Water Res.,* vol. 38, no. 12, pp. 2795-2809, 2004.
[http://dx.doi.org/10.1016/j.watres.2004.03.034] [PMID: 15223273]

[49]   A.A. Alturki, N. Tadkaew, J.A. McDonald, S.J. Khan, W.E. Price, and L.D. Nghiem, "Combining MBR and NF/RO membrane filtration for the removal of trace organics in indirect potable water reuse applications", *J. Membr. Sci.,* vol. 365, no. 1–2, pp. 206-215, 2010.
[http://dx.doi.org/10.1016/j.memsci.2010.09.008]

[50]   A.W. Mohammad, Y.H. Teow, W.L. Ang, Y.T. Chung, D.L. Oatley-Radcliffe, and N. Hilal, "Nanofiltration membranes review: Recent advances and future prospects", *Desalination,* vol. 356, pp. 226-254, 2015.
[http://dx.doi.org/10.1016/j.desal.2014.10.043]

[51]   Z. Qiao, Z. Wang, C. Zhang, S. Yuan, Y. Zhu, and J. Wang, "PVAm–PIP/PS composite membrane with high performance for CO2/N2 separation", *AIChE J.,* vol. 59, no. 4, pp. 215-228, 2012.
[http://dx.doi.org/10.1002/aic]

[52]   N. Bolong, A.F. Ismail, M.R. Salim, and T. Matsuura, "A review of the effects of emerging contaminants in wastewater and options for their removal", *Desalination,* vol. 239, no. 1–3, pp. 229-246, 2009.
[http://dx.doi.org/10.1016/j.desal.2008.03.020]

[53]   Y. Yoon, P. Westerhoff, S.A. Snyder, and E.C. Wert, "Nanofiltration and ultrafiltration of endocrine disrupting compounds, pharmaceuticals and personal care products", *J. Membr. Sci.,* vol. 270, no. 1–2, pp. 88-100, 2006.
[http://dx.doi.org/10.1016/j.memsci.2005.06.045]

[54]   M.J. López-Muñoz, A. Sotto, and J.M. Arsuaga, "Nanofiltration removal of pharmaceutically active compounds", *Desalination Water Treat.,* vol. 42, no. 1–3, pp. 138-143, 2012.
[http://dx.doi.org/10.1080/19443994.2012.683099]

[55]   F.N. Ruíz, A.A. Arévalo, J.C.D. Álvarez, and B.J. Cisneros, "Operating conditions and membrane selection for the removal of conventional and emerging pollutants from spring water using nanofiltration technology: The Tula Valley case", *Desalination Water Treat.,* vol. 42, no. 1–3, pp. 117-124, 2012.
[http://dx.doi.org/10.1080/19443994.2012.682969]

[56]   L.D. Nghiem, A.I. Schäfer, and M. Elimelech, "Removal of natural hormones by nanofiltration membranes: measurement, modeling, and mechanisms", *Environ. Sci. Technol.,* vol. 38, no. 6, pp. 1888-1896, 2004.
[http://dx.doi.org/10.1021/es034952r] [PMID: 15074703]

[57]   F. Qu, "Ultrafiltration membrane fouling caused by extracellular organic matter (EOM) from Microcystis aeruginosa: Effects of membrane pore size and surface hydrophobicity", *J. Membr. Sci.,* vol. 449, pp. 58-66, 2014.
[http://dx.doi.org/10.1016/j.memsci.2013.07.070]

[58]   R. Krüger, D. vil, D. Arifin, M. Weber, and M. Heijnen, "Novel ultrafiltration membranes from low-fouling copolymers for RO pretreatment applications", *Desalination Water Treat.,* vol. 57, no. 48–49, pp. 23185-23195, 2016.
[http://dx.doi.org/10.1080/19443994.2016.1153906]

[59]   C. Hong, Z. Xiaode, G. Mengjing, W. Wei, and P. Info, *Onl in Onl in* vol. 37. , 2016, pp. 603-609.

[60]   J.M. Laîné, D. vial, and P. Moulart, "Status after 10 years of operation - overview of UF technology today", *Desalination,* vol. 131, no. 1–3, pp. 17-25, 2000.
[http://dx.doi.org/10.1016/S0011-9164(00)90002-X]

[61]   J.L. Acero, F.J. Benitez, F. Teva, and A.I. Leal, "Retention of emerging micropollutants from UP water and a municipal secondary effluent by ultrafiltration and nanofiltration", *Chem. Eng. J.,* vol. 163, no. 3, pp. 264-272, 2010.
[http://dx.doi.org/10.1016/j.cej.2010.07.060]

[62]   F. Javier Benitez, J.L. Acero, F.J. Real, G. Roldán, and E. Rodriguez, "Ultrafiltration and

nanofiltration membranes applied to the removal of the pharmaceuticals amoxicillin, naproxen, metoprolol and phenacetin from water", *J. Chem. Technol. Biotechnol.,* vol. 86, no. 6, pp. 858-866, 2011.
[http://dx.doi.org/10.1002/jctb.2600]

[63]   J.L. Acero, F.J. Benitez, F.J. Real, and F. Teva, "Removal of emerging contaminants from secondary effluents by micellar-enhanced ultrafiltration", *Separ. Purif. Tech.,* vol. 181, pp. 123-131, 2017.
[http://dx.doi.org/10.1016/j.seppur.2017.03.021]

[64]   X. Qu, P.J.J. Alvarez, and Q. Li, "Applications of nanotechnology in water and wastewater treatment", *Water Res.,* vol. 47, no. 12, pp. 3931-3946, 2013.
[http://dx.doi.org/10.1016/j.watres.2012.09.058] [PMID: 23571110]

[65]   M. Torki, N. Nazari, and T. Mohammadi, "Evaluation of biological fouling of RO/MF membrane and methods to prevent it", *Eur. J. Adv. Eng. Technol.,* vol. 4, no. 9, pp. 707-710, 2017.

[66]   K. Fischer, M. Grimm, J. Meyers, C. Dietrich, R. Gläser, and A. Schulze, "Photoactive microfiltration membranes *via* directed synthesis of TiO_2 nanoparticles on the polymer surface for removal of drugs from water", *J. Membr. Sci.,* vol. 478, pp. 49-57, 2015.
[http://dx.doi.org/10.1016/j.memsci.2015.01.009]

[67]   S. Rodriguez-Mozaz, M. Ricart, M. Köck-Schulmeyer, H. Guasch, C. Bonnineau, L. Proia, M.L. de Alda, S. Sabater, and D. Barceló, "Pharmaceuticals and pesticides in reclaimed water: Efficiency assessment of a microfiltration-reverse osmosis (MF-RO) pilot plant", *J. Hazard. Mater.,* vol. 282, pp. 165-173, 2015.
[http://dx.doi.org/10.1016/j.jhazmat.2014.09.015] [PMID: 25269743]

[68]   D. Ragab, H.G. Gomaa, R. Sabouni, M. Salem, M. Ren, and J. Zhu, "Micropollutants removal from water using microfiltration membrane modified with ZIF-8 metal organic frameworks (MOFs)", *Chem. Eng. J.,* vol. 300, pp. 273-279, 2016.
[http://dx.doi.org/10.1016/j.cej.2016.04.033]

[69]   R. Owen, and S. Jobling, "Environmental science: The hidden costs of flexible fertility", *Nature,* vol. 485, no. 7399, pp. 441-441, 2012.
[http://dx.doi.org/10.1038/485441a] [PMID: 22622553]

[70]   M. Huerta-Fontela, M.T. Galceran, and F. Ventura, "Occurrence and removal of pharmaceuticals and hormones through drinking water treatment", *Water Res.,* vol. 45, no. 3, pp. 1432-1442, 2011.
[http://dx.doi.org/10.1016/j.watres.2010.10.036] [PMID: 21122885]

[71]   J. Han, S. Meng, Y. Dong, J. Hu, and W. Gao, "Capturing hormones and bisphenol A from water via sustained hydrogen bond driven sorption in polyamide microfiltration membranes", *Water Res.,* vol. 47, no. 1, pp. 197-208, 2013.
[http://dx.doi.org/10.1016/j.watres.2012.09.055] [PMID: 23127621]

[72]   S. Kim, "Removal of contaminants of emerging concern by membranes in water and wastewater: A review", In: *Chem. Eng. J* vol. 335. , 2018, pp. 896-914.
[http://dx.doi.org/10.1016/j.cej.2017.11.044]

CHAPTER 12

Catalytic Processes for Removal of Emerging Water Pollutants

Shabnam Taghipour[1,2], Marziyeh Jannesari[3,4], Behzad Ataie-Ashtiani[1,5], Seiyed Mossa Hosseini[6] and Mohammadhossein Taghipour[7]

[1] *Department of Civil Engineering, Sharif University of Technology, P.O. Box 11155-9313, Tehran, Iran*

[2] *Department of Chemical and Biological Engineering, The Hong Kong University of Science and Technology, Clear Water Bay, Kowloon, Hong Kong*

[3] *Institute for Nanoscience and Nanotechnology, Sharif University of Technology, P.O. Box 14588 89694, Tehran, Iran*

[4] *Department of Physics, Sharif University of Technology, P.O. Box 11155-9161, Tehran, Iran*

[5] *National Centre for Groundwater Research & Training and College of Science & Engineering, Flinders University, GPO Box 2100, Adelaide, South Australia 5001, Australia*

[6] *Physical Geography Department, University of Tehran, P.O. Box 14155-6465, Tehran, Iran*

[7] *Department of Materials Engineering, University of Tabriz, P.O. Box 51666-16471, Tabriz, Iran*

Abstract: An unprecedented increase in urbanization and industrialization ignited by an upsurge in the development of consumer goods. This has been steadily destroying the environmental balance and ecosystem and diminishing the water quality. Inevitably, we are facing one of the biggest challenges of the time, which needs to be resolved with proper remediation strategies to provide clean water as one of the essential components for human beings and agriculture, livestock, and several industrial survivals. With the growing demand for water and sustainable improvement, utilizing unconventional water supplies such as contaminated fresh water, brackish water, and wastewater is required. Although some of the traditional water treatment and purification methods still retain their importance. However, there is a need to provide faster and more efficient technologies beyond conventional methods for treating various contaminated water sources, including emerging pollutants. Recently, catalytic processes such as ozonation and electrocatalysis, including electrocatalytic oxidation, electro-Fenton process, photo electro-Fenton process, photocatalysis, and reduction by hydrodehalogenation, exhibited unique features and have opened wide opportunities in the field of water treatment. This chapter describes various types of

[*] **Corresponding author Shabnam Taghipour & Marziyeh Jannesari:** Department of Civil Engineering, Sharif University of Technology, P.O. Box 11155-9313, Tehran, Iran and Department of Chemical and Biological Engineering, The Hong Kong University of Science and Technology, Clear Water Bay, Kowloon, Hong Kong; E-mails: sh.taghipour70@student.sharif.edu, staghipour@connect.ust.hk and Institute for Nanoscience and Nanotechnology, Sharif University of Technology, P.O. Box 14588 89694, Tehran, Iran, and Department of Physics, Sharif University of Technology, P.O. Box 11155-9161, Tehran, Iran; E-mail: mjannesari2000@gmail.com

emerging contaminants, their effect on human health and the ecosystem, and analytical methods of ECs quantification. Moreover, the features, mechanisms, and potential applications of catalytic processes in treating emerging pollutants are discussed in detail.

Keywords: Catalytic processes, Emerging pollutants, Environmental application, Fenton processes, Oxidation processes, Remediation, Water treatment.

INTRODUCTION

The industrial revolution resulted in changes in lifestyle and the patterns of consumption and production, causing serious environmental hazards due to the release of new substances into the environment. These materials may impact differently from the traditional recognized environmental contaminants. However, traditional investigations into water quality mainly focus on microbial pollutants, heavy metals, and nutrients as the essential factors in providing safe drinking water to human beings, animals, and wildlife. Traditional analytical methods encounter instrumental restrictions in detecting and measuring the organic contaminants from the abovementioned sources. As an illustrative example, perfluorinated compounds (PFAS) as water repellent dates back to the 1950s, while this class of pollutants was recognized in the environment until the 2000s due to the lack of appropriate monitoring instruments [1]. As a result, the term "emerging" does not necessarily refer to a new class of substances. Still, it also indicates the materials that have been recently recognized as environmentally hazardous materials with potentially eco-toxicity effects.

Recently, several groups of chemical substances have been recognized as emerging contaminants (EC), and their number is continuously increasing. Among others, pharmaceutical residuals [2], personal care products [3, 4], food additives and artificial sweeteners [5, 6], synthetic organic materials, and recently nanoparticles [7, 8] excited more interest. Unfortunately, the lack of available data about occurrence routes, risk assessments, and, more importantly, the eco-toxicity effect of the materials makes the prediction of their health effects difficult. So, this is imperative to different types of ECs, the route of occurrence, their health effects, analytical methods for their quantification, and, more importantly, technological options for their removal.

The catalytic techniques include ozonation, electrocatalysis (including electro-catalytic oxidation, electro-Fenton, photo electro-Fenton), photocatalysis, sonocatalysis, and reduction by hydrodehalogenation have gained a reputation for removal or degradation of emerging water pollutants. The ozonation method benefits from the powerful oxidizer of ozone. In the electro-Fenton method, Fenton reagents oxidize the contaminants on the surface of the anode in the presence of •OH radicals (generated *via* ferrous ions (Fe^{2+})). At present, many photocatalytic materials responsive to visible light have been developed. However, applying UV/visible light radiations can effectively increase the efficiency of Fenton reactions. Photocatalysis is based on the formation of electron/hole pairs on the surface of the catalyst and a series of oxidation and reduction reactions for photodegradation of the contaminants. In the sonocatalysis method, the generation of hotspots due to sequential nucleation, growth, and collision of micro-bubbles is the main factor for the degradation of pollutants.

In this chapter, we aim to discuss different kinds of ECs, the route of their occurrence, their health effect, the analytical methods for their quantification, and in-detail mechanism and procedures of each of the catalytic techniques for removal of ECs from water. In addition, the overall pros and cons of each of these technologies are mentioned and briefly debated.

EMERGING WATER CONTAMINANTS

Researchers have realized that some organic pollutants profoundly impact the quality of the water bodies. These pollutants occur in relatively low concentrations (ng/L-µg/L) and have lately been recognized as considerable water pollutants [9]. The following sections describe different types of ECs, their effects, pathways, and various existing analytical detection methods in detail. The operations of various industries release directly and indirectly various substances which enter the waterbodies. Investigation of these compounds revealed their toxicity and detrimental effects on both human health and the environment in the long term. For this reason, several researchers have investigated their monitoring mechanisms. However, there is no exact agreement on the compounds that should be monitored.

Pharmaceuticals are widely used by humans and poultry, livestock, and fish farming to reduce illnesses and diseases. However, these chemicals are one of the most important emerging contaminants, which may cause a huge concern due to their frequent occurrence by remaining residual chemical groups such as steroids in the ecosystem (*i.e.*, drinking, surface, ground, waste, and storm waters) potentially endangering all human beings [10]. It has been estimated that more than 3000 different pharmaceutical ingredients (such as antibiotics, illicit drugs, hormones, antineoplastic compounds, beta-blockers, antidiabetics, and anti-inflammatory) [1, 11]. However, research studies have recognized just a small ratio of these contaminants released in aquatic systems. Beyond the potential adverse effect on the quality of potable water supplies and the health of human beings, bioaccumulation of these compounds interrupting hormonal control and inducing breast and testosterone cancers can worsen circumstances. The major concern that corresponds to the drugs such as antibiotics is the appearance and the development of bacterial drug resistance phenomena favoring superbugs.

The chemicals such as antiseptics, sunscreens, fragrances, per-fluorinated compounds, antimicrobials, and preservative products (the chemicals with inhibitory effects on the growth of microbes such as bacteria, fungi, viruses, and protozoa) and insect repellents widely used in urban areas are known as personal care contaminants when inevitably excreted and released in the environment [1, 10]. Unfortunately, due to the design for external usage, most of these materials are undegradable and cannot be metabolized in the environment [12], where they show bioactivity potential at low concentrations [4]. The bad news is that the presence of these types of contaminants has been continuously growing in recent years. Among others, antimicrobials and preservative compounds such as triclosan (as an additive used in cosmetics products of soaps, toothpaste, and body shampoos) and parabens (widely used in commercial moisturizers, makeup, personal lubricants, shampoo, and many other cosmetic products) are the main classes of personal care products (PCPs).

The occurrence of sweeteners (such as sucralose and saccharin) in domestic water was reported in 2007 [13]. Sucralose is one of the most favorable artificial sweeteners, which is a multistep- chemically processed sucrose in which tree oxygen functional groups are substituted by chlorine atoms. This chlorinated sucrose-based sugar cannot be completely digested by the body resulting in its widespread usage as a noncaloric sweetener [5]. Although sucralose consumption has been considered safe for humans, concerns remain about adverse environmental effects, such as their influence on aquatic organisms (such as crustaceans). Thermal stability (even at elevated temperatures) and its long life make sucralose favorable for baking and soft drink products [6]. On the other hand, its higher stability may endanger the ecosystem. This artificial sweetener in the environment has been extensively reported after the first discovery of sucralose in river and ocean waters [5].

Nanomaterials (NMs) as materials with sizes ranging between 0-100 nm (for at least one dimension) have exponentially attracted much attention due to their extraordinary physio-chemical and biological properties [14 - 16]. Various types of nanomaterials can be simply classified into organic (graphene-based [17], fullerene [18], and carbons [7]) and inorganic (such as metals (Ag [8], Cu [19], Au [20], and Zn [21]), and metal oxides (CuO_2 [15] and ZnO) based nanomaterials. The increased use of nanomaterials in different fields (such as biomedical [22, 23], agricultural [24], industrial [25], and environmental [26 - 34] applications) due to their unique properties raises environmental concerns about their toxic effect (which has not been completely understood) on biological systems. Hence, attempts have been made to study nanomaterials' transport, fate, stability, and long-term eco-toxicology in the aquatic environment. For this reason, various categories have been developed for the classification of NMs based on their characteristics. Among all, some of the well-known categories of different NMs are carbonaceous nanostructures (including fullerene, carbon nanotubes, and graphene), nanocomposites (including metal-metal, metal-metal oxide, metal-Oxyacids, Oxyacids- Oxyacids, metal oxide- metal oxide, and nanoscale supports), and nanoparticles (including metal oxides and Oxyacids). Some metallic NMs such as Ag, TiO_2, and carbonaceous aqueous fullerene (nC60) and carbon nanotubes (CNTs) have also exhibited considerable antimicrobial properties in water treatment [15].

Effects of Emerging Water Pollutants

The modern compounds containing emerging pollutants are improperly disposed of or inefficiently managed. The release of emerging contaminants into the environment may experience different transformation processes of degradation or persist in their initial structures. Although the concentrations of ECs found in the environment (such as ground and surface waters) are not necessarily high (maybe in ng/L), however their long term exposure to these active chemicals declines the LC_{50} and lethal concentration and therefore, they can induce damaging and toxic effects on the ecosystems [35]. Rachel Carson [36], by describing the effect of pesticides on human beings, initially raised public concerns about the environmentally adverse effect of the chemical released in nature. To understand the toxic effects of the ECs, different experimental investigations such as cytotoxicity, genotoxicity, and estrogenicity tests on different models of organisms or cells have been conducted [35, 37]. It has been established that in addition to the chemical structures of the ECs, the physiochemical properties of the polluted environment or the living organisms determine the level of toxicity [35, 38, 39]. It is worth knowing that the presence of a complex mixture of the ECs (either in the under-graded original or their constituent forms) can induce direct or indirect undesirable synergistic effects on human health or ecosystems. Although ethical challenges restrict the direct studies on the toxicological effects of the ECs on human beings, scientists generalize strong evidence of animal tests to human beings based on similar mechanisms. Various studies indicate adverse effects of ECs, including reproductive malfunction, a decrease in the number of male sperm, and an increase in breast and prostate cancers.

The environmentally toxic effects of pharmaceuticals and personal care products (PPCPs) have attracted more attention due to their high usage in human lives. Among others, estrogens and androgens (as the sex hormones), antibiotics, and antineoplastic drugs (with the ability to cure abnormalities in growth tissues) have been considered the greatest potential concerns [40]. Recent research has focused on the contaminants interfering with the function of the endocrine system [41]. The main findings from the initial research on the sewage treatment plants demonstrated stimulating the feminization of the male fish due to the existence of 17-β estradiol [42, 43]. Estrogenic hormones possessing the most endocrine-disrupting chemicals can disrupt the organism's growth, reproduction, and pregnancy at low concentrations of µg/L or even ng/L. These chemical contaminants can change endogenous hormones' synthesis, metabolism, transportation, and expression and influence antagonizing and mimicking of these hormones [44].

Extensive applications of antibiotics in human and animal therapy remain residues in the environment with serious adverse effects on the terrestrial and aquatic organisms and their populations and cells. Long-term exposure to antibiotics has resulted in the ecological risk of antibiotic resistance (the top focus of the toxicological effect of the antibiotics [44]). This effect has been assigned to genetic variations and transfer [45]. Antibiotics also inflict adverse effects on the agroecosystems, including seed germination inhibition and growth prevention, restriction in the microorganisms (such as bacteria) activity in the soil, bioaccumulation of the antibiotics in crops, and leaching into the ground and surface waters [46].

The chemicals used in personal care products, such as triclosan and triclocarban, as other compounds with a considerable toxic effect on microbes and alga (as preservatives), have attracted increasing attention. Experiencing photo-transformation in aqueous solutions (which results in a toxic by-product of 2,8-dichlorodibenzo-p-dioxin (2,8-DCDD)), these chemicals exhibit potential for estrogenicity and bacterial drug-resistant [47 - 49]. Beyond the direct effects of the ECs on the health of living organisms, their impacts may also grow due to climate change, as it introduces increased droughts and lowers water levels making their concentrations high in aquatic systems [50]. The main sources of ECs and their human health hazards are listed in Table **1**.

Table 1. Health risks and hazards of various emerging contaminants.

Category	Contaminant	Health Risk
Pharmaceuticals	Antibiotics	Prevalence of antibiotic resistance, inducing genetic variations and affecting populations of organisms [51]
	Steroidal estrogen	Inducing estrogen hormone effect on the non-target (feminization in male fish) [52]
	Pain reliever	Disrupt the tissues, kidneys and gills, inducing chronic toxic effects suggesting regeneration inhibition and reproduction reduction in aquatic organisms such as fish [53]
PCPs	Paraben	Immune dysfunction, weak estrogenic activity, behavioral problems, possible carcinogenesis effect and endocrine-disrupting potential [54]
	Bisphenol	the estrogenic and hormonal effect, inducing breast cancer in women and feminizing side-effects in men [55 - 57]
	Phthalates	Increasing the risk of miscarriage and pregnancy complications [58]
	Preservatives	Toxic effect killing microorganisms and inducing bacteria resistance toward antibacterial drugs [59]
Textile industry	Azo, nitro and acridine dyes	Mutagenicity, carcinogenesis [60 - 62]

Pathway of the Entrance of Emerging Contaminants into the Ecosystem

Different types of ECs directly or indirectly affecting the biosphere are daily discharged into ecosystems through various pathways, mainly divided into two main classes; point (local) and off-point (diffuse) sources [63, 64]. For example, many pharmaceuticals cannot be completely metabolized after administration; they remain in their original form. These compounds are released from the body through excretion (such as renal or biliary) systems to the sewage and then ecosystems [65]. According to the U.S. Environmental Protection Agency (USEPA), a point-source pollutant is defined as "any single identifiable source of pollution from which pollutants are discharged, such as a pipe, ditch, ship or factory smokestack" [52]. The concentration of these pollutants is inversely proportional to the distance of the point source (such as pipe end) locations. It means that pollutant concentration decreases by increasing the distance away from the sources [66]. There exist different types of point sources of ECs, including hospitals, agricultural soil, industries, municipal sewage, accidental leakages, and wastewater treatment plants (WWTPs), which directly discharge the pollutants into the water streams. Fish farming, landfills, power stations, oil, and agricultural industries are the other main activities potentially local contaminant environments [63, 67]. Some of these pathways are illustrated in Fig. (**1**).

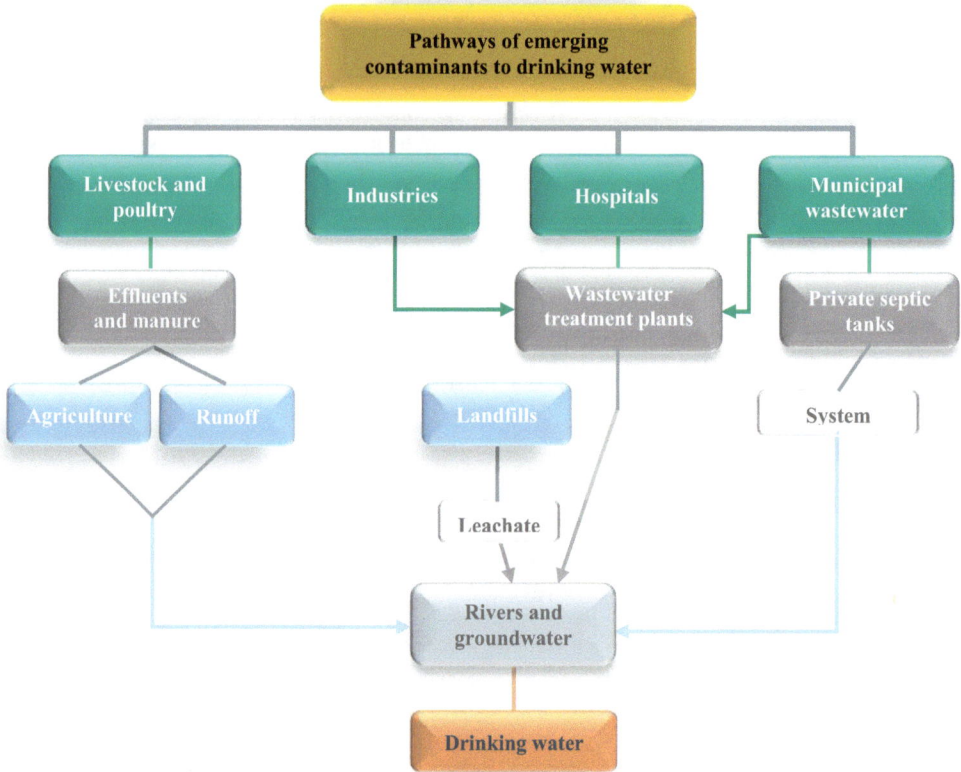

Fig. (1). Different pathways for drinking water contamination by emerging contaminants.

On the other hand, non-point sources contaminate the ecosystem through multiple sources rather than a given discrete source, influencing a large area of the environment [63]. It has been shown that most of the environmental pollution originated from the non-point sources of ECs [68]. These pollutants can adversely affect the ecosystem by decreasing oxygen concentration and the physical confinement of streambeds [69]. The most important examples of these sources include the irrigation or runoff of biosolids (agrochemicals, chemicals, nutrients, fertilizers, and pesticides), livestock excrement, river floods, electronic-waste deposal, and recycling sites, bacteria and wastewaters from septic systems in agricultural or urban areas. In addition, leaching and percolation of sewage from treatment facilities and farming and forestry may cause dispersed environmental pollution [70].

Analytical Methods of ECs Quantification

Low detection limit requirements and separation difficulty of the samples from the interferences (such as their matrices) and the complexity of these substances introduce serious challenges for detecting the ECs through traditional methods. Before analyzing the ECs, several pretreatment steps such as filtration, adjustment of pH, and isolation from matrices are required [71]. An efficient extraction step is critical for proper isolation and purification of the target analyte from its matrices. However, the classical extraction techniques such as shaking (either through hand agitation, orbital or vortex shakers) and Soxhlet techniques are still being applied. Pressurized-liquid extraction (PLE), microwave-assisted extraction, and ultrasound-assisted extraction (UAE) are the most preferred extraction technique recently reported [72]. In addition, liquid-liquid extraction (LLE), solid-phase extraction (SPE), and more novel techniques of solid-phase microextraction (SPME) are the most commonly used techniques [73, 74]. Instrumental analysis for qualification as well as quantification detection of emerging organic contaminants is commonly based on chromatographic separation coupled with mass spectrometry (MS) in a single MS or tandem (MS/MS) modality [75, 76]. In general, the separation mechanisms in chromatography are based on differential partitioning between a mobile and a stationary phase [76]. While excellent separation can be achieved through gas chromatography (GC) for volatile and semi-volatile nonpolar/lipophilic contaminants, nonvolatile thermolabile contaminants with wide polarity ranges are well identified by liquid chromatography (LC). Besides the MS, other detection approaches such as electron capture detector (ECS) and flame ionization detector (FID) have also been used with GC [72]. However, the best option for quantitative analysis based on selected reaction monitoring is triple quadrupole (QqQ) [77]. Quadrupole to time-of-flight coupling (Q-TOF) has also been reported for high-resolution target analysis. Ultra-high-performance liquid chromatography (UHPLC) with a remarkable enhancement in sensitivity, resolution, and speed has been introduced as an improved modality [71].

CATALYTIC WASTEWATER TREATMENT

Various wastewater treatment technologies exist for the novel and conventional pollutants from the wastewater. The catalytic processes efficiently break down the organic waste into value-added products or manageable water waste. The catalytic processes for wastewater treatment include oxidation processes (ozonation, electrocatalysis, electrocatalytic oxidation, electro-Fenton, photo electro-Fenton), photocatalysis, sonocatalysis, and reduction by reduction hydrodehalogenation *etc*. Processes like activated persulfate-based oxidation are also common for removing organic pollutants from wastewater.

Oxidation Processes

Due to emerging pollutants at very low concentrations, conventional wastewater treatment plants do not function well in removing these contaminants. Eventually, these substances enter the aquatic systems and affect the whole life cycle [78]. Diverse technologies have been explored for the remediation of ECs [79]. Biological treatment methods do not have enough efficiency in purifying pollutants even at such low concentrations [80 - 82]. Among the investigated technologies for the degradation of emerging contaminants, oxidation and reduction processes have demonstrated outstanding efficiency. By using various oxidants and catalysts, oxidation processes are likely to be the most applied for wastewater purification. These oxidants can be categorized based on their standard oxidation potential (Table **2**), from the most powerful agent to the weak ones [83, 84].

Table 2. Oxidation potential of well-known oxidizer [83, 84].

Oxidizer	Oxidation Potential (eV)
Fluorine (F_2)	3.03
Hydroxyl radical ($^\cdot HO$)	2.8
Sulfate radical ($^\cdot SO_4^-$)	2.5–3.1
Atomic oxygen	2.42
Ozone (O_3)	2.1
Persulfate ($S_2O_8^{2-}$)	2.1
Peroxymonosulfate (HSO_5^-)	1.8
Hydrogen Peroxide (H_2O_2)	1.78
Hydroperoxyl	1.7
Permanganate (MnO_4^-)	1.68
Chlorine dioxide (ClO_2)	1.5
Hypochlorite (ClO^-)	1.49
Hypochlorous acid ($HClO$)	1.49
Chlorine (Cl_2)	1.4
Molecular oxygen (O_2)	1.23

According to the experimental conditions and the utilized oxidants, the oxidation processes can be classified into the following categories: 1) processes conducted under or near ambient conditions such as AOPs in which $^\cdot HO$ is the dominant specie (*e.g.*, sonocatalysis and photocatalysis) and also processes that utilizes powerful oxidizers such as O_3 and H_2O_2,) and the processes conducted at pressure and temperature above the ambient condition (*e.g.*, catalytic wet air oxidation) [85].

Ozonation

Ozone is an electrophilic molecule that can directly react with high electronic density molecules such as unsaturated bonds and aromatic rings of the emerging pollutants. Under special conditions, this powerful oxidizer can also decompose water to generate ˙OH radicals. Upon dissolution of ozone in water and based on the pH and the type of the contaminants in water, O_3 can react *via* two various mechanisms, including 1) molecular ozone (through cycloaddition, nucleophilic or electrophilic reactions) and 2) indirect reactions (by the formation of hydroxyl radicals). The indirect reaction mechanism of ozonation consists of three different stages, including initiation, propagation, and termination (Eqs. 1-6) [86, 87]:

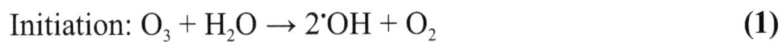

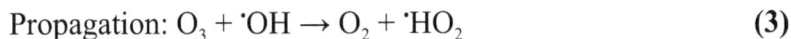

$$\text{Initiation: } O_3 + H_2O \rightarrow 2\dot{O}H + O_2 \tag{1}$$

$$O_3 + OH^- \rightarrow \dot{O}_2^- + \dot{H}O_2 \tag{2}$$

$$\text{Propagation: } O_3 + \dot{O}H \rightarrow O_2 + \dot{H}O_2 \tag{3}$$

$$O_2 + \dot{H}O_2 \rightarrow \dot{O}_2^- + H^+ \tag{4}$$

$$O_3 + \dot{H}O_2 \rightarrow 2O_2 + \dot{O}H \tag{5}$$

$$\text{Termination: } 2\,\dot{H}O_2 \rightarrow O_2 + H_2O_2 \tag{6}$$

Aqueous ozone solutions are unstable. To reduce instability of the ozonation method, various attempts such as acidification and injection of •OH scavengers (*e.g.*, HCO3-) are carried out [88]. High energy requirements and generation of oxidative byproducts, and interference of radical scavengers during the experiment are other disadvantages of this technology [89]. One of the most important advantages of ozonation is that this method is relatively harmless to human health and the ecosystem [88]. Moreover, this method can efficiently improve the biodegradability of the contaminants and reduce the toxicity of the target pollutants. The presence of hydrogen peroxide can also contribute to the high removal of ECs; some other oxidants can conduct disinfection and sterilization [90].

Generally, approximately all ECs are efficiently degraded by ozonation (efficiency > 90%). This powerful oxidant has removed ECs *via* reaction with their double bonds and aromatic rings. Various ECs, such as endocrine-disrupting compounds (bisphenol A, 4-n-nonylphenol), triclosan, Atenolol, metoprolol, propranolol, sotalol, salbutamol, naproxen, ibuprofen, diclofenac, carbamazepine, gemfibrozil, metronidazole, atrazine, codeine, paracetamol, erythromycin, ofloxacin, trimethoprim, ciprofloxacin, levofloxacin, roxithromycin, diazepam, ranitidine have been successfully degraded by ozonation method with high efficiency (> 80%) [91 - 94].

Electrocatalysis Process

Electrocatalysis is one of the effective strategies in water treatment. Electrochemical reactions usually occur at low rates; therefore, the presence of an efficacious electrocatalyst can potentially reduce high energy consumption. Furthermore, multiple electron transfer *via* this method can lead to the formation of intended products relying on selecting appropriately applied catalysts [95]. A proper electrocatalyst is expected to comply with items such as high catalytic activity (to increase the energy efficiency), ascendance selectivity (to catalyze the production of a specific product), suitable conductivity (to enhance the electron transfer and reaction rate), high stability for long-run usage, and cost-effectiveness in real-scale application [96].

Electrocatalytic Oxidation

In electrocatalytic oxidation, electrodes affect activity, selectivity, and current efficiency. Oxidation mostly occurs on the anode side and can be categorized as direct and indirect oxidation. In direct oxidation, contaminants can be degraded to be harmless or biodegradable components (favorably CO_2 and H_2O) without the participation of other substances. At the same time, indirect oxidation (also known as mediate oxidation) refers to the formation of intermediates on the surface of the electrode (*e.g.*, OH, H_2O_2, O_3, and OCL$^-$) for later degradation of contaminants [97]. In the anodic process, a high O_2-overpotential anode (A) will participate in the oxidization of water to O_2 and generate hydroxyl radical A($^.$OH) (Eqs. 7 and 8) as proposed by Feng and Johnson [98, 99]:

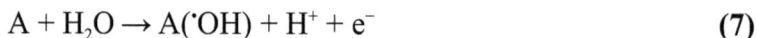

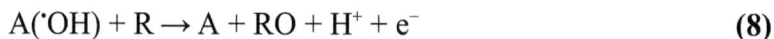

$$A + H_2O \rightarrow A(^.OH) + H^+ + e^- \qquad (7)$$

$$A(^.OH) + R \rightarrow A + RO + H^+ + e^- \qquad (8)$$

The appropriate material of electrodes has to meet three main prerequisites: low cost, resistance in the electrolysis medium, high activity for organic oxidation, and simultaneously slight activity for secondary reactions [100]. The functionality of catalytic reactions on the electrode surfaces can be boosted *via* modification of specified metal oxides such as PbO_2, TiO_2, RuO_2, SnO_2, IrO_2, MnO_2, *etc* [101, 102]. Electrocatalytic oxidation has been successfully applied for industrial wastewater effluents (*e.g.*, textile, food, and pharmaceutical industries) [103, 104]. Kaur *et al.* (2019) fabricated Ti/RuO_2 anodes for electro-catalytic oxidation of amoxicillin trihydrate (AM) and assessed the effect of various variables, including pH (2-9), AM concentration (10–50 mg/L), electrolyte concentration (1–2 g/L), and current density (1.47–5.88 mA/cm^2) on the removal efficiency. At the optimum conditions, including neutral pH, AM concentration of 50 mg/L, NaCl concentration of 2 g/L, and current density of 5.88 mA cm^{-2}, the results showed that increasing current density adversely affects the mineralization in such a way that increasing current density from 1.47 to 5.88 mA/cm^2 reduced TOC removal from 11.77 to 7.67%, respectively. In this study, NaCl was selected as the electrolyte for two main reasons, including 1) increasing electrical conductivity and 2) transferring active Cl$^-$ species as oxidizers of contaminants [105].

Sun *et al.* (2019) prepared Ti–Ag/γ-Al$_2$O$_3$ electrodes to oxidize pesticide wastewater (COD=2300 mg/L). At the optimized conditions, at the pH of 2.0, the electrical conductivity of 4000 μS/cm^{-1}, the current density of 30 mA/cm^2, electrode interval of 3.0 cm, and airflow of 3.0 L/min, the utmost removal efficiency and energy consumption was 82.50% and 10.91 (kW.h)/kg COD after 3 h, respectively. In addition, electrocatalytic oxidation potentially improved the biodegradability of the pollutants. The electrode interval greatly decreased and increased the reaction's functionality and operational cost. The reason can be attributed to the fact that diffusion of organic contaminants did not occur in the low distance and consequently resulted in concentration polarization of wastewater. UV–vis analysis indicated that organic large molecules were transformed into small molecular components within 30 min [106]. Yao *et al.* (2019) developed Bi-doped PbO$_2$ anodes for electrocatalytic degradation of 10-90 mg/L hexazinone insecticide. To optimize the experiment, the concentration of Na$_2$SO$_4$, current density, and pH varied from 0.05 to 0.25 M, 40 to 200 mA/cm^2, and 3.0 to 11.0, respectively. After 120 min, in the presence of 10 mg/L hexazinone, 0.1 M Na$_2$SO$_4$, and the applied current density of 200 mA/cm^2, 99.9% of hexazinone was indirectly degraded in the presence of hydroxyl radicals. The kinetic studies illustrated that contaminant degradation has the best-fitted pseudo-first-order model with a high reaction rate equal to 32.37 × 10^{-3} 1/min [107].

Salazar *et al.* [98] investigated the effect of electrode material and electrolyte medium (Na$_2$SO$_4$ and NaCl) on the degradation of yellow 3 textile dye. Various anodes were prepared, including the boron-doped diamond (BDD), Ti/Ru0.3Ti0.7O2, and Ti/Pt. In the presence of 50 mM Na$_2$SO$_4$ and initial TOC equal to 100 mg/L, regardless of pH and the current density, the BDD anode exhibited the most appropriate function (over 90% removal of TOC and color). In contrast, the corresponding amount for other anodes was 50%. By using NaCl as the medium, the obtained mineralization from various anodes was in the following order (removal efficiency): Ti/Ru$_{0.3}$Ti$_{0.7}$O$_2$ (~100%) > BDD (~98%) > Ti/Pt (~90%). A pseudo-first-order reaction was conducted by kinetic studies and a linear correlation for pollutant degradation at neutral pH. Energy consumption studies exhibited that utilizing BDD anode at neutral pH leads to approximately USD\$ 0.025–0.027, while this quantity for $Ti/Ru_{0.3}Ti_{0.7}O_2$ anode was USD\$ 0.015–0.020.

Electro-Fenton

Electro-Fenton, one of the cathodic processes, has attracted widespread attention across many fields and has been applied especially in wastewater treatment. Electro-Fenton is conducted in two diverse methods, including 1) injection of Fenton reagents to the reactor from outside and utilization of greatly-reactive inert anode electrodes inside the reactor and 2) addition of H_2O_2 from outside and reduction of ferric ion to ferrous ion from sacrificial cast iron anode inside the reactor [108]. The steps are presented in Eqs. 9-11 [109, 110]:

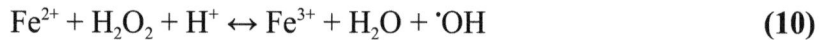

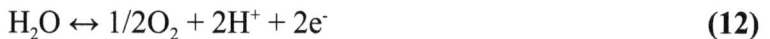

$$O_2 + 2H^+ + 2e^- \rightarrow H_2O_2 \tag{9}$$

$$Fe^{2+} + H_2O_2 + H^+ \leftrightarrow Fe^{3+} + H_2O + {}^{\cdot}OH \tag{10}$$

$$Fe^{3+} + e^- \leftrightarrow Fe^{2+} \tag{11}$$

$$H_2O \leftrightarrow 1/2O_2 + 2H^+ + 2e^- \tag{12}$$

Electro-Fenton technology suffers from several drawbacks, such as the production of secondary sludge, precipitation of hydroxide due to Fe^{2+} and Fe^{3+} ions in the treated stream, and consequently increasing operational costs and reducing the removal efficiency. To efficiently control the production of solid $Fe(OH)_2$ and $Fe(OH)_3$, continuous pH control of the input stream (< 4) is necessary. Furthermore, the ongoing presence of the ${}^{\cdot}OH$ radicals requires pure oxygen and H_2O_2 provision [111, 112]. To overcome these disadvantages, many kinds of research have been made by focusing on the material of the cathodes (to boost the production of hydrogen peroxide), utilization of metal catalysts (*e.g.*, Pd [113], Ru [114], Au [115], and Pt [108]) to enhance conductivity and chemical stability [116]. In the following, the application of electro-Fenton technology for abatement of emerging pollutant-contaminated waters is expressed.

Ghasemi *et al.* (2019) investigated the degradation of 20 mg/L gentamicin by 1.25 g/L Cu-Fe-double hydroxide (DH) as the catalyst in the electro-Fenton process. This study used the graphite plate to generate H_2O_2, and a platinum sheet applicated as the anode. Results indicated that at optimum condition (pH= 6, current= 400 mA, and $C_{Na2SO4} = 0.05$ mol/L), the maximum removal efficiency of different processes after 100 min was in the following order: Cu-Fe-DH-elecro-Fenton (91.3%) > Fenton (50%) > electro-oxidation (25.6%). ${}^{\cdot}OH$ was the main free radical responsible for the degradation of gentamicin. The amount of electro-generated H_2O_2 in the absence and presence of Cu-Fe-DH was equal to 91.6 and 21.1 µ mol/L (due to the accumulation of the hydrogen peroxide in the absence of the Cu-Fe-DH). The initial COD (168 mg/L) was reduced by 77% after 300 min. By increasing the current from 150 to 450 mA, the degradation efficiency increased from 21.7 to 88.2%, respectively, which was attributed to the enhancement of electron transfer through the cathode in higher currents and improvement of H_2O_2 and 'OH radical generation [117].

Ghanbarlou *et al.* (2020) developed nitrogen-doped catalytic electrodes (NCEs) for abatement of 50 mg/L of mixed 2,6-dichlorobenzamide (BAM) 2-methyl-4-chlorophenoxy acetic acid (MCPA), and 2-methyl-4-chlorophenoxy propionic acid (MCPP) pesticides at pH 5.6. Under 15 V voltage, catalyst loading of 1 g/L, after 480 min highest removal efficiency for both MCPA and MCPP and BAM were 93 and 84%, respectively. The existence of nitrogen potentially promoted oxygen reduction activity [118].

Yang *et al.* (2020) applied electro Fenton to remove 10 mg/L diisobutyl phthalate (DP) from water. The effect of various parameters such as pH (2–9), used current (0.5–2.0 A), cathode distance (3−9 cm), 40 µL H_2O_2, the concentration of humic acid and calcium ion (1−40 mg/L and 0−20 mmol/L, respectively) on the removal efficiency was evaluated. At pH 5, electrode's distance of 5 cm, initial current of 1 A, Na_2SO4 concentration of 2 g/L, and hydrogen peroxide amount of 40 µL, 93.07% degradation efficiency was obtained. Adding further hydrogen peroxide led to decreased degradation efficiency due to the consumption of 'OH radicals. The Addition of humic acid and calcium ions reduced the degradation of DP. The presence of humic acid increased the final production of the reaction due to the variation in the attack sites of the pollutant. Moreover, the elimination rate was gently reduced by increasing the plates' distance due to resistance of the reaction system and lower mass transfer [119].

Photo-electro-Fenton Process

The photo electro-Fenton procedure is one of the widely applied developments of Fenton reaction in water and wastewater treatment. In this method, wastewater undergoes UV/visible light irradiation simultaneously with or after the electro-Fenton procedure. As a kind of carboxylic acid, Oxalic acids act as photo-active compounds in Fe(III) presence at acidic pH. As a result, the photo-decarboxylation of hard-degradable Fe^{3+}/carboxylate complexes will occur (Eq. 13) [120]:

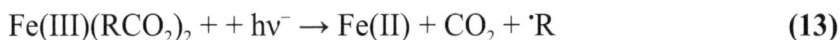

$$Fe(III)(RCO_2)_2 + + hv^- \rightarrow Fe(II) + CO_2 + \ ^.R \quad\quad (13)$$

In the next step, more Fe^{2+} besides ˙OH radicals are produced from further photoreduction of ferrous hydroxide (Eq. 14) [121]:

$$Fe(OH)^{2+} + hv^- \rightarrow Fe^{2+} + \text{˙OH} \tag{14}$$

The photo-decarboxylation can efficiently contribute to carbon mineralization, specifically in the presence of aromatic contaminants. It is noteworthy that the generation of ferrous ions promises a continuation of the Fenton reaction in the existence of hydrogen peroxide. For this reason, ions from the carboxylates family, such as oxalate and citrate, are utilized in photo Fenton and photo-electr--Fenton methods [122]. Several studies have investigated the efficiency of electro-Fenton and photo-electro-Fenton in the decontamination of emerging pollutants from aqueous solutions. Generally, scholars have expressed that photo-electro-Fenton has indicated higher and more rapid elimination of pollutants than that of electro-Fenton [123 - 126]. Despite the mentioned advantages, this method suffers from drawbacks such as high energy consumption and the need for UV lamps, which are, in turn, expensive [127, 172 - 175].

Photocatalysis

When photons are associated with the reaction system during a catalytic reaction, the process is called photocatalysis. The photoreaction accelerates due to the catalyst's interaction with a substrate and the primary photoproduct. The photocatalysis process provides an efficient alternative to the wastewater treatment and degradation of bacterial substances and dissolved organics. Generally, a photocatalytic reaction undergoes five main stages (Eqs. 15-19) [128, 129]: 1) photon absorption for the generation of the electron (e^-) and hole (h^+) in the conduction band (CB) and valence band (VB), respectively 2) separation of photoexcited e^-, 3) transmission of e^- and hole to the surface of photocatalysts, 4) recombination of e^- and hole, and 5) application of e^-/h^+ on the surface for redox reactions. A schematic of a photocatalytic reaction is shown in Fig. (2).

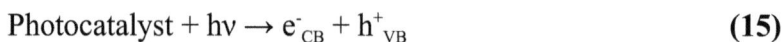

$$\text{Photocatalyst} + hv \rightarrow e^-_{CB} + h^+_{VB} \tag{15}$$

$$H_2O + h^+_{VB} \rightarrow \text{˙OH} + H^+ \tag{16}$$

$$O_2(\text{adsorbed}) + e^-_{CB} \rightarrow \text{˙O}^-_2 \tag{17}$$

$$H^+ + \text{˙O}^-_2 \rightarrow \text{˙OOH} \tag{18}$$

$$\text{Contaminant} + \text{˙OH} + \text{˙OOH} \rightarrow CO_2 + H_2O \tag{19}$$

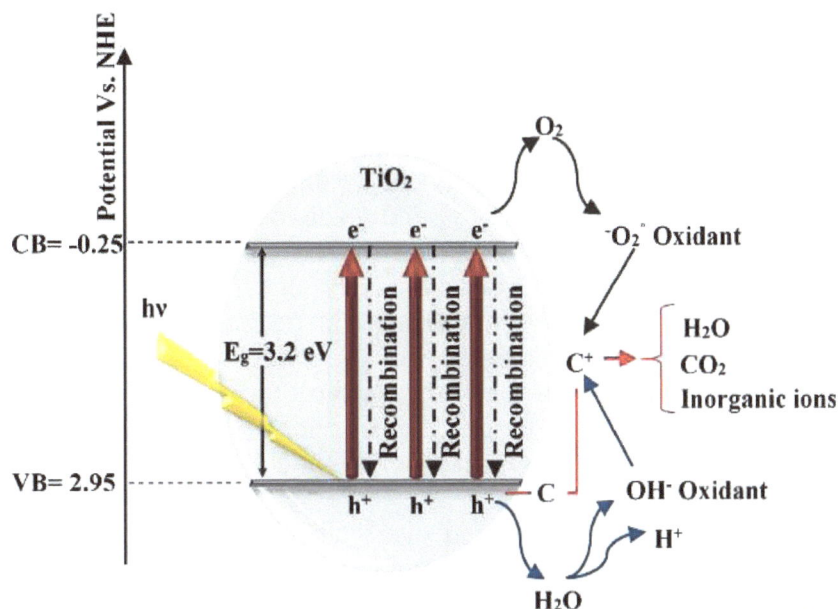

Fig. (2). Photocatalytic removal of water contaminant (C) in the presence of photocatalyst (TiO_2 as a sample). CB= conduction band, VB= valence band. Adapted with permission from Ref [14]. Copyright 2019, Royal Society of Chemistry.

The photo-generated electron carriers can be recombined, trapped on the surface, or migrated to the surface of the photocatalyst for oxidization of contaminants into reactive intermediates or final products [130]. These photocatalysts are often semiconductors of metal oxides with a relatively narrow energy bandgap (E_g) and are applied to improve the reaction rate without changing [10]. Numerous photocatalysts such as TiO_2 [131], ZnO [132], Ag_3PO_4 [133], g-C_3N_4 based materials [134], $BaTiO_3$ [135], $BiVO_4$ [136], *etc.*, individually or alongside other nanomaterials have been used so far for the treatment of emerging pollutants. Fig. (2) shows photocatalytic removal of water contaminants using TiO_2.

Carbuloni *et al.* [138] constructed TiO_2–ZrO_2 for photocatalytic degradation of 10 mg/L metformin at room temperature. Effect of different parameters including catalyst concentration (0.5, 1.0, and 1.5 g/L), operational pH (5.4, 8, and 10), and various percentage of TiO_2-ZrO_2 (5-95% and 95-5%) was evaluated. At the optimized pH 8 and catalyst concentration of 1 g/L, the removal efficiency was highest for TiO_2–ZrO_2 (95%–5%), approximately 55% after 30 min. Methylbiguanide and guanylurea were two oxidation products. Pseudo-first orders kinetic studies revealed a 9.9×10^{-3} 1/min rate constant for TiO_2–ZrO_2 (95–5) catalyst. A Toxicity study proved 100% germination for the treated samples [137]. Majumder and Gupta (2020) synthesized polythiophene modified Al-doped ZnO for photodegradation of 17β-estradiol. Polythiophene was used in this study due to its high light adsorption and charge carrying capacity. The achieved k_{app}, and electrical energy per order of the experiment were 0.4451 1/hour and 1604 kW h/m^3/order, respectively. The optimum percentage of material contained in the catalyst was 0.47 wt% polythiophene and 3.14 mol% aluminum. The optimum experimental parameters were pH 6.5 and a calcination temperature of 200 °C. The results demonstrated that a further increase in the Al content (> 3%) decreases the capturing sites' interval and changes them to a place for e^-/h^+ recombination, and consequently, degradation declines. It is noteworthy that calcination temperature can remarkably improve the particle's crystallinity and photocatalytic activity by reducing the surface moisture. The Photocatalysis process can degrade a wide range of resistant ECs, plus using catalyst can improve the reaction rate. However, this method has drawbacks such as low ability to purify a large volume of wastewater, high overall costs of the process when using artificial UV lamps and electricity, and difficulty separating and recycling the catalysts in aqueous solutions.

Sonocatalysis

Benefiting several privileges such as simple and safe operability, high degradation rate, and eco-friendliness without secondary pollutant sonolysis is one of the promising technologies in wastewater treatment [139]. This acoustic-cavitatio--based method consists of sequential steps of nucleation, growth (expansion), the collision of micro-bubbles (implosion), emission of light (sonoluminescence), and formation of hot spots due to the energy accumulation with high pressure and temperature (~1000 atm and 5000 °K, respectively) [140, 141]. Afterward, H_2O decomposes to produce ˙OH radicals to oxidize contaminants to CO_2 and H_2O [142]. In addition, volatile or hydrophobic pollutants penetrating the cavity interior will be immediately subjected to a pyrolytic reaction [143]. Furthermore, this method has exhibited high degradation and mineralization efficiency for various ECs, including pesticides and personal care products, in a short time. Besides the advantages mentioned above, this technology suffers from high reaction time, low efficiency for organic and emerging contaminants, high energy consumption, and operational and maintenance expenses. As a result, taking efficient actions such as using catalysts alongside this technology to overcome the existed drawbacks are required [139]. The changes may enhance the removal efficiency and reduce the overall energy consumption in practical applications.

In the sonocatalysis method, target compounds diffuse and sorb on the catalyst's surface, and chemical reactions occur on the available active sites. Moreover, the microjet can cause the disintegration of the catalysts to finer particles [144]. This can enhance mass transfer by providing extensive surface area and operating as a nucleation spot to enhance the number of cavitation bubbles [145]. Various factors affect the nucleation step in the sonocatalysis method, including material type (*e.g.*, pellet [146], particles [147], wire mesh [148], *etc.*) and material properties (*e.g.*, particle size, roughness, pore size and shape, and wettability). The critical radii of cavitation nuclei are expected to be 10-100 nm at 20 kHz–1 MHz. The effect of micro-jet on catalysts is related to the particle size. Above all, particles can spread the input energy of the ultrasound wave in the aqueous solution. Thus, studying particle size's impact on the nucleation procedure is undeniable [139, 149].

According to the Wenzel model (Eq. 20), if it is assumed that water complies with the surface topography, the roughness is strongly correlated with the contact angle [150, 151]:

$$\cos \theta_m = r . \cos \theta_\gamma \tag{20}$$

Where θ_m stands for the contact angle, r stands for the roughness ratio (the actual surface area of the rough surface to the geometric projected area), and θ_γ stands for the Young contact angle. Pursuant to Eq. 18, by adding surface roughness, the hydrophobicity of a hydrophobic surface $(\theta_\gamma \geq \frac{\pi}{2})$ will be improved, and similarly, the addition of surface roughness will intensify the hydrophilicity of a hydrophilic surface $(\theta_\gamma \leq \frac{\pi}{2})$ [152].

The pore size has been proved that pores with a suitable aspect ratio can accelerate the nucleation compared to a planar surface constructed from the same material [153]. Studies have revealed that the surface's hydrophobicity can cause a lesser required energy barrier to generate cavitation bubbles [154]. All the studies aim to improve the quality and efficiency of the sonocatalysis method to expand its applicability in the treatment of various contaminants such as emerging pollutants.

Gholami *et al.* (2020) synthesized novel Fe-Cu layered double hydroxide/biochar (Fe-Cu-LDH/BC) for sonocatalytic removal of 0.2 mM cefazolin sodium (CFZ). Using 300 W ultrasonic power (at natural pH of 6.5), the maximum removal efficiency (97.6%) was obtained in the presence of 1.0 g/L catalyst and 0.1 mM CFZ after 80 min contact time. The stability evaluation of sonocatalyst exhibited a 9% removal efficiency drop after 5 cycles. The kinetic studies proved that the removal efficiency best fitted the pseudo-first-order model with a rate constant of 332.4×10^{-4} 1/min. Adding dissolved gas to the reactor leads to more nucleation sites and facile cavitation generation. Moreover, the utilization of gas with low thermal conductivity may lead to fewer bubble collisions, and consequently, minimum heat loss and a drop in degradation efficiency may result [155]. In another study, Lekshmi *et al.* (2018) investigated sonocatalytic elimination of indigo carmine (IC) using MnO_2 as the catalyst. Various ions had diverse effects on the removal efficiency, such as PO_4^{3-} (inhibition), SO_4^{2-} (without notable impact), and NO_3^- (improvement). Results indicated that an increase in the catalyst dosage would increase degradation due to the rise in adsorption sites. Increasing the initial concentration to 40 mg/L increased the degradation efficiency, and further increments reversely affected the removal. Increasing the concentration of IC did not affect the concentration of hydrogen peroxide. At the optimum condition (1.4 g/L catalysts and 40 mg/L IC), the removal efficiency was approximately 80%. Kinetic studies proved the best correlation of the data with the modified Langmuir-Hinshelwood (L–H) model (k_{app}= 1.6 × 10^{-1} 1/min) [156].

Bampos *et al.* (2019) prepared Pd/carbon black nanoparticles (Pd/C) to degrade butylparaben-contaminated water. 25 mg/L catalysts with 10 wt% Pd considerably improved the k_{app} from 0.0126 to 0.071 1/min, confirming pseudo-first-order kinetics. By increasing the pollutant concentration from 0.5 to 2 mg/L, k_{app} dropped from 0.071 to 0.030 1/min. After five cycles, the fabricated catalyst had high stability, soo the k_{app} reduced gently from 0.071 to 0.059 1/min. Adding 250 mg/L of bicarbonates reduced the kinetic constant, while the addition of 250 mg/L chlorides (as one of the most common substrates in water bodies) did not affect k_{app} [157].

Although this method has various privileges, issues such as the dimension of bubbles and hydroxyl radicals can impact the process. Moreover, it is hard to correlate all the involved factors to attain an optimum condition in ultrasonic frequency.

Reduction by Hydrodehalogenation

Catalytic hydrodehalogenation (CHDH) is another alternate purification method for converting hazardous compounds to harmless and facile-degradable substances with higher commercial importance [158]. The halogenated compounds such as dioxines, polychlorinated biphenyls, *etc.*, are among the highly toxic by-products of diverse industrial proceedings [159]. In this case, the CHDH method enables the degradation of halogenated organic substances (HOSs) and, most of the time, attain beneficial products. CHDH of HOSs generates non-halogenated organic substrates under 300 °C and is one of the promising remediation technologies [160]. The catalytic hydrochlorination method is mostly sufficient for purifying chlorinated volatile organic compounds at 30–50 °C and atmospheric pressure. As shown in Fig. (**3**), the releasement of halides (H) in this method reduces carbon and, consequently, the generation of C-H bonds. Hydrodehalogenation of C-X (X= F, Br, Cl) (Eq. 21) occurs on the surface of supported palladium nanoparticles as catalyst [161] in the following order: C-I >C-Br> C-Cl>>C-F. The reaction moves forward *via* consecutive elimination of halogen atoms with hydrogen and the formation of a halogen acid (HX) [85].

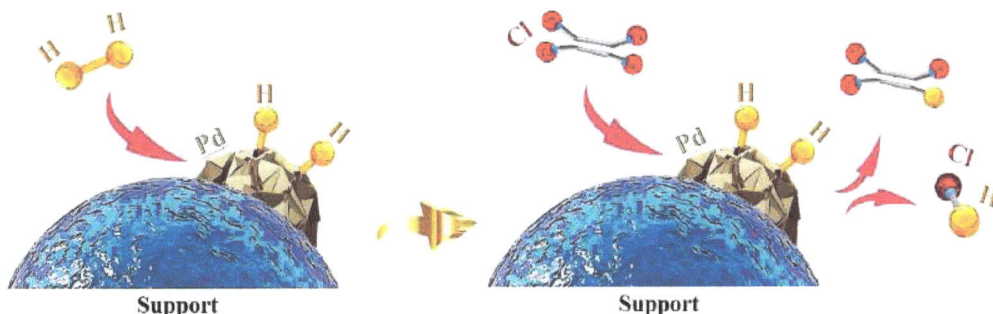

Fig. (3). Schematic of catalytic hydrodehalogenation reaction on the surface of the palladium-supported catalyst.

$$R–X + \text{reducing agent} (H_2) \rightarrow R–H + HX \qquad (21)$$

Fig. (**3**) shows the catalytic hydrodehalogenation reaction on the surface of a palladium-supported catalyst.

In prior studies on CHDH, noble metals such as Pd [162, 163], Rh [164], Pt [165], and Ru [166] have been utilized for wastewater treatment. To decrease operational cost, non-noble metals such as Ni [167], Co [168], Fe [169], Mn [170], *etc.*, have also been applicated in catalytic reactions. Some of the prominences of the CHDH method over conventional oxidation and separation methods are as follows: (I) Cost-effectiveness due to conduction in normal conditions (low temperatures) and energy-saving, (II) excellent selectivity and production of non-toxic or harmless compounds, (III) absence of free radicals and therefore no need further treatment, (IV) feasible in batch and continuous flow conditions, (V) applicable as a pre-treatment phase for detoxification of input flows before the entrance to other technologies (*e.g.*, biological methods) [161].

Considering these advantages, CHDH has been widely applied in the remediation of various emerging pollutants. For instance, Biswas *et al.* (2019) experimented with polyionic resin-Pd/Fe2O3 composite to purify haloaromatic substances. The results demonstrated superior catalytic function in the CHDH of various haloaromatics. Powerful interactions of Fe_2O_3 and Pd nanoparticles simplify the dissociation of the B-H bond and hydride generation. Selecting 9,10-dibromoanthracene as the model pollutant, at optimum conditions (catalyst= 50 mg/L, 70 °C, tetrahydrofuran: H_2O= 2:1), after 5 hr, the hydrodebrominated product was achieved the highest yield of 96%. Studying the stability of the catalyst revealed that after five successive cycles, the catalyst's efficiency did not change [162].

Tarach *et al.* (2018) experimented efficiency of Ni-zeolites Y and ZSM-5 catalysts in the hydrochlorination of 6.4×10^{-5} M trichloroethylene. Ni-zeolite, including 0.79 wt% Ni and Si/Al equal to 31 (pore size: 4–8 nm), was the most active catalyst with a reaction rate constant of 0.024 1/min and removal efficiency of 68% after 150 min. The Ni@ZSM-5 catalyst (consisting of 1.24% wt % Ni) removed only 48% of the pollutant after 150 min. Researchers emphasize that the acidic properties of zeolitic support, besides the pore distribution of zeolite grains, can undeniably impact the catalytic performance of the catalyst [167]. In another study, 0.5 mmol nitrobenzenes were used as the target pollutant to evaluate removal efficiency by FeS_2/N,S-codoped porous carbon (FeS_2/NSC) catalyst. Reports imply that FeS_2/NSC has had outstanding ability in chemo-selection and reduction of nitro groups with various reducible groups. At 120 °C, 20 bar, FeS_2/NSC (with 6.8 wt% equal to 2.3 mol% Fe, 3.07 wt% N, and 26.75 wt% S), almost 100% of the contaminant was degraded within 5 h. Researchers reported that reducing the reaction temperature or H_2 pressure can strongly decrease reaction efficiency. Moreover, the high pore volume, the great specific surface area, and the presence of N and S besides carbon undeniably improve the material's catalytic performance. After eight cycles, the catalyst exhibited considerable stability without significantly dropping catalytic performance [171].

CONCLUSION

The release of emerging contaminants into the environment has raised much concern due to the continuous disruption of the ecosystem balance. This may result in jeopardizing the health of all the organisms, including human beings, through the water cycle. To address these problems, attempts have been made to decontaminate the ECs with high toxic potential to provide clean and safe water as a basic need for sustaining life. The catalytic process has played a leading role in removing emerging water pollutants on the laboratory scale. Some major catalytic processes involve oxidation (electrocatalytic oxidation, electro-Fenton, and photo-electro-Fenton), photocatalysis, sonocatalysis processes, and reduction. Many ECs can be degraded through powerful oxidizers like ozone. However, instability, high energy requirement, and oxidative byproducts restrict the ozonation method for water treatment. Electrocatalytic oxidation is successfully applied for different industrial wastewater treatments. In the electro-Fenton method, Fenton reagents accompanied are by oxidation on the surface of the anode to decontaminate water. Though this method overcomes several challenges attributed to the conventional Fenton reaction in which a highly reactive and strong oxidizing agent of OH radical (the ability to react with most organic compounds) is generated *via* ferrous ions (Fe^{2+}) (as a catalyst) from hydrogen peroxide. This method encounters drawbacks such as secondary sludge generation, slow production of H_2O_2, low current efficiency at higher pH, and precipitation of hydroxide ions resulting in an efficiency drop and increasing the cost of the process. Concurrently with or after the electro-Fenton method, wastewater can be irradiated with UV/visible light radiations to improve Fenton reactions. However, photo-accompanied with Fenton reactions has shown more efficiency than the processes without the photo-irradiation step. However, high energy consumption and the need for UV lamps make this method expensive for industrial applications.

Photocatalysis involves electron/hole production through irradiation of incident light and migration of the photogenerated electron/hole to the surface of the photocatalysts (parts of them can be recombined) followed by simultaneous reduction (production of hydrogen and ˙OH radicals) and oxidation reactions through photoelectrons and holes, respectively. The photocatalysis method has demonstrated degrading a wide variety of ECs. However, challenges such as high overall costs of the process separation and recycling of the catalysts and limitations attributed to the purification of large volumes of wastewater need solutions. Sonocatalysis is another comparatively eco-friendly process, a straightforward method that works based on sequential steps of nucleation, growth, and collision of micro-bubbles. On the other hand, controlling parameters such as the size of microbubbles and OH radical generation affecting the procedure is hard to obtain a suitable efficiency. As a result, all the presented methodologies possess advantages and disadvantages, showing promising opportunities for improvement. In addition, despite suitable efficiency on an experimental scale, a continued need remains for introducing more effective, inexpensive, and convenient technologies for industrial and large-scale water remediation.

CONSENT FOR PUBLICATION

Not applicable.

CONFLICT OF INTEREST

The author declares no conflict of interest, financial or otherwise.

ACKNOWLEDGEMENTS

Declared none.

REFERENCES

[1]   S.D. Richardson, and S.Y. Kimura, "Emerging environmental contaminants: Challenges facing our next generation and potential engineering solutions", *Environ. Technol. Innov.,* vol. 8, pp. 40-56, 2017.
        [http://dx.doi.org/10.1016/j.eti.2017.04.002]

[2]   P.R. Rout, T.C. Zhang, P. Bhunia, and R.Y. Surampalli, "Treatment technologies for emerging contaminants in wastewater treatment plants: A review", *Sci. Total Environ.,* vol. 753, p. 141990, 2021.
        [http://dx.doi.org/10.1016/j.scitotenv.2020.141990] [PMID: 32889321]

[3]   C.C. Montagner, F.F. Sodré, R.D. Acayaba, C. Vidal, I. Campestrini, M.A. Locatelli, I.C. Pescara, A.F. Albuquerque, G.A. Umbuzeiro, and W.F. Jardim, "Ten years-snapshot of the occurrence of emerging contaminants in drinking, surface and ground waters and wastewaters from São Paulo State, Brazil", *J. Braz. Chem. Soc.,* vol. 30, pp. 614-632, 2019.
        [http://dx.doi.org/10.21577/0103-5053.20180232]

[4]   B.M. Sharma, J. Bečanová, M. Scheringer, A. Sharma, G.K. Bharat, P.G. Whitehead, J. Klánová, and L. Nizzetto, "Health and ecological risk assessment of emerging contaminants (pharmaceuticals, personal care products, and artificial sweeteners) in surface and groundwater (drinking water) in the Ganges River Basin, India", *Sci. Total Environ.,* vol. 646, pp. 1459-1467, 2019.
        [http://dx.doi.org/10.1016/j.scitotenv.2018.07.235] [PMID: 30235631]

[5]   R. Loos, B.M. Gawlik, K. Boettcher, G. Locoro, S. Contini, and G. Bidoglio, "Sucralose screening in European surface waters using a solid-phase extraction-liquid chromatography-triple quadrupole mass spectrometry method", *J. Chromatogr. A,* vol. 1216, no. 7, pp. 1126-1131, 2009.
        [http://dx.doi.org/10.1016/j.chroma.2008.12.048] [PMID: 19131070]

[6]   A.K.E. Wiklund, M. Breitholtz, B.E. Bengtsson, and M. Adolfsson-Erici, "Sucralose - an ecotoxicological challenger?", *Chemosphere,* vol. 86, no. 1, pp. 50-55, 2012.
        [http://dx.doi.org/10.1016/j.chemosphere.2011.08.049] [PMID: 21955350]

[7]   M.S. Mauter, and M. Elimelech, "Environmental applications of carbon-based nanomaterials", *Environ. Sci. Technol.,* vol. 42, no. 16, pp. 5843-5859, 2008.
        [http://dx.doi.org/10.1021/es8006904] [PMID: 18767635]

[8]   D.H. Wang, J.K. Kim, G.H. Lim, K.H. Park, O.O. Park, B. Lim, and J.H. Park, "Enhanced light harvesting in bulk heterojunction photovoltaic devices with shape-controlled Ag nanomaterials: Ag nanoparticles versus Ag nanoplates", *RSC Advances,* vol. 2, pp. 7268-7272, 2012.
        [http://dx.doi.org/10.1039/c2ra20815f]

[9]   S. Zgheib, R. Moilleron, M. Saad, and G. Chebbo, "Partition of pollution between dissolved and particulate phases: what about emerging substances in urban stormwater catchments?", *Water Res.,* vol. 45, no. 2, pp. 913-925, 2011.
        [http://dx.doi.org/10.1016/j.watres.2010.09.032] [PMID: 20970821]

[10]   M.B. Ahmed, J.L. Zhou, H.H. Ngo, W. Guo, N.S. Thomaidis, and J. Xu, "Progress in the biological and chemical treatment technologies for emerging contaminant removal from wastewater: A critical review", *J. Hazard. Mater.,* vol. 323, no. Pt A, pp. 274-298, 2017.
        [http://dx.doi.org/10.1016/j.jhazmat.2016.04.045] [PMID: 27143286]

[11]   A. Liu, P. Egodawatta, Y. Guan, and A. Goonetilleke, "Influence of rainfall and catchment characteristics on urban stormwater quality",
        [http://dx.doi.org/10.1016/j.scitotenv.2012.11.053]

[12]   M.B. Campanha, A.T. Awan, D.N.R. de Sousa, G.M. Grosseli, A.A. Mozeto, and P.S. Fadini, "A 3-year study on occurrence of emerging contaminants in an urban stream of São Paulo State of Southeast Brazil", *Environ. Sci. Pollut. Res. Int.,* vol. 22, no. 10, pp. 7936-7947, 2015.
        [http://dx.doi.org/10.1007/s11356-014-3929-x] [PMID: 25516246]

[13]   I.J. Buerge, H.R. Buser, M. Kahle, M.D. Müller, and T. Poiger, "Ubiquitous occurrence of the artificial sweetener acesulfame in the aquatic environment: an ideal chemical marker of domestic

wastewater in groundwater", *Environ. Sci. Technol.,* vol. 43, no. 12, pp. 4381-4385, 2009.
[http://dx.doi.org/10.1021/es900126x] [PMID: 19603650]

[14]   S. Taghipour, S.M. Hosseini, and B. Ataie-Ashtiani, "Engineering nanomaterials for water and wastewater treatment: review of classifications, properties and applications", *New J. Chem.,* vol. 43, pp. 7902-7927, 2019.
[http://dx.doi.org/10.1039/C9NJ00157C]

[15]   M. Jannesari, O. Akhavan, H.R. Madaah Hosseini, and B. Bakhshi, "Graphene/CuO2 Nanoshuttles with Controllable Release of Oxygen Nanobubbles Promoting Interruption of Bacterial Respiration", *ACS Appl. Mater. Interfaces,* vol. 12, no. 32, pp. 35813-35825, 2020.
[http://dx.doi.org/10.1021/acsami.0c05732] [PMID: 32664715]

[16]   M. Jannesari, O. Akhavan, and H.R. Madaah Hosseini, "Graphene oxide in generation of nanobubbles using controllable microvortices of jet flows", *Carbon N. Y.,* vol. 138, pp. 8-17, 2018.
[http://dx.doi.org/10.1016/j.carbon.2018.05.068]

[17]   O. Akhavan, M. Saadati, and M. Jannesari, "Graphene Jet Nanomotors in Remote Controllable Self-Propulsion Swimmers in Pure Water", *Nano Lett.,* vol. 16, no. 9, pp. 5619-5630, 2016.
[http://dx.doi.org/10.1021/acs.nanolett.6b02175] [PMID: 27483134]

[18]   T. Benn, P. Herckes, and P. Westerhoff, *Fullerenes in environmental samples: C 60 in atmospheric particulate matter.* Compr. Anal. Chem, 2012, pp. 291-303.
[http://dx.doi.org/10.1016/B978-0-444-56328-6.00010-4]

[19]   S.H. Wu, and D.H. Chen, "Synthesis of high-concentration Cu nanoparticles in aqueous CTAB solutions", *J. Colloid Interface Sci.,* vol. 273, no. 1, pp. 165-169, 2004.
[http://dx.doi.org/10.1016/j.jcis.2004.01.071] [PMID: 15051447]

[20]   D. Šojić Merkulov, M. Lazarević, A. Djordjevic, M. Náfrádi, T. Alapi, P. Putnik, Z. Rakočević, M. Novaković, B. Miljević, S. Bognár, and B. Abramović, "Potential of TiO$_2$ with various au nanoparticles for catalyzing mesotrione removal from wastewaters under sunlight", *Nanomaterials (Basel),* vol. 10, no. 8, pp. 1-15, 2020.
[http://dx.doi.org/10.3390/nano10081591] [PMID: 32823509]

[21]   K. Chen, H. Lu, G. Li, J. Zhang, Y. Tian, Y. Gao, Q. Guo, H. Lu, and J. Gao, "Surface functionalization of porous In$_2$O$_3$ nanofibers with Zn nanoparticles for enhanced low-temperature NO$_2$ sensing properties", *Sens. Actuators B Chem.,* vol. 308, 2020.
[http://dx.doi.org/10.1016/j.snb.2020.127716]

[22]   M. Zamani, M. Morshed, J. Varshosaz, and M. Jannesari, "Controlled release of metronidazole benzoate from poly ε-caprolactone electrospun nanofibers for periodontal diseases", *Eur. J. Pharm. Biopharm.,* vol. 75, no. 2, pp. 179-185, 2010.
[http://dx.doi.org/10.1016/j.ejpb.2010.02.002] [PMID: 20144711]

[23]   M. Jannesari, J. Varshosaz, M. Morshed, and M. Zamani, "Composite poly(vinyl alcohol)/poly(vinyl acetate) electrospun nanofibrous mats as a novel wound dressing matrix for controlled release of drugs", *Int. J. Nanomedicine,* vol. 6, pp. 993-1003, 2011.
[http://dx.doi.org/10.2147/IJN.S17595] [PMID: 21720511]

[24]   L. Zhao, L. Lu, A. Wang, H. Zhang, M. Huang, H. Wu, B. Xing, Z. Wang, and R. Ji, "Nano-Biotechnology in Agriculture: Use of Nanomaterials to Promote Plant Growth and Stress Tolerance", *J. Agric. Food Chem.,* vol. 68, no. 7, pp. 1935-1947, 2020.
[http://dx.doi.org/10.1021/acs.jafc.9b06615] [PMID: 32003987]

[25]   H.J. Fecht, K. Brühne, and P. Gluche, "Carbon-based nanomaterials and hybrids - synthesis, properties and commercial applications",

[26]   S.M. Hosseini, B. Ataie-Ashtiani, and M. Kholghi, "Bench-scaled nano-Fe0 permeable reactive barrier for nitrate removal", *Ground Water Monit. Remediat.,* vol. 31, pp. 82-94, 2011.
[http://dx.doi.org/10.1111/j.1745-6592.2011.01352.x]

[27]   S. Mossa Hosseini, B. Ataie-Ashtiani, and M. Kholghi, "Nitrate reduction by nano-Fe/Cu particles in packed column", *Desalination,* vol. 276, pp. 214-221, 2011.
[http://dx.doi.org/10.1016/j.desal.2011.03.051]

[28]   S.M. Hosseini, T. Tosco, B. Ataie-Ashtiani, and C.T. Simmons, "Non-pumping reactive wells filled with mixing nano and micro zero-valent iron for nitrate removal from groundwater: Vertical, horizontal, and slanted wells", *J. Contam. Hydrol.,* vol. 210, pp. 50-64, 2018.
[http://dx.doi.org/10.1016/j.jconhyd.2018.02.006] [PMID: 29519731]

[29]   S.M.H. Tiziana Tosco, "Bimetallic Fe/Cu nanoparticles for groundwater remediation: optimized injection strategies *via* transport modelling in porous media", vol. 31, pp. 2013-2014, 2013,

[30]   A.S. Kazemi, A.A. Noroozi, A. Khamsavi, A. Mazaheri, S.M. Hosseini, and Y. Abdi, "Engineering Water and Solute Dynamics and Maximal Use of CNT Surface Area for Efficient Water Desalination", *ACS Omega,* vol. 4, no. 4, pp. 6826-6847, 2019.
[http://dx.doi.org/10.1021/acsomega.9b00188] [PMID: 31459801]

[31]   A. Saberinasr, M. Rezaei, M. Nakhaei, and S.M. Hosseini, "Transport of CMC-Stabilized nZVI in Saturated Sand Column: the Effect of Particle Concentration and Soil Grain Size", *Water Air Soil Pollut.,* vol. 227, 2016.
[http://dx.doi.org/10.1007/s11270-016-3097-3]

[32]   S.M. Hosseini, and T. Tosco, "Integrating NZVI and carbon substrates in a non-pumping reactive wells array for the remediation of a nitrate contaminated aquifer", *J. Contam. Hydrol.,* vol. 179, pp. 182-195, 2015.
[http://dx.doi.org/10.1016/j.jconhyd.2015.06.006] [PMID: 26142547]

[33]   S.M. Hosseini, and T. Tosco, "Transport and retention of high concentrated nano-Fe/Cu particles through highly flow-rated packed sand column", *Water Res.,* vol. 47, no. 1, pp. 326-338, 2013.
[http://dx.doi.org/10.1016/j.watres.2012.10.002] [PMID: 23141767]

[34]   A.S. Kazemi, S.M. Hosseini, and Y. Abdi, "Large total area membrane of suspended single layer graphene for water desalination", *Desalination,* pp. 160-171, 2019.
[http://dx.doi.org/10.1016/j.desal.2017.12.050]

[35]   L.C. Pereira, A.O. de Souza, M.F. Franco Bernardes, M. Pazin, M.J. Tasso, P.H. Pereira, and D.J. Dorta, "A perspective on the potential risks of emerging contaminants to human and environmental health", *Environ. Sci. Pollut. Res. Int.,* vol. 22, no. 18, pp. 13800-13823, 2015.
[http://dx.doi.org/10.1007/s11356-015-4896-6] [PMID: 26201652]

[36]   R. Carson, "Silent spring", In: *Key Readings Journal*, 2012, pp. 290-298.
[http://dx.doi.org/10.9774/GLEAF.978-1-907643-44-6_4]

[37]   O.J. Bandele, M.F. Santillo, M. Ferguson, and P.L. Wiesenfeld, "In vitro toxicity screening of chemical mixtures using HepG2/C3A cells", *Food Chem. Toxicol.,* vol. 50, no. 5, pp. 1653-1659, 2012.
[http://dx.doi.org/10.1016/j.fct.2012.02.016] [PMID: 22381260]

[38]   D.M. Leme, F.L. Primo, G.G. Gobo, C.R.V. da Costa, A.C. Tedesco, and D.P. de Oliveira, "Genotoxicity assessment of reactive and disperse textile dyes using human dermal equivalent (3D cell culture system)", *J. Toxicol. Environ. Health A,* vol. 78, no. 7, pp. 466-480, 2015.
[http://dx.doi.org/10.1080/15287394.2014.999296] [PMID: 25785560]

[39]   J.D. Wolt, "Advancing environmental risk assessment for transgenic biofeedstock crops", *Biotechnol. Biofuels,* vol. 2, no. 1, p. 27, 2009.
[http://dx.doi.org/10.1186/1754-6834-2-27] [PMID: 19883509]

[40]   H. Sanderson, R.A. Brain, D.J. Johnson, C.J. Wilson, and K.R. Solomon, "Toxicity classification and evaluation of four pharmaceuticals classes: antibiotics, antineoplastics, cardiovascular, and sex hormones", *Toxicology,* vol. 203, no. 1-3, pp. 27-40, 2004.
[http://dx.doi.org/10.1016/j.tox.2004.05.015] [PMID: 15363579]

[41]   C.G. Campbell, S.E. Borglin, F.B. Green, A. Grayson, E. Wozei, and W.T. Stringfellow, "Biologically directed environmental monitoring, fate, and transport of estrogenic endocrine disrupting compounds in water: A review", *Chemosphere,* vol. 65, no. 8, pp. 1265-1280, 2006.
[http://dx.doi.org/10.1016/j.chemosphere.2006.08.003] [PMID: 16979218]

[42]   W. Witte, "Medical consequences of antibiotics use in agriculture", *Science (80-.),* vol. 279, pp. 996997-, 1998.
[http://dx.doi.org/10.1126/science.279.5353.996]

[43]   M.R. Servos, D.T. Bennie, B.K. Burnison, A. Jurkovic, R. McInnis, T. Neheli, A. Schnell, P. Seto, S.A. Smyth, and T.A. Ternes, "Distribution of estrogens, 17β-estradiol and estrone, in Canadian municipal wastewater treatment plants", *Sci. Total Environ.,* vol. 336, no. 1-3, pp. 155-170, 2005.
[http://dx.doi.org/10.1016/j.scitotenv.2004.05.025] [PMID: 15589256]

[44]   D.M. Bila, and M Dezotti, "Desreguladores endócrinos no meio ambiente: efeitos e conseqüências, Quim. Nova",
[http://dx.doi.org/10.1590/s0100-40422007000300027]

[45]   L. Du, and W. Liu, "Occurrence, fate, and ecotoxicity of antibiotics in agro-ecosystems. A review", *Agron. Sustain. Dev.,* vol. 32, pp. 309-327, 2012.
[http://dx.doi.org/10.1007/s13593-011-0062-9]

[46]   A.R. Gomes, C. Justino, T. Rocha-Santos, A.C. Freitas, A.C. Duarte, and R. Pereira, "Review of the ecotoxicological effects of emerging contaminants to soil biota", *J. Environ. Sci. Health Part A Tox. Hazard. Subst. Environ. Eng.,* vol. 52, no. 10, pp. 992-1007, 2017.
[http://dx.doi.org/10.1080/10934529.2017.1328946] [PMID: 28598770]

[47]   "The handbook of environmental chemistry, Environ. Pollut. Ser. A", *Ecol. Biol.,* vol. 27, p. 166, 1982.
[http://dx.doi.org/10.1016/0143-1471(82)90111-8]

[48]   D. van de Meent, "Simplebox: a generic multimedia fate evaluation model", 1993.

[49]   J. Klasmeier, and M. Matthies, "Application of the Geography-Referenced Environmental Assessment Tool for European Rivers (GREAT-ER) in the Catchment of the River MAIN (Germany)", 1993.

[50]   J.S. LaKind, J. Overpeck, P.N. Breysse, L. Backer, S.D. Richardson, J. Sobus, A. Sapkota, C.R. Upperman, C. Jiang, C.B. Beard, J.M. Brunkard, J.E. Bell, R. Harris, J.P. Chretien, R.E. Peltier, G.L. Chew, and B.C. Blount, "Exposure science in an age of rapidly changing climate: challenges and opportunities", *J. Expo. Sci. Environ. Epidemiol.,* vol. 26, no. 6, pp. 529-538, 2016.
[http://dx.doi.org/10.1038/jes.2016.35] [PMID: 27485992]

[51]   J. Wang, R. Zhuan, and L. Chu, "The occurrence, distribution and degradation of antibiotics by ionizing radiation: An overview", *Sci. Total Environ.,* vol. 646, pp. 1385-1397, 2019.
[http://dx.doi.org/10.1016/j.scitotenv.2018.07.415] [PMID: 30235624]

[52]   T. Heberer, "Occurrence, fate, and removal of pharmaceutical residues in the aquatic environment: a review of recent research data", *Toxicol. Lett.,* vol. 131, no. 1-2, pp. 5-17, 2002.
[http://dx.doi.org/10.1016/S0378-4274(02)00041-3] [PMID: 11988354]

[53]   B. Hoeger, B. Köllner, D.R. Dietrich, and B. Hitzfeld, "Water-borne diclofenac affects kidney and gill integrity and selected immune parameters in brown trout (Salmo trutta f. fario)", *Aquat. Toxicol.,* vol. 75, no. 1, pp. 53-64, 2005.
[http://dx.doi.org/10.1016/j.aquatox.2005.07.006] [PMID: 16139376]

[54]   W. Li, L. Gao, Y. Shi, Y. Wang, J. Liu, and Y. Cai, "Spatial distribution, temporal variation and risks of parabens and their chlorinated derivatives in urban surface water in Beijing, China", *Sci. Total Environ.,* vol. 539, pp. 262-270, 2016.
[http://dx.doi.org/10.1016/j.scitotenv.2015.08.150] [PMID: 26363399]

[55]   "Molecular structure in relation to oestrogenic activity. Compounds without a phenanthrene nucleus", *Proc. R. Soc. London. Ser. B - Biol. Sci,* vol. 125, pp. 222-232, 1938.

[http://dx.doi.org/10.1098/rspb.1938.0023]

[56]   A.V. Krishnan, P. Stathis, S.F. Permuth, L. Tokes, and D. Feldman, "Bisphenol-A: an estrogenic substance is released from polycarbonate flasks during autoclaving", *Endocrinology,* vol. 132, no. 6, pp. 2279-2286, 1993.
[http://dx.doi.org/10.1210/endo.132.6.8504731] [PMID: 8504731]

[57]   P. Sohoni, and J.P. Sumpter, "Several environmental oestrogens are also anti-androgens", *J. Endocrinol.,* vol. 158, no. 3, pp. 327-339, 1998.
[http://dx.doi.org/10.1677/joe.0.1580327] [PMID: 9846162]

[58]   I. for E. and Health, Environmental Oestrogens: Consequences To Human Health and Wildlife, 2000. papers2://publication/uuid/87A4C585-82D9-4732-8F5F-F71780357676.

[59]   L.M. McMurry, M. Oethinger, and S.B. Levy, "Triclosan targets lipid synthesis", *Nature,* vol. 394, no. 6693, pp. 531-532, 1998. [4].
[http://dx.doi.org/10.1038/28970] [PMID: 9707111]

[60]   K.T. Chung, "Azo dyes and human health: A review", *J. Environ. Sci. Health Part C Environ. Carcinog. Ecotoxicol. Rev.,* vol. 34, no. 4, pp. 233-261, 2016.
[http://dx.doi.org/10.1080/10590501.2016.1236602] [PMID: 27635691]

[61]   C.E. Zubieta, P.V. Messina, and P.C. Schulz, "Photocatalytic degradation of acridine dyes using anatase and rutile TiO2", *J. Environ. Manage.,* vol. 101, pp. 1-6, 2012.
[http://dx.doi.org/10.1016/j.jenvman.2012.02.014] [PMID: 22387324]

[62]   F.M. Drumond Chequer, G.A.R. de Oliveira, E.R. Anastacio Ferraz, J. Carvalho, M.V. Boldrin Zanoni, and D.P. de Oliveir, "Textile Dyes: Dyeing Process and Environmental Impact", In: *Eco-Friendly Text. Dye. Finish,* 2013.
[http://dx.doi.org/10.5772/53659]

[63]   D.P. Mohapatra, M. Cledón, S.K. Brar, and R.Y. Surampalli, "Application of Wastewater and Biosolids in Soil: Occurrence and Fate of Emerging Contaminants", *Water Air Soil Pollut.,* vol. 227, 2016.
[http://dx.doi.org/10.1007/s11270-016-2768-4]

[64]   M.K. Hill, "Understanding environmental pollution", 1997.
[http://dx.doi.org/10.5860/CHOICE.48-3900]

[65]   N. Ratola, A. Cincinelli, A. Alves, and A. Katsoyiannis, "Occurrence of organic microcontaminants in the wastewater treatment process. A mini review", *J. Hazard. Mater.,* vol. 239-240, pp. 1-18, 2012.
[http://dx.doi.org/10.1016/j.jhazmat.2012.05.040] [PMID: 22771351]

[66]   G.W. Olsen, S.C. Chang, P.E. Noker, G.S. Gorman, D.J. Ehresman, P.H. Lieder, and J.L. Butenhoff, "A comparison of the pharmacokinetics of perfluorobutanesulfonate (PFBS) in rats, monkeys, and humans", *Toxicology,* vol. 256, no. 1-2, pp. 65-74, 2009.
[http://dx.doi.org/10.1016/j.tox.2008.11.008] [PMID: 19059455]

[67]   V. Novotny, "Diffuse pollution from agriculture - A worldwide outlook", In: *Water Sci. Technol,* 1999, pp. 1-13.
[http://dx.doi.org/10.2166/wst.1999.0124]

[68]   Y. Darradi, E. Saur, R. Laplana, J.M. Lescot, V. Kuentz, and B.C. Meyer, "Optimizing the environmental performance of agricultural activities: A case study in la Boulouze watershed", *Ecol. Indic.,* vol. 22, pp. 27-37, 2012.
[http://dx.doi.org/10.1016/j.ecolind.2011.10.011]

[69]   D. Norse, "Non-point pollution from crop production: Global, regional and national issues", *Pedosphere,* vol. 15, pp. 499-508, 2005.

[70]   P.M. Melia, A.B. Cundy, S.P. Sohi, P.S. Hooda, and R. Busquets, "Trends in the recovery of phosphorus in bioavailable forms from wastewater", *Chemosphere,* vol. 186, pp. 381-395, 2017.
[http://dx.doi.org/10.1016/j.chemosphere.2017.07.089] [PMID: 28802130]

[71] J. Jernberg, "Novel analytical methods for the identification of emerging contaminants in aquatic environments", 2013.

[72] W. Giger, "Hydrophilic and amphiphilic water pollutants: Using advanced analytical methods for classic and emerging contaminants", *Anal. Bioanal. Chem.,* vol. 393, no. 2009, pp. 37-44, .

[73] L. Martín-Pozo, B. de Alarcón-Gómez, R. Rodríguez-Gómez, M.T. García-Córcoles, M. Çipa, and A. Zafra-Gómez, "Analytical methods for the determination of emerging contaminants in sewage sludge samples. A review", *Talanta,* vol. 192, pp. 508-533, 2019.
[http://dx.doi.org/10.1016/j.talanta.2018.09.056] [PMID: 30348425]

[74] A. Llop, F. Borrull, and E. Pocurull, "Pressurised hot water extraction followed by headspace solid-phase microextraction and gas chromatography-tandem mass spectrometry for the determination of N-nitrosamines in sewage sludge", *Talanta,* vol. 88, pp. 284-289, 2012.
[http://dx.doi.org/10.1016/j.talanta.2011.10.042] [PMID: 22265500]

[75] K. Wille, H.F. De Brabander, L. Vanhaecke, E. De Wulf, P. Van Caeter, and C.R. Janssen, "Coupled chromatographic and mass-spectrometric techniques for the analysis of emerging pollutants in the aquatic environment, TrAC -", *Trends Analyt. Chem.,* vol. 35, pp. 87-108, 2012.
[http://dx.doi.org/10.1016/j.trac.2011.12.003]

[76] S.D. Richardson, and S.Y. Kimura, "Water analysis: Emerging contaminants and current issues", *Anal. Chem.,* vol. 92, no. 1, pp. 473-505, 2020.
[http://dx.doi.org/10.1021/acs.analchem.9b05269] [PMID: 31825597]

[77] N. Pérez-Lemus, R. López-Serna, S.I. Pérez-Elvira, and E. Barrado, "Analytical methodologies for the determination of pharmaceuticals and personal care products (PPCPs) in sewage sludge: A critical review", *Anal. Chim. Acta,* vol. 1083, pp. 19-40, 2019.
[http://dx.doi.org/10.1016/j.aca.2019.06.044] [PMID: 31493808]

[78] P. Rajasulochana, and V. Preethy, "Comparison on efficiency of various techniques in treatment of waste and sewage water – A comprehensive review", *Resour. Technol.,* vol. 2, pp. 175-184, 2016.
[http://dx.doi.org/10.1016/j.reffit.2016.09.004]

[79] H. Khorsandi, M. Teymori, A.A. Aghapour, S.J. Jafari, S. Taghipour, and R. Bargeshadi, "Photodegradation of ceftriaxone in aqueous solution by using UVC and UVC/H_2O_2 oxidation processes", *Appl. Water Sci.,* vol. 9, 2019.
[http://dx.doi.org/10.1007/s13201-019-0964-2]

[80] S. Taghipour, and B. Ayati, "Cultivation of aerobic granules through synthetic petroleum wastewater treatment in a cyclic aerobic granular reactor, Desalin", *Water Treat.,* vol. 76, pp. 134-142, 2017.
[http://dx.doi.org/10.5004/dwt.2017.20779]

[81] S. Taghipour, B. Ayati, and M. Razaei, "Study of the SBAR performance in COD removal of Petroleum and MTBE", *Modares Civ. Eng. J,* 2017pp. 17-27.https://mcej.modares.ac.ir/article--6-7139-en.html

[82] S. Taghipour, and B. Ayati, "Petroleum Wastewater Treatment Using Granular Sequencing Batch Reactor: Physical Characteristics and Capabilities of the Aerobic Granules", In: *Wastewater Treat. Process. Uses Importance* NOVA SCIENCE PUBLISHERS, INC, 2019, pp. 145-179.

[83] S. Parsons, "Advanced Oxidation Processes for Water and Wastewater Treatment", In: *Water Intell. Online.,* S. Parsons, Ed., IWA Publishing: London, 2015, pp. 9781780403076-9781780403076.
[http://dx.doi.org/10.2166/9781780403076]

[84] S. Guerra-Rodríguez, E. Rodríguez, D.N. Singh, and J. Rodríguez-Chueca, "Assessment of sulfate radical-based advanced oxidation processes for water and wastewater treatment: A review", *Water (Switzerland),* vol. 10, 2018.
[http://dx.doi.org/10.3390/w10121828]

[85] C. Descorme, "Catalytic wastewater treatment: Oxidation and reduction processes. Recent studies on chlorophenols", *Catal. Today,* vol. 297, pp. 324-334, 2017.

[http://dx.doi.org/10.1016/j.cattod.2017.03.039]

[86]   B. Dolly, *Singh, A. Singh, S. Sharma, Ozone as a Shelf-Life Extender of Fruits.* Emerg. Technol. Shelf-Life Enhanc. Fruits, 2020, pp. 289-312.
[http://dx.doi.org/10.1201/9780429264481-11]

[87]   M.G. Nagarkatti, *Ozone in Water Treatment: Application and Engineering.* Routledge, 1991.
[http://dx.doi.org/10.2134/jeq1991.00472425002000040040x]

[88]   C. von Sonntag, and U. von Gunten, "Chemistry of Ozone in Water and Wastewater Treatment: From Basic Principles to Applications", 2015.
[http://dx.doi.org/10.2166/9781780400839]

[89]   K. Dhangar, and M. Kumar, "Tricks and tracks in removal of emerging contaminants from the wastewater through hybrid treatment systems: A review", *Sci. Total Environ.,* vol. 738, p. 140320, 2020.
[http://dx.doi.org/10.1016/j.scitotenv.2020.140320] [PMID: 32806367]

[90]   K. Ikehata, M.G. El-Din, and S.A. Snyder, "Ozonation and advanced oxidation treatment of emerging organic pollutants in water and wastewater", In: *Ozone Sci. Eng,* 2008, pp. 21-26.
[http://dx.doi.org/10.1080/01919510701728970]

[91]   J. Rivera-Utrilla, M. Sánchez-Polo, M.Á. Ferro-García, G. Prados-Joya, and R. Ocampo-Pérez, "Pharmaceuticals as emerging contaminants and their removal from water. A review", *Chemosphere,* vol. 93, no. 7, pp. 1268-1287, 2013.
[http://dx.doi.org/10.1016/j.chemosphere.2013.07.059] [PMID: 24025536]

[92]   A. Rodríguez, R. Rosal, J.A. Perdigón-Melón, M. Mezcua, A. Agüera, M.D. Hernando, P. Letón, A.R. Fernández-Alba, and E. García-Calvo, "Ozone-based technologies in water and wastewater treatment", *Handb. Environ. Chem,* vol. 5, pp. 127-175, 2008.
[http://dx.doi.org/10.1007/698_5_103]

[93]   J. Reungoat, B.I. Escher, M. Macova, F.X. Argaud, W. Gernjak, and J. Keller, "Ozonation and biological activated carbon filtration of wastewater treatment plant effluents", *Water Res.,* vol. 46, no. 3, pp. 863-872, 2012.
[http://dx.doi.org/10.1016/j.watres.2011.11.064] [PMID: 22172561]

[94]   S. Esplugas, D.M. Bila, L.G.T. Krause, and M. Dezotti, "Ozonation and advanced oxidation technologies to remove endocrine disrupting chemicals (EDCs) and pharmaceuticals and personal care products (PPCPs) in water effluents", *J. Hazard. Mater.,* vol. 149, no. 3, pp. 631-642, 2007.
[http://dx.doi.org/10.1016/j.jhazmat.2007.07.073] [PMID: 17826898]

[95]   "X.B.F.W.H.L.Z.S.W.Z. Han Y, Recent Progress on Two-dimensional Electrocatalysis", *Chem. Res. Chin. Univ.,* vol. 36, pp. 611-621, 2020.
[http://dx.doi.org/10.1007/s40242-020-0182-3]

[96]   M.B. Ross, P. De Luna, Y. Li, C.T. Dinh, D. Kim, P. Yang, and E.H. Sargent, "Designing materials for electrochemical carbon dioxide recycling", *Nat. Catal.,* vol. 2, pp. 648-658, 2019.
[http://dx.doi.org/10.1038/s41929-019-0306-7]

[97]   Y. Wang, *Development of Novel Bioelectrochemical Wastewater Treatment Technologies for Membrane Separation and Resource Recovery.* Springer Nature, 2020.
[http://dx.doi.org/10.1007/978-981-15-3078-4]

[98]   R. Salazar, M.S. Ureta-Zañartu, C. González-Vargas, C.D.N. Brito, and C.A. Martinez-Huitle, "Electrochemical degradation of industrial textile dye disperse yellow 3: Role of electrocatalytic material and experimental conditions on the catalytic production of oxidants and oxidation pathway", *Chemosphere,* vol. 198, pp. 21-29, 2018.
[http://dx.doi.org/10.1016/j.chemosphere.2017.12.092] [PMID: 29421732]

[99]   J. Feng, and D.C. Johnson, "Electrocatalysis of Anodic Oxygen-Transfer Reactions: Fe-Doped Beta-Lead Dioxide Electrodeposited on Noble Metals", *J. Electrochem. Soc.,* vol. 137, pp. 507-510, 1990.

[http://dx.doi.org/10.1149/1.2086488]

[100]   C.A. Martínez-Huitle, and L.S. Andrade, "Electrocatalysis in wastewater treatment: Recent mechanism advances", *Quim. Nova,* vol. 34, pp. 850-858, 2011.
[http://dx.doi.org/10.1590/S0100-40422011000500021]

[101]   "J. QU, The progress of catalytic technologies in water purification: A review", *J. Environ. Sci. (China),* 2009.
[http://dx.doi.org/10.1016/S1001-0742(08)62329-3]

[102]   T.A. Kenova, G.V. Kornienko, and V.L. Kornienko, "Electrocatalytic Oxidation of Aromatic Ecopollutants on Composite Anodic Materials", *Russ. J. Electrochem.,* vol. 56, pp. 337-348, 2020.
[http://dx.doi.org/10.1134/S1023193520040047]

[103]   "P. PARTHEEBAN, R. GANESAN, Treatment of Textile Wastewater by Electrochemical Method", *Int. J. Earth Sci. Eng.,* vol. 10, pp. 146-149, 2017.
[http://dx.doi.org/10.21276/ijee.2017.10.0124]

[104]   A.P. Borole, C. Tsouris, S.G. Pavlostathis, S. Yiacoumi, A.J. Lewis, X. Zeng, and L. Park, "Efficient conversion of aqueous-waste-carbon compounds into electrons, hydrogen, and chemicals via separations and microbial electrocatalysis", *Front. Energy Res.,* vol. 6, 2018.
[http://dx.doi.org/10.3389/fenrg.2018.00094]

[105]   R. Kaur, J.P. Kushwaha, and N. Singh, "Amoxicillin electro-catalytic oxidation using Ti/RuO2 anode: Mechanism, oxidation products and degradation pathway", *Electrochim. Acta,* vol. 296, pp. 856-866, 2019.
[http://dx.doi.org/10.1016/j.electacta.2018.11.114]

[106]   Y. Sun, S. Zhu, W. Sun, and H. Zheng, "Degradation of high-chemical oxygen demand concentration pesticide wastewater by 3D electrocatalytic oxidation", *J. Environ. Chem. Eng.,* vol. 7, 2019.
[http://dx.doi.org/10.1016/j.jece.2019.103276]

[107]   Y. Yao, M. Li, Y. Yang, L. Cui, and L. Guo, "Electrochemical degradation of insecticide hexazinone with Bi-doped PbO2 electrode: Influencing factors, intermediates and degradation mechanism", *Chemosphere,* vol. 216, pp. 812-822, 2019.
[http://dx.doi.org/10.1016/j.chemosphere.2018.10.191] [PMID: 30404074]

[108]   P.V. Nidheesh, and R. Gandhimathi, "Trends in electro-Fenton process for water and wastewater treatment: An overview", *Desalination,* vol. 299, pp. 1-15, 2012.
[http://dx.doi.org/10.1016/j.desal.2012.05.011]

[109]   T. Yu, and C.B. Breslin, "Graphene-modified composites and electrodes and their potential applications in the electro-fenton process", *Materials (Basel),* vol. 13, no. 10, p. E2254, 2020.
[http://dx.doi.org/10.3390/ma13102254] [PMID: 32422892]

[110]   S. Wang, Q. Wang, Y. Fang, and Y. Huang, "Degradation of organic pollutants by visible light synergistic electro-Fenton oxidation process", *Sci. China Chem.,* vol. 56, pp. 813-820, 2013.
[http://dx.doi.org/10.1007/s11426-012-4809-1]

[111]   Z. Qiang, J.H. Chang, and C.P. Huang, "Electrochemical regeneration of Fe2+ in Fenton oxidation processes", *Water Res.,* vol. 37, no. 6, pp. 1308-1319, 2003.
[http://dx.doi.org/10.1016/S0043-1354(02)00461-X] [PMID: 12598195]

[112]   A.F. Jahromi, M. Elektorowicz, and N. Zhukovskaya, "Enhanced electro-oxidation for tkn removal from highly polluted industrial wastewater", *Proceedings, Annu. Conf. - Can. Soc. Civ. Eng,* 2016pp. 957-964

[113]   R. Nazari, L. Rajić, A. Ciblak, S. Hernández, I.E. Mousa, W. Zhou, D. Bhattacharyya, and A.N. Alshawabkeh, "Immobilized palladium-catalyzed electro-Fenton's degradation of chlorobenzene in groundwater", *Chemosphere,* vol. 216, pp. 556-563, 2019.
[http://dx.doi.org/10.1016/j.chemosphere.2018.10.143] [PMID: 30390586]

[114]   X. Liu, B. Lv, G. Liu, H. Hua, X. Wang, S. Zhou, Y. Wang, and X. Mao, "Affordable Polymer-Carbon

Composite Electrodes for Electroosmotic Dehydration and Electro-Fenton Processes", *Ind. Eng. Chem. Res.,* vol. 58, pp. 19917-19925, 2019.
[http://dx.doi.org/10.1021/acs.iecr.9b03818]

[115] F. Liu, Y. Liu, Q. Yao, Y. Wang, X. Fang, C. Shen, F. Li, M. Huang, Z. Wang, W. Sand, and J. Xie, "Supported Atomically-Precise Gold Nanoclusters for Enhanced Flow-through Electro-Fenton", *Environ. Sci. Technol.,* vol. 54, no. 9, pp. 5913-5921, 2020.
[http://dx.doi.org/10.1021/acs.est.0c00427] [PMID: 32271550]

[116] J. Song, and S. Cho, "Catalytic materials for efficient electrochemical production of hydrogen peroxide", *APL Mater.,* vol. 8, p. 050701, 2020.
[http://dx.doi.org/10.1063/5.0002845]

[117] M. Ghasemi, A. Khataee, P. Gholami, and R.D. Cheshmeh Soltani, "Template-free microspheres decorated with Cu-Fe-NLDH for catalytic removal of gentamicin in heterogeneous electro-Fenton process", *J. Environ. Manage.,* vol. 248, p. 109236, 2019.
[http://dx.doi.org/10.1016/j.jenvman.2019.07.007] [PMID: 31306926]

[118] H. Ghanbarlou, B. Nasernejad, M. Nikbakht Fini, M.E. Simonsen, and J. Muff, "Synthesis of an iron-graphene based particle electrode for pesticide removal in three-dimensional heterogeneous electro-Fenton water treatment system", *Chem. Eng. J.,* vol. 395, 2020.
[http://dx.doi.org/10.1016/j.cej.2020.125025]

[119] Z. Yang, H. Chen, J. Wang, R. Yuan, F. Wang, and B. Zhou, "Efficient degradation of diisobutyl phthalate in aqueous solution through electro-Fenton process with sacrificial anode", *J. Environ. Chem. Eng.,* vol. 8, 2020.
[http://dx.doi.org/10.1016/j.jece.2020.104057]

[120] S.M. Aramyan, "Advances in Fenton and Fenton Based Oxidation Processes for Industrial Effluent Contaminants Control-A Review, Int", *J. Environ. Sci. Nat. Resour.,* vol. 2, 2017.
[http://dx.doi.org/10.19080/IJESNR.2017.02.555594]

[121] W. Huang, *Homogeneous and heterogeneous Fenton and photo-Fenton processes : impact of iron complexing agent ethylenediamine-N, N ' -disuccinic acid (EDDS) Wenyu Huang To cite this version : Diplômée de Master Homogeneous and heterogeneous Fenton and photo- Fento, Univ.* Blaise Pascal - Clermont-Ferrand, 2012, pp. 1-198.

[122] J. Casado, "Towards industrial implementation of Electro-Fenton and derived technologies for wastewater treatment: A review", *J. Environ. Chem. Eng.,* vol. 7, 2019.
[http://dx.doi.org/10.1016/j.jece.2018.102823]

[123] M.D.G. de Luna, M.L. Veciana, C.C. Su, and M.C. Lu, "Acetaminophen degradation by electro-Fenton and photoelectro-Fenton using a double cathode electrochemical cell", *J. Hazard. Mater.,* vol. 217-218, pp. 200-207, 2012.
[http://dx.doi.org/10.1016/j.jhazmat.2012.03.018] [PMID: 22480705]

[124] W. Wang, Y. Li, Y. Li, M. Zhou, and O.A. Arotiba, "Electro-Fenton and photoelectro-Fenton degradation of sulfamethazine using an active gas diffusion electrode without aeration", *Chemosphere,* vol. 250, p. 126177, 2020.
[http://dx.doi.org/10.1016/j.chemosphere.2020.126177] [PMID: 32114336]

[125] S.O. Ganiyu, M. Zhou, and C.A. Martínez-Huitle, "Heterogeneous electro-Fenton and photo-electr--Fenton processes: A critical review of fundamental principles and application for water/wastewater treatment", *Appl. Catal. B,* vol. 235, pp. 103-129, 2018.
[http://dx.doi.org/10.1016/j.apcatb.2018.04.044]

[126] D. Martínez-Pachón, M. Ibáñez, F. Hernández, R.A. Torres-Palma, and A. Moncayo-Lasso, "Photo-electro-Fenton process applied to the degradation of valsartan: Effect of parameters, identification of degradation routes and mineralization in combination with a biological system", *J. Environ. Chem. Eng.,* vol. 6, pp. 7302-7311, 2018.
[http://dx.doi.org/10.1016/j.jece.2018.11.015]

[127] E. Brillas, *A review on the degradation of organic pollutants in waters by UV photo-electro-fenton and solar photo-electro-fenton.* J. Braz. Chem. Soc., 2014, pp. 393-417.
[http://dx.doi.org/10.5935/0103-5053.20130257]

[128] G. Vilé, "Photocatalytic materials and light-driven continuous processes to selectively remove emerging pharmaceutical pollutants from water: recent developments and future trends in catalysis and reactor engineering", *Catal. Sci. Technol.,* 2020.
[http://dx.doi.org/10.1039/D0CY01713B]

[129] S. Zhu, and D. Wang, "Photocatalysis: Basic principles, diverse forms of implementations and emerging scientific opportunities", *Adv. Energy Mater.,* vol. 7, 2017.
[http://dx.doi.org/10.1002/aenm.201700841]

[130] L. Li, P.A. Salvador, and G.S. Rohrer, "Photocatalysts with internal electric fields", *Nanoscale,* vol. 6, no. 1, pp. 24-42, 2014.
[http://dx.doi.org/10.1039/C3NR03998F] [PMID: 24084897]

[131] M. Norouzi, A. Fazeli, and O. Tavakoli, "Phenol contaminated water treatment by photocatalytic degradation on electrospun Ag/TiO2 nanofibers: Optimization by the response surface method", *J. Water Process Eng.,* vol. 37, 2020.
[http://dx.doi.org/10.1016/j.jwpe.2020.101489]

[132] A.L.N. Beraldo, P.J.M.T. Abreu, L.G. Gonçalves, and S.O.A. Andreo, "Degradation of caffeine by heterogeneous photocatalysis using ZnO with Fe and Ag", *Braz. Arch. Biol. Technol.,* vol. 63, pp. 1-9, 2020.
[http://dx.doi.org/10.1590/1678-4324-2020180614]

[133] A. Sharma, N. Liu, Q. Ma, H. Zheng, N. Kawazoe, G. Chen, and Y. Yang, "PEG assisted P/Ag/Ag2O/Ag3PO4/TiO2 photocatalyst with enhanced elimination of emerging organic pollutants in salinity condition under solar light illumination", *Chem. Eng. J.,* vol. 385, 2020.
[http://dx.doi.org/10.1016/j.cej.2019.123765]

[134] K. Zhao, I. Khan, K. Qi, Y. Liu, and A. Khataee, "Ionic liquid assisted preparation of phosphorus-doped g-C3N4 photocatalyst for decomposition of emerging water pollutants", *Mater. Chem. Phys.,* vol. 253, 2020.
[http://dx.doi.org/10.1016/j.matchemphys.2020.123322]

[135] A. Ali, Y. Liang, S. Ahmed, B. Yang, B. Guo, and Y. Yang, "Mutual contaminants relational realization and photocatalytic treatment using Cu2MgSnS4 decorated BaTiO3", *Appl. Mater. Today,* vol. 18, 2020.
[http://dx.doi.org/10.1016/j.apmt.2019.100534]

[136] L. Wang, and Z. Bian, "Photocatalytic degradation of paracetamol on Pd-BiVO4 under visible light irradiation", *Chemosphere,* vol. 239, p. 124815, 2020.
[http://dx.doi.org/10.1016/j.chemosphere.2019.124815] [PMID: 31526994]

[137] C.F. Carbuloni, J.E. Savoia, J.S.P. Santos, C.A.A. Pereira, R.G. Marques, V.A.S. Ribeiro, and A.M. Ferrari, "Degradation of metformin in water by TiO2-ZrO2 photocatalysis", *J. Environ. Manage.,* vol. 262, p. 110347, 2020.
[http://dx.doi.org/10.1016/j.jenvman.2020.110347] [PMID: 32250822]

[138] A. Majumder, and A.K. Gupta, "Enhanced photocatalytic degradation of 17β-estradiol by polythiophene modified Al-doped ZnO: Optimization of synthesis parameters using multivariate optimization techniques", *J. Environ. Chem. Eng.,* vol. 8, p. 104463, 2020.
[http://dx.doi.org/10.1016/j.jece.2020.104463]

[139] N. Zhang, G. Xian, X. Li, P. Zhang, G. Zhang, and J. Zhu, "Iron Based Catalysts Used in Water Treatment Assisted by Ultrasound: A Mini Review", *Front Chem.,* vol. 6, p. 12, 2018.
[http://dx.doi.org/10.3389/fchem.2018.00012] [PMID: 29473033]

[140] M. Ashokkumar, *The characterization of acoustic cavitation bubbles - An overview.* Ultrason.

Sonochem, 2011, pp. 864-872.
[http://dx.doi.org/10.1016/j.ultsonch.2010.11.016]

[141] K.S. Suslick, Y. Didenko, M.M. Fang, T. Hyeon, K.J. Kolbeck, W.B. McNamara, M.M. Mdleleni, and M. Wong, "Acoustic cavitation and its chemical consequences, Philos. Trans. R. Soc. A Math", *Phys. Eng. Sci.,* vol. 357, pp. 335-353, 1999.
[http://dx.doi.org/10.1098/rsta.1999.0330]

[142] P. Qiu, B. Park, J. Choi, B. Thokchom, A.B. Pandit, and J. Khim, "A review on heterogeneous sonocatalyst for treatment of organic pollutants in aqueous phase based on catalytic mechanism", *Ultrason. Sonochem.,* vol. 45, pp. 29-49, 2018.
[http://dx.doi.org/10.1016/j.ultsonch.2018.03.003] [PMID: 29705323]

[143] M. Pirsaheb, and N. Moradi, "Sonochemical degradation of pesticides in aqueous solution: investigation on the influence of operating parameters and degradation pathway - a systematic review", *RSC Advances,* vol. 10, no. 13, pp. 7396-7423, 2020.
[http://dx.doi.org/10.1039/C9RA11025A] [PMID: 35492163]

[144] H.C. Yap, Y.L. Pang, S. Lim, A.Z. Abdullah, H.C. Ong, and C.H. Wu, "A comprehensive review on state-of-the-art photo-, sono-, and sonophotocatalytic treatments to degrade emerging contaminants", *Int. J. Environ. Sci. Technol.,* vol. 16, pp. 601-628, 2019.
[http://dx.doi.org/10.1007/s13762-018-1961-y]

[145] E. Kuna, R. Behling, S. Valange, G. Chatel, and J.C. Colmenares, "Sonocatalysis: A Potential Sustainable Pathway for the Valorization of Lignocellulosic Biomass and Derivatives", *Top. Curr. Chem. (Cham),* vol. 375, no. 2, p. 41, 2017.
[http://dx.doi.org/10.1007/s41061-017-0122-y] [PMID: 28337669]

[146] M.T. Taghizadeh, and R. Abdollahi, "Sonolytic, sonocatalytic and sonophotocatalytic degradation of chitosan in the presence of TiO_2 nanoparticles", *Ultrason. Sonochem.,* vol. 18, no. 1, pp. 149-157, 2011.
[http://dx.doi.org/10.1016/j.ultsonch.2010.04.004] [PMID: 20466578]

[147] S. Chong, G. Zhang, Z. Wei, N. Zhang, T. Huang, and Y. Liu, "Sonocatalytic degradation of diclofenac with FeCeOx particles in water", *Ultrason. Sonochem.,* vol. 34, pp. 418-425, 2017.
[http://dx.doi.org/10.1016/j.ultsonch.2016.06.023] [PMID: 27773264]

[148] J.S. Park, N. Her, J. Oh, and Y. Yoon, "Sonocatalytic degradation of bisphenol A and 17α-ethinyl estradiol in the presence of stainless steel wire mesh catalyst in aqueous solution", *Separ. Purif. Tech.,* vol. 78, pp. 228-236, 2011.
[http://dx.doi.org/10.1016/j.seppur.2011.02.007]

[149] M.E.A. Mohammed, and M.R. Alhajhoj, "Importance and Applications of Ultrasonic Technology to Improve Food Quality", In: *Food Process,* 2020.
[http://dx.doi.org/10.5772/intechopen.88523]

[150] C. Zhang, J. Lin, H. Sun, and X. Shi, "Influence of Characteristics and Roughness of Nucleation in Conical Pits on Surface Heterogeneous Nucleation", In: *Procedia Eng,* 2015, pp. 1281-1288.
[http://dx.doi.org/10.1016/j.proeng.2015.09.004]

[151] M. Ramiasa-MacGregor, A. Mierczynska, R. Sedev, and K. Vasilev, "Tuning and predicting the wetting of nanoengineered material surface", *Nanoscale,* vol. 8, no. 8, pp. 4635-4642, 2016.
[http://dx.doi.org/10.1039/C5NR08329J] [PMID: 26854095]

[152] B. Wang, Y. Zhang, L. Shi, J. Li, and Z. Guo, "Advances in the theory of superhydrophobic surfaces", *J. Mater. Chem.,* vol. 22, pp. 20112-20127, 2012.
[http://dx.doi.org/10.1039/c2jm32780e]

[153] L.O. Hedges, and S. Whitelam, "Patterning a surface so as to speed nucleation from solution", *Soft Matter,* vol. 8, pp. 8624-8635, 2012.
[http://dx.doi.org/10.1039/c2sm26038g]

[154] V. Belova, D.A. Gorin, D.G. Shchukin, and H. Möhwald, "Controlled effect of ultrasonic cavitation on hydrophobic/hydrophilic surfaces", *ACS Appl. Mater. Interfaces,* vol. 3, no. 2, pp. 417-425, 2011.
[http://dx.doi.org/10.1021/am101006x] [PMID: 21280665]

[155] P. Gholami, L. Dinpazhoh, A. Khataee, A. Hassani, and A. Bhatnagar, "Facile hydrothermal synthesis of novel Fe-Cu layered double hydroxide/biochar nanocomposite with enhanced sonocatalytic activity for degradation of cefazolin sodium", *J. Hazard. Mater.,* vol. 381, p. 120742, 2020.
[http://dx.doi.org/10.1016/j.jhazmat.2019.120742] [PMID: 31204019]

[156] K.P. Vidya Lekshmi, S. Yesodharan, and E.P. Yesodharan, "MnO_2 efficiently removes indigo carmine dyes from polluted water", *Heliyon,* vol. 4, no. 11, p. e00897, 2018.
[http://dx.doi.org/10.1016/j.heliyon.2018.e00897] [PMID: 30450434]

[157] G. Bampos, and Z. Frontistis, "Sonocatalytic degradation of butylparaben in aqueous phase over Pd/C nanoparticles", *Environ. Sci. Pollut. Res. Int.,* vol. 26, no. 12, pp. 11905-11919, 2019.
[http://dx.doi.org/10.1007/s11356-019-04604-5] [PMID: 30820921]

[158] X. Ma, S. Liu, Y. Liu, G. Gu, and C. Xia, "Comparative study on catalytic hydrodehalogenation of halogenated aromatic compounds over Pd/C and Raney Ni catalysts", *Sci. Rep.,* vol. 6, p. 25068, 2016.
[http://dx.doi.org/10.1038/srep25068] [PMID: 27113406]

[159] V.V. Lunin, and E.S. Lokteva, "Catalytic hydrodehalogenation of organic compounds", *Russ. Chem. Bull.,* vol. 45, pp. 1519-1534, 1996.
[http://dx.doi.org/10.1007/BF01431781]

[160] B. Coq, G. Ferrat, and F. Figueras, "Conversion of chlorobenzene over palladium and rhodium catalysts of widely varying dispersion", *J. Catal.,* vol. 101, pp. 434-445, 1986.
[http://dx.doi.org/10.1016/0021-9517(86)90271-X]

[161] M. Hu, Y. Liu, Z. Yao, L. Ma, and X. Wang, "Catalytic reduction for water treatment", *Front. Environ. Sci. Eng.,* vol. 12, 2018.
[http://dx.doi.org/10.1007/s11783-017-0972-0]

[162] K. Biswas, S. Chattopadhyay, Y. Jing, R. Che, G. De, B. Basu, and D. Zhao, Polyionic Resin Supported Pd/Fe_2O_3 Nanohybrids for Catalytic Hydrodehalogenation: Improved and Versatile Remediation for Toxic Pollutants., *Ind. Eng. Chem. Res.,* vol. 58, pp. 2159-2169, 2019.
[http://dx.doi.org/10.1021/acs.iecr.8b04464]

[163] Y.L. Chen, L. Xiong, X.N. Song, W.K. Wang, Y.X. Huang, and H.Q. Yu, "Electrocatalytic hydrodehalogenation of atrazine in aqueous solution by Cu@Pd/Ti catalyst", *Chemosphere,* vol. 125, pp. 57-63, 2015.
[http://dx.doi.org/10.1016/j.chemosphere.2015.01.052] [PMID: 25697805]

[164] F. Qing, C. Zhang, and H. Quan, "Synthesis of hydrofluorocyclopentanes by vapor-phase catalytic hydrodehalogenation", *J. Fluor. Chem.,* vol. 213, pp. 61-67, 2018.
[http://dx.doi.org/10.1016/j.jfluchem.2018.06.010]

[165] M. Hegedus, P. Lacina, M. Plotěný, J. Lev, B. Kamenická, and T. Weidlich, "Fast and efficient hydrodehalogenation of chlorinated benzenes in real wastewaters using Raney alloy", *J. Water Process Eng.,* vol. 38, 2020.
[http://dx.doi.org/10.1016/j.jwpe.2020.101645]

[166] B. Qiu, Y. Hu, C. Liang, L. Wang, Y. Shu, Y. Chen, and J. Cheng, "Enhanced degradation of diclofenac with Ru/Fe modified anode microbial fuel cell: Kinetics, pathways and mechanisms", *Bioresour. Technol.,* vol. 300, p. 122703, 2020.
[http://dx.doi.org/10.1016/j.biortech.2019.122703] [PMID: 31911312]

[167] K.A. Tarach, A. Śrębowata, E. Kowalewski, K. Gołąbek, A. Kostuch, K. Kruczała, V. Girman, and K. Góra-Marek, "Nickel loaded zeolites FAU and MFI: Characterization and activity in water-phase hydrodehalogenation of TCE", *Appl. Catal. A Gen.,* vol. 568, pp. 64-75, 2018.
[http://dx.doi.org/10.1016/j.apcata.2018.09.026]

[168] T. Liu, J. Luo, X. Meng, L. Yang, B. Liang, M. Liu, C. Liu, A. Wang, X. Liu, Y. Pei, J. Yuan, and J. Crittenden, "Electrocatalytic dechlorination of halogenated antibiotics via synergistic effect of chlorine-cobalt bond and atomic H", *J. Hazard. Mater.,* vol. 358, pp. 294-301, 2018.
[http://dx.doi.org/10.1016/j.jhazmat.2018.06.064] [PMID: 29990817]

[169] J.M. Yan, X.B. Zhang, S. Han, H. Shioyama, and Q. Xu, "Iron-nanoparticle-catalyzed hydrolytic dehydrogenation of ammonia borane for chemical hydrogen storage", *Angew. Chem. Int. Ed. Engl.,* vol. 47, no. 12, pp. 2287-2289, 2008.
[http://dx.doi.org/10.1002/anie.200704943] [PMID: 18266252]

[170] X. Zhou, D. Xu, Y. Chen, and Y. Hu, "Enhanced degradation of triclosan in heterogeneous E-Fenton process with MOF-derived hierarchical Mn/Fe@PC modified cathode", *Chem. Eng. J.,* vol. 384, 2020.
[http://dx.doi.org/10.1016/j.cej.2019.123324]

[171] Y. Duan, X. Dong, T. Song, Z. Wang, J. Xiao, Y. Yuan, and Y. Yang, "Hydrogenation of Functionalized Nitroarenes Catalyzed by Single-Phase Pyrite FeS2 Nanoparticles on N,S-Codoped Porous Carbon", *ChemSusChem,* vol. 12, no. 20, pp. 4636-4644, 2019.
[http://dx.doi.org/10.1002/cssc.201901867] [PMID: 31411806]

[172] S. Taghipour, A. Khadir, and M. Taghipour, "Carbon Nanotubes Composite Membrane for Water Desalination. InSustainable Materials and Systems for Water Desalination", *Springer, Cham.,* pp. 163-184, 2021.

[173] A. Khadir, A.M. Ramezanali, S. Taghipour, and K. Jafari, "Insights of the Removal of Antibiotics From Water and Wastewater: A Review on Physical, Chemical, and Biological Techniques", *Applied Water Science: Remediation Technologies,* vol. 2, pp. 1-47, 2021.

[174] S. Taghipour, and B. Ayati, "Study of SBAR Capability in Petroleum Wastewater Treatment", *Water Reuse,* vol. 2, no. 2, pp. 119-28, 2015.

[175] S. Taghipour, B. Ataie-Ashtiani, S.M. Hosseini, and K.L. Yeung, "Graphitic carbon nitride-based composites for photocatalytic abatement of emerging pollutants: In Nano structured Carbon Nitrides for Sustainable Energy and Environmental Applications", *Elsevier,* pp. 175-214, 2022.

CHAPTER 13

Integrated Processes for Removal of Emerging Water Pollutants

Muhammad Saud Baig[1], Siraj Ahmed[1], Ghulam Mujtaba[1,*], Muhammad Rizwan[2], Naveed Ahmed[2] and Sheeraz Ahmed[3]

[1] *Department of Energy & Environment Engineering, Dawood University of Engineering & Technology, Karachi, Pakistan*

[2] *U.S.-Pakistan Center for Advanced Studies in Water, Mehran University of Engineering & Technology, Jamshoro, Pakistan*

[3] *Institute of Environmental Engineering, Mehran University of Engineering & Technology, Jamshoro, Pakistan*

Abstract: As the world advances rapidly in technology, industries are experiencing rampant growth, and the healthcare sector is reaching new heights; however, novel challenges are emerging that threaten humanity in entirely new ways. Industrial development, large-scale urbanization, and hazardous effluent from healthcare facilities increase concentrations of emerging pollutants in our surface waters. Emerging pollutants have puzzled the researchers as they are relatively in smaller quantities than other pollutants, yet they pass through the conventional water treatment processes unscathed. Innovative integrated methods must be employed to enhance the water quality by significantly removing these persistent emerging pollutants. This chapter dives deeper into modern research to remove emerging water pollutants effectively. Integrated methods such as integrated electrocoagulation, activated sludge with membrane technology, and construction of wetlands are thoroughly presented.

Keywords: Integrated wastewater processes, Removal of hazardous materials, Wastewater, Wastewater treatment.

INTRODUCTION

Water and oxygen are the essential components to sustain life on planet Earth. Water bodies cover nearly 70% of the total Earth. Unfortunately, as little as 1% of total water is available as fresh water for human, plant, and animal consumption in rivers, lakes, ponds, and wells. The influx of masses towards urban areas in search of a better life, increasing industrialization, and growing economy are the

---

* **Corresponding author Ghulam Mujtaba:** Department of Energy & Environment Engineering, Dawood University of Engineering & Technology, Karachi, Pakistan; E-mail: ghulam.mujtaba1@duet.edu.pk

Shaukat Ali Mazari, Nabisab Mujawar Mubarak & Nizamuddin Sabzoi (Eds.)

important factors that account for water stress and less freshwater availability. The problem is further exacerbated by anthropogenic activities' addition of harmful pollutants to water resources, which renders the freshwater bodies unpleasant to the environment and hazardous for consumption. This uncontrolled problem adds to the phenomenon known as water scarcity [1]. United Nations World Water Development reported that more than 700 million people around the globe suffer from access to potable water. Additionally, industrial water demand (*e.g.*, pharmaceutical, textile, and agriculture) would possibly be increased by 400% by 2050.

The addition of nutrients in water causes dissolved oxygen to drop as microorganisms begin feeding on them. Still, due to rapid urbanization and industrial growth, researchers have identified new contaminants that are toxic to human health and threaten civilization. The water pollutants like phenols, organic dyes, benzene compounds, antibiotics, Bisphenol A, Clofibric acid, Benzotriazole, Atrazine, Carbamazepine, Diclofenac, Estriol, Gemifibrozil, Iopamidole, Naproxen, Primidone, Tonalide, Triclosan, Tolyltriazole, Sulfamethoxazole, Ibuprofen, Acetaminophen, Estrone, and halogenated hydrocarbon are among some of the known emerging water pollutants [2]. These emerging pollutants disrupt the food chain and significantly hamper the environmental equilibrium. These hazardous pollutants are concentrated in effluents from pharmaceutical, food processing, and polymer industries, particularly [3]. Effluents from the industries mentioned above are largely released in receiving waters without any treatment, posing a danger to human and animal lives and altering both species' fertility [4]. Owing to the immense dependence of the economy on industries, emerging pollutants are the most common contaminant in industrial wastewater. They have a harmful effect on humans. These pollutants can cause reproductive system malfunction, disruption of thyroid function, hypertension, diabetes mellitus, insulin resistance, cardiovascular diseases, blood disorders, and hormone-dependent cancer [5].

Concerned about the rampant growth in water-borne diseases and problems, researchers have focused on eradicating water contamination through cost-effective treatment systems. Novel integrated approaches, which require the combination of two or more technologies to tackle this issue and successfully remove emerging pollutants from the wastewaters, are being studied and developed. Integrated technologies have garnered much interest due to higher removal efficiency and performance than stand-alone methods. Other approaches such as waste volume reduction [6], zero liquid discharge [7, 8], and recovery of salts/minerals [9, 10] are being researched. Removal of emerging pollutants is a challenge that is now being tackled as a top priority. Technologies like electrocoagulation, constructed wetlands with waste stabilization ponds, and activated sludge processes with physical processes like ultrafiltration and reverse osmosis are extensively explored worldwide and discussed in this chapter.

INTEGRATED ELECTROCOAGULATION TREATMENT PROCESS

Electrocoagulation (EC) plays a vital role in wastewater treatment technologies and alternative treatment systems designed for the different types of wastewater due to its low footprint, durability, flexibility, simplicity, and sustainable process for nature [11]. It is an electrochemical process in which wastewater pollutants get eliminated due to the direct current (DC) supply affecting electrode dissolution. Together with all impurities available in the effluent, it can be easily separated through the electrolytic mixture [12]. The geometry of EC is very simple, which relates to the factors like DC power supply and electrodes occupied with the protected containers for wastewater treatment. EC reduces the moving parts during the process due to the simplicity of the system, which decreases the maintenance and proves the hybrid EC with chemical coagulation (CC) for the higher chemical oxygen demand (COD) removal efficiency of the oil industry effluents [12 - 14].

Electricity is used for moving the parts of the EC process, and required chemical addition is made in the CC process. Underdeveloped countries have a shortage of electricity, and chemicals are costly. Due to the electricity shortage in developing countries, alternative energy sources like solar power have been investigated. Similar ideas for running the EC process have been reported with remarkable results [15 - 17]. The utilization of alternative energy sources for the electrocoagulation process plays a vital role in the environmental and economic sector for sustainability in wastewater treatment. EC treatment process studies have been widely conducted for treating various wastewaters like oily, municipal, textile, tannery, and mineral processing, urban laundry, metalworking fluids, palm oil mill effluent (POME), industrial wastewater, *etc* [11]. EC wastewater treatment combines with other technologies and can help reduce the EC limitations and increase the efficiencies of the system. Hybrid technologies can help reduce the scarcity of fresh water and reuse water for further processes in industries. It is reported that the efficiencies of any combined system are higher than the single treatment process; for instance, there are 20% higher efficiencies of a combined processes system as compared to the electrocoagulation alone [18]. Due to limitations associated with EC technology, recently, various research studies have reported on integrated systems with EC to enhance the removal efficiency of pollutants [11].

The combination of peroxidation with EC plays a vital role in the integrated technologies system for wastewater treatment. Peroxidation is broadly used with

EC. EC and advanced oxidation combine wastewater treatment technologies called the electrochemical hydrogen peroxide H_2O_2 while operating the EC process (ECP) [19, 20]. Various research studies on the ECP reported promising results for removing the pollutant from wastewater [21]. The initial results achieved to remove the COD for adding the H_2O_2 concentration of 234 mg/L, integrated technologies achieved 13% more COD removal than the single EC process. Fe electrodes were used at initial pH of 5 the COD removal of 83% of industrial effluent was achieved. Adding 2% of H_2O_2 with the coagulant aid poly aluminum chloride (PAC) achieved the COD removal of up to 94.5% on POME [20]. As per results achieved in various studies, they concluded that the combination of ECP with coagulant proved to be capable hybrid technology, which removes organic pollutants and is beneficial for saving time and energy costs. Electrocoagulation wastewater treatment process combined with Chemical Coagulation (CC) will help remove the total organic carbon from the effluent of the industries. The removal efficiency of the total organic carbon is more than the single process of EC. The integrated process removal of the total organic carbon (TOC) oil sands produced water is 38.7% [22].

EC integrated wastewater technologies with the photovoltaic (PV) cells achieved remarkable results for utilizing solar power instead of conventional power. Using the renewable energy source for carrying out the EC process will benefit cost-saving and environmental sustainability. Very little research has been done to analyze the large-scale implementation of solar power for continuous EC operations. The long-term supply of PV cells and the efficiency of EC will be investigated for continuous operation and sustainability [17]. The combination of photovoltaic EC hybrid technologies has the potential to remove the textile wastewater color and turbidity by 98.5% and 94.5%, respectively [15]. Continuous operation of integrated technologies achieved high removal percentage of total dissolved solids (TDS), Turbidity, and COD, 49%, 93%, and 92%, respectively, from municipal wastewater. Photovoltaic EC integrated technologies are promising alternatives for integrated technologies to increase their efficiency.

EC integrated with membrane technology requires minimum chemicals, common space, and simple integration with EC compared to all above-integrated systems [23]. The integrated system achieved remarkable results, like 100% removing the manganese (Mn) from steel industry effluent. EC integrated with microfiltration has removed up to 100% heavy metals from the wastewater and promising environmental sustainability considering waste reduction by recycling and reuse. Therefore, replacing the chemical coagulation process with EC and integrating it with the membrane process or membrane bioreactor (MBR) is a promising alternative for water/wastewater treatment. However, further investigations are required on its large-scale viability and continuous operations. Table **1** highlights the removal efficiencies of various integrated processes for wastewater treatment.

Table 1. Removal of Emerging Pollutant Through Integrated Process With EC.

Emerging Pollutant	Wastewater Source	Integrated Process	Removal Efficiency	References
COD	Textile Wastewater	Al-Al (Combination of Electrodes)	18.16%	[24]
Turbidity			83.5%	
COD	Municipal Wastewater	Al-Al	92.01%	[15]
Turbidity			93.97%	
TDS			49.97%	
COD	Distillery Industrial Effluent	Photo-EC, Electrochemical Peroxidation (ECP), Photo-ECP	86%	[21]
Color			100%	
COD	Tannery Wastewater	EC-Fungal	96%	[25]
COD	Bilge Water	EC-Nanofiltration	62.38%	[26]
COD	Municipal Effluent	Photovoltaic-EC	92.01%	[15]
Turbidity			93.97%	
TDS			49.78%	
COD	Pharmaceutical Wastewater	EO-EC	80%	[27]
TSS	Textile Wastewater	EO-EC	97%	
Color			97.5%	
COD			93.5%	
COD	Dairy Industry Wastewater	Phytoremediation -Electrocoagulation	86.4%	[28]
Sulfide	Tannery Wastewater	Electrocoagulation-UVC/VUV Photoreactor	98.27%	[29]
Cr(T)			100%	
COD			99.52%	
Calcium	Petroleum Industry	EC-Adsorption	88%	
Strontium			72%	
COD			52%	

INTEGRATED MEMBRANE PROCESSES

Membrane technologies are effectively used for wastewater treatment. In membrane filtration, a semi-permeable membrane separates pollutants from effluent wastewater [30]. The effectiveness of membrane technologies depends on the properties of the membrane used (molecular weight cut off and pore size). In membrane filtration, membranes and their modules are consumed. The most common membrane modules are hollow fiber, spiral wound, and flat sheet [31]. The most common materials used in membrane technology include polyvinylidene fluoride, polysulfone, nylon polyether-sulfone, and ceramics [32]. The membrane technology is improved by introducing novel membrane materials and improving the structure of membrane pores. The most common pressure-driven membrane processes that are used for water and wastewater treatment include microfiltration (MF), nanofiltration (NF), and ultrafiltration (UF) [33]. Table **2** highlights various membrane features.

Table 2. Pressure-driven wastewater treatment membrane technologies and their features (modified from [34 - 36]).

Membrane Technology	Membrane Type	Membrane Pore Size (μm)	Pollutants Blocked	Pressure Required (bar)
Microfiltration (MF)	porous, asymmetric, or symmetric	0.1-10	suspended solids, bacteria, fat, organics, microparticles	0.1-2
Ultrafiltration (UF)	microporous, asymmetric	0.01-0.1	pigments, most bacteria colloids, sugar, some viruses	1-5
Nanofiltration (NF)	tight porous, asymmetric, thin-film composite	0.001	viruses, divalent cations, multivalent ions, NaCl	5-20
Reverse osmosis (RO)	semi-porous, asymmetric, thin-film composite	0.001	all pollutants, including monovalent ions	10-100

In pressure-driven membranes, positive hydrostatic pressure is involved infiltration. Microfiltration, nanofiltration, and ultrafiltration membranes remove the large molecular organic matter. However, reverse osmosis (RO) can remove molecules of small size, *e.g.*, monovalent ions. MF, NF, and UF are used as a pretreatment step before RO due to their ability to remove turbidity, organic matter, and bacteria [37, 38].

Pretreatment of RO wastewater with MF, UF, and NF can significantly improve recovery and flux rate, reducing the overall treatment cost [39]. Micro and ultra-filtration can produce water that can be used in industrial processes. On the other hand, nanofiltration effectively removes multivalent ions, bacteria, and micropollutants. RO is a promising technology for desalination. The recovery rate of the RO process for brine wastewater is up to 70%. However, the treatment cost increases significantly with a higher salt concentration. The pumping cost of the

RO process can be reduced by using energy-efficient equipment [40]. Membrane fouling and scaling is another major issue in RO treatment. Scaling the membrane can be avoided by using anti-scaling material, resulting in higher removal of organic matter [41]. Pressure limitation is another issue in RO treatment. There is work to develop RO systems capable of working at low pressures to overcome pressure limitations. These low-pressure RO systems are energy efficient and capable of producing high permeate flux. Low-pressure reverse osmosis (LPRO) systems have successfully treated agriculture wastewater, red-sea water, and fouling feed water [42 - 44]. Advanced biological treatment has the potential to treat industrial wastewater and saline water [45 - 47]. Membrane bioreactor (MBR) is another emerging technology used to treat industrial wastewaters. MBR combines filtration with biological treatment, which results in increased hydraulic retention time and biomass in reactors. MBR also has other advantages such as easy operation, high efficiency and loading capacity, and less space requirement [48 - 50]. MBR processes have successfully treated municipal and industrial wastewater at the pilot and large scale [51, 52]. The MBR-treated water can be used for non-potable applications.

Osmotic membrane bioreactor (OMBR) is another emerging technology having an integrated membrane and biological system capable of treating wastewater. In OMBR, forward osmosis (FO) is combined with the biological process. OMBR transport biologically treated wastewater through a semi-permeable FO membrane into a highly concentrated draw solution. OMBR is better than the MBR system, producing better quality water and having low fouling [53]. Another upcoming biological technology is electro-biochemical reactors used for salts and organic pollutants removal from wastewater. This technology consists of electrodes (cathode and anode) and an ion-exchange membrane. This process converts the chemical energy of biodegradable matter into electrical energy [54].

Integration of biological treatment methods (aerobic/anaerobic) with conventional physio-chemical treatment methods can better treat wastewater containing organic matter [55]. The conventional activated sludge process is not very efficient in treating saline wastewater. Combining the biological process with the activated sludge process can improve the overall efficiency of conventional activated sludge for saline wastewater treatment [56]. The most common biological process which can be integrated with the activated sludge process includes sequencing batch reactor, fed-batch reactor, moving bed biofilm reactor, and rotating bio-disc contactors [56 - 60]. These systems can treat less saline (<1 wt%) water but are not effective at high salt concentrations [61].

The pressure-driven membrane technologies such as microfiltration, nanofiltration, and Ultrafiltration are recently integrated with RO to treat saline and industrial wastewater. The use of MF and UF treatment before RO can result in better permeate flux, reduce treatment cost and improve reliability [62]. MF–NF–RO membrane systems with membrane distillation/membrane crystallization process have been integrated. The integrated technology improved desalination and reduced the overall desalination cost. The four different combinations were tested in the integrated system. The systems included MF, NF, RO, feed solution (FS), crystalline membrane, and membrane desalination (MD). The FS1, MF, and NF were first combined as a pretreatment step before the RO unit. FS1 reduced energy consumption (up to 25– 30%), decreased discharged volume, and 50% water recovery. In the second and third configurations, NF and RO retentate was passed through membrane crystallization (MCr) in the FS1 system. Calcium ions should be removed from the water before MCr to avoid scaling. In the fourth combination, RO retentate was passed through the MD unit in addition to the FS2 configuration. MCr and MD units can reduce thermal and electrical energy for MF–NF–RO [63]. In another study, refinery wastewater was treated using integrated MF–UF and UF–RO systems. The pilot-scale hybrid system successfully removed toxic pollutants from refinery wastewater. UF–RO hybrid system showed better performance (~98%) for salts and organic pollutants removal than MF–UF system. The treated wastewater was reused in boilers and cooling towers [64]. Another study reported removing organic matter, emerging micropollutants (EMPs), nutrients, and salts from municipal wastewater by integrating NF/RO pilot plant with a membrane bioreactor. The use of flat sheet and hollow fiber modules removed suspended solids and COD 100% and 98%, respectively [65]. Falizi *et al.* [66] recently treated MBR effluent with and without NF/RO treatment. The treated wastewater was reused for agriculture purposes. MBR effluents and RO permeate were combined in different proportions and were analyzed with irrigation standards. The 2:1 ratio of RO permeate, and MBR effluent was suitable for irrigation usage. The proportion not only reduced the toxicity of MBR effluent but reduced salinity and adverse impact of RO permeate infiltration. Table **3** exhibits various combinations of membrane based integrate processes.

Table 3. Integrated membrane other processes for wastewater treatment.

Technology	Type of Pollutant	Removal Efficiency	Wastewater Types	References
bench-scale moving-bed biofilm reactors	ammonia	80%	chemical industry and domestic sewage	[59]
membrane bioreactor	COD	90%	saline water	[49]
membrane bioreactor	COD	81%	vegetable oil refinery wastewaters	[50]
submerged membrane bioreactor	COD	90%	textile wastewater	[52]

(Table 3) cont.....

Technology	Type of Pollutant	Removal Efficiency	Wastewater Types	References
electro biochemical reactor	hardness	100%	saline water	[54]
MF–NF–RO/ membrane distillation/crystallization	salinity	92.8%	seawater	[63]
MF/UF and UF/RO System	total hardness	99.94%	refinery wastewater	[64]
MBR–NF/RO treatment	COD	100%	municipal wastewater	[65]
MBR/NF-RO	salinity	-	industrial wastewater	[66]

INTEGRATED CONSTRUCTED WETLAND FOR MICROPOLLUTANT REMOVAL

Constructed wetland (CW) is a nature-based waste treatment system that is a green engineered and economical wastewater treatment system that requires minimal supervision and works by gravitational force. Many types of wastewaters have been treated by this method, such as municipal, agricultural, industrial, run-offs, and storm wastewaters [67 - 69]. Though it has modest organics removal and ineffective removal of nutrients due to the absence of electron acceptor (*e.g.*, oxygen), it requires a larger area and time to provide reasonable treatment efficiency [70 - 72]. To overcome removal efficiency and deficiency of terminal electron acceptor availability, the process can be integrated with other treatment technologies [69, 73, 74].

Integrated constructed wetland-microbial fuel cell (MFC) is a recent integrated technology developing a new paradigm of water-energy-land nexus. The construction of CW-MFC is based on aerobic (upper part) and anaerobic (lower part) portions corresponding to the MFC's aerobic cathode and anaerobic anode. This integration upscales the process performance and reduces the limitations of individual processes.

Several attempts have optimized this process, construction, and electricity generation. Since this is a budding technology, a lower number of documents are available focusing on the principles of electron transfer, the potential of integration, operational parameters, landscaping, and prospects [75 - 79]. The environmental factors, other than electricity generation and electron transfer, need to be considered, such as heavy metal removal, greenhouse gas removal, enhanced treatment, and utilization as biosensors [80].

The naturally available self-purifying systems (lakes, rivers, *etc.*) also have several redox processes. Similarly, the redox processes in aerobic and anaerobic sections can share the electron transfer and complement each other. Srivastava *et al.* [81]) reviewed the electron transfer mechanisms for pollutant abatement in CW-MFCs. The treatment performance largely depends on the electron flow/transfer, especially in the oxygen-deficient area. However, electron transfer chemistry is complex due to several types of electron donors and acceptors, microbial species, and electrode material working simultaneously.

The lagoons are also being used with the constructed wetland technology in warm and tropical areas [82, 83]. Both systems harbor the natural wastewater treatment systems with great robustness, few expenditures, effortlessness, the lagoon system, and high-rate algal ponds (HRAP). Especially, macrophyte shadings can control the algal population in the effluent, and filtration will be enhanced when effluent passes through the filter bed and plant roots of the constructed wetlands, besides providing a habitat for the invertebrates known for filter-feeding and grazing activity [84]. Additionally, biological removal of nutrients can be achieved due to the large wetland footprint, phosphorus sorption on media, and uptake by plants [84]. Consequently, several integrated systems are currently being utilized in countries like the United States, Brazil, France, Spain, Thailand, and New Zealand [85].

The integration of the lagoon system can be done with all types of constructed wetlands, *i.e.*, 1-free water surface flow, 2-horizontal, and 3-vertical subsurface flow wetlands [86]. Despite being similar in technology, the removal efficiencies of three types of constructed wetlands might be different. The removal efficiency of vertical flow subsurface constructed wetland is comparatively higher due to the greater inflow concentrations [86]. A comparative study on maturation ponds, horizontal, and vertical flow constructed wetlands, conducted in Brazil, concluded that each system has pros and cons. The option can be taken based on suitability [87, 88]. The real drawbacks of the lagoon technology are the smell and accumulation of sludge, while constructed wetlands can easily get clogged by the filtration substrate [89]. The hydrodynamics and clogging of the constructed wetlands are strongly affected by plant species' position and type of substrate, which proposes the analysis of solid characteristics and pore space expansion due to roots [90].

The effectiveness of constructed wetland utilization for the lagoon effluent is reported to remove toxicants and heavy metals from agricultural wastewaters, like dairy farm wastewater [84], piggery wastewater [91], and swine wastewater [92]. The studies' main findings were higher nitrogen removal due to nitrification and mineralization, up to 85 and 49%, respectively. The phosphorus removal was good due to lower loading rates, low plant uptake, or lesser adsorption by media.

A novel lagoon system has been proposed named enhanced pond and wetland system composed of the high-rate algal pond, algae settlers, surface flow constructed wetland, and denitrification filter (woodchip). The system is characterized by the smaller footprint of the lagoon and conventional high-rate algal ponds. The removal efficiencies of ammonium (95%), nitrate (77%), phosphate (65%), and E. coli (3 log reduction) show a good promise. This system can provide further monetary benefits to algae and nutrient recovery as biomass and fertilizer [93].

Normally, the micropollutants are not removed during conventional treatment systems (activated sludge process, lagoons, *etc.*), but the removal is enhanced when integrated processes are used. The removal of antibiotics, antibiotic resistance genes, hormones, biocides, *etc.*, is summarized in Table **4** using lagoon integrated constructed wetlands.

Table 4. The characteristics of the integrated constructed wetlands to remove several types of micropollutants.

S. No.	Target Micropollutants	Influent/Scale	Preceding Treatment	CW Type	Vegetation Type	References
1	Antibiotic resistance genes, Carbamezapene, Carbamazepine, Gemfibrozil, Sulfamethoxazole, Sulfapyridine	Municipal wastewater/full scale	Hybrid lagoons	Surface flow CW	*Typha* spp.	[94]
2	Acetaminophen, Atenolol, Caffeine, Carbamazepine, Cotinine, Fluoxetine, Gemfibrozil, Ibuprofen, Metoprolol, Nadolol, Naproxen, Propranolol, Sotalol, Sulfamethoxazole, Sulfapyridine	Urban wastewater/Full scale	Hybrid aerated lagoons	Surface flow CW		[95]

S. No.	Target Micropollutants	Influent/Scale	Preceding Treatment	CW Type	Vegetation Type	References
3	Caffeine, Carbamazepine, Diclofenac, Ibuprofen, Ketoprofen, Naproxen, Salicylic acid	Urban wastewater/Full scale	Hybrid facultative pond	Surface Flow-CW and Horizontal SubSurface Flow-CW	*Phragmites australis,* Typha angustifolia	[96, 97]
4	Antibiotic Resistance Genes (sul1, sul2, tetM, and tetO), Leucomycin, ofloxacin, lincomycin, and sulfamethazine	Rural wastewater	Regulating pool and stabilization lagoon	Surface Flow-CW and SubSurface Flow-CW	*Myriophyllum verticillatum L., Pontederia cordata,* and *M. verticillatum*	[98]
5	Fourteen androgens, five progestogens, five glucocorticoids, and fifteen biocides	Rural wastewater	Regulating pool and stabilization lagoon	Surface Flow-CW and SubSurface Flow-CW	*Myriophyllum verticillatum L., Pontederia cordata*	[99]

CONCLUSION

Emerging pollutants are the contaminants that need extra attention when wastewater treatment systems are being designed. Due to their toxic nature, these emerging pollutants are hazardous to life and pose a threat to the smooth functioning of society. On top of that, these pollutants pass unscathed through the conventional treatment systems. Integrated wastewater treatment technologies are novel approaches. These technologies combine two or more existing technologies with new advantages over conventional ones. Various integrated wastewater treatment technologies are being developed and installed to tackle the looming danger of these pollutants. Innovative options such as integrated wastewater treatment technologies must be further researched and developed effectively to eradicate these novel micropollutants successfully.

CONSENT FOR PUBLICATION

Not applicable.

CONFLICT OF INTEREST

The author declares no conflict of interest, financial or otherwise.

ACKNOWLEDGEMENTS

Declared none.

REFERENCES

[1]   D. Seckler, R. Barker, and U. Amarasinghe, "Water scarcity in the twenty-first century", *Int. J. Water Resour. Dev.,* vol. 15, no. 1-2, pp. 29-42, 1999.
[http://dx.doi.org/10.1080/07900629948916]

[2]   T. Deblonde, C. Cossu-Leguille, and P. Hartemann, "Emerging pollutants in wastewater: a review of the literature", *Int. J. Hyg. Environ. Health,* vol. 214, no. 6, pp. 442-448, 2011.
[http://dx.doi.org/10.1016/j.ijheh.2011.08.002] [PMID: 21885335]

[3]   M. Gavrilescu, K. Demnerová, J. Aamand, S. Agathos, and F. Fava, "Emerging pollutants in the environment: present and future challenges in biomonitoring, ecological risks and bioremediation", *N. Biotechnol.,* vol. 32, no. 1, pp. 147-156, 2015.
[http://dx.doi.org/10.1016/j.nbt.2014.01.001] [PMID: 24462777]

[4]   V. Belgiorno, L. Rizzo, D. Fatta, C. Della Rocca, G. Lofrano, A. Nikolaou, V. Naddeo, and S. Meric, "Review on endocrine disrupting-emerging compounds in urban wastewater: occurrence and removal by photocatalysis and ultrasonic irradiation for wastewater reuse", *Desalination,* vol. 215, no. 1-3, pp. 166-176, 2007.
[http://dx.doi.org/10.1016/j.desal.2006.10.035]

[5]   Y. Tang, M. Yin, W. Yang, H. Li, Y. Zhong, L. Mo, Y. Liang, X. Ma, and X. Sun, "Emerging pollutants in water environment: Occurrence, monitoring, fate, and risk assessment", *Water Environ. Res.,* vol. 91, no. 10, pp. 984-991, 2019.
[http://dx.doi.org/10.1002/wer.1163] [PMID: 31220374]

[6]   A. Subramani, and J.G. Jacangelo, "Treatment technologies for reverse osmosis concentrate volume minimization: A review", *Separ. Purif. Tech.,* vol. 122, pp. 472-489, 2014.
[http://dx.doi.org/10.1016/j.seppur.2013.12.004]

[7]   D. Xevgenos, P. Michailidis, K. Dimopoulos, M. Krokida, and M. Loizidou, "Design of an innovative vacuum evaporator system for brine concentration assisted by software tool simulation", *Desalination Water Treat.,* vol. 53, no. 12, pp. 3407-3417, 2015.
[http://dx.doi.org/10.1080/19443994.2014.948660]

[8]   D. Xevgenos, K. Moustakas, D. Malamis, and M. Loizidou, "An overview on desalination & sustainability: renewable energy-driven desalination and brine management", *Desalination Water Treat.,* vol. 57, no. 5, pp. 2304-2314, 2016.
[http://dx.doi.org/10.1080/19443994.2014.984927]

[9]   S.A. Abdul-Wahab, and M.A. Al-Weshahi, "Brine management: substituting chlorine with on-site produced sodium hypochlorite for environmentally improved desalination processes", *Water Resour. Manage.,* vol. 23, no. 12, pp. 2437-2454, 2009.
[http://dx.doi.org/10.1007/s11269-008-9389-7]

[10]   D.H. Kim, "A review of desalting process techniques and economic analysis of the recovery of salts from retentates", *Desalination,* vol. 270, no. 1-3, pp. 1-8, 2011.
[http://dx.doi.org/10.1016/j.desal.2010.12.041]

[11]   A. Tahreen, M.S. Jami, and F. Ali, "Role of electrocoagulation in wastewater treatment: A developmental review", *J. Water Process Eng.,* vol. 37, p. 101440, 2020.
[http://dx.doi.org/10.1016/j.jwpe.2020.101440]

[12]   J.N. Hakizimana, B. Gourich, M. Chafi, Y. Stiriba, C. Vial, P. Drogui, and J. Naja, "Electrocoagulation process in water treatment: A review of electrocoagulation modeling approaches", *Desalination,* vol. 404, pp. 1-21, 2017.

[http://dx.doi.org/10.1016/j.desal.2016.10.011]

[13]   P. Aswathy, R. Gandhimathi, S. Ramesh, and P. Nidheesh, "Removal of organics from bilge water by batch electrocoagulation process", *Separ. Purif. Tech.,* vol. 159, pp. 108-115, 2016.
[http://dx.doi.org/10.1016/j.seppur.2016.01.001]

[14]   K. Sardari, P. Fyfe, D. Lincicome, and S.R. Wickramasinghe, "Aluminum electrocoagulation followed by forward osmosis for treating hydraulic fracturing produced waters", *Desalination,* vol. 428, pp. 172-181, 2018.
[http://dx.doi.org/10.1016/j.desal.2017.11.030]

[15]   C. Nawarkar, and V. Salkar, "Solar powered electrocoagulation system for municipal wastewater treatment", *Fuel,* vol. 237, pp. 222-226, 2019.
[http://dx.doi.org/10.1016/j.fuel.2018.09.140]

[16]   B. Khemila, B. Merzouk, A. Chouder, R. Zidelkhir, J-P. Leclerc, and F. Lapicque, "Removal of a textile dye using photovoltaic electrocoagulation", *Sustain. Chem. Pharm.,* vol. 7, pp. 27-35, 2018.
[http://dx.doi.org/10.1016/j.scp.2017.11.004]

[17]   A. Ziouvelou, A.G. Tekerlekopoulou, and D.V. Vayenas, "A hybrid system for groundwater denitrification using electrocoagulation and adsorption", *J. Environ. Manage.,* vol. 249, p. 109355, 2019.
[http://dx.doi.org/10.1016/j.jenvman.2019.109355] [PMID: 31499372]

[18]   Z. Al-Qodah, M. Tawalbeh, M. Al-Shannag, Z. Al-Anber, and K. Bani-Melhem, "Combined electrocoagulation processes as a novel approach for enhanced pollutants removal: A state-of-the-art review", *Sci. Total Environ.,* vol. 744, p. 140806, 2020.
[http://dx.doi.org/10.1016/j.scitotenv.2020.140806] [PMID: 32717462]

[19]   S. Sharma, A. Aygun, and H. Simsek, "Electrochemical treatment of sunflower oil refinery wastewater and optimization of the parameters using response surface methodology", *Chemosphere,* vol. 249, p. 126511, 2020.
[http://dx.doi.org/10.1016/j.chemosphere.2020.126511] [PMID: 32208219]

[20]   M. Nasrullah, L. Singh, Z. Mohamad, S. Norsita, S. Krishnan, N. Wahida, and A.W. Zularisam, "Treatment of palm oil mill effluent by electrocoagulation with presence of hydrogen peroxide as oxidizing agent and polialuminum chloride as coagulant-aid", *Water Resour. Ind.,* vol. 17, pp. 7-10, 2017.
[http://dx.doi.org/10.1016/j.wri.2016.11.001]

[21]   P. Asaithambi, B. Sajjadi, A.R.A. Aziz, and W.M.A.B.W. Daud, "Performance evaluation of hybrid electrocoagulation process parameters for the treatment of distillery industrial effluent", *Process Saf. Environ. Prot.,* vol. 104, pp. 406-412, 2016.
[http://dx.doi.org/10.1016/j.psep.2016.09.023]

[22]   L. Shamaei, B. Khorshidi, B. Perdicakis, and M. Sadrzadeh, "Treatment of oil sands produced water using combined electrocoagulation and chemical coagulation techniques", *Sci. Total Environ.,* vol. 645, pp. 560-572, 2018.
[http://dx.doi.org/10.1016/j.scitotenv.2018.06.387] [PMID: 30029132]

[23]   M. Changmai, P. Das, P. Mondal, M. Pasawan, A. Sinha, and P. Biswas, "Hybrid electrocoagulation–microfiltration technique for treatment of nanofiltration rejected steel industry effluent", *Int. J. Environ. Anal. Chem.,* pp. 1-22, 2020.

[24]   S. Bener, Ö. Bulca, B. Palas, G. Tekin, S. Atalay, and G. Ersöz, "Electrocoagulation process for the treatment of real textile wastewater: Effect of operative conditions on the organic carbon removal and kinetic study", *Process Saf. Environ. Prot.,* vol. 129, pp. 47-54, 2019.
[http://dx.doi.org/10.1016/j.psep.2019.06.010]

[25]   E.Ü. Deveci, C. Akarsu, Ç. Gönen, and Y. Özay, "Enhancing treatability of tannery wastewater by integrated process of electrocoagulation and fungal via using RSM in an economic perspective", *Process Biochem.,* vol. 84, pp. 124-133, 2019.

[http://dx.doi.org/10.1016/j.procbio.2019.06.016]

[26]   C. Akarsu, Y. Ozay, N. Dizge, H. Elif Gulsen, H. Ates, B. Gozmen, and M. Turabik, "Electrocoagulation and nanofiltration integrated process application in purification of bilge water using response surface methodology", *Water Sci. Technol.,* vol. 74, no. 3, pp. 564-579, 2016.
[http://dx.doi.org/10.2166/wst.2016.168] [PMID: 27508361]

[27]   P.V. Nidheesh, J. Scaria, D.S. Babu, and M.S. Kumar, "An overview on combined electrocoagulation-degradation processes for the effective treatment of water and wastewater", *Chemosphere,* vol. 263, p. 127907, 2021.
[http://dx.doi.org/10.1016/j.chemosphere.2020.127907] [PMID: 32835972]

[28]   J. Akansha, P.V. Nidheesh, A. Gopinath, K.V. Anupama, and M. Suresh Kumar, "Treatment of dairy industry wastewater by combined aerated electrocoagulation and phytoremediation process", *Chemosphere,* vol. 253, p. 126652, 2020.
[http://dx.doi.org/10.1016/j.chemosphere.2020.126652] [PMID: 32272308]

[29]   M. Moradi, and G. Moussavi, "Enhanced treatment of tannery wastewater using the electrocoagulation process combined with UVC/VUV photoreactor: Parametric and mechanistic evaluation", *Chem. Eng. J.,* vol. 358, pp. 1038-1046, 2019.
[http://dx.doi.org/10.1016/j.cej.2018.10.069]

[30]   A.S. Rathore, and A. Shirke, "Recent developments in membrane-based separations in biotechnology processes: review", *Prep. Biochem. Biotechnol.,* vol. 41, no. 4, pp. 398-421, 2011.
[http://dx.doi.org/10.1080/10826068.2011.613976] [PMID: 21967339]

[31]   R. Baker, *Membranes and modules.* vol. 3. Membrane Technology and Applications, 2004.

[32]   H.A. Maddah, A.S. Alzhrani, M. Bassyouni, M. Abdel-Aziz, M. Zoromba, and A.M. Almalki, "Evaluation of various membrane filtration modules for the treatment of seawater", *Appl. Water Sci.,* vol. 8, no. 6, pp. 1-13, 2018.
[http://dx.doi.org/10.1007/s13201-018-0793-8]

[33]   P.Y. Apel, O. Bobreshova, A. Volkov, V. Volkov, V. Nikonenko, I. Stenina, A.N. Filippov, Y.P. Yampolskii, and A.B. Yaroslavtsev, "Prospects of membrane science development", *Membranes and Membrane Technologies,* vol. 1, no. 2, pp. 45-63, 2019.
[http://dx.doi.org/10.1134/S2517751619020021]

[34]   V.G. Gude, "Desalination and water reuse to address global water scarcity", *Rev. Environ. Sci. Biotechnol.,* vol. 16, no. 4, pp. 591-609, 2017.
[http://dx.doi.org/10.1007/s11157-017-9449-7]

[35]   R. Singh, and N. Hankins, *Emerging membrane technology for sustainable water treatment.* Elsevier, 2016.

[36]   C. Muro, F. Riera, and M. del Carmen Díaz, "Membrane separation process in wastewater treatment of food industry", In: *Food industrial processes–methods and equipment,* 2012, pp. 253-280.
[http://dx.doi.org/10.5772/31116]

[37]   A. Teuler, K. Glucina, and J. Laine, "Assessment of UF pretreatment prior RO membranes for seawater desalination", *Desalination,* vol. 125, no. 1-3, pp. 89-96, 1999.
[http://dx.doi.org/10.1016/S0011-9164(99)00126-5]

[38]   A. Brehant, V. Bonnelye, and M. Perez, "Assessment of ultrafiltration as a pretreatment of reverse osmosis membranes for surface seawater desalination", *Water Sci. Technol. Water Supply,* vol. 3, no. 5-6, pp. 437-445, 2003.
[http://dx.doi.org/10.2166/ws.2003.0200]

[39]   A. M'nif, S. Bouguecha, B. Hamrouni, and M. Dhahbi, "Coupling of membrane processes for brackish water desalination", *Desalination,* vol. 203, no. 1-3, pp. 331-336, 2007.
[http://dx.doi.org/10.1016/j.desal.2006.04.016]

[40]   F. Esmaeilion, "Hybrid renewable energy systems for desalination", *Appl Water Sci,* vol. 10, no. 84,

2020.

[41]   B.K. Pramanik, Y. Gao, L. Fan, F.A. Roddick, and Z. Liu, "Antiscaling effect of polyaspartic acid and its derivative for RO membranes used for saline wastewater and brackish water desalination", *Desalination,* vol. 404, pp. 224-229, 2017.
[http://dx.doi.org/10.1016/j.desal.2016.11.019]

[42]   R-W. Lee, J. Glater, Y. Cohen, C. Martin, K. Kovac, M.N. Milobar, and D.W. Bartel, "Low-pressure RO membrane desalination of agricultural drainage water", *Desalination,* vol. 155, no. 2, pp. 109-120, 2003.
[http://dx.doi.org/10.1016/S0011-9164(03)00288-1]

[43]   V. Yangali-Quintanilla, Z. Li, R. Valladares, Q. Li, and G. Amy, "Indirect desalination of Red Sea water with forward osmosis and low pressure reverse osmosis for water reuse", *Desalination,* vol. 280, no. 1-3, pp. 160-166, 2011.
[http://dx.doi.org/10.1016/j.desal.2011.06.066]

[44]   H-G. Park, and Y-N. Kwon, "Long-term stability of low-pressure reverse osmosis (RO) membrane operation—A pilot scale study", *Water,* vol. 10, no. 2, p. 93, 2018.
[http://dx.doi.org/10.3390/w10020093]

[45]   E. Semenova, N. Bublienko, T. Shilofost, and A. Bublienko, "Biochemical treatment of petroleum-containing waters", *J. Water Chem. Technol.,* vol. 35, no. 4, pp. 183-188, 2013.
[http://dx.doi.org/10.3103/S1063455X13040073]

[46]   M. El-Sheekh, W. El-Shouny, M. Osman, and E. El-Gammal, "Treatment of sewage and industrial wastewater effluents by the cyanobacteria Nostoc muscorum and Anabaena subcylinderica", *J. Water Chem. Technol.,* vol. 36, no. 4, pp. 190-197, 2014.
[http://dx.doi.org/10.3103/S1063455X14040079]

[47]   C. Narayanan, and V. Narayan, "Biological wastewater treatment and bioreactor design: a review", *Sustain. Environ. Res.,* vol. 29, no. 1, pp. 1-17, 2019.
[http://dx.doi.org/10.1186/s42834-019-0036-1]

[48]   A. Barreiros, C. Rodrigues, J.P. Crespo, and M. Reis, "Membrane bioreactor for drinking water denitrification", *Bioprocess Eng.,* vol. 18, no. 4, pp. 297-302, 1998.
[http://dx.doi.org/10.1007/s004490050445]

[49]   G. Luo, Z. Wang, Y. Li, J. Li, and A-M. Li, "Salinity stresses make a difference in the start-up of membrane bioreactor: performance, microbial community and membrane fouling", *Bioprocess Biosyst. Eng.,* vol. 42, no. 3, pp. 445-454, 2019.
[http://dx.doi.org/10.1007/s00449-018-2048-3] [PMID: 30478779]

[50]   E. Abdollahzadeh Sharghi, A. Shourgashti, and B. Bonakdarpour, "Considering a membrane bioreactor for the treatment of vegetable oil refinery wastewaters at industrially relevant organic loading rates", *Bioprocess Biosyst. Eng.,* vol. 43, no. 6, pp. 981-995, 2020.
[http://dx.doi.org/10.1007/s00449-020-02294-9] [PMID: 31993797]

[51]   R. Alnaizy, A. Aidan, and H. Luo, "Performance assessment of a pilot-size vacuum rotation membrane bioreactor treating urban wastewater", *Appl. Water Sci.,* vol. 1, no. 3-4, pp. 103-110, 2011.
[http://dx.doi.org/10.1007/s13201-011-0013-2]

[52]   S.A. Deowan, W. Korejba, J. Hoinkis, A. Figoli, E. Drioli, R. Islam, and L. Jamal, "Design and testing of a pilot-scale submerged membrane bioreactor (MBR) for textile wastewater treatment", *Appl. Water Sci.,* vol. 9, no. 3, pp. 1-7, 2019.
[http://dx.doi.org/10.1007/s13201-019-0934-8]

[53]   X. Wang, V.W. Chang, and C.Y. Tang, "Osmotic membrane bioreactor (OMBR) technology for wastewater treatment and reclamation: Advances, challenges, and prospects for the future", *J. Membr. Sci.,* vol. 504, pp. 113-132, 2016.
[http://dx.doi.org/10.1016/j.memsci.2016.01.010]

[54]   P. Jegathambal, R. Nisha, K. Parameswari, and M. Subathra, "Desalination and removal of organic pollutants using electrobiochemical reactor", *Appl. Water Sci.,* vol. 9, no. 4, pp. 1-10, 2019.
[http://dx.doi.org/10.1007/s13201-019-0990-0]

[55]   O. Lefebvre, and R. Moletta, "Treatment of organic pollution in industrial saline wastewater: a literature review", *Water Res.,* vol. 40, no. 20, pp. 3671-3682, 2006.
[http://dx.doi.org/10.1016/j.watres.2006.08.027] [PMID: 17070895]

[56]   G. Wu, Y. Guan, and X. Zhan, "Effect of salinity on the activity, settling and microbial community of activated sludge in sequencing batch reactors treating synthetic saline wastewater", *Water Sci. Technol.,* vol. 58, no. 2, pp. 351-358, 2008.
[http://dx.doi.org/10.2166/wst.2008.675] [PMID: 18701785]

[57]   F. Kargi, and A.R. Dincer, "Effect of salt concentration on biological treatment of saline wastewater by fed-batch operation", *Enzyme Microb. Technol.,* vol. 19, no. 7, pp. 529-537, 1996.
[http://dx.doi.org/10.1016/S0141-0229(96)00070-1]

[58]   F. Kargi, and A. Uygur, "Biological treatment of saline wastewater in a rotating biodisc contactor by using halophilic organisms", *Bioprocess Eng.,* vol. 17, no. 2, pp. 81-85, 1997.
[http://dx.doi.org/10.1007/s004490050357]

[59]   J.P. Bassin, M. Dezotti, and G.L. Sant'anna Jr, "Nitrification of industrial and domestic saline wastewaters in moving bed biofilm reactor and sequencing batch reactor", *J. Hazard. Mater.,* vol. 185, no. 1, pp. 242-248, 2011.
[http://dx.doi.org/10.1016/j.jhazmat.2010.09.024] [PMID: 20933327]

[60]   J.P. Bassin, R. Kleerebezem, G. Muyzer, A.S. Rosado, M.C. van Loosdrecht, and M. Dezotti, "Effect of different salt adaptation strategies on the microbial diversity, activity, and settling of nitrifying sludge in sequencing batch reactors", *Appl. Microbiol. Biotechnol.,* vol. 93, no. 3, pp. 1281-1294, 2012.
[http://dx.doi.org/10.1007/s00253-011-3428-7] [PMID: 21744134]

[61]   S. Abdelkader, F. Gross, D. Winter, J. Went, J. Koschikowski, S.U. Geissen, and L. Bousselmi, "Application of direct contact membrane distillation for saline dairy effluent treatment: performance and fouling analysis", *Environ. Sci. Pollut. Res. Int.,* vol. 26, no. 19, pp. 18979-18992, 2019.
[http://dx.doi.org/10.1007/s11356-018-2475-3] [PMID: 29948689]

[62]   M. Wilf, and M.K. Schierach, "Improved performance and cost reduction of RO seawater systems using UF pretreatment", *Desalination,* vol. 135, no. 1-3, pp. 61-68, 2001.
[http://dx.doi.org/10.1016/S0011-9164(01)00139-4]

[63]   E. Drioli, E. Curcio, G. Di Profio, F. Macedonio, and A. Criscuoli, "Integrating membrane contactors technology and pressure-driven membrane operations for seawater desalination: energy, exergy and costs analysis", *Chem. Eng. Res. Des.,* vol. 84, no. 3, pp. 209-220, 2006.
[http://dx.doi.org/10.1205/cherd.05171]

[64]   M. Nadjafi, A. Reyhani, and S. Al Arni, "Feasibility of treatment of refinery wastewater by a pilot scale MF/UF and UF/RO system for reuse at boilers and cooling towers", *J. Water Chem. Technol.,* vol. 40, no. 3, pp. 167-176, 2018.
[http://dx.doi.org/10.3103/S1063455X18030098]

[65]   P. Cartagena, M. El Kaddouri, V. Cases, A. Trapote, and D. Prats, "Reduction of emerging micropollutants, organic matter, nutrients and salinity from real wastewater by combined MBR–NF/RO treatment", *Separ. Purif. Tech.,* vol. 110, pp. 132-143, 2013.
[http://dx.doi.org/10.1016/j.seppur.2013.03.024]

[66]   N.J. Falizi, M.C. Hacıfazlıoğlu, İ. Parlar, N. Kabay, T.Ö. Pek, and M. Yüksel, "Evaluation of MBR treated industrial wastewater quality before and after desalination by NF and RO processes for agricultural reuse", *J. Water Process Eng.,* vol. 22, pp. 103-108, 2018.
[http://dx.doi.org/10.1016/j.jwpe.2018.01.015]

[67]   S. Gupta, Y. Mittal, P. Tamta, P. Srivastava, and A.K. Yadav, "Textile wastewater treatment using microbial fuel cell and coupled technology: a green approach for detoxification and bioelectricity generation", In: *Integrated Microbial Fuel Cells for Wastewater Treatment* Elsevier, 2020, pp. 73-92. [http://dx.doi.org/10.1016/B978-0-12-817493-7.00004-7]

[68]   J. Vymazal, "The use constructed wetlands with horizontal sub-surface flow for various types of wastewater", *Ecol. Eng.,* vol. 35, no. 1, pp. 1-17, 2009. [http://dx.doi.org/10.1016/j.ecoleng.2008.08.016]

[69]   S. Wu, P. Kuschk, H. Brix, J. Vymazal, and R. Dong, "Development of constructed wetlands in performance intensifications for wastewater treatment: a nitrogen and organic matter targeted review", *Water Res.,* vol. 57, pp. 40-55, 2014. [http://dx.doi.org/10.1016/j.watres.2014.03.020] [PMID: 24704903]

[70]   J. Garcia, D.P. Rousseau, J. Morato, E. Lesage, V. Matamoros, and J.M. Bayona, "Contaminant removal processes in subsurface-flow constructed wetlands: a review", *Crit. Rev. Environ. Sci. Technol.,* vol. 40, no. 7, pp. 561-661, 2010. [http://dx.doi.org/10.1080/10643380802471076]

[71]   S. Gupta, P. Srivastava, and A.K. Yadav, "Simultaneous removal of organic matters and nutrients from high-strength wastewater in constructed wetlands followed by entrapped algal systems", *Environ. Sci. Pollut. Res. Int.,* vol. 27, no. 1, pp. 1112-1117, 2020. [http://dx.doi.org/10.1007/s11356-019-06896-z] [PMID: 31820236]

[72]   P. Srivastava, A.K. Yadav, V. Garaniya, and R. Abbassi, "Constructed wetland coupled microbial fuel cell technology: development and potential applications", In: *Microbial Electrochemical Technology* Elsevier, 2019, pp. 1021-1036.

[73]   A. Yadav, "Design and development of novel constructed wetland cum microbial fuel cell for electricity production and wastewater treatment", *Proceedings of 12th international conference on wetland systems for water pollution control (IWA),* pp. 4-10, 2010.

[74]   A.K. Yadav, P. Srivastava, N. Kumar, R. Abbassi, and B.K. Mishra, *Constructed wetland-microbial fuel cell: an emerging integrated technology for potential industrial wastewater treatment and Bio-electricity generation.* Constructed Wetlands for Industrial Wastewater Treatment, 2018, pp. 493-510. [http://dx.doi.org/10.1002/9781119268376.ch22]

[75]   C. Corbella, and J. Puigagut, "Microbial fuel cells implemented in constructed wetlands: Fundamentals, current research and future perspectives", *Contrib. Sci.,* pp. 113-120, 2016.

[76]   L. Doherty, Y. Zhao, X. Zhao, Y. Hu, X. Hao, L. Xu, and R. Liu, "A review of a recently emerged technology: Constructed wetland--Microbial fuel cells", *Water Res.,* vol. 85, pp. 38-45, 2015. [http://dx.doi.org/10.1016/j.watres.2015.08.016] [PMID: 26295937]

[77]   O. Guadarrama-Pérez, T. Gutiérrez-Macías, L. García-Sánchez, V.H. Guadarrama-Pérez, and E.B. Estrada-Arriaga, "Recent advances in constructed wetland-microbial fuel cells for simultaneous bioelectricity production and wastewater treatment: A review", *Int. J. Energy Res.,* vol. 43, no. 10, pp. 5106-5127, 2019. [http://dx.doi.org/10.1002/er.4496]

[78]   C.A. Ramírez-Vargas, A. Prado, C.A. Arias, P.N. Carvalho, A. Esteve-Núñez, and H. Brix, "Microbial electrochemical technologies for wastewater treatment: principles and evolution from microbial fuel cells to bioelectrochemical-based constructed wetlands", *Water,* vol. 10, no. 9, p. 1128, 2018. [http://dx.doi.org/10.3390/w10091128]

[79]   W. Wang, Y. Zhang, M. Li, X. Wei, Y. Wang, L. Liu, H. Wang, and S. Shen, "Operation mechanism of constructed wetland-microbial fuel cells for wastewater treatment and electricity generation: A review", *Bioresour. Technol.,* vol. 314, p. 123808, 2020. [http://dx.doi.org/10.1016/j.biortech.2020.123808] [PMID: 32713782]

[80]   L. Xu, W. Yu, N. Graham, Y. Zhao, and J. Qu, "Application of Integrated Bioelectrochemical-

Wetland Systems for Future Sustainable Wastewater Treatment", In: *Environmental Science & Technology* vol. 53. , 2019, pp. 1741-1743.
[http://dx.doi.org/10.1021/acs.est.8b07159]

[81]   P. Srivastava, R. Abbassi, A. K. Yadav, V. Garaniya, and M. Asadnia, "A review on the contribution of electron flow in electroactive wetlands: Electricity generation and enhanced wastewater treatment", *Chemosphere,* vol. 254, p. 126926, 2020.
[http://dx.doi.org/10.1016/j.chemosphere.2020.126926]

[82]   T. Gschlößl, C. Steinmann, P. Schleypen, and A. Melzer, "Constructed wetlands for effluent polishing of lagoons", *Water Research,* vol. 32, pp. 2639-2645, 1998.

[83]   C. R. Steinmann, S. Weinhart, and A. Melzer, "A combined system of lagoon and constructed wetland for an effective wastewater treatment", *Water Research,* vol. 37, pp. 2035-2042, 2003.
[http://dx.doi.org/10.1016/S0043-1354(02)00441-4]

[84]   C.C. Tanner, and J.P. Sukias, "Linking pond and wetland treatment: performance of domestic and farm systems in New Zealand", *Water Sci. Technol.,* vol. 48, no. 2, pp. 331-339, 2003.
[http://dx.doi.org/10.2166/wst.2003.0138] [PMID: 14510228]

[85]   A. Shilton, ""Pond treatment technology,"",

[86]   J. Vymazal, "Constructed wetlands for wastewater treatment", *Water,* vol. 2, no. 3, pp. 530-549, 2010.
[http://dx.doi.org/10.3390/w2030530] [PMID: 20795704]

[87]   M. von Sperling, F.L. Dornelas, F.A. Assunção, A.C. de Paoli, and M.O. Mabub, "Comparison between polishing (maturation) ponds and subsurface flow constructed wetlands (planted and unplanted) for the post-treatment of the effluent from UASB reactors", *Water Sci. Technol.,* vol. 61, no. 5, pp. 1201-1209, 2010.
[http://dx.doi.org/10.2166/wst.2010.961] [PMID: 20220242]

[88]   M. von Sperling, "Comparison of simple, small, full-scale sewage treatment systems in Brazil: UASB-maturation ponds-coarse filter; UASB-horizontal subsurface-flow wetland; vertical-flow wetland (first stage of French system)", *Water Sci. Technol.,* vol. 71, no. 3, pp. 329-337, 2015.
[http://dx.doi.org/10.2166/wst.2014.496] [PMID: 25714630]

[89]   M.P. Matos, M. von Sperling, A.T. Matos, P.R.A. Aranha, M.A. Santos, F.D.B. Pessoa, and P.D.D. Viola, "Clogging in constructed wetlands: Indirect estimation of medium porosity by analysis of ground-penetrating radar images", *Sci. Total Environ.,* vol. 676, pp. 333-342, 2019.
[http://dx.doi.org/10.1016/j.scitotenv.2019.04.168] [PMID: 31048164]

[90]   S.T. Miranda, A.T. de Matos, M.P. de Matos, C.B. Saraiva, and D.L. Teixeira, "Influence of the substrate type and position of plant species on clogging and the hydrodynamics of constructed wetland systems", *J. Water Process Eng.,* vol. 31, p. 100871, 2019.
[http://dx.doi.org/10.1016/j.jwpe.2019.100871]

[91]   P.H. Sezerino, V. Reginatto, M.A. Santos, K. Kayser, S. Kunst, L.S. Philippi, and H.M. Soares, "Nutrient removal from piggery effluent using vertical flow constructed wetlands in southern Brazil", *Water Sci. Technol.,* vol. 48, no. 2, pp. 129-135, 2003.
[http://dx.doi.org/10.2166/wst.2003.0103] [PMID: 14510203]

[92]   K. Stone, P. Hunt, A. Szögi, F. Humenik, and J. Rice, "Constructed wetland design and performance for swine lagoon wastewater treatment", *Trans. ASAE,* vol. 45, no. 3, p. 723, 2002.
[http://dx.doi.org/10.13031/2013.8828]

[93]   J.B. Park, R.J. Craggs, and C.C. Tanner, "Eco-friendly and low-cost Enhanced Pond and Wetland (EPW) system for the treatment of secondary wastewater effluent", *Ecol. Eng.,* vol. 120, pp. 170-179, 2018.
[http://dx.doi.org/10.1016/j.ecoleng.2018.05.029]

[94]   J.C. Anderson, J.C. Carlson, J.E. Low, J.K. Challis, C.S. Wong, C.W. Knapp, and M.L. Hanson, "Performance of a constructed wetland in Grand Marais, Manitoba, Canada: Removal of nutrients,

pharmaceuticals, and antibiotic resistance genes from municipal wastewater", *Chem. Cent. J.,* vol. 7, no. 1, p. 54, 2013.
[http://dx.doi.org/10.1186/1752-153X-7-54] [PMID: 23506187]

[95]   J.L. Conkle, J.R. White, and C.D. Metcalfe, "Reduction of pharmaceutically active compounds by a lagoon wetland wastewater treatment system in Southeast Louisiana", *Chemosphere,* vol. 73, no. 11, pp. 1741-1748, 2008.
[http://dx.doi.org/10.1016/j.chemosphere.2008.09.020] [PMID: 18977010]

[96]   M. Hijosa-Valsero, V. Matamoros, R. Sidrach-Cardona, A. Pedescoll, J. Martín-Villacorta, J. García, J.M. Bayona, and E. Bécares, "Influence of design, physico-chemical and environmental parameters on pharmaceuticals and fragrances removal by constructed wetlands", *Water Sci. Technol.,* vol. 63, no. 11, pp. 2527-2534, 2011.
[http://dx.doi.org/10.2166/wst.2011.500] [PMID: 22049744]

[97]   M. Hijosa-Valsero, V. Matamoros, J. Martín-Villacorta, E. Bécares, and J. M. Bayona, "Assessment of full-scale natural systems for the removal of PPCPs from wastewater in small communities", *Water Research,* vol. 44, pp. 1429-1439, 2010.
[http://dx.doi.org/10.1016/j.watres.2009.10.032]

[98]   J. Chen, Y-S. Liu, H-C. Su, G-G. Ying, F. Liu, S-S. Liu, L.Y. He, Z.F. Chen, Y.Q. Yang, and F.R. Chen, "Removal of antibiotics and antibiotic resistance genes in rural wastewater by an integrated constructed wetland", *Environ. Sci. Pollut. Res. Int.,* vol. 22, no. 3, pp. 1794-1803, 2015.
[http://dx.doi.org/10.1007/s11356-014-2800-4] [PMID: 24687794]

[99]   J. Chen, Y.-S. Liu, W.-J. Deng, and G.-G. Ying, "Removal of steroid hormones and biocides from rural wastewater by an integrated constructed wetland", *Science of The Total Environment,* vol. 660, pp. 358-365, 2019.
[http://dx.doi.org/10.1016/j.scitotenv.2019.01.049]

SUBJECT INDEX

A

Abnormalities 14, 61, 86
  endocrine gland 61
  eye 86
Absorption 84, 86, 88
Acetate, trenbolone 187
Acid(s) 10, 53, 58, 63, 65, 88, 106, 109, 110, 111, 113, 161, 162, 180, 188, 209, 215, 256, 299, 304, 310, 327
  amino 161, 162
  ascorbic 162
  carboxylic 304
  citric 162
  clofibric 10, 180, 188, 256, 327
  fatty 58
  ferulic 162
  fulvic 88
  halogen 310
  humic 63, 304
  hydrofluoric 109
  Hypochlorous 299
  Iothalamic 188
  perflurooctanoic 113
  rain 209
  sulfuric 215
  threnodic 162
Acidification 300
Activated sludge process 270, 276, 328, 332, 336
Activity 15, 82, 83, 90, 91, 120, 123, 163, 190, 191, 295, 301, 307, 327
  anthropogenic 120, 123, 327
  antiestrogenic 15
  cellular metabolic 91
  dehydrogenase 163
  manganese peroxidase 83
  microbial enzymes 82
  photocatalytic 307
Acute respiratory pathologies 104
Adsorbents 64, 185, 204, 217, 218, 219, 221
  natural 217, 218

  synthetic 217, 218
Adsorption 78, 79, 156, 185, 204, 217, 221, 252, 253, 279, 280, 281, 282, 283
  filtration 221
  macrolide antibiotics 156
Advanced oxidation processes (AOPs) 42, 43, 187, 197, 230, 279, 299
Agencies 2, 11, 12, 209, 296
  environmental protection 2, 11, 12, 296
Agents 32, 39, 41, 53, 65, 104, 180, 188, 299, 310
  antibacterial 39
  anti-infection 41
  disinfecting 53, 65
  infectious 32
  reducing 310
Agricultural 65, 146, 159, 210, 217, 219, 220
  environments 159
  food processes 65
  waste 146, 210, 217, 219, 220
Agriculture product processing 58
Agroecosystems 295
Algae 11, 13, 14, 15, 126, 131, 146, 181, 182, 213, 214, 336
  fresh-weight 182
  growth 146
Aliphatic hydrocarbon compounds 102
Alzheimer's disease 206
AMF symbiosis 83
Amino acid conjugates 161
Aminoglycoside 187
Amoxicillin trihydrate 301
Analeptic forms, numerous 88
Analgesics 4, 6, 10, 14, 35, 160, 179, 180, 187
  tetracyline 187
Analysis, economic 190
Analytical 1, 292
  detection methods 292
  techniques 1
Androgens 148, 149, 153, 294, 337
  androgenic hormones 149
Androstenedione 43

www.ingramcontent.com/pod-product-compliance
Lightning Source LLC
Chambersburg PA
CBHW050802220326
41598CB00006B/98